AF774643

IET TELECOMMUNICATIONS SERIES 78

Principles and Applications of Free Space Optical Communications

Other volumes in this series:

Volume 9	**Phase Noise in Signal Sources** W.P. Robins
Volume 12	**Spread Spectrum in Communications** R. Skaug and J.F. Hjelmstad
Volume 13	**Advanced Signal Processing** D.J. Creasey (Editor)
Volume 19	**Telecommunications Traffic, Tariffs and Costs** R.E. Farr
Volume 20	**An Introduction to Satellite Communications** D.I. Dalgleish
Volume 26	**Common-Channel Signalling** R.J. Manterfield
Volume 28	**Very Small Aperture Terminals (VSATs)** J.L. Everett (Editor)
Volume 29	**ATM: The broadband telecommunications solution** L.G. Cuthbert and J.C. Sapanel
Volume 31	**Data Communications and Networks, 3rd Edition** R.L. Brewster (Editor)
Volume 32	**Analogue Optical Fibre Communications** B. Wilson, Z. Ghassemlooy and I.Z. Darwazeh (Editors)
Volume 33	**Modern Personal Radio Systems** R.C.V. Macario (Editor)
Volume 34	**Digital Broadcasting** P. Dambacher
Volume 35	**Principles of Performance Engineering for Telecommunication and Information Systems** M. Ghanbari, C.J. Hughes, M.C. Sinclair and J.P. Eade
Volume 36	**Telecommunication Networks, 2nd Edition** J.E. Flood (Editor)
Volume 37	**Optical Communication Receiver Design** S.B. Alexander
Volume 38	**Satellite Communication Systems, 3rd Edition** B.G. Evans (Editor)
Volume 40	**Spread Spectrum in Mobile Communication** O. Berg, T. Berg, J.F. Hjelmstad, S. Haavik and R. Skaug
Volume 41	**World Telecommunications Economics** J.J. Wheatley
Volume 43	**Telecommunications Signalling** R.J. Manterfield
Volume 44	**Digital Signal Filtering, Analysis and Restoration** J. Jan
Volume 45	**Radio Spectrum Management, 2nd Edition** D.J. Withers
Volume 46	**Intelligent Networks: Principles and applications** J.R. Anderson
Volume 47	**Local Access Network Technologies** P. France
Volume 48	**Telecommunications Quality of Service Management** A.P. Oodan (Editor)
Volume 49	**Standard Codecs: Image compression to advanced video coding** M. Ghanbari
Volume 50	**Telecommunications Regulation** J. Buckley
Volume 51	**Security for Mobility** C. Mitchell (Editor)
Volume 52	**Understanding Telecommunications Networks** A. Valdar
Volume 53	**Video Compression Systems: From first principles to concatenated codecs** A. Bock
Volume 54	**Standard Codecs: Image compression to advanced video coding, 3rd Edition** M. Ghanbari
Volume 59	**Dynamic Ad Hoc Networks** H. Rashvand and H. Chao (Editors)
Volume 60	**Understanding Telecommunications Business** A Valdar and I Morfett
Volume 65	**Advances in Body-Centric Wireless Communication: Applications and state-of-the-art** Q.H. Abbasi, M.U. Rehman, K. Qaraqe and A. Alomainy (Editors)
Volume 67	**Managing the Internet of Things: Architectures, theories and applications** J. Huang and K. Hua (Editors)
Volume 68	**Advanced Relay Technologies in Next Generation Wireless Communications** I. Krikidis and G. Zheng
Volume 69	**5G Wireless Technologies** A. Alexiou (Editor)
Volume 70	**Cloud and Fog Computing in 5G Mobile Networks** E. Markakis, G. Mastorakis, C.X. Mavromoustakis and E. Pallis (Editors)
Volume 71	**Understanding Telecommunications Networks, 2nd Edition** A. Valdar
Volume 72	**Introduction to Digital Wireless Communications** Hong-Chuan Yang
Volume 73	**Network as a Service for Next Generation Internet** Q. Duan and S. Wang (Editors)
Volume 74	**Access, Fronthaul and Backhaul Networks for 5G and Beyond** M.A. Imran, S.A.R. Zaidi and M.Z. Shakir (Editors)
Volume 76	**Trusted Communications with Physical Layer Security for 5G and Beyond** T.Q. Duong, X. Zhou and H.V. Poor (Editors)
Volume 77	**Network Design, Modelling and Performance Evaluation** Q. Vien
Volume 79	**Satellite Communications in the 5G Era** S.K. Sharma, S. Chatzinotas and D. Arapoglou
Volume 80	**Transceiver and System Design for Digital Communications, 5th Edition** Scott R. Bullock
Volume 905	**ISDN Applications in Education and Training** R. Mason and P.D. Bacsich

Principles and Applications of Free Space Optical Communications

Edited by
Arun K. Majumdar, Zabih Ghassemlooy, and
A. Arockia Bazil Raj

The Institution of Engineering and Technology

Published by The Institution of Engineering and Technology, London, United Kingdom

The Institution of Engineering and Technology is registered as a Charity in England & Wales (no. 211014) and Scotland (no. SC038698).

© The Institution of Engineering and Technology 2019

First published 2019

This publication is copyright under the Berne Convention and the Universal Copyright Convention. All rights reserved. Apart from any fair dealing for the purposes of research or private study, or criticism or review, as permitted under the Copyright, Designs and Patents Act 1988, this publication may be reproduced, stored or transmitted, in any form or by any means, only with the prior permission in writing of the publishers, or in the case of reprographic reproduction in accordance with the terms of licences issued by the Copyright Licensing Agency. Enquiries concerning reproduction outside those terms should be sent to the publisher at the undermentioned address:

The Institution of Engineering and Technology
Michael Faraday House
Six Hills Way, Stevenage
Herts, SG1 2AY, United Kingdom

www.theiet.org

While the authors and publisher believe that the information and guidance given in this work are correct, all parties must rely upon their own skill and judgement when making use of them. Neither the authors nor publisher assumes any liability to anyone for any loss or damage caused by any error or omission in the work, whether such an error or omission is the result of negligence or any other cause. Any and all such liability is disclaimed.

The moral rights of the authors to be identified as authors of this work have been asserted by them in accordance with the Copyright, Designs and Patents Act 1988.

British Library Cataloguing in Publication Data
A catalogue record for this product is available from the British Library

ISBN 978-1-78561-415-6 (hardback)
ISBN 978-1-78561-416-3 (PDF)

Typeset in India by MPS Limited
Printed in the UK by CPI Group (UK) Ltd, Croydon

To my family Gargi, Sharmistha and Ben who believe in my continued creativity and imagination under all adverse conditions to make some contributions and to inspire other researchers worldwide in my field.

– Arun K. Majumdar
San Diego, CA, USA

To all my research students, research fellows and colleagues whom have worked with me on the topic of FSO and for their ideas, insight and dedications, which has benefited me enormously.

– Zabih Ghassemlooy
Northumbria University, UK

To my parents A. Jothy Mary and A. AnthoniSamy
To my wife A. Johnsi Rosita and daughter A. Feona Hydee Mitchell
To Mr. J. Niranjan Samuel, who has admirably and unreservedly extended his helping hand and support

– A. Arockia Bazil Raj
Defence Institute of Advanced Technology (DIAT), Pune, India

Contents

List of acronyms

Acronym	Expansion
A	
APD	Avalanche photodiode
AO	Adaptive optics
AWG	Arbitrary waveform generator
ABER	Average bit error rate
ASE	Amplified spontaneous emission
ADC	Analog-to-digital converter
AF	Amplify-and-forward
ANOVA	Analysis of variance
AGC	Automatic gain control
AOAF	All-optical amplify-and-forward
AORF	All-optical regenerate-and-forward
AT	Atmospheric turbulence
AO	All-optical
ASE	Amplified spontaneous emission
AOR	All-optical regenerator
AWGN	Additive white Gaussian noise
ACF	Amplitude comparison function
B	
BER	Bit error rate
BTS	Box of turbulence simulator
BEP	Bit error probability
BERR	Bit error rate ratio
BPSK	Binary phase shift keying
BWA	Broadband wireless access
BPPM	Binary pulse position modulation
bps	Bits per second

BS	Beam splitter
BE	Beam expander

C

CDF	Cumulative distribution function
CSI	Channel-state-information
CDF	Cumulative density function
CCs	Cross correlations
CSS	Chirp spread spectrum
CDMA	Code division multiple access
CW	Continuous wave

D

DF	Decode-and-forward
DOE	Design of experiment
DWDM	Dense wavelength division multiplexing
DF	Decode-and-forward
DSO	Digital storage oscilloscope
DFA	Doped fiber amplifier
DoF	Degree-of-freedom
DF	Distributed feedback
DSSS	Direct-sequence spread spectrum
DS	Direct sequence
DDS	Direct digital synthesizer
DFT	Discrete Fourier transform
DF	**Decision-feedback**
DS	**Differential signaling**
DSP	Digital signal processing

E

ESFL	Electronically steerable flash lidar
EMI	Electro-magnetic interference
EOM	Electro optical modulator
ELPF	Electrical low-pass filter
EO	Electrical-to-optical
EPONs	Ethernet passive optical networks

EDFA	Erbium-doped fiber amplifier
ER	Extinction ratio
EGC	Equal gain combining

F

FOV	Field-of-view
FHSS	Frequency-hopping spread spectrum
FH	Frequency hopping
FBGs	Fiber Bragg gratings
FSOC	Free-space optical communications
FFT	Fast Fourier transform
FDPP	Frequency-dependent power penalty
FS-RDL	Frequency shift recirculating delay line
FOC	Fiber optic cable
FIR	Finite impulse response
FTTH	Fiber to the home
FEC	Forward error correction
FWHM	Full-width at half-maximum

G

G-G	Gamma–gamma
GRIN	Gradient index
Gbps	Gigabits per second
GPS	Global positioning system
GUI	Graphical user interface
GMT	Greenwich mean time

H

HNLF	Highly-nonlinear fiber
HDTV	High-definition television
HAOF	Hybrid acousto-optic feedback

I

IP	Internet protocol
ISDB-T	Integrated services digital broadcasting-terrestrial
IQ	In-phase/quadrature

IFM	Instantaneous frequency measurement
IDFT	Inverse discrete fourier transform
IM/DD	Intensity modulation/direct detection
ICF	Irradiance correlation function
IL	Interleaver
ISI	Inter symbol interference
IM-DD	Intensity modulation-direct detection
IF	Intermediate frequency

L

LE	Lyapunov exponent
LPF	Low-pass filter
LOS	Line-of-sight
LM	Link margin
LG	Laguerre–Gaussian
LDPC	Low-density parity-check
LO	Local oscillator
LOs	Local oscillators
LFM	Linear frequency modulated
LD	Laser diode
LEO	Lower earth orbit
LWC	Liquid water content
LAN	Local area network

M

MRC	Maximal ratio combining
MZM	Mach–Zehnder modulator
MEO	Medium-earth-orbit
MISO	Multiple-input single-output
MIMO	Multiple-input multiple-output
MMSE	Minimum mean square error
MSE	Mean square error
MLD	Maximum-likelihood-decoding
MLD	Maximum likelihood decoder
MP	Mono-pulse
MLSD	Maximum-likelihood sequence detection

ML	Maximum-likelihood
MMF	Multi-mode fiber
MLL	Mode-locking laser
MVR	Meteorological visual range
MSL	Mean sea level

N

NRZ-OOK	Non-return-to-zero on–off keying

O

OSA	Optical spectrum analyzer
OHL	Optical hard-limiter
OSF	Optical spatial filter
OSNR	Optical signal-to-noise-ratio
OE	Optical-to-electrical
OA	Optical amplifier
OAM	Orbital-angular-momentum
OTF	Optical-tunable filter
OC	Optimal combining
OBPF	Optical bandpass filter

P

PSK	Phase shift keying
PSAM	Pilot-symbol-assisted modulation
PIB	Power-in-bucket
PRBS	Pseudorandom binary sequence
PD	Photodiode
PolSK	Polarization shift keying
PC	Personal computer
PNS	Photon number splitting attack
PAAs	Phased-array antennas
PM	Phase modulator
PICs	Photonic integrated chips
PEs	Pointing errors
PPM	Pulse position modulation
PCB	Partially coherent beam

PDF	Probability density function
PSD	Power spectral density
PPS	Phase power spectrum

Q

QPPM	Q-ary pulse-position modulation
QKD	Quantum key distribution
QBER	Quantum bit error ratio
QPSK	Quadrature phase shift keying

R

RF ID	RF identification
RZ	Return-to-zero
RMS	Root mean square
RC	Repetition coding
RD	Relay-to-destination

S

SIMO	Single-input multiple-output
SLM	Spatial light modulator
SMF	Single-mode fiber
STBCs	Space time block codes
SI	Scintillation index
STTCs	Space time trellis codes
SC	Selection combining
SAS	Statistical analysis system
SLC	Submarine laser communication
SAE	Sum of absolute error
SA	Sum of absolute
SIM	Subcarrier intensity modulation
SISO	Single-input single-output
SOA	Semiconductor optical amplifier
SMC	Single-step Markov chain
SPM	Self-phase modulation
SelC	Selection combining
SR	Source-to-relay

SNR	Signal-to-noise ratio
SwP	Size, weight, and power
SDR	Software-defined radio

T

TIA	Trans impedance amplifier
Tbps	Terabits per second
TFF	Transfer function formalism
TRS	Transmit–receive switches
THSS	Time-hopping spread spectrum

U

UAV	Unmanned aerial vehicle

V

VOA	Variable optical attenuator
VLC	Visible light communication
VCSELs	Vertical-cavity surface-emitting lasers

W

WS	Wind speed
WDM	Wavelength division multiplexing
WLAN	Wireless local area network

Z

ZF	Zero-forcing

Chapter 1

Introduction to free space optical (FSO) communications

Zabih Ghassemlooy[1], *Arun Majumdar*[2], *and Arockia Bazil Raj*[3]

1.1 Introduction

The evolution of wireless communication applications over the past decades is enormous, driven by the ever increasing number of wireless broadband internet, mobile phones, smart devices, social web, gaming, and video-centric applications. The number of end users is grown by about 30%–40% per year, i.e., from 16 million to 3.6 billion in 1995 and 2016 [1,2], which has put a tremendous pressure on the network infra-structure thus forcing the service providers to upgrade their current systems for higher wireless access data rate and improve quality of service. Until now, the radio frequency (RF)-based wireless systems have been the prominent and mature technology in a range of fields including wireless local area network (WLAN), global positioning system (GPS), RF identification (RF ID) systems, home satellite network, etc. [3]. In addition, the third and fourth generations (3G/4G) wireless networks have experienced a growing increase in the data traffic due to the wide-spread use of smart devices any-time anywhere. The volume of mobile and wireless users and thus the data traffic are predicted to increase a thousand-fold over the next decade [4], thus resulting in the mobile spectrum congestion (i.e., bandwidth bottleneck) at both the backhaul and last mile access networks [5]. Of course, the situation is going to get even more challenging by the introduction of 5G and beyond wireless technologies.

However, operators consider alternative technologies to overcome the spectrum congestion in certain applications, where the RF-based technology cannot be used or is not suitable. For example, in highly populated indoor environments (train station, airports, etc.), and "the last mile access" network, where the end user, using the RF-based wireless technologies, do experience lower data rates and low quality services due to the spectrum congestion (i.e., bandwidth bottleneck). Thus ensuring

[1]Faculty of Engineering and Environment, Northumbria University, UK
[2]Department of Physics, Colorado State University-Pueblo, USA
[3]Department of Electronics Engineering, Defence Institute of Advanced Technology, India

the most efficient and effective utilization of the RF spectrum in dense-traffic areas. This could include point-to-multipoint links in areas where spectrum for the conventional point-to-point links is becoming scarce and costly. The microwave, millimeter wave and optical fiber-based technologies will continue to retain their importance as a backhaul bearer. To increase the bandwidth and capacity, service providers are considering moving to higher frequencies (i.e., 40 and 80 GHz bands), but at the expense of reduced transmission coverage, which has adverse effects on the cost (i.e., deployment, site rental, maintenance, equipment, etc.).

In a perfect scenario, all end users should have access to the optical fiber-based backbone network with an ultra-high capacity, to benefit from truly high-speed data communications with a very low end-to-end transmission latency. Fiber optical communication systems as the most reliable and high bandwidth transmission technology meet the bandwidth requirements and high quality of services mostly at the backbone network with the potential to move in into the last mile and last meter access networks. However, the cost and challenges associate with installation of optical fiber particularly in rural areas as well as maintenance of such a network is rather high, therefore is not considered for the last mile access network [6]. Of course, for environments where deployment of optical fiber is not economical a combination of satellite communications and optical fiber communications technologies would be the most suitable option. However, this could also be quite costly and therefore may not be feasible in the long run.

1.2 Free space optics

The demand for high bandwidth and secure communication is increasing in future. Therefore, free space optical (FSO) wireless communications technology could be one possible alternative option to the RF technologies that can be adopted in certain application to un-locked the bandwidth bottleneck issue more specifically in the last mile access networks, between mobile base station in RF cellular wireless network, and as of the radio over optical fiber [7–9]. During the last decade, we have seen a growing research and development activities in the FSO communications in the field of high data rates wireless technology applications as well as the emergence of commercial systems. The principle reason behind the increasing popularity of FSO is its capability to meet the user's ever increasing demand for bandwidth, which is not possible with the existing RF-based wireless technologies [6,10].

Note that, most FSO systems are based on the line-of-sight (LoS) intensity modulation/direct detection (IM/DD) laser (single or multiple wavelength) transmission, which offering similar capabilities as optical fiber communications with attractive features including (i) huge bandwidth; (ii) no licensing fee since the optical spectrum bands lies outside the telecommunication regulations; (iii) inherent security at the physical layer for mostly the point-to-point link configurations; (iv) low cost of installation and maintenance [10–13]; (v) lower power consumption; (vi) immunity to the RF-based electromagnetic interference [8–10]; (vii) back-bone network compatibility, where FSO is operating at optical transmission windows of 850, 1,300 and 1,550 nm, which are compatible with optical fiber back bone networks, as well as 10 μm [14]; and (viii) no or very little inter-channel interference due to a narrow laser beams, which guarantees high spatial selectivity. FSO systems employing heterodyne detection techniques are also used in order to increase the sensitivity of the receiver and improve the robustness of the systems against channel induced impairments [15]. Some of the key features of FSO links are as follows:

(i) The global information and communications technology is responsible for 2%–10% of the global energy consumption according to the report smart 2020 [16]. The global warming and the existing concern to reduce the power usage is a critical motivation to replace RF links with FSO in certain applications since the FSO technology is potentially green in terms of energy consumption compared to RF [17,18].

(ii) At the present time RF-based wireless technologies provides 1 to 2 Mbps for unregulated 2.4 GHz ISM bands [19], 20 Mbps 875 Mbps at 5.7 GHz 4G mobile and 60 GHz millimeter wave (MMW), respectively [20]. Potentially FSO can provide bandwidth as large as 2,000 THz, which is far beyond the maximum data rate of RF technologies [21,22]. In addition, FSO offers dense spatial reuse [23].

(iii) A review survey conducted with the operators in Europe and USA companies concluded that FSO is much faster to deploy than any other fixed communication technology [23]. Moreover, the speed of installation of FSO is in hours as compared to the RF wireless technology which can take up to months [10,14].

(iv) The main advantage of the confined beam of FSO communications is the ability to provide a significant degree of covertness. A malicious eavesdropper would need to within the LoS transmission path in order to intercept the light and therefore access the information [24]. This makes the interception almost impossible as the eavesdropper's antenna is also likely to cause link outage for the intended recipient due to beam obstruction. Jamming an FSO is also difficult because of the nature of optical beam is narrow and also invisible [25,26].

1.3 FSO applications

The FSO technology with data rates ranging from multi-Gbit/sec to a few Mbit/sec or less, over typical link spans of a few micro-meters to hundreds of meters have been adopted for civilian applications [27,28] including:

- **In-chip optical interconnections**—with path lengths ranging from hundreds of microns up to ~ 1 cm.
- **Last meter indoor communications**—with path length <100 m—usually using infrared and visible lights.

- **Last mile access network in rural areas** [29] – The bandwidth bottlenecks within the access network is a major issue, which has been depriving the end users with sufficient amount of bandwidth in order to meet their requirements [23]. A number of technologies (wired and wireless) have been developed to bridge the last mile. With the increasing level of deployment of fiber optic technology such as the Ethernet passive optical networks (EPONs), the bandwidth bottleneck is being shifted toward the last mile access network. EPONs are designed to carry Ethernet frames at gigabit Ethernet rates but are not cost effective [30,31]. However, wireless networks capable of offering gigabit per second data rates in the last mile access network still have bandwidth limitations, thus enabling the end users to have a full access to broadband internet [32]. This problem is more acute in the rural areas where access to high-speed broadband using the existing technologies is rather limited [33,34]. In such scenarios the FSO technology could offer gigabit Ethernet to the end users. FSO can replace optical fiber access technologies such as fiber to the home (FTTH) in order to provide connectivity between in-building networks and to broadband and backbone data networks [35,36].
- **LAN-to-LAN inter-connectivity—[27,37] and electronic commerce** [38]—provides high-speed, flexibility and high security connectivity for campus and metropolitan applications.
- **Audio and video streaming** [39]—for video surveillance and monitoring, as well as live broadcasting of sporting events, in emergency situation [40], etc.
- **Unmanned aerial vehicle (UAV) and high attitude platforms [27,41]**—for monitoring traffic and disaster areas, or broadcasting vital data to the emergency services, etc.
- **Disaster and emergency relief network**—where the existing communications networks is no longer operational and therefore FSO systems can be quickly used to establish communications links for emergency services.
- **Inter-satellite communications, ground to/from satellite communications, and deep space communications**—Numerous ultra-long-haul outdoor optical wireless systems have been proposed and implemented in a wide range of applications including terrestrial, earth-to-satellite and satellite-to-earth, earth-to-high altitude platform, intersatellite, and interplanetary communication links [42–45]. Major technical advances in the field of adaptive optics and beam acquisition/tracking, which are the critical and importance functions in ultra-long-haul outdoor FSO links have enabled the deployment of tractable broadband outdoor optical wireless communications systems [46,47]. FSO satellite network can provide a high-bandwidth optical wireless network access to the end users since the satellites can cover large areas on the earth.
- **Underwater communication networks**—Is suitable for various applications like undersea explorations, environmental monitoring, disaster prevention, distributed tactical surveillance (unmanned underwater vehicles), and underwater sensors to monitor the surveillance, targeting, and intrusion detection [48]. FSO can provide broadband communications for underwater wireless sensor networks such as to transmit video streams or downloads a large burst

Table 1.1 Comparison between FSO and RF communication systems

Parameter	FSO link	RF link
Data rate	Up to 10's Gbps using multiple wavelength	<1 Gbps
Devices size	Small	Medium–Large
Bandwidth	License free	Required for most frequency range
Security at the physical layer	Very high	Very low
Network architecture	Scalable	Non-scalable
Cost	Moderate	Low to moderate
Link performance effects	Fog, atmospheric turbulence misalignment or obstruction	Multipath fading, rain, interferences
Transmission range	• Short to medium • Long for ground to space and space to ground	Long
Noise limitation	Background light	Other sources
Installation complexity	Low	Medium

of stored data in a brief time slot. This is very crucial when the data is required from a specific location or within a short polling time slot.

- **Hybrid FSO/RF communications**—FSO and RF communications can be realized as a complementary scheme in order to overcome the limitations of both technologies and achieve 99.999% link availability under all weather conditions [48–50]. The scattering effects due to fog/smoke and atmospheric turbulence degrades the bit error rate (BER) performance of a FSO link. Note that, for a FSO link at 830 nm the measured attenuation due to fog is 37 dB/km at a visibility of 200 m, whereas for a 58 GHz RF link it is 3 dB/km [51]. However, with rain the RF link experience attenuation of 17 dB/km at a rain rate of 40 mm/h, whereas for a FSO link at a wavelength of 830 it is 2 dB/km [51]. Note that, the probability of occurrence of fog and rain simultaneously is very low [52]. Hence, the RF link provides a back-up link to FSO in fog conditions [53]. Table 1.1 summarizes the differences between FSO and RF communications.

1.4 Key features and advantageous

In most cases, the FSO modules, which are compact in size, are installed on the tall buildings, connected from rooftop-to-rooftop, window-to-rooftop or window-to-window. In practice, using a hybrid link the availability of five nines (99.999%) is reported [19]) as well as to reduce the cost of the complete system. The cost effectiveness of FSO system compared to the RF system is more obvious, when the RF system is supposed to deliver the same high data rate connection service [54–58]. With the emergence of powerful and efficient optoelectronic components and advanced communication techniques, current states-of-the-art FSO prototypes have demonstrated transmission data rates of 10 Gbps over a range of few kilometers, which are commercially available with improve the link quality (e.g., it is desirable to achieve ideal 100% link availability in all weather conditions [59], a data rate of 7.5 Gbps up to distance of 40 km [60], and up to 1.6 Tbps over a 80 m single outdoor link based on the dense wavelength division multiplexing technique [60–62].

One option adopted in order to increase the spectral efficiency (i.e., total data rate) and performance of FSO is to adopt spatial division multiplexing (SDM), where multiple independent data carrying optical beams simultaneously are transmitted over the same free space channel [14,16]. Note, a sub-set of SDM use mode division multiplexing, where orthogonal modes from a model basis set are used for transmitting each optical beams [63], in order to ensure multiplexing, transmission, and demultiplexing of different beams with little inherent crosstalk [64]. In [65], an experimental demonstration of 400 Gbps FSO transmission using mode division multiplexing was reported.

Note that, SDM can provide a simple solution to the major setbacks in designing multiple input multiple output (MIMO) systems, i.e., reduced interchannel interference and improved inter-antenna synchronization [16,19]. Space shift keying (SSK)—a special case of SDM [19]—exploits only the spatial positions of transmitters to encode the information bits thereby trading the receiver's complexity off against the data rate. More recently, quantum key distribution has been recognized as the only approach so far to realize secure communications, which can be implemented in FSO links [66–68]. In [69], a high-speed (120 Mbps) four-state continuous-variable quantum key distribution system, based on wavelength-division multiplexing, polarization multiplexing, and orbital angular momentum multiplexing was investigated under atmospheric turbulence.

In order to improve link availability (i.e., 99.999% availability) hybrid FSO-RF systems are proposed and adopted. Indeed, the gap between the backbone and a last mile access network can be filled by adopting the hybrid RF-FSO technology, thereby enabling multiplexing of a number RF users via a single FSO link. More recently, a hybrid dual-hop amplify-and-forward-based RF-FSO link have been considered [70–72], which offers a strong option for the last mile access network connectivity. For such links, the system performance has been investigated assuming Rayleigh and Gamma-Gamma fading on the RF and FSO links, respectively [70], and considering the effect of pointing error [73]. Note, the hybrid FSO-RF system performance can be increased by adopting relay selection approaches, such as the partial relay selection [74,75], the relay is selected based on the CSI of only one hop, in contrast the other relay selection approaches requiring global knowledge of the CSI of all hops. In [76], a unified and general performance evaluation of a dual-hop amplify-and-forward-based RF-FSO link with the relay selection strategy based on outdated CSI estimates is reported.

The FSO technology along with plastic optical fiber/ multimode fiber and indoor optical wireless communication links can also be used as part of heterogeneous optical networks to address the bandwidth bottleneck in communication networks, where ultra-wideband RF signal at 60 GHz are transmitted over fiber (i.e., the radio-over-fiber (RoF)). Note that, RoF approach requires frequent electrical-to-optical and optical-to-electrical conversions and costly optical components. In addition, it is not very efficient in terms of optical bandwidth utilization since it transmits low-speed wireless data over a wide bandwidth optical channel [78]. The FSO-based heterogeneous optical networks can deliver high-speed optical signal, i.e., 40 Gbps and beyond to the end-user, thus providing high-bandwidth, and solving interoperability problem of future optical networks [78]. Furthermore, optical networking facilitates the operation complexity at the IP layer and thus reduce the connection latency and the cost to deploy and operate the networks.

1.5 FSO networks

The LoS FSO-based topologies can be classified into three categories of point-to-point, ring, and mesh networks as shown in Figure 1.1. The first topology is the most basic and most commonly adopted, which only requires two LoS FSO transceiver between two locations typically few hundred meters up to few kilometers (usually $\leq$ 4 km) a part. The second topology is composed of more than two nodes offering improved network and traffic protection. However, it does suffer from the link failure, which can be resolved using the traffic loop back mechanism provided there is a monitoring mechanism to detect the fault as well as protocols to activate the loop-back mechanism, which is usually located at the physical layer near to the transceiver head or at the MAC layer [77]. To cover a longer and wider transmission span and area, the ring network can be linked up with another ring networks.

The mesh topology, which is capable of interconnecting multi-nodes, the communication links are scalable and expandable up to several kilometers-square, therefore is considered as the best solution to interconnect FSO nodes in different terrain. In addition, this topology offers flexibility in mounting the individual FSO module at different conditions as well as the best network and service protection, which is suitable as part of ad-hoc networks provided the LoS is maintained

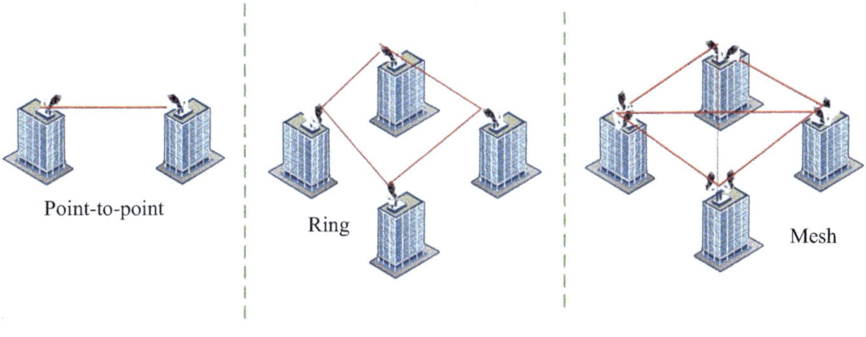

Figure 1.1 FSO topologies

(i.e., high availability, connectivity, increased capacity, and network utilization) [78]. In this topology, one detecting a faulty node, the traffic is rerouted via an alternative path based on the routing algorithms. However, the drawbacks are complexity and increased maintenance cost (requiring traffic capacity, traffic routing, and traffic balancing). As in the ring topology network, this network needs thorough monitoring for fault detection, require traffic balancing so that all nodes deliver the acceptable performances, and also maintain the traffic priorities.

The basic concept of FSO communications is similar to RF communications in terms of data generation, modulation, transmission, reception, and processing of data. A typical IM/DD FSO link is shown in Figure 1.2. Both the transmitter and the receiver must directly point to each other without having any obstruction in their path to ensure the communications link is always established. The unguided channel could be either space, seawater, or the atmosphere.

The Tx composed of four main components of laser modulator, driver, optical source, and transmit telescope. Both laser diodes and light-emitting diodes (LEDs)-based transmitters can be used in FSO systems [79], where laser-based FSO links are mostly used for high data rates (>Gbps) and long range transmission [80] and LED-based FSO systems are used for low data rates and short range inter-building communications [81]. The IM laser beam is transmitted via the transmit telescope, which collimates and directs the optical radiation toward the receiver telescope at the other end through free space channel. The functionality of the driver is to regulate current flowing through the light source and stabilizes its performances as well as neutralizes temperatures and aging effects on the performance of the laser. While there are a number of modulation scheme that can be adopted in FSO systems, the most common modulation formats considered is the binary amplitude-shift-keying or widely known as the on-and-off (OOK) due to its high bandwidth efficiency.

The non-return-to-zero (NRZ) OOK format is the most simplest followed by return-to-zero (RZ), which offers higher sensitivity compared to NRZ OOK [82] and with the clock frequency being part of the signal spectrum. However, in both NRZ- and RZ-based OOK long transmitted bit stream of "1" and "0" can lead to the loss of clock synchronization. This can be avoided by using Manchester coding, and pulse position modulations, the clock can easily be recovered, but at the cost of increased transmission bandwidth. OOK has become a dominant form of

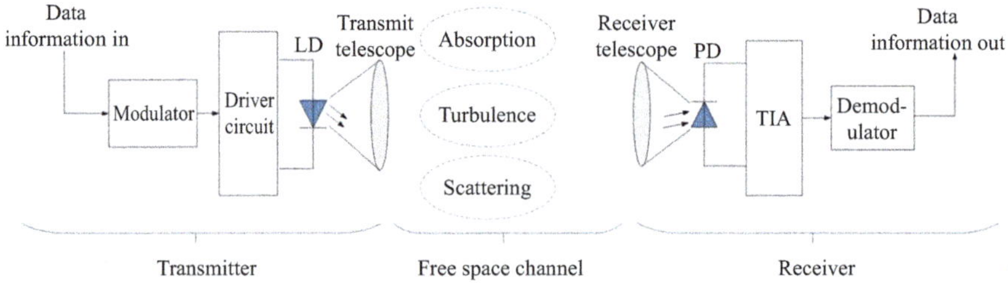

Figure 1.2 The block diagram of an FSO communications system

signaling because of the transmitter and the receiver hardware are relatively simple and fiber optics networks generally operate a high signal-to-noise ratio (SNR) with a lower dynamic range requirement and well controlled signal levels at the receiver [83,84].

In addition, line codes (i.e., 8B10B with 25% more bandwidth requirement than NRZ) can be used to maintain a constant short time average of the signal (i.e., the baseline) in order to reduce inter-symbol interference due to high-pass filtering at the receiver, and to recover the clock signal [83]. Alternatively, modulation can be carried out externally where the transmitted laser beam's phase and frequency is modulated using the external modulator such as the symmetric Mach–Zehnder interferometer. Note, the transmit optics such as telescope or lens is used to focus the optical energy toward the receiver in order to minimize the divergence [84,85].

The commercial FSO systems can be classified into two transmission windows of 780–850 nm and 1,520–1,600 nm [86]. The transmission window of 780–850 nm is preferable for FSO due to low cost, reliable, and high performance transmitter and detector components are readily available and are commonly used in networks and transmission equipment. Moreover, avalanche photodiode and a cheaper vertical-cavity surface-emitting lasers (VCSELs) can be used in this wavelength range offering modulation speeds beyond 3 Gbps [87]. On the other hand, the 1,550 nm wavelength is well suited for free space transmission due to low attenuation (i.e., Rayleigh scattering), as well as the proliferation of high-quality transmitter and detector components. In addition, it facilitate higher transmit power, 50 times greater than 800 nm wavelength [40], due to lower eye safety requirements, thus enabling data transmission over longer distances. According to [86], laser beams at 1,550 nm wavelength are more eye safety since the laser beam at the wavelength above 1,400 nm is absorbed by the lens and the cornea, and thus there is no destructive focal point to create damage on the retina [60]. On the other hand, at 800 nm the retina could be permanently damaged as the collimated light beam entering the eye is concentrated by a factor of 100,000 times when it strikes the retina. Most of the optical devices are also compatible with 1,550 nm. However, the trade-off are less receiver sensitivity and higher component cost.

At the receiver end, a large receiver telescope aperture is desirable in order to collect the uncorrelated flux of optical beams and focusing their average flux onto the photodetector. This is called aperture averaging; however, this will collect more background noise intrinsic to a wide aperture area [88]. Hence, an optical band pass filter is normally used to minimize the magnitude of the background noise. Compared to the transmitter, the receiver choice is much more limited. The two most common photodetectors used are PIN diode and avalanche photodiode (APD). Note, PIN-based optical receivers are widely used in outdoor FSO links compared to APD, which shows reduced SNR by amplifying the ambient noise [89]. In addition, APD require a much higher bias voltages level (i.e., >30 V for InGaAs to 300 V for the silicon-based APDs) [90].

The received optical beam is collected and focused by the receiver telescope to the PD, which converts the optical signal into an electrical form prior to being amplified by trans-impedance amplifier, and demodulator. Semiconductor photodiodes (i.e., p-i-n and avalanche photodiode) are usually preferable because of their compact size, relatively high spectral sensitivity, and a very fast response time (rise and fall time) [12,91]. Normally, an optical band filter is placed before the PD to minimize the effects of background radiation [92]. Following amplification, the original electrical signal is reconstructed by the demodulator from the time-varying current in spite of the channel-induced degradation and the added noise at the receiver. The design of demodulator is depends on the nature of the signal (i.e., analog or digital) and the modulation format [14,93]. Prior to recovery of the information, post detection processor is used where the necessary filtering and signal processing are carried out to guarantee high fidelity data. With IM/DD-based FSO communication systems, effective detection techniques are needed to mitigate the channel-induced performance degradations [94–97]. There are a number of detection schemes including.

(i) **Maximum-likelihood sequence detection (MLSD)**—which outperforms the maximum-likelihood (ML) symbol-by-symbol detection scheme provided the temporal correlation of turbulence τ_0 is known [94]. However, for $\tau_0 \cong 1 - 10\,\text{ms}$ the computational (i.e., implementation) complexity for MLSD at the receiver is relatively high. To overcome, this issue suboptimal MLSD schemes based on the single-step Markov chain (SMC) model could be adopted [98], which require perfect channel state information at the receiving end. Provided, τ_0 is known, a pilot symbol, periodically added to the data frame in pilot-symbol-assisted modulation (PSAM), could be used to mitigate the effects of channel fading, but at the cost reduced system throughputs [99].

(ii) **Decision-feedback detection**—a ML sequence receiver with no requirement for the knowledge of CSI, channel distribution and transmit power where detection is based on the prior knowledge of previous decisions made and on the observation window over τ_0 [100]. However, this scheme has a drawback where the value of τ_0 depends on the data stream, where one needs to use a fast multi-symbol detection scheme based on block-wise decisions and a fast search algorithm [101]. The main drawback of this method is the trade-off between the throughput and performance as well as being too complex to implement.

(iii) **Blind detection**—where there is no CSI considering the case for background-noise limited and a sub-optimum ML detection-based receivers [102,103], but with poor performance over a small observation window.

(iv) **Differential signaling**—which utilizes a pre-fixed threshold level under various channel conditions (rain, turbulence, etc.), and it does not require CSI and neither has extensive computations at the receiver [104]. Note that, in this scheme (a) the system throughput is not reduced since no pilot signals or training sequences are used; (b) offers simplified detection procedure; and (c) mitigates for the background noise (i.e., the ambient noise) at the receive [105].

(v) **Spatial diversity and multiplexing [96,97]**—this technique offers substantial link performance improvement in spatially uncorrelated channels by employing multiple apertures at the transmitter and/or the receiver that are sufficiently spaced as in single-input multiple-output (SIMO), multiple-input single-output (MISO) or multiple-input multiple-output (MIMO) systems [12,106–108]. In FSO systems, the most commonly adopted spatial diversity techniques are repetition coding, which achieves transmit diversity by simultaneous transmission of the data via all transmitters, and orthogonal space time block codes. Note that, in IM/DD FSO links repetition coding outperforms orthogonal space time block codes [109,110]. Multiplexing of multiple orbital-angular-momentum (OAM) beams is another possible approach for increasing system capacity (up to 100 Tbps combined with wavelength division multiplexing) and spectral efficiency in FSO systems [111].

(vi) **Relay-assisted or multi-hop FSO**—a powerful fading mitigation tool as an alternate option in realizing the spatial diversity scheme advantages, which is based on the broadcast nature of the RF wireless technology [112–116]. An all-optical FSO relay-assisted system can be adopted to mitigate the destructive effects due to distance dependent atmospheric turbulence-induced fading [117]. It offers an efficient and low-cost solution compared to the MIMO systems as it does not need an additional transmitter and receiver aperture.

(vii) **Hybrid FSO/RF**—the hybrid FSO/RF link refers to a single antenna unit with dual functionalities at the transmitter and receiver for both optical and RF signals transmission [118]. The key features of the hybrid system are (i) reduced power consumption and costs, which is achieved by means of incorporating the optical aperture as part of the RF antenna and only utilizing FSO or RF path at any given time depending on the weather conditions; (ii) link alignment, where both FSO and RF could be used to establish the link alignment and maintain it via auto-tracking system within a certain degree; (iii) high link availability, which ensures full link availability under all weather conditions with higher data rates capability; and (iv) installation cost and complexity, which is much lower in the hybrid antenna-based wireless link than the dual antenna base systems [116–122].

FSO systems with coherent receivers and the benefit of adopting spatial diversity techniques are extensively reported in the literature [15,123,124]. Coherent detection (homodyne and heterodyne) is employed in less reliable FSO links in order to increase the sensitivity of the receiver and improve the robustness of the systems against channel-induced impairments such as turbulence [15,125]. In homodyne-based detection, which uses a local optical oscillator synchronized to the transmitted optical signal carrier frequency, the optical signal is directly demodulated to the baseband. However, optical synchronization is a bit unstable in practice, therefore heterodyne detection is adopted, which simplifies the receiver design by converting the optical signal back into an electrical signal with an intermediate frequency followed by a phase noise compensation technique for the IF signal phase noise tracking. A FSO link with a coherent receiver, which mitigates the degradation performance caused by phase fluctuations due to turbulence phase compensation schemes, have been proposed [126,127].

Note that, in coherent FSO systems with relatively higher system complexity compared with IM/DD offer features such (i) the signal dependent shot noise limited SNR, provided the optical local oscillator has a sufficiently high power; (ii) the extraction of phase information allows for a large number of modulation schemes compared IM/DD; (iii) excellent background noise rejection compared to IM/DD; and (iv) higher sensitivity, and improved spectral efficiency [128,129].

Note, in LoS FSO systems, the link performance highly depends on the number of received photons. For DD-based FSO systems, there is an optimal receiver's field of view to ensure improve link performance. If the radial angle-of-arrival of the optical signal is within the receiver's field of view, then the entire received optical beam will be collected at the receiver. However, if the receiver's field-of-view is small, then the number of received photons is strongly related to the turbulence-induced angular spread, which needs considering in order to properly investigate the performance of FSO systems. Thus, the receiver's aperture acts as a spatial filter only collecting photons within its field of view. In [128], the benefit of coherent detection over DD in the presence of angular spread was investigated. Moreover, spatial diversity receivers, which are able to significantly improve the performance of atmospheric optical systems are discussed in detail.

In optical communications including optical wireless communications, where the transmitted data rates are very high, the bit duration is very short compared to the channel coherence time. Therefore, in a turbulence channel the FSO link performance is best measure by the outage probability instead of the most commonly used bit error probability [123]. Note that, outage occurs when the error probability is above a threshold level, which indicates how often the link performance falls below the given threshold level.

1.6 Factors affecting FSO systems

There has been tremendous technical advancement of available components such as laser/LED transmitters, high sensitivity optical receivers offering extremely high bandwidth, efficient modulation techniques, improvement in low power consumption, weight, and size. In spite of many such technological development, the major limitation of FSO communications performance is the atmosphere conditions. The terrestrial LoS FSO link operating in the troposphere layer will experience a medium, which is continuously changing in chemical composition, humidity, pressure, temperature, and air movements. As a result, the FSO link performance is hampered by the atmospheric channel, which is highly variable, unpredictable and vulnerable to different weather conditions such as smoke, fog, haze, sandstorm, low clouds, snow, rain, atmospheric turbulence, and pointing errors [1,29,130], which may result in noticeable distance-related power loss, and phase distortion at the receiving end [29,106,131]. With conditions of the earth's atmosphere, only a few atmospheric windows are suitable for FSO due to selective absorption by gases and water vapor [61]. The interaction with solid and liquid water particles in adverse weather can also generates signal fades which can lead to link outage [132]. Moreover, even in clear sky conditions, the turbulence induced by temperature and pressure gradients results in random fluctuations and the loss of wave-front coherence [7]. Fog, haze, and dust-induced atmospheric attenuation are critical and can result in link failures. Scattering due to collision of photons with the scatterers, which is wavelength dependent, leads to reduced light intensity over a longer transmission span. Physical obstructions due to tall buildings, flying birds, trees, etc. can temporarily block the propagating beam, thus resulting in burst errors or link failures; whereas geometric losses due to the beam spreading reduces the power level, i.e., lower SNR at the receiver [133]. Absorption due to water molecules and carbon dioxide, reduces the power density of the propagating optical beam and therefore directly affecting the FSO link availability [134].

Atmospheric condition thus ultimately determines the FSO communication systems performance not only of terrestrial applications but also for space (satellite) links involving uplink-downlink communications (e.g., between ground and satellite, aircraft or UAV terminals), because a portion of the atmospheric path always includes turbulence and multiple scattering effects. There has been many research published during the last two decades on the subject of effects of atmosphere on optical communication channel and therefore are not repeated in this section. Interested readers can review them for understanding the details of the atmospheric channel for establishing communications between a transmitter and a receiver.

The FSO link visibility, attenuation in dB/km, and its effective link range under various weather conditions are discussed in [12,135]. For telecommunication applications, FSO systems will need to meet very high availability requirements. For example, carrier-class availability is considered to be 99.999% ("5 nines") for very high data rate communications. This reliability of 99.999% is equivalent to the link availability as the percentage of time over a year that a FSO link will be operational is the same as "down 5 minute/year." The FSO link ranges in the worst measured conditions for fog, snow, and rain in order to extrapolate 99.999% availability link ranges, as well as the actual 99.999% link ranges for Phoenix, San Juan, Las Vegas, and Honolulu are given in [135].

1.7 FSO link reliability

The connection in a FSO system is accomplished by means of a narrow optical beam with low divergence. For a successful and reliable installation of optical link, it is therefore necessary to know the steady parameters for standard atmosphere and statistical character of the weather in a given locality. The performance of a FSO communications system is generally quantified by the "link margin" (LM), which is defined as ration of the signal power received to the signal power required to achieve a specified data rate with a specified acceptable probability of error. The

LM calculation is therefore essential in order to design an acceptable system. The atmospheric conditions affecting the FSO link performance needs to be considered in the calculation. A link budget model, which includes dependence on the atmospheric channel and on the transmitter and the receiver [9,136–138], will aid designers in optimizing the FSO base station main parameters in order to be able to establish a data link with adequate performance. The link budget includes the transmit optical power P_t [dBm], the received power P_r [dBm], which is related to the receiver sensitivity N_b [dBm] and all propagation losses L_p [dBm]. If we express the LM in dBm, then we can write [9]:

$$LM = P_t - L_p - N_b \tag{1.1}$$

The LM value shows how much margin a FSO system has at a given range to compensate for scattering, absorption (due to fog, snow, and rain) and scintillation (turbulence). A simple calculation shows that, using a laser with a transmit power of 30 mW, a detector sensitivity of 25 nW, a mis-pointing loss of 3 dB, and an optical loss of 4 dB the FSO link margin is estimated to be 54 dB. Note, for this link budget the FSO range even in the heaviest fog with 350 dB/km of attenuation, which corresponds to 99.999% link availability, could be up to 140 m.

We can define the dynamic range as the interval of acceptable power in which the link function is guaranteed with a definite error rate. The receiver is saturated when $P_r > P_S$ (a specified saturation value), and the required SNR is not provided when $P_r < N_b$. Link reliability is quantified by the availability, which is the percentage of time T_{av} (%), when the data transmission bit rate is higher than its required value. The link availability can equivalently be defined as the probability that additional power losses L_A caused by atmospheric effects (including absorption and scattering) are less (in dB) than the LM, which is defined as:

$$T_{av} = 100\% \cdot \int_0^{\alpha_A \lim} p(\alpha_A) \cdot d\alpha_A \tag{1.2}$$

where $p(\alpha_A)$ is the probability density of an attenuation coefficient α_A (dB/km), L is the range and the limiting attenuation coefficient value is given by:

$$\alpha_{A_{\mathrm{LIM}}} = \mathrm{LM}(L) \cdot 1{,}000/L \tag{1.3}$$

$p(\alpha_A)$ can be determined from long-time monitoring of the received signal level of a real link, or using the data collected over a long time in a specific location. In practical network design, the concept of link availability in presence of atmospheric turbulence is an important consideration. In many FSO systems, an automatic tracking system is used so that the pointing error is not taken into consideration when defining the link reliability.

Practical solution to extend the high availability link range can be (i) a hybrid FSO/RF communications system [76,118]; and (ii) optical network topologies employing relayed FSO links [9] particularly attractive for the metro-access market offered by the carriers. To evaluate a FSO as an access technology for a particular location with a given link range, estimating the link availability is essential. Enterprise-class availabilities can also extend the possible FSO link ranges to much longer distances depending on the geographical locations. The researchers have suggested that, for high reliability the optimum network architecture is a meshed network, because it combines the advantages of ring and star architectures.

A reliability analysis of FSO communications link using an aberrated divergent rectangular partially coherent flat-topped beam is been reported in [139], which is based on numerical values for power-in-bucket (PIB), SNR and BER considering atmospheric losses due to absorption, scattering and turbulence. Reliability of FSO links including the laser link stability can be improved using auto-track subsystems in presence of different beam divergences considering the atmospheric effects including absorption, scattering and turbulence [140]. An innovative approach based on all-optical relaying technique was proposed to mitigate the degrading effects of atmospheric turbulence-induced fading by relaying data from the source to the destination using intermediate terminals [27,70,95,112,117]. The proposed techniques of optical amplify-and-forward relaying and optical regenerate-and-forward relaying were applied to multihop FSO systems in order to extend the maximum accessible distance for high data rates, which improved the link reliability. Reliability of the optical communications system was improved using polarization shift keying modulation-based FSO systems [12,141–143]. There are other techniques to improve the reliability of the FSO systems such as hybrid FSO/RF communications with channel coding [144,145], soft-switching of FSO/RF links using field-programmable gate arrays, modulation schemes combined with LDPC forward correction scheme and multipath diversity [48,70,72,118,144]. Optimization of the FSO-based network for cellular backhauling in order to improve reliability was achieved by introducing a novel integer linear programming model [142,145,146].

References

[1] "Internet growth statistics," 2016. [Online]. Available: http://www.internetworldstats.com/emarketing.htm. [Accessed: 4-Sep-2016].

[2] S. Wabnitz and B. J. Eggleton, All-optical Signal Processing: Data Communication and Storage. Springer Series in Optical Sciences. Cham: Springer, Cham: Springer, 2015.

[3] K.-L. Du and M. N. S. Swamy, Wireless Communication Systems from RF Subsystems to 4G Enabling Technologies. United Kingdom: Cambridge University Press, 2010.

[4] A. Osseiran, F. Boccardi, V. Braun, *et al.*, "Scenarios for the 5G mobile and wireless communications: the vision of the METIS project," IEEE Communications Magazine, vol. 52, no. 5, pp. 26–35, 2014.

[5] A. K. Majumdar, Advanced Free Space Optics (FSO): A System Approach, vol. 140. USA: Springer, 2015.

[6] S. Arnon, J. R. Barry, G. K. Karagiannidis, R. Schober, and M. Uysal, Advanced Optical Wireless Communication Systems. New York, USA: Cambridge University Press, 2012.
[7] L. Andrews, R. Phillips, and C. Hopen, Laser Beam Scintillation with Applications. New York: USA: SPIE Press, 2001.
[8] H. Henniger and O. Wilfert, "An introduction to free-space optical communications," Radio Engineering, vol. 19, no. 2, pp. 203–212, 2010.
[9] A. K. Majumdar and J. C. Ricklin, Free-Space Laser Communications: Principles and Advances, Springer, New York, 2008.
[10] A. Bazil Raj, Free Space Optical Communication: System Design, Modeling, Characterization and Dealing with Turbulence, de Gruyter, Gmbh, 2015.
[11] http://www.laseroptronics.com/index.cfm/id/57-66.htm. Last visited 12 Aug. 2018.
[12] Z. Ghassemlooy, W. Popoola, and S. Rajbhandari, Optical wireless communications: System and channel modelling with MATLAB. United Kingdom: CRC Press Taylor and Francis Group, 2013.
[13] M. A. Khalighi and M. Uysal, "Survey on free space optical communication: A communication theory perspective," IEEE Communications Surveys and Tutorials, vol. 16, no. 4, pp. 2231–2258, 2014.
[14] H. Willebrand and B. Ghuman, Free Space Optics: Enabling Optical Connectivity in Today's Networks, 1st ed. US: Sams Publishing, 2001.
[15] V. W. S. Chan, "Free-space optical communications," Journal of Lightwave Technology, vol. 24, no. 12, pp. 4750–4762, 2006.
[16] G. e.-S. Initiative, "SMART 2020: Enabling the low carbon economy in the information age," press release, Brussels, Belgium, June, vol. 20, 2008.
[17] E. Bonetto, S. Buzzi, D. Cuda, G. A. G. Castillo, and F. Neri, "Optical technologies can improve the energy efficiency of networks," in Optical Communication, 2009. ECOC '09. 35th European Conference on, 2009, pp. 1–4.
[18] E. Gulsen, E. Olivetti, L. C. Kimerling, and R. Kirchain, "Energy concerns in information and communication technology and the potential for photonics integration," in Sustainable Systems and Technology (ISSST), 2010 IEEE International Symposium on, 2010, p. 1.
[19] A. Sikora and V. F. Groza, "Coexistence of IEEE 802. 15.4 with other Systems in the 2.4 GHz-ISM-Band," in Instrumentation and Measurement Technology Conference, 2005. IMTC 2005. Proceedings of the IEEE, 2005, pp. 1786–1791.
[20] L. Rakotondrainibe, Y. Kokar, G. Zaharia, and G. El-Zein, "60 GHz high data rate wireless communication system," in Vehicular Technology Conference, 2009. VTC Spring 2009. IEEE 69th, 2009, pp. 1–5.
[21] D. Sinefeld, D. Shayovitz, O. Golani, and D. M. Marom, "Adaptive rate and bandwidth WDM optical sampling pulse streams with LCoS-based photonic spectral processor," in Optical Fiber Communication Conference, 2013.
[22] J. I. Saari, M. M. Krause, B. R. Walsh, and P. Kambhampati, "Terahertz bandwidth all-optical modulation and logic using multiexcitons in semiconductor nanocrystals," Nano Letters, 2013.

[23] A. K. Rahman, M. S. Anuar, S. A. Aljunid, and M. N. Junita, "Study of rain attenuation consequence in free space optic transmission," in Proceedings of the 2nd Malaysia Conference on Photonics Telecommunication Technologies (NCTT-MCP '08), pp. 64–70, IEEE, Putrajaya, Malaysia, August 2008.
[24] X. Liu, "Secrecy capacity of wireless channels subject to log-normal fading," Security and Communication Networks, pp. n/a-n/a, 2013.
[25] L. F. Abdulameer, U. Sripati, and M. Kulkarni, "Enhancement of security for free space optics based on reconfigurable chaotic technique," in International Conference on Communication and Electronics System Design, 2013, pp. 87600O–87600O-7.
[26] R. McClintock, A. Haddadi, and M. Razeghi, "Free-space optical communication using mid-infrared or solar-blind ultraviolet sources and detectors," in Proceedings of SPIE Vol, 2012, pp. 826810-1.
[27] F. Ahdi and S. Subramaniam, "Optimal placement of FSO relays for network disaster recovery," in Proceedings of IEEE International Conference on Communications (ICC), pp. 3921–3926, June 2013, Budapest, Hungary.
[28] R. Peach, G. Burdge, F. Reitberger, *et al.*, "Performance of a 10 Gbps QoS-based buffer in a FSO/RF IP network," in SPIE Optical Engineering + Applications, 2010, pp. 781402-781402-12.
[29] K. P. Peppas and C. K. Datsikas, "Average symbol error probability of general-order rectangular quadrature amplitude modulation of optical wireless communication systems over atmospheric turbulence channels," Journal of Optical Communications and Networking, vol. 2, no. 2, pp. 102–110, 2010.
[30] S. Ahmad Anas, F. Hamat, S. Hitam, and R. K. Sahbudin, "Hybrid fiber-to-the-x and free space optics for high bandwidth access networks," Photonic Network Communications, pp. 1–7, 2012.
[31] G. Kramer and G. Pesavento, "Ethernet passive optical network (EPON): Building a next-generation optical access network," Communications Magazine, IEEE, vol. 40, no. 2, pp. 66–73, 2002.
[32] H. A. Fadhil, A. Amphawan, H. A. Shamsuddin, *et al.*, "Optimization of free space optics parameters: An optimum solution for bad weather conditions," Optik-International Journal for Light and Electron Optics, 2013.
[33] E. Leitgeb, J. Bregenzer, P. Fasser, and M. Gebhart, "Free space optics-extension to fiber-networks for the," in Lasers and Electro-Optics Society, 2002. LEOS 2002. The 15th Annual Meeting of the IEEE, 2002, pp. 459–460.
[34] E. Leitgeb, M. Gebhart, and U. Birnbacher, "Optical networks, last mile access and applications," Journal of Optical and Fiber Communications Research, vol. 2, no. 1, pp. 56–85, 2005.
[35] A. Acampora, "Last mile by laser," Scientific American, vol. 287, no. 1, pp. 32–7, 2002.
[36] N. Blaunstein, S. Arnon, A. Zilberman, and N. Kopeika, "Applied aspects of optical communication and LIDAR," Boca Raton: CRC Press, London, pp. 53–55, 2010.

[37] A. Bazil Raj, J. A. V. Selvi, and S. Raghavan, "Real-time measurement of meteorological parameters for estimating low altitude atmospheric turbulence strength (C_n^2)," IET–Science Measurement & Technology, vol. 8, no. 6, pp. 459–469, 2014.

[38] W. O. Popoola and Z. Ghassemlooy, "BPSK subcarrier intensity modulated free-space optical communications in atmospheric turbulence," Journal of Lightwave Technology, vol. 27, no. 8, pp. 967–973, 2009.

[39] E. Bayaki, R. Schober, and R. K. Mallik, "Performance analysis of MIMO free-space optical systems in gamma-gamma fading," IEEE Transactions on Communications, vol. 57, no. 11, pp. 3415–3424, 2009.

[40] W. Popoola, Z. Ghassemlooy, J. Allen, E. Leitgeb, and S. Gao, "Freespace optical communication employing subcarrier modulation and spatial diversity in atmospheric turbulence channel," IET Optoelectronics, vol. 2, no. 1, pp. 16–23, 2008.

[41] N. Letzepis and A. G. Fabregas, "Outage probability of the Gaussian MIMO free-space optical channel with PPM," IEEE Transactions on Communications,vol. 57, no. 12, pp. 3682–3690, 2009.

[42] H. Hemmati, Deep Space Optical Communications. New York: John Wiley & Sons, 2006.

[43] H. Hemmati and B. J. Thompson, Near-Earth Laser Communications. London: Taylor & Francis Group, 2008.

[44] H. Kaushal, V. K. Jain, and S. Kar, "Improvement of ground to satellite FSO link performance using transmit diversity in weak atmospheric turbulence," in International Conference on Intelligent and Advanced Systems (ICIAS) 2010, 2010, pp. 1–6.

[45] M. Toyoshima, Y. Takayama, T. Takahashi, *et al.*, "Ground-to-satellite laser communication experiments," IEEE Aerospace and Electronic Systems Magazine, vol. 23, no. 8, pp. 10–18, 2008.

[46] A. Bazil Raj, J. Arputha Vijaya Selvi, and S. Durairaj, Comparison of different models for ground-level atmospheric turbulence strength (C_n^2) prediction with a new model according to local weather data for FSO applications, Applied Optics, vol. 54, no. 4, pp. 802–815, 2015.

[47] A. Bazil Raj, J. A. V. Selvi, D. Kumar, and N. Sivakumaran, "Mitigation of beam fluctuation due to atmospheric turbulence and prediction of control quality using intelligent decision-making tools," Applied Optics, vol. 53, no. 17, pp. 3796–3806, 2014.

[48] O. Awwad, A. Al-Fuqaha, B. Khan, and G. B. Brahim, "Topology Control Schema for Better QoS in Hybrid RF/FSO Mesh Networks," IEEE Transactions on Communications, vol. 60, no. 5, pp. 1398–1406, 2012.

[49] V. Vishnevskii, O. Semenova, and S. Y. Sharov, "Modeling and analysis of a hybrid communication channel based on free-space optical and radio-frequency technologies," Automation and Remote Control, vol. 74, no. 3, pp. 521–528, 2013.

[50] J. Libich, M. Mudroch, P. Dvorak, and S. Zvanovec, "Performance analysis of hybrid FSO/RF link," in Antennas and Propagation (EUCAP), 2012 6th European Conference on, 2012, pp. 1235–1238.
[51] F. Nadeem, V. Kvicera, M. S. Awan, E. Leitgeb, S. Muhammad, and G. Kandus, "Weather effects on hybrid FSO/RF communication link," IEEE Journal on Selected Areas in Communications, vol. 27, no. 9, pp. 1687–1697, 2009.
[52] Z. Kolka, Z. Kincl, V. Biolkova, and D. Biolek, "Hybrid FSO/RF test link," in Ultra Modern Telecommunications and Control Systems and Workshops (ICUMT), 2012 4th International Congress on, 2012, pp. 502–505.
[53] Y. Tang, M. Brandt-Pearce, and S. Wilson, "Link adaptation for throughput optimization of parallel channels with application to hybrid FSO/RF systems," IEEE Transactions on Communications, vol. 60, no. 9, pp. 2723–2732, 2012.
[54] Guowei Yang, Mohammad-Ali Khalighi, Zabih Ghassemlooy, and Salah Bourennane, "Performance evaluation of receive-diversity free-space optical communications over correlated gamma–gamma fading channels," Applied Optics, vol. 52, no. 24, pp. 5903–5911, 2013.
[55] X. Liu, "Free-space optics optimization models for building sway and atmospheric interference using variable wavelength," IEEE Transactions on Communications, vol. 57, no. 2, pp. 492–498, 2009.
[56] W. Gappmair, "Further results on the capacity of free-space optical channels in turbulent atmosphere," IET Communications, vol. 5, no. 9, pp. 1262–1267, 2011.
[57] I. E. Lee, Z. Ghassemlooy, W. P. Ng, and M.-A. Khalighi, "Joint optimization of a partially coherent Gaussian beam for free-space optical communication over turbulent channels with pointing errors," Optics Letters, vol. 38, no. 3, pp. 350–352, 2013.
[58] F. Yang, J. Cheng, and T. A. Tsiftsis, "Free-space optical communication with nonzero boresight pointing errors," IEEE Transactions on Communications, vol. 62, no. 2, pp. 713–725, 2014.
[59] A. Paraskevopoulos, J. Vucic, S. H. Voss, R. Swoboda, and K.-D. Langer, "Optical wireless communication systems in the Mb/s to Gb/s range, suitable for industrial applications," IEEE/ASME Transactions on Mechatronics, vol. 15, no. 4, pp. 541–547, 2010.
[60] G. Parca, "Optical wireless transmission at 1.6-Tbit/s (16×100 Gbit/s) for next generation convergent urban infrastructures," Optical Engineering, vol. 52, no. 11, p. 116102, 2013.
[61] M. Uysal, C. Capsoni, Z. Ghassemlooy, and A. Bousouvalas, Optical Wireless Communications: An Emerging Technology, Switzerland: Springer International Publishing, 2016.
[62] E. Ciaramella, Y. Arimoto, G. Contestabile, *et al.*, "1.28 Terabit/s (32×40 Gbit/s) WDM transmission system forfree space optical communications," IEEE Journal on Selected Areas in Communications, vol. 27, no. 9, pp. 1639–1645, 2009.

[63] I. B. Djordjevic, "Deep-space and near-Earth optical communications by coded orbital angular momentum (OAM) modulation," Optics Express, vol. 19, no. 15, pp. 14277–14289, 2011.

[64] G. Gibson, J. Courtial, M. J. Padgett, *et al.*, "Free-space information transfer using light beams carrying orbital angular momentum," Optics Express, vol. 12, no. 22, 5448–5456, 2004.

[65] K. Pang, H. Song, Z. Zhao, *et al.*, "Experimental Demonstration of 400-Gbit/s Free-Space Mode-Division-Multiplexing by Varying Both Indices when using Four Laguerre–Gaussian Modes or Four Hermite-Gaussian Modes," 2018 Conference on Lasers and Electro-Optics (CLEO), San Jose, CA, USA, 2018, pp. 1–2.

[66] B. Heim, C. Peuntinger, N. Killoran, *et al.*, "Atmospheric continuous-variable quantum communication," New Journal of Physics, vol. 16, no. 11, p. 113018, 2014.

[67] X. Sun, I. B. Djordjevic, and M. A. Neifeld, "Multiple spatial modes based QKD over marine free-space optical channels in the presence of atmospheric turbulence," Optics Express, vol. 24, no. 24, pp. 27663–27673, 2016.

[68] C. Paterson, "Atmospheric turbulence and orbital angular momentum of single photons for optical communication," Physical Review Letters, vol. 94, no. 15, p. 153901, 2005.

[69] Z. Qu and I. B. Djordjevic, "High-speed free-space optical continuous-variable quantum key distribution enabled by three-dimensional multiplexing," Optics Express, vol. 25, no. 7, 7919–7928, 2017.

[70] E. Lee, J. Park, D. Han, and G. Yoon, "Performance analysis of the asymmetric dual-hop relay transmission with mixed RF/FSO links," IEEE Photonics Technology Letters, vol. 23, no. 21, pp. 1642–1644, 2011.

[71] H. AlQuwaiee, I. S. Ansari, and M.-S. Alouini, "On the performance of free-space optical communication systems over double Generalized Gamma channel," IEEE Journal on Selected Areas in Communications, vol. 33, no. 9, pp. 1829–1840, 2015.

[72] P. V. Trinh, T. C. Thang, and A. T. Pham, "Mixed mm wave RF/FSO relaying systems over generalized fading channels with pointing errors," IEEE Photonics Journal, vol. 9, no. 1, pp. 1–14, 2017.

[73] B. Asharafzadeh, E. Soleimani-Nasab, and M. Kamandar, "Performance analysis of mixed DGG and generalized Nakagami-m dual-hop FSO/RF transmission systems," in Proceedings of IEEE TELFOR'16, Nov. 2016, pp. 1–4.

[74] D. B. da Costa and S. Aıssa, "End-to-end performance of dual-hop semi-blind relaying systems with partial relay selection," IEEE Transactions on Wireless Communications, vol. 8, no. 8, pp. 4306–4315, 2009.

[75] M. Soysa, H. A. Suraweera, C. Tellambura, and H. K. Garg, "Partial and opportunistic relay selection with outdated channel estimates," IEEE Transactions on Communications, vol. 60, no. 3, pp. 840–850, 2012.

[76] G. N. Kamga and S. Aissa, "Relay selection based hybrid RF/FSO transmission over double generalized gamma channels under outdated CSI

and pointing errors," 2018 IEEE International Conference on Communications (ICC), Kansas City, MO, USA, 2018, pp. 1–6. doi: 10.1109/ICC.2018.8422434.
[77] S. V. Kartalopoulos, Free Space Optical Networks for Ultra-Broad Band Services. Hoboken, NJ: Wiley-IEEE Press, 2011.
[78] P. C. Gurumohan and J. Hui, "Topology design for free space optical networks," in IEEE International Conference on Computer Communications and Networks, 2003, pp. 576–579.
[79] J. Liao, A. Mirvakili, A. Boryssenko, V. Joyner, and Z. R. Huang, "Integration of LED chip within patch antenna geometry for hybrid FSO/ RF communication," Electronics Letters, vol. 46, no. 19, pp. 1332–1333, 2010.
[80] A. Jabeena, B. Praneeth, and P. Arulmozhivarman, "Laser based optical transceiver for data transfer of free space optical communication," European Journal of Scientific Research, vol. 67, no. 2, pp. 294–300, 2012.
[81] Y. H. Kho, L. Yong, K. L. Lau, and K. P. Kiu, "Design of an indoor wireless optical transceiver system with source and channel coding," in Industrial Electronics and Applications (ISIEA), 2012 IEEE Symposium on, 2012, pp. 45–49.
[82] M. Pauer, P. Winter, and W. Leeb, "Bit error probability reduction in direct detection optical receivers using RZ coding," Journal of Lightwave Technology, vol. 19, no. 9, pp. 1255–1262, 2001.
[83] A. Street, K. Samaras, D. Obrien, and D. Edwards, "Closed form expressions for baseline wander effects in wireless IR applications," Electronics Letters, vol. 33, no. 12, pp. 1060–1062, 1997.
[84] S. Bloom, E. Korevaar, J. Schuster, and H. Willebrand, "Understanding the performance of free-space optics," Journal of Optical Networking, vol. 2, no. 6, pp. 178–200, 2003.
[85] J. C. Palais, Fiber Optic Communication. Upper Saddle River, NJ: Pearson Education Inc. 2005.
[86] D. Rockwell and S. Mecherle, "Wavelength selection for optical wireless communications systems." Sona Communication Corporation, 2001.
[87] A. K. Majumdar, "Free space laser communication performance in the atmospheric channel," in Free space laser communications: Principles and advances, vol. 2, pp. 57–108, 2005.
[88] M. Khalighi, N. Aitamer, N. Schwartz, and S. Bourennane, "Turbulence mitigation by aperture averaging in wireless optical systems," in Telecommunications, 2009. ConTEL 2009. 10th International Conference on, 2009, pp. 59–66.
[89] F. Xu, M. Khalighi, and S. Bourennane, "Impact of different noise sources on the performance of PIN-and APD-based FSO receivers," in Telecommunications (ConTEL), Proceedings of the 2011 11th International Conference on, 2011, pp. 211–218.
[90] K. Kiasaleh, "Performance of APD-based, PPM free-space optical communication systems in atmospheric turbulence," IEEE Transactions on Communications, vol. 53, no. 9, pp. 1455–1461, 2005.

[91] G. P. Agrawal, Lightwave Technology: Telecommunication Systems. New Jersey: Wiley-Interscience, 2005.

[92] Z. Ghassemlooy and W. O. Popoola, "Terrestrial free-space optical communications," in Mobile and Wireless Communications: Network layer and circuit level design, S. A. Fares and F. Adachi, Eds. InTech, 2010, pp. 355–392.

[93] L. Ibbotson, The Fundamentals of Signal Transmission. London: Arnold, 1999.

[94] X. Zhu and J. M. Kahn, "Free-space optical communication through atmospheric turbulence channels," IEEE Transactions on Communications, vol. 50, no. 8, pp. 1293–1300, 2002.

[95] M. Safari, M. Rad, and M. Uysal, "Multi-hop relaying over the atmospheric Poisson channel: Outage analysis and optimization," IEEE Transactions on Communications, vol. 60, no. 3, pp. 817–829, 2012.

[96] E. Lee and V. W. S. Chan, "Part 1: Optical communication over the clear turbulent atmospheric channel using diversity," IEEE Journal on Selected Areas in Communications, vol. 22, no. 9, pp. 1896–1906, 2004.

[97] M. Safari and S. Hranilovic, "Diversity and multiplexing for near-field atmospheric optical communication," IEEE Transactions on Communications, vol. 61, no. 5, pp. 1988–1997, 2013.

[98] X. Zhu and J. M. Kahn, "Markov chain model in maximum-likelihood sequence detection for free-space optical communication through atmospheric turbulence channels," IEEE Transactions on Communications, vol. 51, no. 3, pp. 509–516, 2003.

[99] X. Zhu and J. M. Kahn, "Pilot-symbol assisted modulation for correlated turbulent free-space optical channels," in Proceedings of SPIE 4489, Free-Space Laser Communication and Laser Imaging, 2002, pp. 138–145.

[100] M. L. B. Riediger, R. Schober, and L. Lampe, "Decision-feedback detection for free-space optical communications," in IEEE 66th Vehicular Technology Conference Fall. 2007, 2007, pp. 1193–1197.

[101] M. L. B. Riediger, R. Schober, and L. Lampe, "Fast multiple-symbol detection for free-space optical communications," IEEE Transactions on Communications, vol. 57, no. 4, pp. 1119–1128, 2009.

[102] M. L. B. Riediger, R. Schober, and L. Lampe, "Blind detection of on-off keying for free-space optical communications," in Electrical and Computer Engineering, 2008. CCECE 2008. Canadian Conference on, 2008, pp. 001361–001364.

[103] N. D. Chatzidiamantis, G. K. Karagiannidis, and M. Uysal, "Generalized maximum-likelihood sequence detection for photon-counting free space optical systems," IEEE Transactions on Communications, vol. 58, no. 12, pp. 3381–3385, 2010.

[104] M. Mansour Abadi, Z. Ghassemlooy, M. A. Khalighi, S. Zvanovec, and M. R. Bhatnagar, "FSO detection using differential signaling in outdoor correlated-channels condition," IEEE Photonics Technology Letters, vol. 28, no. 1, pp. 55–58, 2016.

[105] M. A. Khalighi, F. Xu, Y. Jaafar, and S. Bourennane, "Double-laser differential signaling for reducing the effect of background radiation in free-space optical systems," IEEE/OSA Journal of Optical Communications and Networking, vol. 3, no. 2, pp. 145–154, 2011.
[106] G. Yang, M. A. Khalighi, S. Bourennane, and Z. Ghassemlooy, "Fading correlation and analytical performance evaluation of the space-diversity free-space optical communications system," Journal of Optics, vol. 16, no. 3, pp. 1–10, 2014.
[107] R. Priyadarshani, M. R. Bhatnagar, Z. Ghassemlooy, and S. Zvanovec, "Effect of correlation on BER performance of the FSO-MISO system with repetition coding over gamma-gamma turbulence," IEEE Photonics Journal, vol. 9, no. 5, pp. 1–15, 2017.
[108] M. Abaza, R. Mesleh, A. Mansour, and E. H. M. Aggoune, "Spatial diversity for FSO communication systems over atmospheric turbulence channels," IEEE Wireless Communications and Networking Conference (WCNC), pp. 382–387, Apr. 2014.
[109] M. Safari and M. Uysal, "Do we really need OSTBCs for free-space optical communication with direct detection?" IEEE Transactions on Wireless Communications, vol. 7, no. 11, pp. 4445–4448, 2008.
[110] X. Song and J. Cheng, "Subcarrier intensity modulated MIMO optical communications in atmospheric turbulence," IEEE Journal of Optical Communications and Networking, vol. 5, no. 9, pp. 1001–1009, 2013.
[111] Y. Ren, H. Huang, G. Xie, *et al.*, "Atmospheric turbulence effects on the performance of a free space optical link employing orbital angular momentum multiplexing," Optics Letters, vol. 38, no. 20, 2013.
[112] M. Safari and M. Uysal, "Relay-assisted free-space optical communication," IEEE Transactions on Wireless Communications, vol. 7, no. 12, pp. 5441–5449, 2008.
[113] A. Sendonaris, E. Erkip, and B. Aazhang, "User cooperation diversity-Part I: System description," IEEE Transactions on Communications, vol. 51, no. 11, pp. 1927–1938, 2003.
[114] A. Sendonaris, E. Erkip, and B. Aazhang, "User cooperation diversity-Part II: Implementation aspects and performance analysis," IEEE Transactions on Communications, vol. 51, no. 11, pp. 1939–1948, 2003.
[115] J. N. Laneman, D. N. C. Tse, and G. W. Wornell, "Cooperative diversity in wireless networks: Efficient protocols and outage behavior," IEEE Transactions on Information Theory, vol. 50, no. 12, pp. 3062–3080, 2004.
[116] G. Kramer, M. Gastpar, and P. Gupta, "Cooperative strategies and capacity theorems for relay networks," IEEE Transactions on Information Theory, vol. 51, no. 9, pp. 3037–3063, 2005.
[117] N. A. M. Nor, Z. Ghassemlooy, J. Bohata, *et al.*, "Experimental investigation of all-optical relay-assisted 10 Gb/s FSO link over the atmospheric turbulence channel," Journal of Lightwave Technology, vol. 35, no. 1, pp. 45–53, 2017.

[118] M. Mansour Abadi, Z. Ghassemlooy, S. Zvanovec, D. Smith, M. R. Bhatnagar, and Y. Wu, "Dual purpose antenna for hybrid free space optics/RF communication systems," IEEE Journal of Lightwave Technology (JLT), vol. 34, no. 14, 3432–3439, 2016.

[119] Z. Wenzhe, H. Steve, and S. Ce, "Soft-switching hybrid FSO/RF links using short-length Raptor codes," IEEE Journal on Selected Areas in Communication, vol. 27, no. 9, pp. 1698–1708, 2009.

[120] S. Bloom and W. S. Hartley, "The Last-Mile Solution: Hybrid FSO Radio," AirFiber, Inc., 2002

[121] M. Tatarko, L. Ovsenik, and J. Turan, "Management of switching in hybrid FSO/RF link," in Carpathian Control Conference (ICCC), 2015 16th International, 2015, pp. 532–536.

[122] F. Nadeem, B. Geiger, E. Leitgeb, M. S. Awan, and G. Kandus, "Evaluation of switch-over algorithms for hybrid FSO-WLAN systems," in Wireless Communication, Vehicular Technology, Information Theory and Aerospace & Electronic Systems Technology, 2009. Wireless VITAE 2009. 1st International Conference on, 2009, pp. 565–570.

[123] E. Lee and V. W. S. Chan, "Diversity coherent and incoherent receivers for free-space optical communication in the presence and absence of interference," IEEE/OSA Journal of Optical Communications and Networking, vol. 1, no. 5, pp. 463–483, 2009.

[124] A. Puryear and V.W. S. Chan, "Coherent optical communication over the turbulent atmosphere with spatial diversity and wavefront predistortion," in Global Telecommunications Conference, 2009. GLOBECOM 2009. IEEE, Nov. 2009, pp. 1–8.

[125] Karp S, Gagliardi R, Moran S E, Stotts L B, Optical Channels. New York: Plenum, 1988.

[126] A. Belmonte and J. M. Khan, "Performance of synchronous optical receivers using atmospheric compensation techniques," Optics Express, vol. 16, no. 18, pp. 14, 151–14 162, 2008.

[127] A. Belmonte and J. Khan, "Capacity of coherent free-space optical links using atmospheric compensation techniques," Optics Express, vol. 17, no. 4, pp. 2763–2773, 2009.

[128] Win M Z, Chen C -C, Scholtz R A, "Optical phase-locked loop (OPLL) for an amplitude modulated communications link using solid-state lasers," IEEE Journal on Selected Areas in Communications, vol. 13, no. 3, pp. 569–576, 1995.

[129] Jafar M, O'Brien D C, Stevens C J, and Edwards D J, "Evaluation of coverage area for a wide line-of-sight indoor optical free-space communication system employing coherent detection," IET Communications, vol. 2, no. 1, pp. 18–26, 2008.

[130] S. Hitam, M. Abdullah, M. Mahdi, H. Harun, A. Sali, and M. Fauzi, "Impact of increasing threshold level on higher bit rate in free space optical communications," Journal of Optical and Fiber Communications Research, vol. 6, no. 1–6, pp. 22–34, 2009.

[131] M. R. Bhatnagar and Z. Ghassemlooy, "Performance evaluation of FSOMIMO links in gamma-gamma fading with pointing errors," 2015 IEEE International Conference on Communications (ICC), pp. 5084–5090, Jun. 2015.
[132] T. Plank, E. Leitgeb, and M. Loeschnigg, "Recent developments on free space optical links and wavelength analysis," in International Conference on Space Optical Systems and Applications (ICSOS), 2011, pp. 14–20.
[133] S. Vigneshwaran, I. Muthumani, and A. S. Raja, "Investigations on free space optics communication system," in Proceedings of the International Conference on Information Communication & Embedded Systems (ICICES '13), pp. 819–824, IEEE, Chennai, India, February 2013.
[134] J. Singh and N. Kumar, "Performance analysis of different modulation format on free space optical communication system," Optik, vol. 124, no. 20, pp. 4651–4654, 2013.
[135] I. I. Kim and E. Korevaar, "Availability of free space optics (FSO) and hybrid FSO/RF systems," Proceedings of SPIE, Optical Wireless Communications IV, Eric Korevaar (Editor), 27 November 2002.
[136] L. C. Andrews and R. Phillips, Laser Beam Propagation through Random Media. Bellingham, WA: SPIE Press, 2005.
[137] A. Bazil Raj and S. Padmavathi, "Quality metrics and reliability analysis of laser communication system," Defence Science Journal, vol. 66, no. 2, pp. 175–185, 2017.
[138] A. Bazil Raj and U. Darusalam, "Performance improvement of terrestrial free-space optical communications by mitigating the focal-spot wandering," Journal of Modern Optics, vol. 63, no. 21, pp. 2339–2347, 2016.
[139] B. Ghafary, F. D. Kashani, and E. Kazemian, "Reliability analysis of FSO communication links using aberrated divergent rectangular partially coherent flat-topped beam," Iranian Journal of Electrical & Electronic Engineering, vol. 9, no. 1, pp. 13–19, 2013.
[140] F. D. Kashani, M. R. Hedayati Rad, M. R. Mahzoun, E. Kazemian, and A. Kahrizi, "Investigation to reliability of optical communication links using auto-track subsystems in presence of different beam divergences," Iranian Journal of Electrical & Electronic Engineering, vol. 10, no. 1, pp. 1–9, 2014.
[141] X. Tang, "Polarisation shift keying modulated free-space optical communication systems," Ph.D. Thesis, University of Northumbria at Newcastle, February 2012.
[142] Y. Li, N. Pappas, V. Angelakis, M. Píoro, and D. Yuan, "Optimization of free space optical wireless network for cellular backhauling," IEEE Journal On Selected Areas in Communications, vol. 33, no. 9, pp. 1841–1854, 2015.
[143] Y. Ai, Z. Xiong, J. Chen, *et al.*, "The analysis of 7.5 Gbps 40 km FSO experiments," in Photonics Society Summer Topical Meeting Series, 2012 IEEE, 2012, pp. 128–129.

[144] A. Bazil Raj, A. J. V. Selvi, K. D. Durai, and R. S. Singaravelu, “Intensity feedback-based beam wandering mitigation in free-space optical communication using neural control technique,” EURASIP Journal on Wireless Communications and Networking, vol. 2014, no. 1, pp. 1–18, 2014.
[145] M. A. Kashani, M. Uysal, and M. Kavehrad, “A novel statistical channel model for turbulence-induced fading in free-space optical systems,” Journal of Lightwave Technology, vol. 33, no. 11, pp. 2303–2312, 2015.
[146] A. Bazil Raj, J. A. V. Selvi, D. Kumar, and S. Raghavan, “Design of cognitive decision making controller for autonomous online adaptive beam steering in free space optical communication system,” Wireless Personal Communications, vol. 84, no. 1, pp. 765–799, 2015.

Chapter 2

Free-space optical communication over strong atmospheric turbulence channels

Zhen Qu[1] *and Ivan B. Djordjevic*[1]

2.1 Introduction

The continuously evolving optical/RF devices have been pushing forward the development of optical communication market [1–4]. Optical communication can be deployed over a fiber-optic link [5–8] or a free-space optical (FSO) link [9–12]. Fiber-optic links are based on a closed two waveguide layers (core and cladding), protected by the jacket, for signal transmission, which are characterized by low-loss, extremely large bandwidth, small distortions, and represent relatively stable channels. FSO links are known for the "free" communication media, and have been widely used in shot-reach optical transmission systems. Despite the advantages of FSO channels, the performance of FSO transmission systems may be degraded by the atmospheric effects such as absorption, scattering effects, and atmospheric turbulence [13]. Atmospheric turbulence is the main reason for channel fluctuation, as a result, the received beam will be affected by the intensity scintillation and spatial phase distortion. When spatial-mode multiplexing is used for FSO communication, mode crosstalk may become the limiting factor for the system capacity.

In this chapter, we first analyze the atmospheric turbulence, and build a model to emulate the atmospheric turbulence channel. Then orbital angular momentum (OAM) multiplexing/demultiplexing will be introduced, followed by the description of approaches to mitigate the turbulence effects. The concluding remarks will be provided at the end of chapter.

2.2 Turbulence model

It is quite difficult and also costly to configure and/or field test outdoor FSO communication system. Optical path alignment is very challenging for outdoor experiment; expanding telescope and compressing telescope with large aperture are

[1]Department of Electrical and Computer Engineering, University of Arizona, USA

expensive and fragile for outdoor deployment. In order to reduce the cost and easily build an FSO link, it is strongly suggested to emulate the outdoor atmospheric turbulence environment on the indoor optical table. In general, three types of turbulence models are usually used, i.e., temperature-adjustable turbulence cell [14], rotating phase plates [15], and fast-changing phase screens of spatial light modulator (SLM) [11]. Temperature-adjustable turbulence cell is a cost-efficient solution, and more suitable to emulate the weak-to-medium turbulence channel; rotating phase plates and changing the phase screens of SLM are able to preciously emulate the real atmospheric turbulence environment if a sufficiently large number of slabs are positioned along with the FSO link. It is worthwhile to mention that it is not an easy task to switch between different turbulence conditions by rotating phase plates.

In this chapter, the continuously upgrading SLM phase patterns are used to emulate the atmospheric turbulence environment [13], as we proposed in [11],

$$\Phi_n(\kappa) = 0.033C_n^2\left[1 + 1.802\left(\frac{\kappa}{k_l}\right) - 0.254\left(\frac{\kappa}{k_l}\right)^{7/6}\right]\frac{e^{-\kappa^2/k_l^2}}{\left(\kappa^2 + k_0^2\right)^{11/6}} \qquad (2.1)$$

where C_n^2 represents the refractive structure parameter, $k_l = 3.3/l_0$, $k_0 = 1/L_0$, l_0 and L_0 are the inner and outer scales, respectively.

The conceptual diagram of the experimentally used atmospheric turbulence emulator is depicted in Figure 2.1(a). The corresponding continuous atmospheric turbulence FSO channel is modeled by the well-known split-step solution, as shown in Figure 2.1(b), in which the L-length FSO path is divided into 100 slabs, and each lab follows Andrews' spectra. The precision of the used atmospheric turbulence emulator will be then verified in terms of intensity distributions. We select pinhole captured optical power fluctuation as a figure of merit to verify our emulator, which is expected to follow the well-known gamma–gamma distribution [13], over a large-scale atmospheric turbulence strength. Rytov variance is used to represent the turbulence condition, which is given by $\sigma_R^2 = 1.23C_n^2k^{7/6}L^{11/6}$, where k is the wave number, and L is the link distance. As shown in Figure 2.2(a), our measured power distributions are similar to the theoretical gamma–gamma distributions.

We then use irradiance correlation function (ICF) to investigate the irradiance fluctuations, denoted by $\Gamma(r_1, r_2, L) = E\{[I_1 - E(I_1)][I_2 - E(I_2)]\}$, where I_1 and I_2 are the irradiances measured at L-length apart locations r_1 and r_2, respectively; $E(\cdot)$ represents the ensemble average. The atmospheric turbulence model is designed as per Rytov variance of $\sigma_R^2 = 4.06$, with the key specs listed below: inner scale of 1 mm, outer scale of 10 m, path length of $L = 30$ km, and turbulence strength of $C_n^2 = 10^{-14}\mathrm{m}^{-2/3}$[13]. Figure 2.2(b) shows the experimentally measured ICF, the numerical obtained ICF, and the corresponding ICF obtained by the continuous atmospheric turbulence environment. It is clear that, the experimentally obtained ICF matches well with the ICF obtained by the continuous atmospheric turbulence environment.

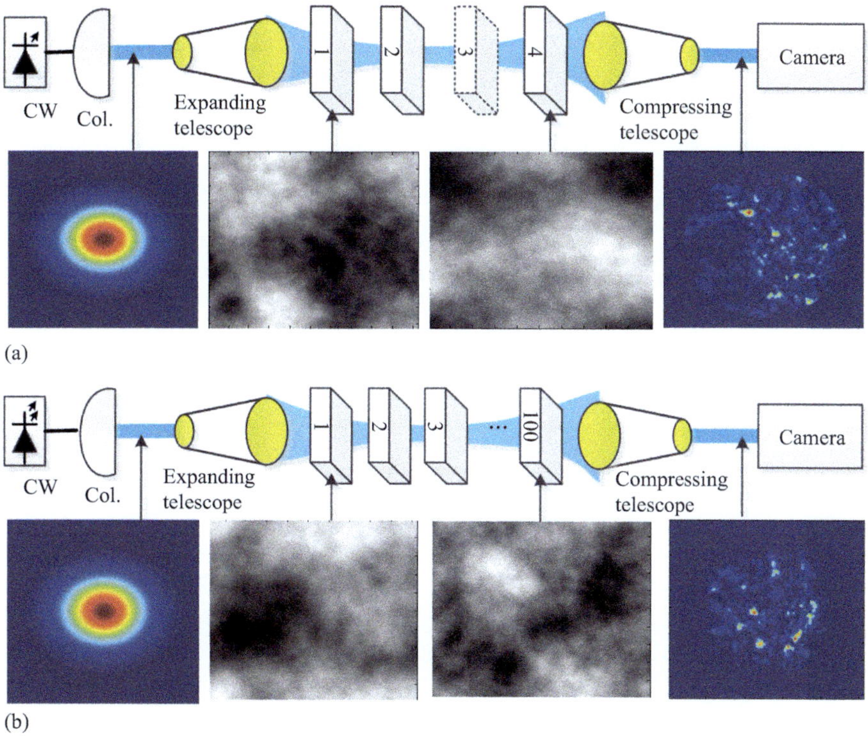

Figure 2.1 The conceptual diagrams: (a) the experimentally implemented atmospheric turbulence emulator and (b) the continuous-like atmospheric turbulence channel

2.3 OAM multiplexing

Historically, the OAM mode was founded decades ago, featured by a helical phase front [16]. The optical beams carrying helical phase front have the OAM of $m\hbar$ per photon, in which m is a "topological charge," also known as index of the OAM beam. The discrete unbounded topological charge, possessed by OAM beams, potentially provides an inexhaustible source for modulation and multiplexing. Among various optical beams that can carry OAM, the Laguerre–Gaussian (LG) beams can easily be implemented with the help of SLMs. The electric field of a LG beam can be expressed in cylindrical coordinates as follows:

$$u(r,\phi,z) = \sqrt{\frac{2p!}{\pi(p+|m|)!}}\frac{1}{w(z)}\left[\frac{r\sqrt{2}}{w(z)}\right]^{|m|} L_p^m\left[\frac{2r^2}{w^2(z)}\right]\exp\left[\frac{-r^2}{w^2(z)}\right]\exp\left[\frac{-ikr^2z}{2(z^2+z_{\mathrm{R}}^2)}\right]$$
$$\times\exp\left[i(2p+|m|+1)\tan^{-1}\left(\frac{z}{z_R}\right)\right]\exp(-im\phi) \tag{2.2}$$

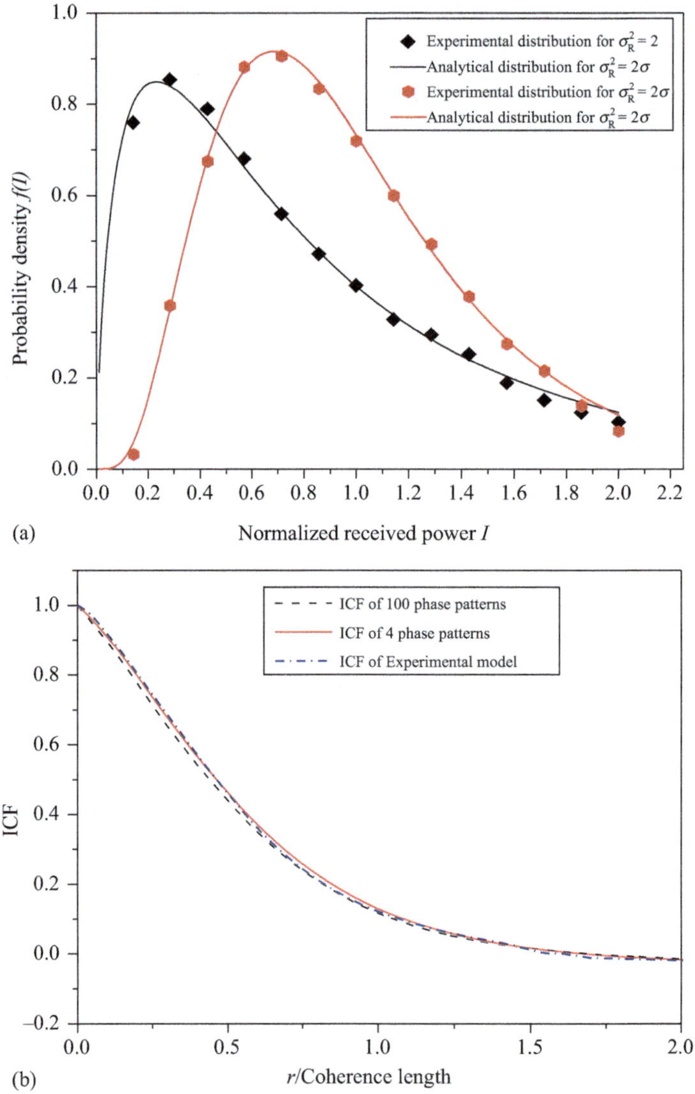

Figure 2.2 (a) Probability density function of the pinhole captured optical power fluctuations and the theoretical gamma–gamma distribution and (b) the experimentally measured ICF, the numerical obtained ICF, and the corresponding ICF obtained by the continuous atmospheric turbulence environment

where r denotes the radial distance from the propagation axis, ϕ is the azimuthal angle, and z is the propagation distance; $w(z) = w_0\sqrt{1 + (z/z_R)^2}$ is the beam radius at distance z, and w_0 is the zero-order Gaussian radius at beam waist; $z_R = \pi w_0^2/\lambda$ is the Rayleigh range, and λ is the optical wavelength. The term $L_p^m(\cdot)$ in (2.2)

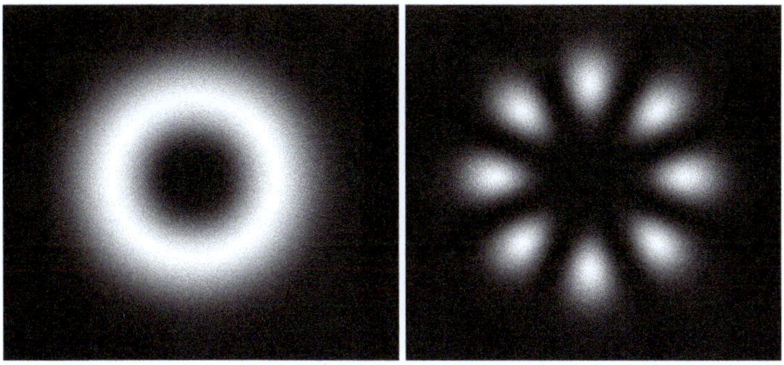

Figure 2.3 The intensity profile of: (left) the OAM mode m = 2 and (right) the superposition of OAM states 4 and −4, wherein the transmitted power is equally divided

represents the generalized Laguerre polynomial, and p and m are the radial and angular mode numbers, respectively, and determine the order of the mode $N = 2p + |m|$. When $p = m = 0$, the field is identified by a zero-order Gaussian beam, i.e., TEM_{00} mode. The radial number p determines the ring index in density distribution. For $p = 0$, the associate Legendre polynomials become 1 for all m's so that the intensity of an LG mode is a ring with radius proportional to $|m|^{1/2}$, as illustrated in Figure 2.3. Without special statement, p is assumed to be 0 in the following content.

The distinguishing characteristic of a LG beam is found in the phase term of $\exp(-im\phi)$, which makes it a natural choice for OAM carrier. For any two OAM modes sharing the fixed radial number p, the following principle of orthogonality is satisfied

$$\begin{aligned}&\langle u_m(r,\phi,z), u_l(r,\phi,z)\rangle \\ &= \int u_m^*(r,\phi,z)u_l(r,\phi,z)rdrd\phi \\ &= \begin{cases} 0, \forall m \neq l \\ \int |u_m(r,\phi,z)|^2 rdrd\phi, m = l \end{cases}\end{aligned} \tag{2.3}$$

where the operator $\langle\cdot\rangle$ denotes the scalar (dot) product (in Dirac notation). Equation (2.3) is the fundamental for OAM-based modulation and/or multiplexing.

A common method to visualize the phase front of an OAM beam is to interfere it with a reference beam whose phase front is already known. In experiments, a Mach–Zehnder interferometer is usually implemented to detect the interferogram by interfering the OAM beam with a collimated Gaussian beam, as shown in Figure 2.4(a)–(d). Theoretically, when an OAM beam interferes with a co-propagating Gaussian beam, a spiral interference pattern is recorded, as shown in Figure 2.4(e)–(h). The OAM state is then identified by the number of arms in the spiral, while the sign of the OAM state is shown by the handedness the spiral pattern.

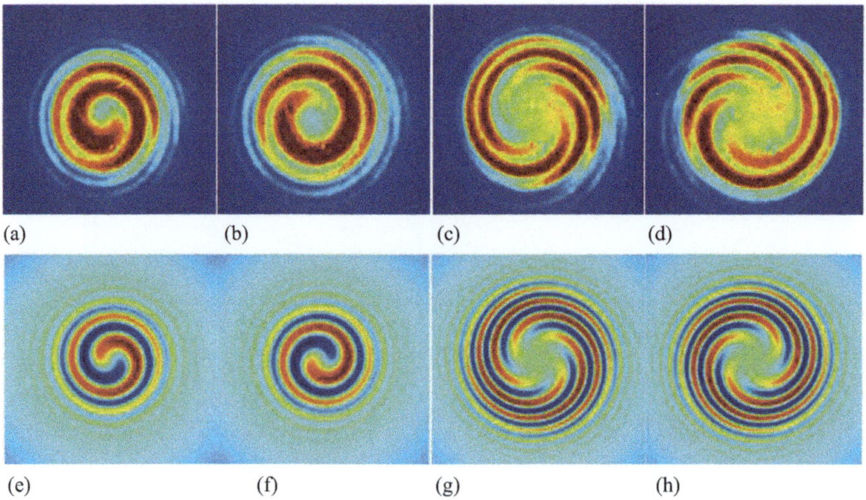

Figure 2.4 Experimental (top) and numerical (bottom) interferograms obtained from the interference of a Gaussian beam and the OAM beams of (a), (e), state 1; (b), (f), state −1; (c), (g), state 4; (c), (h), state −4

The mutual orthogonality among the superimposed OAM modes enables the subsequent demultiplexing via optical means. Theoretically, an OAM beam can be identified and converted to zero-order Gaussian beam by using a complex conjugated version of its electric field. A phase pattern identical to that of the OAM mode but with opposite OAM state sign, can be implemented for detection. The zero-order beam will be obtained at the far field. In Figure 2.5, the detection efficiency of individual OAM mode is shown in terms of the intensity profiles. Note that blazed phase patterns are used for detection and the intensity patterns are observed at the center of the diffraction orders +1. The numerical results, shown in Figure 2.5(a)–(d), are based on the scalar diffraction theory. By comparing the upper and lower patterns, we can see that the experimental results as shown in Figure 2.5(e)–(h), agree well with the numerical ones. The minor discrepancies are mainly due to beam wandering, image saturation, and beam rotation.

2.4 Dealing with atmospheric turbulence effects by adaptive optics and LDPC coding

Basically, the turbulence-induced inter-mode crosstalk can be mitigated by adaptive optics (AO) [15], multi-input multi-output (MIMO) processing [17], spatial misalignment [18], and channel coding [19]. It is worthwhile to notice that MIMO processing is suitable to be applied in weak-to-middle turbulence conditions, while spatial misalignment will introduce extra power loss when the aperture size of the compressing telescope is limited. In our experiment [11], we use AO to compensate the phase distortion in the optical domain first, followed by the

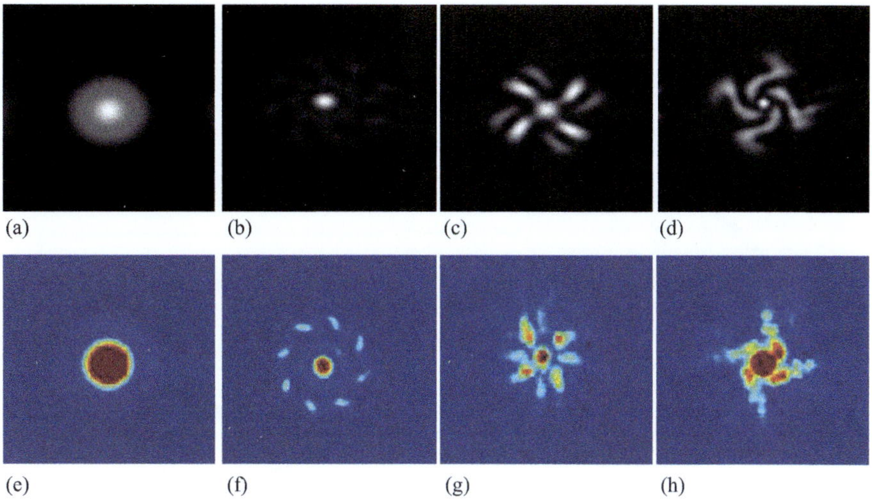

Figure 2.5 Numerical and experimental detection of individual OAM states from the superimposed OAM modes. (a), (e), OAM state 1 is detected by using OAM state −1; (b), (f), superposition of OAM state 4, −4, is detected by using OAM state 4; (c), (g), superposition of OAM state 2, −2, 6, −6, is detected by using OAM state 2; (d), (h), superposition of OAM state 2, −2, 6, −6, is detected by using OAM state −6

low-density parity-check (LDPC)-based forward error correction (FEC) coding to realize an error-free FSO communication.

Figure 2.6 shows the experimental setup for AO-enabled FSO communication system is. The five continuous wave (CW) laser beams are generated with the inter-channel spacing of 50 GHz (1,549.32–1,550.92 nm). The wavelength channels are multiplexed together and used as the optical input of an I/Q modulator. The pseudorandom binary sequence (PRBS) signals are encoded using a binary LDPC code with the code rate of 0.8. The data streams pass to the arbitrary waveform generator (AWG) and drive the I/Q modulator to generate 15.6 GBaud optical QPSK signal. An optical interleaver (IL) is applied to separate the odd and even channel signals, which are then decorrelated by 350 symbols and re-combined together. The resulting WDM QPSK optical signals with the decorrelated adjacent wavelength channels, are boosted by an Erbium-doped fiber amplifier (EDFA), followed by an optical-tunable filter (OTF) to suppress amplified spontaneous emission (ASE) noise. The optical signals are separated by a coupler, and one path is later decorrelated before recombination. The optical Gaussian modes are collimated by fixed fiber optic collimator and converted to OAM modes (OAM states ±2 and ±6) by using a high-resolution SLM. Then the resulting OAM modes are centrally aligned by a beam splitter (BS). Another 1,548.9 nm Gaussian probe beam is used to assist AO compensation. The Gaussian probe beam is expanded by a beam expander (BE) so that its diameter is larger than the widest OAM beams (OAM

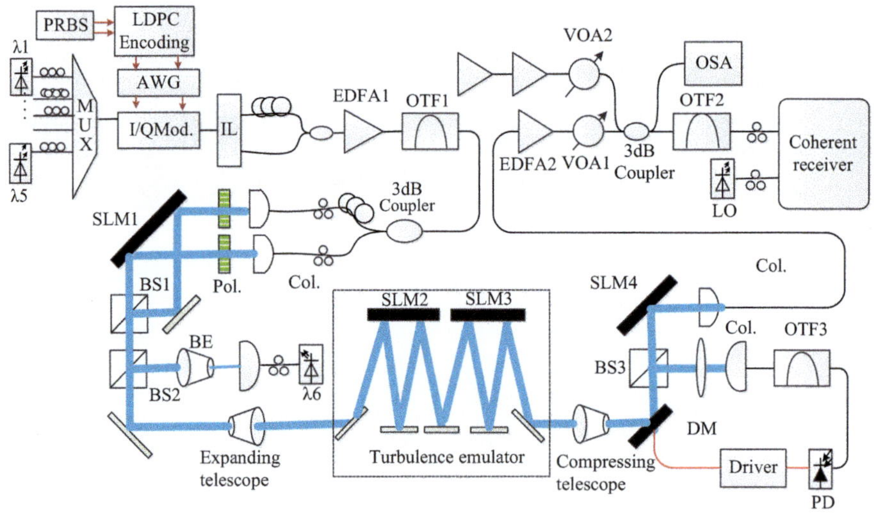

Figure 2.6 Experimental setup

states ±6) generated in this experiment. Then the probe beam is also centrally aligned with the data-carrying OAM modes.

The expanding telescope is applied to adjust the diameters of the collimated OAM beams and Gaussian probe beam, which are then sent to the turbulence emulator. The SLM2 and SLM3 are continuously and randomly updating the phase patterns to be modeled on the dynamic atmospheric turbulence environment. The atmospheric turbulence emulator used in this experiment is designed according to the Rytov variance of $\sigma_R^2 = 2$.

The beam sizes of the distorted OAM beams and probe beam are decreased by a compressing telescope. The distorted beams then pass to the DM for wavefront correction. The distorted Gaussian probe beam is utilized as a stimulus in this wavefront sensorless AO setup, with the assumption that the probe beam and OAM beams are affected by the similar wavefront distortion. Partial optical beams are segregated via a BS and collected by a single-mode fiber (SMF) patch cable. In this experiment, the OTF3 with the central wavelength of 1,548.9 nm is implemented to only capture the Gaussian probe beam, followed by a photodiode (PD) for power monitoring. The detected analog voltage is digitized by an analog-to-digital converter (ADC) to update the DM pixels according to stochastic parallel gradient descent algorithm.

After the AO compensation, the less distorted OAM modes are detected by SLM4 to convert one OAM mode back to the Gaussian-like mode, which is then collected by another SMF patch cable. The collected optical signals are pre-amplified by EDFA2, followed by a variable optical attenuator (VOA) for optical power tuning. Additional ASE noise is generated and adjusted by a sub-system configured by EDFA3, EDFA4, and VOA2. Such ASE noise is added to the optical signal via an optical coupler, after which an OTF with the central wavelength of 1,550.12 nm is applied to single out the corresponding optical wavelength channel.

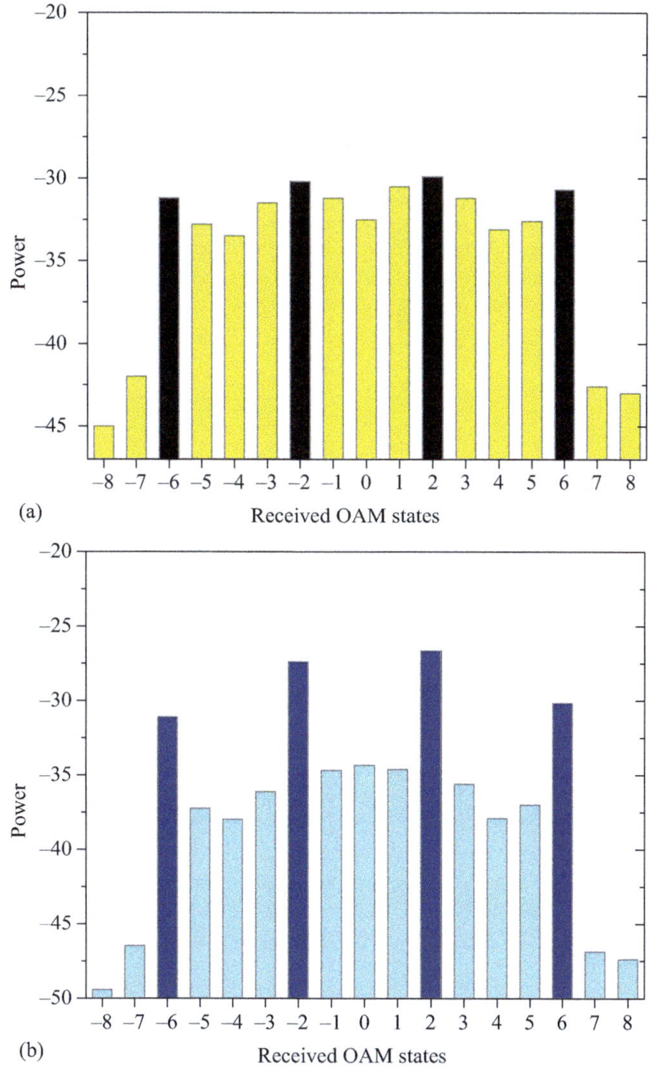

Figure 2.7 Optical power distributions of the OAM models: (a) without AO wavefront correction and (b) with AO wavefront correction

In the coherent receiver, the local oscillator (LO) light and the optical signal are mixed in an optical 90° hybrid, detected by two PDs, and digitized by a real-time oscilloscope. After the off-line digital signal processing (DSP) signal recovery, the sum-product algorithm is used in the LDPC decoding procedure with a maximum of 50 iterations.

The atmospheric turbulence-induced mode crosstalk and the merits of AO-assisted wavefront correction are investigated first. The power ratio between the target OAM modes (OAM states ±2, ±6) and their adjacent modes are used here as the metric to evaluate the effect of mode crosstalk. As illustrated in Figure 2.7(a),

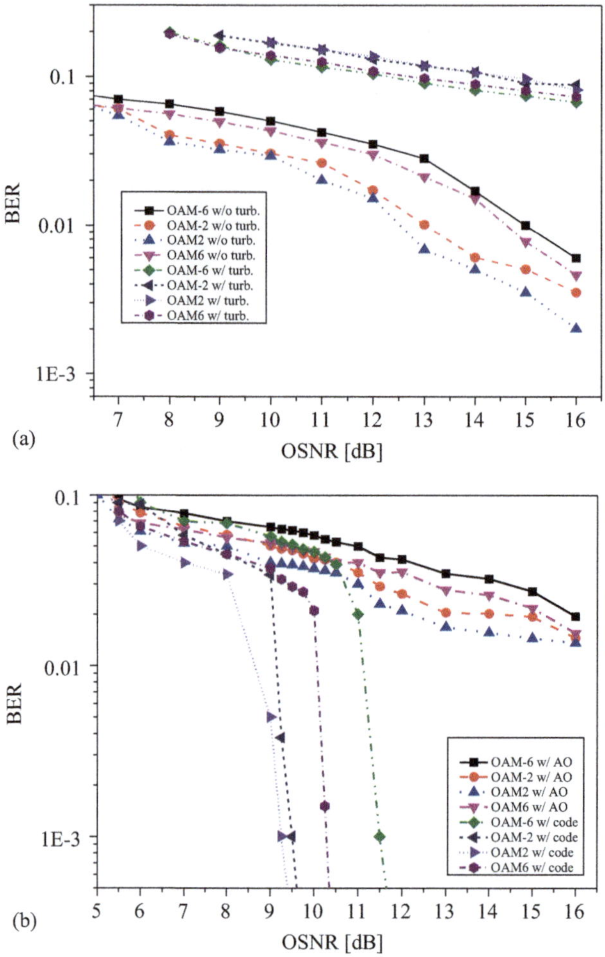

Figure 2.8 (a) Measured BER versus OSNR by turning on and off the atmospheric turbulence emulator and (b) measured BER vs. OSNR if AO wavefront correction and LDPC-based FEC coding are applied

the power of target OAM mode is measured to be similar to the adjacent OAM modes, which means that the data coming from mode crosstalk will severely interfere the desirable data after mode detection. While it can be found from Figure 2.7(b) that the average extinction ratio (ER) after AO-assisted wavefront correction reaches a 6 dB improvement.

Figure 2.8(a) shows the average bit-error rate (BER) vs. OSNR performance in cases of with and without atmospheric turbulence effects. It is clear that the BER curves will not drop quickly even with the increasing OSNR values. It is caused by the unperfect mode generation and detection patterns, which will also introduce unwanted inter-mode crosstalk effects. Note that the worse BER performance of

OAM states ±6, compared to OAM state ±2, is because that high-order OAM mode is more sensitive to the boundary effect.

Figure 2.8(b) shows that distinct average OSNR gain can be reached after AO-assisted wavefront correction and LDPC coding. We can find that the BER curves after AO-assisted wavefront correction can be lower than 0.04, which is the error correction threshold of the used LDPC code. It can also be observed that the performance of the AO-assisted wavefront correction on OAM states ±2 are better than that of OAM states ±6. It is not only because of the boundary effect brought by the limited sizes of the used optical components, but also the features of the used probe beam, which is able to fully cover the small-size OAM modes (OAM states ±2) after the turbulent FSO transmission, rather than the large-size OAM modes (OAM states ±6). After the AO correction is performed, LDPC coding/decoding can be then applied to efficiently eliminate the post-FEC error floor. Figure 2.8(b) shows that the BER curves of all OAM modes can drop quickly as long as the OSNR is higher than 8 dB. More specifically, the coding gains of 3.9, 4.1, 5.2, and 5 dB are reached for OAM states 2, −2, 6, and −6, respectively, when the BER is 2×10^{-2}.

2.5 Concluding remarks

We built an indoor turbulence emulator to study the realistic turbulence effects. The atmospheric turbulence emulator was validated in terms of PDF of the experimentally captured optical power fluctuation and ICF. We have theoretically analyzed the OAM, and experimentally shown the generation and detection of OAM beams. In order to mitigate the turbulence, we experimentally demonstrated a robust OAM multiplexing-based FSO transmission system. The inter-mode crosstalk has been first relieved by using the low-cost wavefront sensorless AO system, and the remaining inter-mode crosstalk was successfully solved by LDPC coding. After turbulence mitigation, 500 Gb/s WDM/OAM multiplexing, per single polarization, based FSO communication has been be realized in strong turbulence regime.

References

[1] M. A. Mestre, H. Mardoyan, C. Caillaud, *et al.*, "Compact InP-based DFB-EAM enabling PAM-4 112 Gb/s transmission over 2 km," J. Lightw. Technol., vol. 34, no. 7, pp. 1572–1578, 1, 2016.

[2] Y. Xie, L. Zhuang, C. Zhu, and A. Lowery, "Nyquist pulse shaping using arrayed waveguide grating routers," Opt. Express, vol. 24, no. 20, pp. 22357–22365, 2016.

[3] Y. Loussouarn, M. Song, E. Pincemin, G. Miller, A. Gibbemeyer, and B. Mikkelsen, "100 Gbps coherent digital CFP interface for short reach, regional, and ultra long-haul optical communications," 2015 European Conference on Optical Communication (ECOC), Valencia, 2015, pp. 1–3.

[4] Y. Xie, L. Zhuang, R. Broeke, *et al.*, "Compact 4 × 5 Gb/s silicon-on-insulator OFDM transmitter," 2017 Optical Fiber Communications Conference and Exhibition (OFC), Los Angeles, CA, 2017, pp. 1–3.
[5] Z. Qu, Y. Li, W. Mo, *et al.*, "Performance optimization of PM-16QAM transmission system enabled by real-time self-adaptive coding," Opt. Lett., vol. 42, no. 20, pp. 4211–4214, 2017.
[6] J. Cho, X. Chen, S. Chandrasekhar, *et al.*, "Trans-Atlantic field trial using probabilistically shaped 64-QAM at high spectral efficiencies and single-carrier real-time 250-Gb/s 16-QAM," Optical Fiber Communication Conference (OFC), 2017, paper Th5B.3.
[7] J. Sakaguchi, "Space division multiplexed transmission of 109-Tb/s data signals using homogeneous seven-core fiber," J. Lightw. Technol., vol. 30, no. 4, pp. 658–665, 2012.
[8] Z. Qu, S. Fu, M. Zhang, M. Tang, P. Shum, and D. Liu, "Analytical investigation on self-homodyne coherent system based on few-mode fiber," IEEE Photon. Technol. Lett., vol. 26, no. 1, pp. 74–77, 2014.
[9] P. Polynkin, Peleg, A., Klein, L., Rhoadarmer, T., and Moloney, J. "Optimized multi emitter beams for free-space optical communications through turbulent atmosphere," Opt. Lett., vol. 32, no. 8, pp. 885–887, 2007.
[10] Z. Qu and I. B. Djordjevic, "Coded orbital angular momentum based free-space optical transmission in the presence of atmospheric turbulence", Asia Commun. Photonics Conf. (ACP), 2015, paper AS3D.3.
[11] Z. Qu and I. B. Djordjevic, "500 Gb/s free-space optical transmission over strong atmospheric turbulence channels," Opt. Lett., vol. 41, no. 14, pp. 3285–3288, 2016.
[12] V. W. S. Chan, "Free-space optical communications," J. Lightwave Technol., vol. 24, no. 12, pp. 4750–4762, 2006.
[13] L. C. Andrews, R. L. Phillips; C. Y. Hopen, Laser Beam Scintillation with Applications. SPIE Press, 2001.
[14] B. J. Pors, C. H. Monken, E. R. Eliel, and J. P. Woerdman, "Transport of orbital-angular-momentum entanglement through a turbulent atmosphere," Opt. Express, vol. 19, no. 7, pp. 6671–6683, 2011.
[15] Y. Ren, G. Xie, H. Huang, *et al.*, "Turbulence compensation of an orbital angular momentum and polarization-multiplexed link using a data-carrying beacon on a separate wavelength," Opt. Lett., vol. 40, no. 10, pp. 2249–2252, 2015.
[16] L. Allen, M. W. Beijersbergen, R. J. C. Spreeuw, and J. P. Woerdman, "Orbital angular-momentum of light and the transformation of Laguerre–Gaussian laser modes," Phys. Rev. A, vol. 45, no. 11, pp. 8185, 1992.
[17] H. Huang, Y. Cao, G. Xie, *et al.*, "Crosstalk mitigation in a free-space orbital angular momentum multiplexed communication link using 4 × 4 MIMO equalization," Opt. Lett., vol. 39, no. 15, pp. 4360–4363, 2014.
[18] Z. Qu and I. Djordjevic, "Two-stage crosstalk mitigation in an orbital angular momentum-based free-space optical communication system," Opt. Lett., vol. 42, no. 10, pp. 3125–3128, 2017.
[19] Z. Qu and I. B. Djordjevic, "Approaching terabit optical transmission over strong atmospheric turbulence channels," Proc. ICTON 2016, Jul. 2016.

Chapter 3

Performance analysis and mitigation of turbulence effects using spatial diversity techniques in FSO systems over combined channel

Prabu Krishnan[1]

3.1 Introduction

Free space optical (FSO) communication is an emerging technology in wireless optical communication systems. The key advantages of FSO systems include huge bandwidth and high data rate and also copious advantages like license free huge bandwidth, high security at low installation and maintenance cost. The potential applications of FSO systems are cellular communication back haul, optical fiber backup, last mile access, disaster recovery, and temporary links, etc. [1–4]. However, the link range, reliability and data rate of FSO communication systems are affected by various atmospheric phenomena, viz. rain, haze, fog, snow, scintillation, and pointing errors.

The atmospheric turbulence fading and misalignment fading are the main deficiencies affecting light signal when transmitted through the atmospheric channel. It degrades the performance of the FSO system. Therefore, the atmospheric turbulences and pointing errors are the foremost limitations in FSO technology. The atmospheric turbulence is modeled using the various channel models in both weak and strong fading regimes include log normal, negative exponential, gamma–gamma, double gamma, and K-distribution, etc., Pointing errors are mainly due to building sway. Thermal expansion, dynamic wind loads and weak earthquakes result in the sway of high-rise buildings. This causes vibrations of the transmitter beam and therefore, misalignment between the transmitter and receiver. The misalignment fading affects the performance of an FSO system rigorously [5,6].

The combined channel model which models the attenuation due to atmospheric weather conditions and fading due to atmospheric turbulence and misalignment. It gives the complete effects of FSO link impairments and useful to analyze the exact

[1]Department of Electronics and Communication Engineering, National Institute of Technology Karnataka, India

performance of FSO communication systems. The impact of pointing errors on FSO system is analyzed over combined channel model is analyzed and improved using various mitigation techniques in literature [7–9].

The various techniques available to mitigate the effects of atmospheric turbulence fading comprise adaptive optics, error control, aperture averaging, and spatial diversity. Each mitigation technique has its own advantages and disadvantages. The major drawback of adaptive optics is high cost and complexity. Random errors and burst errors are the problems exist in coding technique [10,11]. In aperture averaging, keeping the spatial coherence distance in order of centimeters is the challenging one [12]. Spatial diversity is one of the power full mitigation techniques to improve the performance of the system over atmospheric turbulence channel.

In this chapter, the performance of a polarization shift keying (PolSK)-based FSO system with time and wavelength diversity in a strong turbulent atmospheric channel is analyzed. At the receiver side, the various linear combining schemes such as optimal combining (OC), equal gain combining (EGC), and selection combining (SC) are introduced.

The closed-form expressions of average bit error rate (ABER) are derived for single-input multiple-output (SIMO) FSO links with spatial diversity and multiple input multiple output (MIMO) FSO systems over moderate and strong turbulence channel with pointing errors. The results are analyzed and compared between SISO, MIMO, SIMO with OC, EGC, and SC diversity techniques.

3.2 Combined channel model

The combined channel model introduces the complete fading effects due to atmospheric attenuation, atmospheric turbulence, and misalignment or pointing errors. The complete effects of fading on FSO link and its mathematical models are discussed in this section.

3.2.1 Atmospheric attenuation

The atmospheric channel consists of different gases methane, nitrous oxide, oxygen, ozone, carbon dioxide, and aerosols – tiny particles deferred in the atmosphere. The other forms of precipitation existing in the atmosphere are fog, rain, and haze. The amount of precipitation existing in the atmosphere depends on the location and the season. An optical field that passes through the atmospheric channel is absorbed or scattered by the particles present in the atmosphere resulting in power loss.

Absorption: When an optical light signal transmitting through the free space is subject to atmospheric attenuation due to the photon absorption by the molecular elements (fog, water vapor, ozone, etc.) and various gases in the atmosphere [13]. Due to this absorption part of the optical energy is converted into heat. The absorption of atmospheric gases versus wavelength is shown in Figure 3.1 [14].

In Figure 3.1, the first five graphs show the absorptance of methane, nitrous oxide, oxygen and ozone, carbon dioxide, and water molecules, respectively. The latest grid in Figure 3.1 shows the complete absorption of the atmosphere.

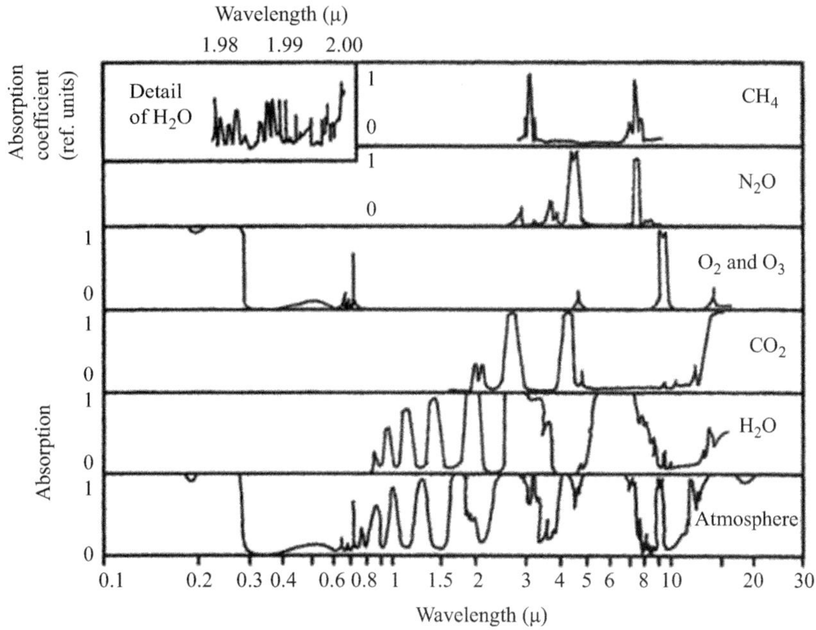

Figure 3.1 Absorption of wavelengths by atmospheric gases (after J. N. Howard, Proceedings of I.R.E. 47, 1451, 1959; and R. M. Goody and G. D. Robinson, Quarterly Journal of the Royal Meteorological Society, 7.7, 153, 1951)

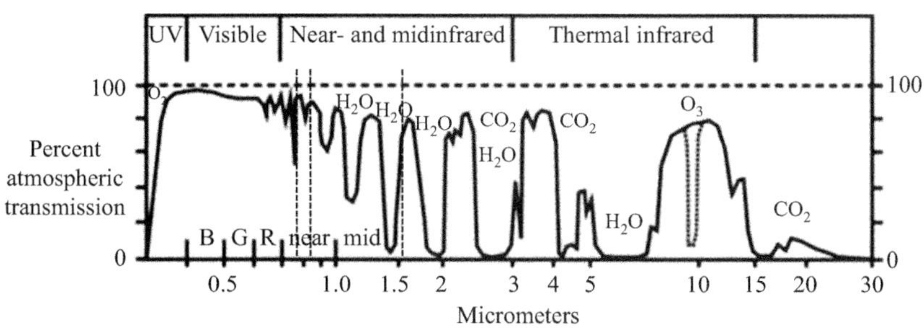

Figure 3.2 Atmospheric transmittance

Figure 3.2 gives the transmittance of the atmosphere versus wavelength. It can be noticed that the major limitation of transmission is caused by water molecules. The dotted lines marked in the figure epitomize the operating wavelengths of 800, 850, and 1,550 nm used in FSO systems. For these wavelengths the atmospheric transmittance is high [15].

Scattering: The optical signal transmitted over the atmosphere is subject to the scattering by atmospheric constituents other than absorption. Scattering is happening due to the atmospheric molecules which have similar order or smaller dimensions of the wavelength of the incident light. The atmospheric factors fog, haze, aerosols, and rain play a major role in FSO communications because, the size of these particles is very close to the wavelengths used in FSO systems. The three forms of scattering are Raman, Rayleigh, and Mie scattering [16].

Mathematical model: The atmospheric attenuation loss is modeled by Beers-Lambert law is given as [17]

$$h_l = e^{-\sigma L} \tag{3.1}$$

where σ denotes an attenuation coefficient in dB/km and L is the propagation distance in kilometer.

3.2.2 Atmospheric turbulence

For long distance communication (i.e. more than 1 km distance) the free-space medium is vastly susceptible to attenuation, scattering, and turbulence. The sunlight is absorbed by the earth, due to this the surface near to the earth get heated up. The heated air rises over the land and mixed intensely with the cooler air at higher altitude. This makes a random change in the atmospheric temperature and pressure along the propagation path. Inhomogeneity caused by turbulence can be viewed as turbulent eddies of different temperature and they act like refractive prisms of different sizes and refractive index. The interaction between the transmitted light signal and the turbulent medium produces random variations in phase and amplitude of the transmitted information bearing optical signal. This phenomenon is called as scintillation. It degrades the performance of FSO links ultimately. Atmospheric turbulence depends on various parameters viz. altitude, wind speed, atmospheric pressure, and refractive index variation due to temperature inhomogeneity. The effects of atmospheric turbulence include polarization fluctuation, beam steering, beam scintillation, spatial coherence degradation, beam spread, and image dancing [18].

Based on the variation of the magnitude of index of refraction and inhomogeneity, the atmospheric turbulence is categorized into four regimes. They are weak, moderate, strong, and saturation. The various models describing the probability density function (PDF) of the irradiance fluctuation in a turbulent channel are log-normal, negative exponential, and gamma–gamma [17]. The log-normal distribution can support only weak turbulence and the gamma–gamma distribution is valid for weak to strong regimes [19,20].

Lognormal turbulence model: In this model, the statistics of the irradiance fluctuations obeys the log-normal distribution. This model is characterized by a single scattering event and is best suited for weak turbulence regime. The PDF can be given as [21,22],

$$f(h) = \frac{1}{\sqrt{2\pi\sigma_h^2}}\frac{1}{h}\exp\left(-\frac{\left(\ln\left(\frac{h}{h_0}\right) - E[h]\right)^2}{2\sigma_h^2}\right), h \geq 0 \tag{3.2}$$

where σ_{h}^2 is the Rytov variance, no unit, h be the field irradiance in turbulent medium while the intensity in free-space (no turbulence) represented as h_0 and $E[h]$ is the mean log intensity.

Negative exponential turbulence model: In this model, the number of independent scattering is very high and it can support for saturation regime. Therefore, the irradiance fluctuation follows the Rayleigh distribution entailing negative exponential statistics for the irradiance. The negative exponential PDF can be given as [21,22],

$$f(h) = \frac{1}{h_0}\exp\left(-\frac{h}{h_0}\right), h_0 > 0 \tag{3.3}$$

where $E[h] = h_0$ is the mean received irradiance.

Gamma–gamma turbulence model: The atmospheric turbulence is modeled by gamma–gamma distribution with scintillation parameters α and β, which are indicated as functions of the Rytov variance and a geometry factor. The PDF of the gamma–gamma channel model is given by [17]

$$f_{h_{\mathrm{s}}}(h_{\mathrm{s}}) = \frac{2(\alpha\beta)^{(\alpha+\beta)/2}}{\Gamma(\alpha)\Gamma(\beta)} {h_{\mathrm{s}}}^{(\alpha+\beta)/2-1} K_{(\alpha-\beta)}\left(2\sqrt{\alpha\beta h_{\mathrm{s}}}\right) \tag{3.4}$$

where α and β are the effective number of large- and small-scale turbulent eddies, $\Gamma(\cdot)$ is the gamma function, and $K_{(\alpha-\beta)}$ is the modified Bessel function of the second kind of order $(\alpha-\beta)$. The effective number of large- and small-scale turbulent eddies α and β for a spherical wave is given by [17],

$$\alpha = \left[\exp\left(\frac{0.49\delta_{\mathrm{n}}^2}{\left(1 + 0.18d^2 + 0.56\delta_{\mathrm{n}}^{12/15}\right)^{7/6}}\right) - 1\right]^{-1} \tag{3.5}$$

$$\beta = \left[\exp\left(\frac{0.51\delta_{\mathrm{n}}^2\left(1 + 0.69\delta_{\mathrm{n}}^{12/15}\right)^{-5/6}}{\left(1 + 0.9d^2 + 0.62d^2\delta_{\mathrm{n}}^{12/15}\right)^{5/6}}\right) - 1\right]^{-1} \tag{3.6}$$

where $d = \sqrt{kD^2/4L}$, $k = 2\pi/\lambda$, is the optical wave number with λ being the operational wavelength in nm, L is the length of the optical link in kilometer and D is the receiver's aperture diameter in centimeter. The parameter δ_{n}^2 is the Rytov variance no unit and is given as, $\delta_{\mathrm{n}}^2 = 0.5C_{\mathrm{n}}^2k^{7/6}L^{11/6}$ and the C_{n}^2 represents the refractive index structure parameter in $m^{-2/3}$. This model is valid for all turbulence regimes from weak to strong and the gamma–gamma distribution approaches negative exponential distribution when it approaches saturation regime [21,22].

3.2.3 Misalignment fading or pointing errors

FSO is a line of sight communication. So, the system performance and reliability of an FSO link is determined by the alignment between the transmitter and receiver.

However, misalignment due to weak earthquakes, thermal expansion, sway of buildings, and wind loads cause pointing errors and signal fading at the receiver [23]. By considering a circular detection aperture of radius r and a Gaussian beam, the PDF of h_p is given by [24]

$$f_{h_p}(h_p) = \frac{\xi^2}{{A_0}^{\xi^2}} {h_p}^{\xi^2-1}, 0 \le h_p \le A_0 \tag{3.7}$$

where $A_0 = [\mathrm{erf}(v)]^2$ is the fraction of the collected power at $r = 0$. The Gauss error function $\mathrm{erf}(\cdot)$ is defined as $\mathrm{erf}(x) = \frac{2}{\sqrt{\pi}} \int_0^x \mathrm{e}^{-t^2}\, \mathrm{d}t$. The radial distance is denoted as r and ξ is the ratio between the equivalent beam radius at the receiver and the pointing error displacement (jitter) standard deviation at the receiver.

3.2.4 Combined channel model

The PDF of the irradiance intensity, h for weak atmospheric turbulence channel can be expressed as [25,26],

$$f_h(h) = \frac{\xi^2}{(A_0 h_l)^{\xi^2}} h^{\xi^2-1} \times \int_{h/A_0 h_l}^{\infty} \frac{1}{h_s^{\xi^2+1} \sigma_I(D)\sqrt{2\pi}} \exp\left(-\frac{\left[\ln(h_s) + 0.5{\sigma_I}^2(D)\right]^2}{2{\sigma_I}^2(D)}\right) \mathrm{d}h_s \tag{3.8}$$

where ${\sigma_I}^2(D)$ is the aperture averaged scintillation index, $\xi = w_{z_{eq}}/\sigma_s$ is the ratio between the equivalent beam radius at the receiver and the pointing errors displacement standard deviation at the receiver and A_0 is the fraction of the received power at zero radial distance.

The PDF of the irradiance intensity, h for strong atmospheric turbulence channel can be expressed as [25,26],

$$f_h(h) = \frac{2\xi^2 (\alpha\beta)^{(\alpha+\beta)/2}}{(A_0 h_l)^{\xi^2} \Gamma(\alpha)\Gamma(\beta)} h^{\xi^2-1} \times \int_{h/A_0 h_l}^{\infty} {h_s}^{(\alpha+\beta)/2-1-\xi^2} K_{(\alpha-\beta)}\left(2\sqrt{\alpha\beta h_s}\right) \mathrm{d}h_s \tag{3.9}$$

where α and β are the effective number of large- and small-scale turbulent eddies, $\Gamma(\cdot)$ is the gamma function, and $K_{(\alpha-\beta)}$ is the modified Bessel function of the second kind of order $(\alpha - \beta)$ which can be simplified using the Meijer G function $K_v(x) = \frac{1}{2} G_{0,2}^{2,0}\left[\frac{x^2}{4}\middle| \begin{matrix} - & - \\ v/2 & -v/2 \end{matrix}\right]$ [27].

$$f_h(h) = \frac{2\xi^2 (\alpha\beta)^{(\alpha+\beta)/2}}{(A_0 h_l)^{\xi^2} \Gamma(\alpha)\Gamma(\beta)} h^{\xi^2-1} \times \frac{1}{2} \times \int_{h/A_0 h_l}^{\infty} {h_s}^{(\alpha+\beta)/2-1-\xi^2} G_{0,2}^{2,0}\left[\alpha\beta h_s \middle| \begin{matrix} - & - \\ (\alpha-\beta)/2 & (\beta-\alpha)/2 \end{matrix}\right] \mathrm{d}h_s \tag{3.10}$$

Then [28, Eq. (07.34.21.0085.01)] is used to solve the above integral followed by further simplification using [28, Eq. (07.34.16.0001.01)], a closed-form expression for the channel model is obtained as [29]

$$f_{\mathrm{h}}(h) = \frac{\alpha\beta\xi^2}{A_0 h_{\mathrm{l}}\Gamma(\alpha)\Gamma(\beta)} G_{1,3}^{3,0}\left[\frac{\alpha\beta h}{A_0 h_{\mathrm{l}}}\middle| \begin{matrix} \xi^2 \\ \xi^2 - 1, \alpha - 1, \beta - 1 \end{matrix}\right] \tag{3.11}$$

3.3 Techniques for improving the reliability of FSO systems

A keen active research is going on FSO communication through atmospheric turbulence and various methods have been proposed to mitigate the effects of atmospheric turbulence fading. BER is the important performance metric used to measure the performance of FSO systems. Atmospheric turbulence is most vulnerable to a BER of FSO system. It can be improved through aperture averaging, diversity techniques, and relay-assisted FSO systems.

3.3.1 *Aperture averaging*

In the 1950s, aperture averaging was used in astronomical measurements [30]. Recently, the aperture averaging effects are analyzed over atmospheric turbulence channel in [31]. Aperture averaging is a simple technique and it does not require any additional power, bandwidth, size or weight overhead to the overall design for mitigating the effects of atmospheric turbulence in FSO systems. The aperture averaging factor A_{avg} is defined as the ratio between the normalized intensity variance of fluctuations of a receiver and a point receiver. The aperture averaging factor may be approximated by [32]

$$A_{\mathrm{avg}} = \frac{\sigma_{\mathrm{I}}^2(A)}{\sigma_{\mathrm{I}}^2(0)} = \left[1 + 1.062\left(\frac{kA^2}{4D}\right)\right]^{-7/6} \tag{3.12}$$

where A is the antenna aperture in m^2, D is the propagation length in km, and $k = 2\pi/\lambda$ is the wave-number in cm^{-1}.

3.3.2 *Diversity techniques*

The different diversity techniques used to mitigate the effects of atmospheric turbulence include temporal diversity, space diversity and wavelength and wavelength diversity [33–37]. Since, the atmospheric absorption of wavelengths has been fixed, it has less significance in FSO communication systems. In temporal diversity, single transmitter transmitting the signal with different time slots. Therefore, it demands longer signal processing time [37]. In spatial diversity technique, multiple receivers are used to receive the same bits of data. The spatial correlation between a pair of constituent transmitters in a MIMO system is investigated in [38].

There are several combining techniques for the spatial diversity, which are OC, EGC, and maximal ratio combining (MRC) [39]. In the case of MRC, the interference in each of the received signals is assumed to be different; thus, it is used to provide the maximal signal-to-noise ratio (SNR). But, the interference will be similar in all of the received signals in which case MRC may not perform well. Instead, we can go to OC, in which, by combining all of the received signals, we can effectively reduce the effects of atmospheric turbulence and improve the FSO system performance relatively [39]. In OC, the signals from multiple receiving apertures are weighted and combined such that the output signal-to-interference-noise ratio is maximized. Spatial diversity is used not only to reduce the co-channel interference, but also to counter the fading of the signal; it also reduces the power of interfering signals [40].

3.3.3 Relaying techniques

The main undignified factor in FSO links is the atmospheric turbulences particularly for long distance communication, more than a few kilometers [19,20]. Relay-assisted FSO system is an effective method to extend coverage, lessen the effects of fading and increase the link range. There are two types of relay-assisted schemes broadly used in FSO systems are serial relaying and parallel relaying. Serial relay is used to improve the link range and parallel relay is used to improve the performance of the system. The widely used relaying strategies are decode-and-forward (DF) and amplify-and-forward (AF). In DF mode, the relay decodes, re-modulate and retransmit the received signal, as AF the relay simply amplify and retransmit the signal without decoding. The outage probability of serial and parallel relay-assisted FSO system over log-normal channel using both DF and AF modes is analyzed in [41]. The end-to-end performance of serial relay-assisted FSO system with AF relays over gamma–gamma channel is studied in [42].

3.4 Transmitter diversity in strong atmospheric turbulence channel using Polsk scheme

The transmitter diversity is one of the mitigation technique to reduce the effects of atmospheric turbulence fading. The average BER and outage probability of an PolSK-based FSO system is analyzed and compared for no diversity, time diversity, and wavelength diversity cases in this section.

3.4.1 The FSO System with wavelength or time diversity

An FSO system with multiple transmitters and receivers are considered here. The transmitters transmit the signal with different wavelengths and each receiver detects the signal at a specified wavelength. Assume, an FSO system consists of Ndiverse transceivers $n = 1 \ldots N$. The transmitters are transmitting the signal simultaneously at different wavelengths. Each nth copy of the transmitted signal can be detected only by the *n*th receiver provided it recognizes only the *n*th wavelength. The signal (y_n) received by the *n*th receiver [35] can be expressed as

$$y_n = h_n \gamma_n x + \vartheta \tag{3.13}$$

where h_n is the channel state, γ_n is the responsivity of the nth detector in A/W, x is the transmitted signal, and ϑ is the noise caused by various sources. The channel state h_n models the optical intensity fluctuations resulting from atmospheric loss, turbulence, and fading as [24]

$$h_n = h_l h_s h_p \tag{3.14}$$

where h_l is the attenuation due to beam extinction and path loss, h_s due to scintillation effects, and h_p due to pointing errors.

The SNR of the received signal [43] can be expressed as

$$\mathrm{SNR}(h_n) = \frac{\gamma_n{}^2 h_n P_n}{N\sigma^2} \tag{3.15}$$

where σ^2 is the variance of the channel noise, N is the number of transceivers, and P_n is the local oscillator power.

For time diversity case, a single transmitter, transmits n copies of the signal at different time slots and a single receiver, receives the signal. OC technique is used to at the receiver to retrieve the original message signal.

3.4.2 Channel model

The combined channel includes the effects of atmospheric attenuation, turbulence, and pointing errors. This expression is written based on the combined channel model expression in (3.11) for multiple n number of channels as follows,

$$f_{h_n}(h_n) = \frac{\alpha_n \beta_n \xi_n^2}{A_{0_n} h_{l_n} \Gamma(\alpha_n)\Gamma(\beta_n)} G_{1,3}^{3,0}\left[\frac{\alpha_n \beta_n h_n}{A_{0_n} h_{l_n}} \middle| \begin{matrix} \xi_n^2 \\ \xi_n^2 - 1, \alpha_n - 1, \beta_n - 1 \end{matrix}\right] \tag{3.16}$$

3.4.3 Average BER

The ABER performance of PolSK-based FSO system is analyzed with and without diversity cases in this section.

No diversity: In this case a single transmitter and receiver are used. For a PolSK-based FSO system, the conditional BER probability [43] can be expressed as

$$P_{ec}(h) = 0.5 \times \mathrm{erfc}\left(\sqrt{\frac{\gamma^2 h P}{2\sigma^2}}\right) \tag{3.17}$$

where γ is the photo detector responsivity, σ^2 is the variance of the channel noise, and P is the local oscillator power.

The ABER of PolSK-based FSO system is obtained using the following expression,

$$P_e = \int_0^\infty P_{ec}(h) f_h(h) \mathrm{d}h \tag{3.18}$$

By substituting (3.16) and (3.17) in (3.18), we get

$$P_e = \frac{\alpha\beta\xi^2}{A_0 h_l \Gamma(\alpha)\Gamma(\beta)} \times \int_0^\infty 0.5 \times \text{erfc}\left(\sqrt{\frac{\gamma^2 hP}{2\sigma^2}}\right) \times G_{1,3}^{3,0}\left[\frac{\alpha\beta h}{A_0 h_l}\middle| \begin{matrix} \xi^2 \\ \xi^2-1, \alpha-1, \beta-1 \end{matrix}\right] \mathrm{d}h \tag{3.19}$$

The complementary error function erfc(·) can be expressed as Meijer G function using [44, Eq. (8.4.14.2)], we get

$$P_e = \frac{\alpha\beta\xi^2}{2\sqrt{\pi} A_0 h_l \Gamma(\alpha)\Gamma(\beta)} \times \int_0^\infty G_{1,2}^{2,0}\left[\frac{\gamma^2 hP}{2\sigma^2}\middle| \begin{matrix} 1 \\ 0, 1/2 \end{matrix}\right] \times G_{1,3}^{3,0}\left[\frac{\alpha\beta h}{A_0 h_l}\middle| \begin{matrix} \xi^2 \\ \xi^2-1, \alpha-1, \beta-1 \end{matrix}\right] \mathrm{d}h \tag{3.20}$$

By using [27, (3.21)] the above integral is simplified as [45],

$$P_{e,PolSK} = \frac{\xi^2}{2\sqrt{\pi}\Gamma(\alpha)\Gamma(\beta)} G_{4,3}^{2,3}\left[\frac{A_0 h_l P\gamma^2}{2\alpha\beta\sigma^2}\middle| \begin{matrix} 1, 1-\xi^2, 1-\alpha, 1-\beta \\ 0, 0.5, -\xi^2 \end{matrix}\right] \tag{3.21}$$

Wavelength diversity: The conditional bit error rate of FSO system with wavelength diversity over the multiple atmospheric channels, $n = 1, 2, \ldots N$ is given by [43]

$$P_{\text{ec}}(h_{\text{n}}) = Q\left(\sqrt{\text{SNR}}\right) \tag{3.22}$$

By substituting (3.15) in (3.22), we get the conditional BER probability as

$$P_{\text{ec}}(h_{\text{n}}) = Q\left(\frac{\gamma_{\text{n}}\sqrt{P_{\text{n}}}}{N\sigma}\sqrt{\sum_{n=1}^{N} h_{\text{n}}}\right) \tag{3.23}$$

where γ_{n} is the responsivity of the nth photo detector, σ^2 is the variance of the channel noise, and $Q(\cdot)$ is the Gaussian Q function. The ABER of PolSK-based FSO system is calculated using the following equation:

$$P_{\text{e}} = \int_0^\infty P_{\text{ec}}(h_{\text{n}}) f_{h_{\text{n}}}(h_{\text{n}}) \mathrm{d}h_n \tag{3.24}$$

By substituting (3.16) and (3.23) in (3.24), we get

$$P_{\text{e}} = \frac{\alpha_{\text{n}}\beta_{\text{n}}\xi_{\text{n}}^2}{A_{0_{\text{n}}} h_{l_{\text{n}}} \Gamma(\alpha_{\text{n}})\Gamma(\beta_{\text{n}})} \times \int_0^\infty Q\left(\frac{\gamma_{\text{n}}\sqrt{P_{\text{n}}}}{N\sigma}\sqrt{\sum_{n=1}^{N} h_{\text{n}}}\right) \times G_{1,3}^{3,0}\left[\frac{\alpha_{\text{n}}\beta_{\text{n}} h_{\text{n}}}{A_{0_{\text{n}}} h_{l_{\text{n}}}}\middle| \begin{matrix} \xi_{\text{n}}^2 \\ \xi_{\text{n}}^2-1, \alpha_{\text{n}}-1, \beta_{\text{n}}-1 \end{matrix}\right] \mathrm{d}h_{\text{n}} \tag{3.25}$$

Equation (3.25) is modified by using the approximation of Q-function [46] to obtain the ABER as

$$P_e \approx \frac{1}{12}\prod_{n=1}^{N}\int_0^{\infty}\frac{\alpha_n\beta_n\xi_n^2}{A_{0_n}h_{l_n}\Gamma(\alpha_n)\Gamma(\beta_n)}\times G_{1,3}^{3,0}\left[\frac{\alpha_n\beta_n h_n}{A_{0_n}h_{l_n}}\middle|\begin{matrix}\xi_n^2\\ \xi_n^2-1,\alpha_n-1,\beta_n-1\end{matrix}\right]e^{-\left(\frac{\gamma_n^2 h_n P_n}{2N\sigma^2}\right)}dh_n$$
$$+\frac{1}{4}\prod_{n=1}^{N}\int_0^{\infty}\frac{\alpha_n\beta_n\xi_n^2}{A_{0_n}h_{l_n}\Gamma(\alpha_n)\Gamma(\beta_n)}\times G_{1,3}^{3,0}\left[\frac{\alpha_n\beta_n h_n}{A_{0_n}h_{l_n}}\middle|\begin{matrix}\xi_n^2\\ \xi_n^2-1,\alpha_n-1,\beta_n-1\end{matrix}\right]e^{-\left(\frac{2\gamma_n^2 h_n P_n}{3N\sigma^2}\right)}dh_n \tag{3.26}$$

By expressing the exp(·) as Meijer G function [27], the average BER of PolSK based FSO system can be evaluated using [27] as

$$P_e \approx \frac{1}{12}\prod_{n=1}^{N}\frac{\xi_n^2}{\Gamma(\alpha_n)\Gamma(\beta_n)}G_{3,2}^{1,3}\left[\frac{\gamma_n^2 P_n A_{0_n}h_{l_n}}{2N\sigma^2\alpha_n\beta_n}\middle|\begin{matrix}1-\xi_n^2,1-\alpha_n,1-\beta_n\\ 0,-\xi_n^2\end{matrix}\right]$$
$$+\frac{1}{4}\prod_{n=1}^{N}\frac{\xi_n^2}{\Gamma(\alpha_n)\Gamma(\beta_n)}G_{3,2}^{1,3}\left[\frac{2\gamma_n^2 P_n A_{0_n}h_{l_n}}{3N\sigma^2\alpha_n\beta_n}\middle|\begin{matrix}1-\xi_n^2,1-\alpha_n,1-\beta_n\\ 0,-\xi_n^2\end{matrix}\right] \tag{3.27}$$

Equation (3.27) is rewritten in terms of average SNR $= \frac{\gamma_n{}^2 h_n P_n}{N\sigma^2}$ as follows [45],

$$P_e \approx \frac{1}{12}\prod_{n=1}^{N}\frac{\xi_n^2}{\Gamma(\alpha_n)\Gamma(\beta_n)}G_{3,2}^{1,3}\left[\frac{A_{0_n}\overline{\text{SNR}}}{2N\alpha_n\beta_n}\middle|\begin{matrix}1-\xi_n^2,1-\alpha_n,1-\beta_n\\ 0,-\xi_n^2\end{matrix}\right]$$
$$+\frac{1}{4}\prod_{n=1}^{N}\frac{\xi_n^2}{\Gamma(\alpha_n)\Gamma(\beta_n)}G_{3,2}^{1,3}\left[\frac{2A_{0_n}\overline{\text{SNR}}}{3N\alpha_n\beta_n}\middle|\begin{matrix}1-\xi_n^2,1-\alpha_n,1-\beta_n\\ 0,-\xi_n^2\end{matrix}\right] \tag{3.28}$$

Time diversity: In this case a single transmitter and receiver is used. Therefore $\lambda_1=\lambda_2=\ldots\ \ldots=\lambda_n=\lambda$, where $\alpha_1=\alpha_2=\ldots\ \ldots=\alpha_n=\alpha$ and $\beta_1=\beta_2=\ldots\ \ldots=\beta_n=\beta$. We obtain ABER for time diversity by substituting the above assumptions in (3.28) as [45]

$$P_e \approx \frac{1}{12}\left[\frac{\xi^2}{\Gamma(\alpha)\Gamma(\beta)}\right]^N\left\{\left[G_{3,2}^{1,3}\left[\frac{A_0\overline{\text{SNR}}}{2N\alpha\beta}\middle|\begin{matrix}1-\xi^2,1-\alpha,1-\beta\\ 0,-\xi^2\end{matrix}\right]\right]^N\right.$$
$$\left.+3\left[G_{3,2}^{1,3}\left[\frac{2A_0\overline{\text{SNR}}}{3N\alpha\beta}\middle|\begin{matrix}1-\xi^2,1-\alpha,1-\beta\\ 0,-\xi^2\end{matrix}\right]\right]^N\right\} \tag{3.29}$$

Comparison of with and without diversity cases: The various system parameters considered for the simulation include the responsivity of the photo detector is 0.5A/W, beam radius is 2.5 m at 1 km distance and jitter standard deviation of 30 cm, the noise standard deviation is 5×10^{-7}A/Hz [24].

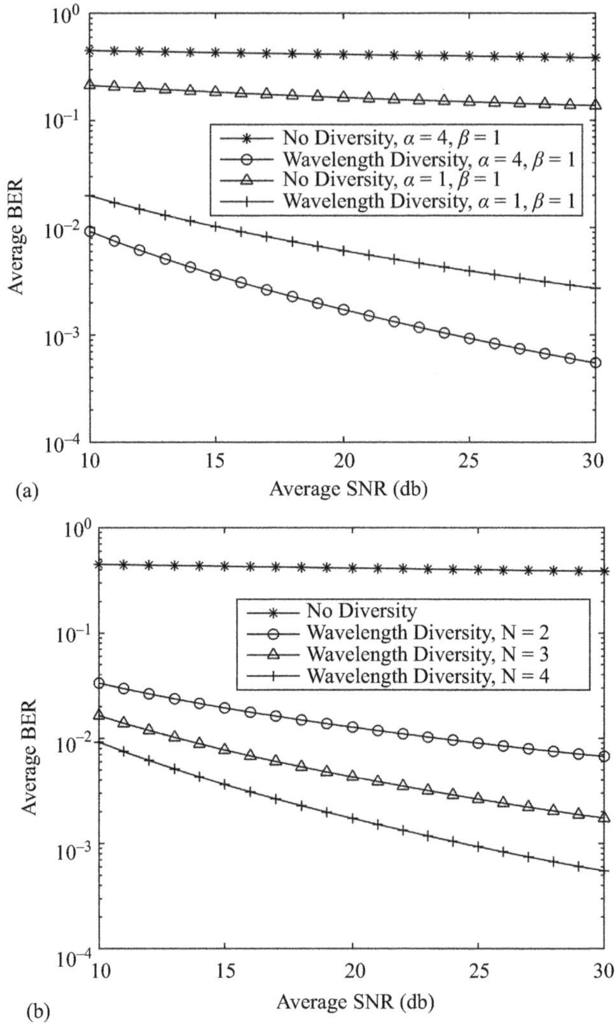

Figure 3.3 ABER against average SNR (a) Over moderate and strong turbulence regimes with and without wavelength diversity (b) with different transmitter apertures $N = 2$, 3 and 4

Figure 3.3 gives the ABER comparison between no diversity and wavelength diversity cases using (3.21) and (3.28). The ABER as a function of average SNR is illustrated in Figure 3.3(a). The ABER is analyzed for weak and moderate turbulence regimes with third order wavelength diversity. The different wavelengths $\lambda_1 = 1{,}550$ nm, $\lambda_2 = 850$ nm, and $\lambda_3 = 1{,}310$ nm [47] are used for the simulation. The moderate and strong turbulences are obtained using the values of $\alpha = 4$ and $\beta = 1$, $\alpha = 1$ and $\beta = 1$, respectively. From the figure, the performance of ABER of FSO system with wavelength diversity provides better performance compared to without diversity case for both moderate and strong turbulences. That is a wavelength diversity technique improves the reliability of the system significantly. Because, multiple copies of the signal are transmitted through different atmospheric turbulence channels, reducing the effect of several atmospheric phenomena, resulting in a decreased ABER [35,48].

The ABER variation with respect to the order of wavelength diversity is shown in Figure 3.3(b). It indicates that the performance of the system is improved with increment in wavelength diversity parameter. Since different wavelengths are used in signal transmission and reception, the absorption of signal in the channel varies with the wavelength of the signal. Therefore, the BER performance can be improved at the receiver [35].

The comparison between no diversity and time diversity cases using (3.21) and (3.29) are illustrated in Figure 3.4. Figure 3.4(a) depicts the ABER performance of the PolSK-based FSO system with time diversity. The single transmitter and receiver used in this case. The wavelength considered for time diversity case is 1,550 nm [35]. It retransmits the signal more than once in different time slots. It improves the performance of the system compared to without diversity case.

The variation in diversity parameter versus ABER is illustrated in Figure 3.4 (b). From the results, the BER of less than 10^{-3} is achieved at SNR of 25 dB compared with other cases.

The ABER as a function of average SNR for with and without diversity technique is plotted in Figure 3.5. It clearly shows that the ABER performance is improved significantly using wavelength and time diversity techniques.

3.4.4 Outage probability

The outage probability is the performance metric to represent the probability of receiving SNR less than the threshold SNR. The minimum required SNR to maintain the acceptable quality of service is referred as threshold SNR. The outage probability over the slow fading channel is expressed as

$$P_{\text{out}} = P(\text{SNR}(h) \leq \text{SNR}_{\text{th}}) \tag{3.30}$$

where SNR_{th} is the threshold SNR below which the signal strength of the receiver is less than acceptable limits.

No diversity: The average SNR of PolSK-based FSO system can be expressed as [43]

$$\text{SNR}(h) = \frac{\gamma^2 h P}{\sigma^2} \tag{3.31}$$

The outage probability of PolSK-based FSO system can be estimated by substituting (3.31) in (3.30) as follows:

$$P_{\text{out}} = P\left(h \leq \frac{\sigma^2 \text{SNR}_{\text{th}}}{\gamma^2 P}\right) = F_h\left(\frac{\sigma^2 \text{SNR}_{\text{th}}}{\gamma^2 P}\right) \tag{3.32}$$

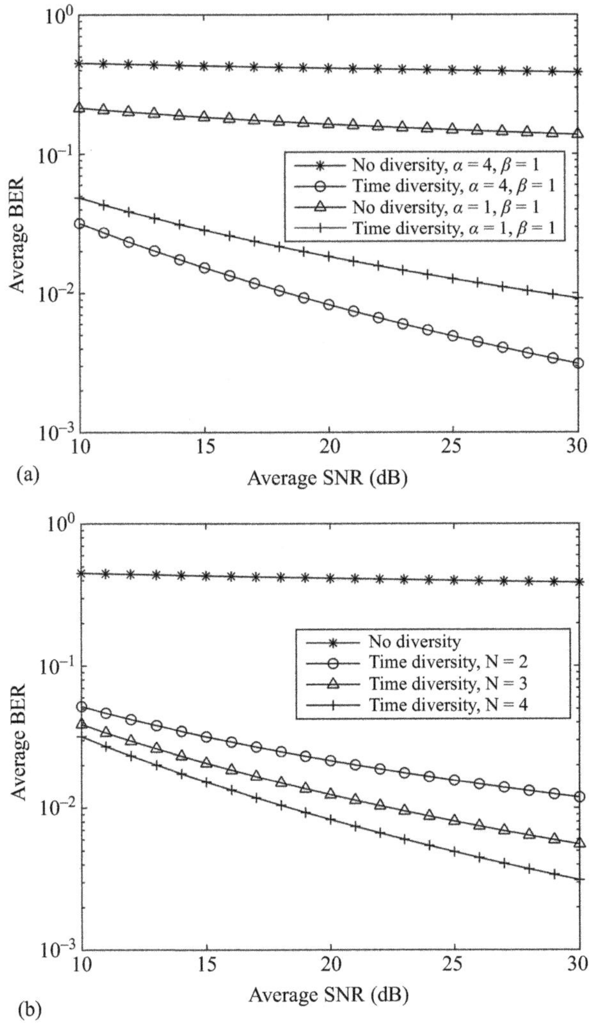

Figure 3.4 ABER against average SNR (a) Over moderate and strong turbulence regimes with and without time diversity (b) with no diversity and time diversity with $N = 2, 3, 4$

where $F_{\mathrm{h}}(\cdot)$ is the cumulative distribution function (CDF) of the channel model. The CDF of $f_{\mathrm{h}}(h)$ is expressed as [30]

$$F_{\mathrm{h}}(h) = \frac{\xi^2}{\Gamma(\alpha)\Gamma(\beta)} G_{2,4}^{3,1}\left[\frac{\alpha\beta}{A_0 h_l} h \middle| \begin{matrix} 1, 1+\xi^2 \\ \xi^2, \alpha, \beta, 0 \end{matrix}\right] \tag{3.33}$$

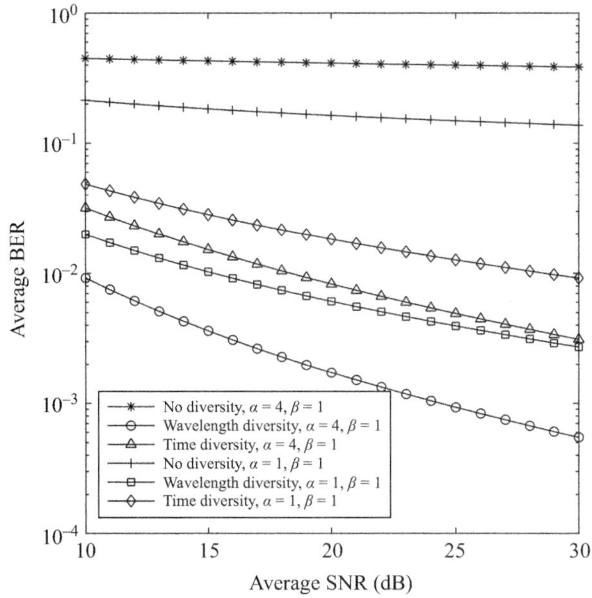

Figure 3.5 ABER comparison of no diversity, time diversity, and wavelength diversity cases

By substituting (3.33) in (3.32), the outage probability of the considered system is obtained as [45],

$$P_{\text{out}} = \frac{\xi^2}{\Gamma(\alpha)\Gamma(\beta)} G_{2,4}^{3,1}\left[\frac{\alpha\beta\sigma^2\text{SNR}_{\text{th}}}{A_0 h_l \gamma^2 P}\middle| \begin{matrix} 1, 1+\xi^2 \\ \xi^2, \alpha, \beta, 0 \end{matrix}\right] \tag{3.34}$$

Wavelength diversity: An FSO system, the wavelength diversity is achieved by using multiple transmitters transmitted the signal simultaneously at different wavelengths over different atmospheric channels. The outage probability of PolSK-based FSO system with wavelength diversity depends on the each FSO link. The outage probability of each n different channel using wavelength diversity is given by [45],

$$P_{\text{out,n}} = \frac{\xi_{\text{n}}^2}{\Gamma(\alpha_{\text{n}})\Gamma(\beta_{\text{n}})} \text{G}_{2,4}^{3,1}\left[\frac{\alpha_{\text{n}}\beta_{\text{n}}\sigma^2\text{SNR}_{\text{th}}}{\gamma_{\text{n}}^2 P_{\text{n}} A_{0_{\text{n}}} h_{l_{\text{n}}}}\middle| \begin{matrix} 1, 1+\xi_{\text{n}}^2 \\ \xi_{\text{n}}^2, \alpha_{\text{n}}, \beta_{\text{n}}, 0 \end{matrix}\right] \tag{3.35}$$

Assuming that the outage probability of each one of the N optical channels $P_{\text{out,n}}$ is independent and then the total outage probability of all the N links can be expressed as [35]

$$P_{\text{out,N}} = \prod_{n=1}^{N} P_{\text{out,n}} \tag{3.36}$$

By substituting (3.35) in (3.36), we obtain the total outage probability of PolSK-based FSO system with wavelength diversity as

$$P_{\text{out,N}} = \prod_{n=1}^{N} \left\{ \frac{\xi_{\text{n}}^2}{\Gamma(\alpha_{\text{n}})\Gamma(\beta_{\text{n}})} G_{2,4}^{3,1} \left[\frac{\alpha_{\text{n}}\beta_{\text{n}}\sigma^2 \text{SNR}_{\text{th}}}{\gamma_{\text{n}}^2 P_{\text{n}} A_{0_{\text{n}}} h_{l_{\text{n}}}} \middle| \begin{matrix} 1, 1+\xi_{\text{n}}^2 \\ \xi_{\text{n}}^2, \alpha_{\text{n}}, \beta_{\text{n}}, 0 \end{matrix} \right] \right\} \tag{3.37}$$

Time diversity: A single transmitter and receiver is used in PolSK-based FSO system with time diversity case. The transmitter retransmits the signal more than once in different time slots to the receiver over a strong atmospheric turbulence channel. The wavelength used in this case is 1,550 nm only. The total outage probability of the FSO system using time diversity can be derived based on the assumptions, $\lambda_1 = \lambda_2 = \ldots \ \ldots = \lambda_{\text{n}} = \lambda, \alpha_1 = \alpha_2 = \ldots \ \ldots = \alpha_{\text{n}} = \alpha$, and $\beta_1 = \beta_2 = \ldots \ \ldots = \beta_{\text{n}} = \beta$. By considering the above assumptions on (3.37), the outage probability of PolSK-based FSO system is expressed as [45],

$$P_{\text{out,n}} = \left\{ \frac{\xi^2}{\Gamma(\alpha)\Gamma(\beta)} G_{2,4}^{3,1} \left[\frac{\alpha\beta\sigma^2 \text{SNR}_{\text{th}}}{A_0 h_l \gamma^2 P} \middle| \begin{matrix} 1, 1+\xi^2 \\ \xi^2, \alpha, \beta, 0 \end{matrix} \right] \right\}^N \tag{3.38}$$

Comparison of with and without diversity cases:

Figure 3.6 narrates the outage performance of the system in terms of threshold SNR over moderate and strong turbulences for with and without time diversity using (3.34) and (3.38). The time diversity provides a significant improvement of outage probability order of 10^{-10} compared to without diversity.

The outage probability against threshold SNR with time diversity is plotted in Figure 3.7 using (3.38). From the figure, it is inferred that the outage performance

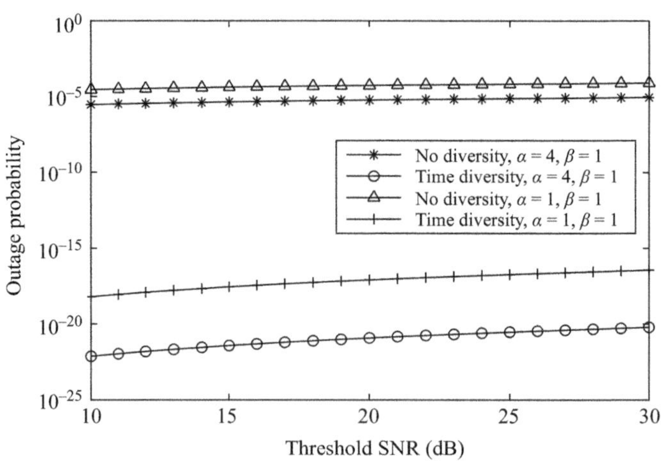

Figure 3.6 Outage probability over moderate and strong turbulences for no diversity and time diversity cases

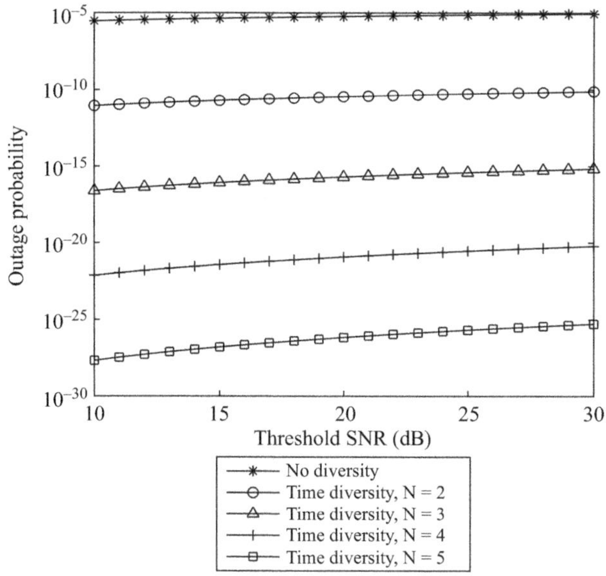

Figure 3.7 Outage probability in terms of SNR for N = 2, 3, 4, and 5

is improved with increment in the order of time diversity. That is the outage probability of 10^{-28} is achieved in the time diversity order of 5.

3.5 Multiple input multiple output

One of the efficient solutions to mitigate the effects of atmospheric turbulences and misalignment fading is diversity. That is the performance of wireless systems can be enhanced by means of spatial diversity, especially the MIMO technique [49,33]. MIMO systems can exploit the diversity in spatial and time domains, and the capacity of the systems increases significantly. The combining techniques used to lessen the effects of atmospheric turbulence fading, where the channel diversity can provide significant improvement in system BER performance [40,50].

MIMO system is useful to reduce the link blockage and misalignment and improves the reliability of an FSO system. The various diversity techniques can be employed to enhance the link performance with low cost. It is employed by creating several independent paths between the transmitter and receiver. Since, each path fades independently, there is a low chance of fading together. The combination of receiver diversity with wide divergence optical sources can avoid the requirement of an active tracking system [36,51,52]. In receiver diversity, a single receiver with a large FOV is replaced by multiple receivers with small surface area and narrow FOV. So, the possibility of fading effects for the entire receiver is much reduced. Also, it provides higher data rate transmission and limits the amount of background light from unwanted sources due to the low capacitance of small area photodetectors [51,53].

The ABER of PolSK-based SISO and MIMO with OC FSO system over atmospheric turbulence channel with pointing errors is analyzed in Section 3.5.1. The ABER of SISO and MIMO FSO system using MRC and EGC is analyzed and the results are compared with and without pointing errors in Section 3.5.2.

3.5.1 ABER analysis of PolSK

The spatial diversity is one of the powerful technique to mitigate effects of atmospheric turbulence. It can be implemented either at transmitter side or receiver side or both the sides defined as SIMO, MISO, and MIMO, respectively. In this section MIMO FSO system with OC is analyzed and compared with SISO FSO system.

SISO FSO link: The conditional BER of PolSK-based FSO system can be expressed as

$$P_{\mathrm{ec,SISO}}(h) = 0.5 \times \exp\left(\frac{-\gamma^2 hP}{2\sigma^2}\right) \tag{3.39}$$

where γ is the photo detector responsivity in A/W, h is the received power irradiance, P is the power of the local oscillator field, and σ^2 is the variance of the channel noise.

The average BER can be computed using (3.40) as follows,

$$P_{\mathrm{e,SISO}} = \int_0^\infty P_{\mathrm{ec,SISO}}(h) f_{\mathrm{h}}(h)\mathrm{d}h \tag{3.40}$$

By using (3.11) and (3.39) in (3.40), the average BER can be obtained as

$$\begin{aligned} P_{\mathrm{e,SISO}} = {} & \frac{\alpha\beta\xi^2}{A_0 h_l \Gamma(\alpha)\Gamma(\beta)} \times \int_0^\infty 0.5 \times \exp\left(\frac{-\gamma^2 hP}{2\sigma^2}\right) \\ & \times G_{1,3}^{3,0}\left[\frac{\alpha\beta h}{A_0 h_l}\middle| \begin{matrix} \xi^2 \\ \xi^2 - 1, \alpha - 1, \beta - 1 \end{matrix}\right] \mathrm{d}h \end{aligned} \tag{3.41}$$

By expressing the exp(·) as Meijer G function [27], the ABER of PolSK-FSO system can be expressed in a closed-form by utilizing [27]

$$P_{\mathrm{e,SISO}} = \frac{2^{\alpha+\beta-4}\xi^2}{\pi\Gamma(\alpha)\Gamma(\beta)} \times G_{6,3}^{1,6}\left[\frac{8\gamma^2 P A_0^2 h_l^2}{\sigma^2\alpha^2\beta^2}\middle| \begin{matrix} \frac{1-\xi^2}{2}, \frac{2-\xi^2}{2}, \frac{1-\alpha}{2}, \frac{2-\alpha}{2}, \frac{1-\beta}{2}, \frac{2-\beta}{2} \\ 0, \frac{-\xi^2}{2}, \frac{1-\xi^2}{2} \end{matrix}\right] \tag{3.42}$$

MIMO FSO link: The ABER for MIMO FSO links can be calculated from [24],

$$P_{\mathrm{e,MIMO}} = \int_0^\infty f_{\mathrm{h}}(h) Q\left(\frac{\gamma^2 P}{\mathrm{MN}\sqrt{2}\sigma_{\mathrm{n}}}\sqrt{\sum_{n=1}^{N}\left(\sum_{m=1}^{M} h_{\mathrm{mn}}\right)^2}\right)\mathrm{d}h \tag{3.43}$$

where $f_{\mathrm{h}}(h)$ is the joint PDF of vector $h = (h_{11}, h_{12}, \ldots h_{\mathrm{MN}})$ of length MN.

The electrical SNR of the FSO system with OC can be expressed as

$$\mathrm{SNR_{OC}} = \frac{\gamma^2 P}{2N\sigma_{\mathrm{n}}^2}\sum_{n=1}^{N} h_{\mathrm{n}}^2 \tag{3.44}$$

The conditional BER of the PolSK-based FSO system can be calculated as

$$P_{\mathrm{ec,OC}}(h) = Q\left(\sqrt{\mathrm{SNR_{OC}}}\right) = Q\left(\frac{\gamma P}{\sqrt{2N}\sigma_{\mathrm{n}}}\sqrt{\sum_{n=1}^{N} h_{\mathrm{n}}^2}\right) \tag{3.45}$$

The OC diversity is applied at the receiver; then the ABER of SIMO FSO is written as

$$P_{\mathrm{e,OC}} = \int_0^\infty Q\left(\frac{\gamma P}{\sqrt{2N}\sigma_{\mathrm{n}}}\sqrt{\sum_{n=1}^{N} h_{\mathrm{n}}^2}\right) f_{\mathrm{h}}(h)\mathrm{d}h \tag{3.46}$$

Equation (3.46) is simplified using the approximation of Q-function [46] to obtain the ABER as

$$P_{\mathrm{e,OC}} \approx \frac{1}{12}\prod_{n=1}^{N}\int_0^\infty f_{\mathrm{h_n}}(h_{\mathrm{n}})\mathrm{e}^{-\left(\frac{\gamma^2 P h_{\mathrm{n}}^2}{4N\sigma_{\mathrm{n}}^2}\right)}\mathrm{d}h_{\mathrm{n}} + \frac{1}{4}\prod_{n=1}^{N}\int_0^\infty f_{\mathrm{h_n}}(h_{\mathrm{n}})\mathrm{e}^{-\left(\frac{\gamma^2 P h_{\mathrm{n}}^2}{3N\sigma_{\mathrm{n}}^2}\right)}\mathrm{d}h_{\mathrm{n}} \tag{3.47}$$

By applying (3.11) in (3.47) and express the exp(·) as Meijer G function [27], the average BER of PolSK-FSO system can be evaluated using (3.21) in [27] as [54]

$$\begin{aligned} P_{\mathrm{e,OC}} \approx & \frac{1}{12}\prod_{n=1}^{N}\frac{2^{\alpha+\beta-3}\xi^2}{\pi\Gamma(\alpha)\Gamma(\beta)} \\ & \times G_{6,3}^{1,6}\left[\frac{4\gamma^2 P A_0^2 h_1^2}{N\sigma_{\mathrm{n}}^2\alpha^2\beta^2}\left|\begin{matrix}\frac{1-\xi^2}{2},\frac{2-\xi^2}{2},\frac{1-\alpha}{2},\frac{2-\alpha}{2},\frac{1-\beta}{2},\frac{2-\beta}{2}\\ 0,\frac{-\xi^2}{2},\frac{1-\xi^2}{2}\end{matrix}\right.\right] + \frac{1}{4}\prod_{n=1}^{N}\frac{2^{\alpha+\beta-3}\xi^2}{\pi\Gamma(\alpha)\Gamma(\beta)} \\ & \times G_{6,3}^{1,6}\left[\frac{16\gamma^2 P A_0^2 h_1^2}{3N\sigma_{\mathrm{n}}^2\alpha^2\beta^2}\left|\begin{matrix}\frac{1-\xi^2}{2},\frac{2-\xi^2}{2},\frac{1-\alpha}{2},\frac{2-\alpha}{2},\frac{1-\beta}{2},\frac{2-\beta}{2}\\ 0,\frac{-\xi^2}{2},\frac{1-\xi^2}{2}\end{matrix}\right.\right] \end{aligned} \tag{3.48}$$

Figures 3.8 and 3.9 are plotted using (3.42) and (3.48). The simulation parameters include, the noise standard deviation $\sigma_{\mathrm{n}} = 10^{-7}\mathrm{A/Hz}$, photo detector responsivity $\gamma = 0.5\mathrm{A/W}$, beam radius $w_{\mathrm{L}} \cong 2.5\,\mathrm{m}$ at 1 km distance, and jitter standard deviation $\sigma_{\mathrm{s}} \cong 30\,\mathrm{cm}$. The effective number of large- and small-scale turbulent eddies for moderate and strong atmospheric turbulences are $\alpha = 4, \beta = 1.9$ and $\alpha = 4.2, \beta = 1.4$, respectively [54].

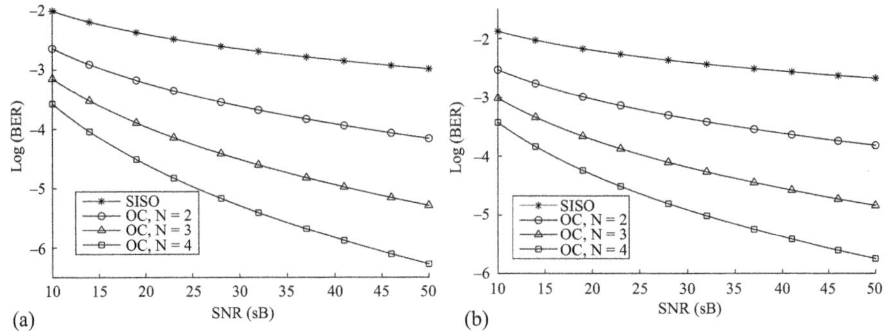

Figure 3.8 BER versus SNR for PolSK-based FSO in (a) moderate turbulence and (b) strong turbulence

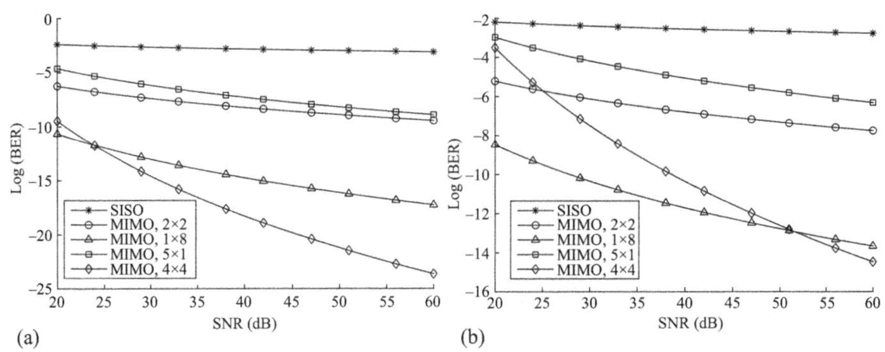

Figure 3.9 BER versus SNR for PolSK-based MIMO-FSO in (a) moderate turbulence and (b) strong turbulence

Figure 3.8(a) illustrates the average BER of SISO and MIMO-FSO systems with OC under moderate turbulence conditions, as a function of the average SNR. The ABER performance is compared with different number of receiver apertures using OC technique. It is inferred from the graph, increase in receiver aperture increases the performance of the system. A better ABER of 10^{-4} is achieved at SNR of 15 dB MIMO FSO system using an OC technique with 4 receiver apertures.

The ABER performance of MIMO FSO system with OC against the average SNR over strong atmospheric turbulence is shown in Figure 3.8(b). Since the high absorption and laser radiation, the performance of an FSO system over strong turbulence is worse compared with moderate turbulence. From the figure, the ABER performance is improved from 10^{-2} to 10^{-4} using 4 MIMO with OC.

Figure 3.9 represents the ABER of PolSK-based SISO and MIMO systems over moderate and strong turbulences, respectively. In both the cases, the ABER improves as the SNR and the order of the MIMO (number of transmitting and

Table 3.1 Comparison of ABER for SISO and MIMO systems

System	Average BER at SNR = 50 dB	
	Moderate turbulence channel $\alpha = 4, \beta = 1.9$	**Strong turbulence channel** $\alpha = 4.2, \beta = 1.4$
SISO	10^{-3}	$\approx 10^{-2}$
MIMO with OC ($N = 4$)	10^{-6}	$\approx 10^{-5}$
MIMO (4 × 4)	10^{-24}	$\approx 10^{-14}$

receiving apertures) increases. In weak turbulence, it is inferred that the ABER performance improved from 10^{-5} to 10^{-22} at SNR of 50 dB by adding an extra receiver as 4 × 4 MIMO. It is observed that the ABER performance is improved from 10^{-2} to 10^{-8} at SNR of 20 dB using 4 × 4 MIMO.

The comparison of ABER of SISO and MIMO FSO systems over moderate and strong turbulence channel at SNR of 50 dB is given in Table 3.1. From the table, the effects of turbulences can be mitigated through MIMO systems. We can reduce the error rate to 10^{-24}, 10^{-14} using MIMO (4 × 4) systems for moderate and strong turbulent cases, respectively.

3.5.2 BER of Polsk with and without pointing errors

The misalignment fading is one of the major challenges in FSO system. This section discusses the influence of pointing errors in an FSO system based on the comparison of ABER performance using SISO and MIMO with MRC and EGC over the atmospheric channel with pointing errors.

SISO FSO link: The ABER of SISO FSO system over atmospheric channel without pointing error is calculated using (3.4), (3.17), and (3.18). To further simplify it, exponential (exp(·)) and Bessel functions $K_v(\cdot)$ using [27, (3.11) and (3.14)] are rewritten in terms of Meijer's G function [27, (3.21)] is given as,

$$P_e = \frac{1}{8\pi\Gamma(\alpha)\Gamma(\beta)}\left(\frac{\sqrt{2}\alpha\beta}{\sqrt{\mu}}\right)^{\frac{(\alpha+\beta)}{2}} \times G_{1,4}^{4,1}\left[\frac{(\alpha\beta)^2}{8\mu}\middle|\begin{matrix} 1-\dfrac{\alpha+\beta}{4} \\ \dfrac{\alpha-\beta}{4}, \dfrac{\alpha-\beta+2}{4}, \dfrac{\beta-\alpha}{4}, \dfrac{\beta-\alpha+2}{4}\end{matrix}\right] \tag{3.49}$$

where μ is the average SNR. The ABER of SISO FSO system over atmospheric turbulence channel with pointing error is previously derived in (3.42). It can be expressed in terms of average SNR as follows,

$$P_e = \frac{2^{\alpha+\beta-4}\xi^2}{\pi\Gamma(\alpha)\Gamma(\beta)} G_{6,3}^{1,6}\left[\frac{8\mu A_0^2}{\alpha^2\beta^2}\middle|\begin{matrix}\dfrac{1-\xi^2}{2}, \dfrac{2-\xi^2}{2}, \dfrac{1-\alpha}{2}, \dfrac{2-\alpha}{2}, \dfrac{1-\beta}{2}, \dfrac{2-\beta}{2} \\ 0, \dfrac{-\xi^2}{2}, \dfrac{1-\xi^2}{2}\end{matrix}\right] \tag{3.50}$$

MMO FSO link: A MIMO PolSK-FSO system is achieved using Mtransmitters and Nreceivers operating through atmospheric turbulence channel. At the receiver, the total SNR at the input of the coherent demodulator is obtained as [53]

$$\gamma_T = \frac{(\gamma P)^2 \left(\sum_{i=1}^{N} G_i \sum_{j=1}^{M} h_{ij} \right)^2}{M^2 N \sum_{i=1}^{N} G_i^2 \sigma_{\rm n}^2} \tag{3.51}$$

where h_{ij} denotes the received signal amplitude through the optical turbulence channel between the ith transmitter and the jth receive aperture.

The ABER is obtained by averaging conditional BER over gamma–gamma distribution

$$P_{\rm e,MIMO} = \int_0^\infty \frac{1}{2} \exp\left(\frac{-\gamma_T}{2} \right) f_{\rm h}(h) {\rm d}h \tag{3.52}$$

3.5.2.1 ABER without pointing errors

The ABER of MIMO PolSK-based FSO system using maximal ratio combining and EGC without pointing error is analyzed in this section.

Maximal Ratio Combining: The average SNR of MIMO PolSK-based FSO system using maximal ratio combining scheme is given as [53],

$$\gamma_{\rm MRC} = \frac{\sum_{i=1}^{N} \sum_{j=1}^{M} \gamma_{ij}}{M^2 N} \tag{3.53}$$

The ABER of MIMO FSO system using EGC is derived using (3.54) as follows

$$P_{\rm e,MRC} = \int_0^\infty \frac{1}{2} \exp\left(\frac{-\gamma_{\rm MRC}}{2} \right) f_{\rm h}(h) {\rm d}h \tag{3.54}$$

By substituting (3.4) and (3.53) into (3.54) and expressing exponential $(\exp(\cdot))$ and Bessel functions $K_v(\cdot)$ using [27, (3.11) and (3.14)] Meijer's G function, the ABER for MIMO is given as [55],

$$P_{\rm e,MRC} = \prod_{i=1}^{N} \prod_{j=1}^{M} \frac{\left(\sqrt{2M^2 N \alpha\beta} \right)^{\frac{\alpha+\beta}{2}}}{8\pi \Gamma(\alpha) \Gamma(\beta) \left(\mu_{ij} \right)^{\frac{\alpha+\beta}{4}}} \times G_{1,4}^{4,1} \left[\frac{M^2 N (\alpha\beta)^2}{8\mu_{ij}} \middle| \begin{matrix} 1 - \frac{\alpha+\beta}{4} \\ \frac{\alpha-\beta}{4}, \frac{\alpha-\beta+2}{4}, \frac{\beta-\alpha}{4}, \frac{\beta-\alpha+2}{4} \end{matrix} \right] \tag{3.55}$$

Equal gain combining: The SNR for MIMO-PolSK-FSO system with EGC technique is given as [53],

$$\gamma_{\text{EGC}} = \frac{\left(\sum_{i=1}^{N}\sum_{j=1}^{M}\sqrt{\gamma_{ij}}\right)^2}{M^2N^2} \tag{3.56}$$

The ABER for MIMO-PolSK-FSO system over gamma–gamma distribution is obtained as

$$P_{\text{e,EGC}} = \int_0^{\infty} \frac{1}{2}\exp\left(\frac{-\gamma_{\text{EGC}}}{2}\right) f_{\text{h}}(h)\text{d}h \tag{3.57}$$

By substituting (3.4) and (3.56) in (3.57) and it is simplified as (3.58) by replacing exponential $(\exp(\cdot))$ and Bessel functions $K_v(\cdot)$ using (3.11) and (3.14) in [27], Meijer's G function [56] is given as [55]

$$P_{\text{e,EGC}} = \frac{2^{\alpha_{\gamma T}+\beta_{\gamma T}-3}}{\pi\Gamma(\alpha_{\gamma T})\Gamma(\beta_{\gamma T})} \times G_{1,4}^{4,1}\left[\frac{\left(\frac{M}{N}\alpha_{\gamma T}\beta_{\gamma T}\right)^2}{8\mu_{\gamma T}}\middle|\begin{matrix}1\\ \frac{\alpha_{\gamma \text{T}}}{2}, \frac{\alpha_{\gamma T}+1}{2}, \frac{\beta_{\gamma T}}{2}, \frac{\beta_{\gamma T}+1}{2}\end{matrix}\right] \tag{3.58}$$

3.5.2.2 ABER with pointing errors

The ABER of MIMO PolSK-based FSO system using maximal ratio combining and EGC with pointing error is analyzed in this section.

Maximal ratio combining: The ABER for MIMO-PoSK-FSO system with pointing errors is obtained by substituting (3.11) and (3.53) into (3.54) and expressing exp(·) using (3.11) in [27], Meijer's G function, the closed form expression can be derived as [55],

$$P_{\text{e,MRC}} = \prod_{i=1}^{N}\prod_{j=1}^{M}\frac{2^{\alpha+\beta-4}\xi^2}{\pi\Gamma(\alpha)\Gamma(\beta)} \times \text{G}_{6,3}^{1,6}\left[\frac{8\mu A_0^2}{M^2N\alpha^2\beta^2}\middle|\begin{matrix}\frac{1-\xi^2}{2}, \frac{2-\xi^2}{2}, \frac{1-\alpha}{2}, \frac{2-\alpha}{2}, \frac{1-\beta}{2}, \frac{2-\beta}{2}\\ 0, \frac{-\xi^2}{2}, \frac{1-\xi^2}{2}\end{matrix}\right] \tag{3.59}$$

Equal gain combining: The ABER for MIMO-PolSK-FSO system with EGC over gamma–gamma distribution with pointing errors is obtained by substituting (3.11) and (3.56) into (3.57) and expressing exponential exp(·) using (3.11) in [27],

Meijer's G function, the closed form expression can be derived as [55],

$$P_{\mathrm{e,EGC}} = \prod_{i=1}^{N} \frac{2^{\alpha+\beta-4}\xi^2}{\pi\Gamma(\alpha)\Gamma(\beta)} G_{6,3}^{1,6}\left[\frac{8\mu A_0^2}{N^2\alpha^2\beta^2}\left|\begin{matrix}\frac{1-\xi^2}{2},\frac{2-\xi^2}{2},\frac{1-\alpha}{2},\frac{2-\alpha}{2},\frac{1-\beta}{2},\frac{2-\beta}{2}\\ 0,\frac{-\xi^2}{2},\frac{1-\xi^2}{2}\end{matrix}\right.\right] \tag{3.60}$$

Numerical results and discussions: Figure 3.10 depicts the ABER in terms of SNR over moderate and strong turbulences with and without pointing errors for SISO PolSK-FSO system. It is plotted using (3.49) and (3.50). From the figure, it is inferred that the pointing errors degrades the ABER of the system for both the turbulence cases. From moderate to strong turbulence, 10 dB additional SNR is required to maintain the same ABER. That is, the result shows that the importance of pointing error compensation in optical wireless communication systems.

Equations (3.49), (3.50), (3.55), (3.58), (3.59), and (3.60) are plotted in Figures 3.11 and 3.12. FSO system link with beam radius $w_L \cong 2.5$ m at 1 km distance, jitter standard deviation $\sigma_s \cong 30$ cm, photo detector responsivity $\gamma = 0.5\text{A/W}$, and noise standard deviation $\sigma_n = 10^{-7}\text{A/Hz}$ are considered.

Figure 3.11 illustrates the ABER comparison of SISO and MIMO (2 × 2) FSO system using MRC and EGC combining techniques over moderate and strong turbulence regime, respectively. It is noticed that to reduce the BER from 10^{-2} to 10^{-3}, an extra ~ 20 dB of SNR is needed for with and without pointing errors, respectively, in moderate and strong turbulences.

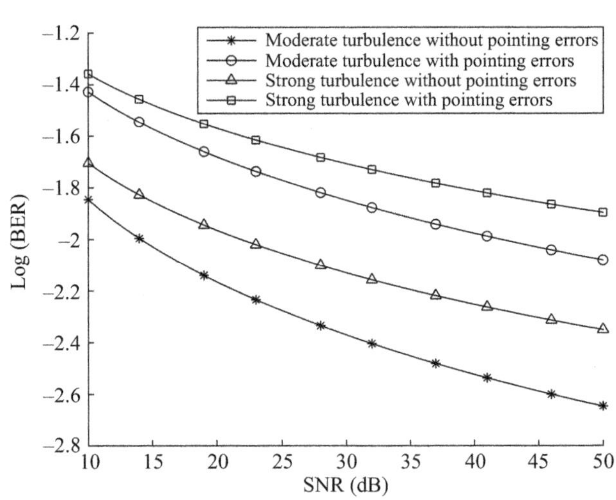

Figure 3.10 BER against SNR over moderate and strong turbulences

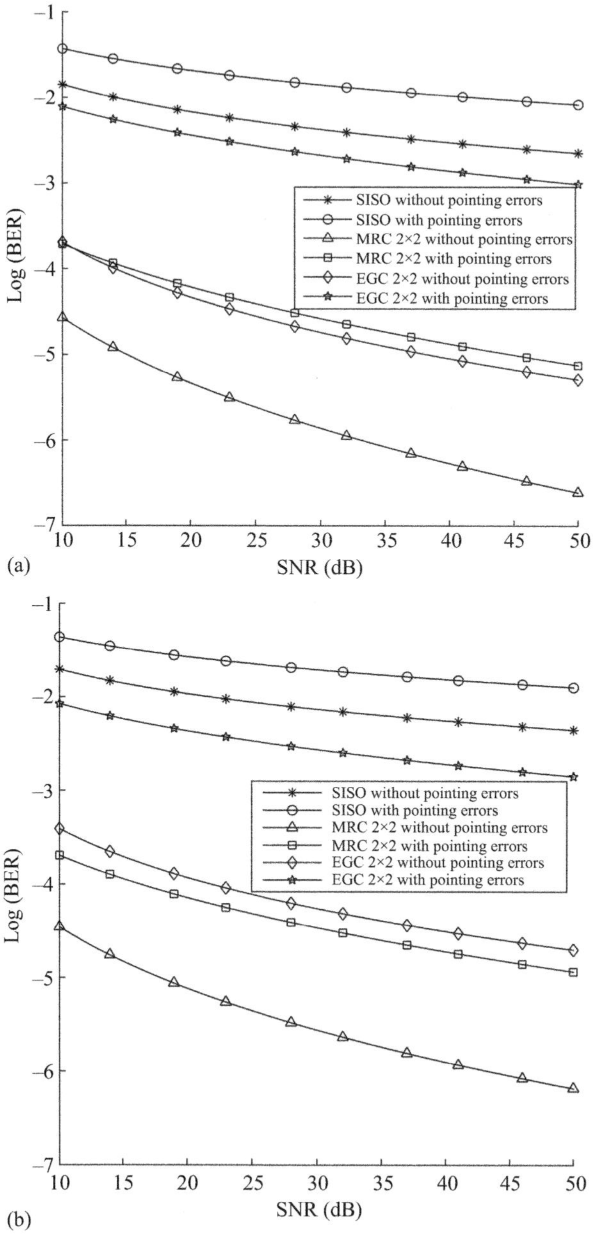

Figure 3.11 BER comparison of SISO and MIMO with EGC, MRC, and OC combining schemes over (a) moderate turbulence and (b) strong turbulence

The ABER comparison of SISO and MIMO with MRC and EGC reception over moderate and strong turbulences is illustrated in Figure 3.12. It is observed that, the BER performances of MIMO systems are significantly improved as the transmitter/receiver numbers increase, compared to the SISO system. It can be observed that 2 × 2 MIMO MRC outperform well compared to other combining schemes.

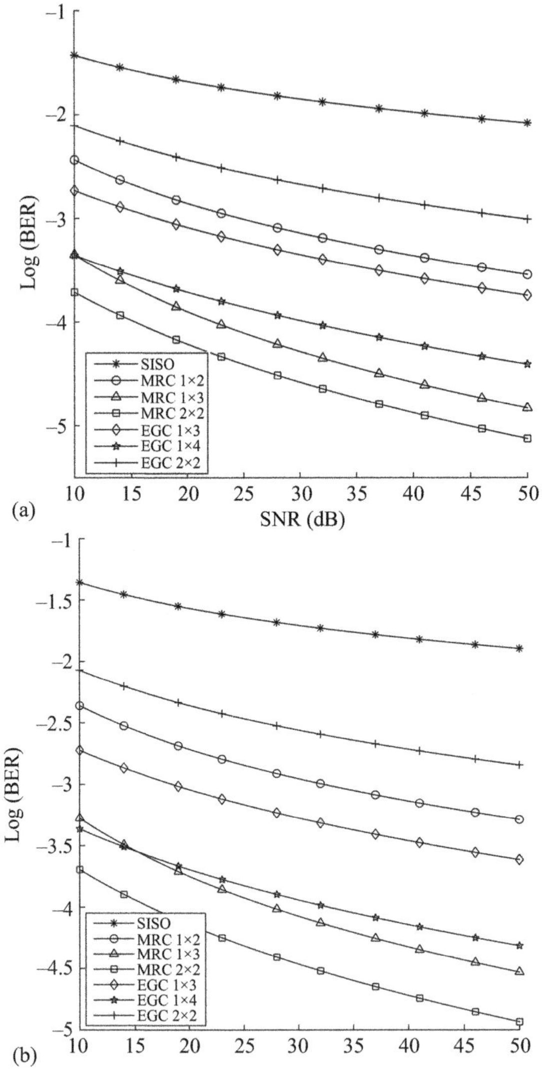

Figure 3.12 BER comparison of SISO and MIMO-FSO system with pointing errors against the average SNR for different N and M over (a) moderate turbulence channel and (b) strong turbulence channel

3.6 Summary

In this chapter, the ABER and outage performance of PolSK-based FSO system in analyzed over moderate and strong atmospheric turbulences. By using wavelength and time diversity techniques, the performance of the system is enhanced. Also, the PolSK-based system extended with spatial diversity and MIMO techniques using OC, EGC, and MRC combining schemes. The performance of SISO and MIMO-FSO systems are analyzed in moderate and strong turbulence regimes, the results are compared and plotted. The better BER performance of 10^{-7} at SNR = 30dB is achieved by using MIMO with MRC combining in weak turbulence without pointing errors. The performance degradation effect of atmospheric turbulence and pointing errors can be reduced by using MIMO scheme, that is by increasing the number of transmit and/or receive apertures. Through the analytical and numerical studies, it is observed that MIMO systems provide the better BER of 10^{-13} and 10^{-22} for moderate and strong turbulences at SNR at 60 dB, respectively.

Appendix A

This complete chapter presents the numerical results only. The sample MuPAD® codes for outage probability of PolSK-based FSO system with no diversity (3.34), with time diversity (3.38) as follows [i.e. Code for Figure 4.7].

Program 4.1: MuPAD Code to Calculate the outage probability of PolSK-based FSO system (3.34) and (3.38)

```
%****************SIMULATION PARAMETERS*********************
a := 4; % Effective number of large-scale turbulent eddies b := 1; %
Effective number of small-scale turbulent eddies wza :=6;

v :=sqrt(PI/2)/wza; A0 :=(erfc(v))^2; % Fraction of the collected
power at r = 0 wz :=12.5;% beam waist wzeq :=sqrt(wz^2*sqrt(PI)
*erfc(v)/2*v*exp(-v^2)); % Equivalent beam width ss :=30*10^-2; %
Jitter standard deviation t :=(wzeq/(2*ss))^2; % ratio between the
equivalent beam radius at

the receiver and the pointing errors displacement standard devia-
tion r :=0.5; % Radial vector from the beam center hl :=0.008; %
Light fog weather condition w :=1550*10^-9; % Wavelength L :=
5*10^3; % Link distance e :=0.8; % transmitter / Receiver optical
efficiency d :=8*10^-2; % Aperture diameter A :=(PI*d^2)/4; %
Aperture area sn2 :=(5*10^-7)^2; % Noise standard deviation

%*********************Outage Calculation*******************
Out_Nodiv:=(t/(gamma(a)*gamma(b)))*meijerG([[1],[t+1]],[[t,a,
b],[0]],((a*b)/(A0*hl))*sqrt((2*sn2*SNRTH)/r^2)):

Out_2 := ((t/(gamma(a)*gamma(b)))*meijerG([[1],[t+1]],[[t,a,
b],[0]],((a*b)/(A0*hl))*sqrt((2*sn2*SNRTH)/r^2)))^2:
```

```
Out_3:= ((t/(gamma(a)*gamma(b)))*meijerG([[1],[t+1]],[[t,a,b],
[0]],((a*b)/(A0*hl))*sqrt((2*sn2*SNRTH)/r^2)))^3:

%*********************Plot Fig. 4.7********************

plot( plot::Function2d(Out_Nodiv, SNRTH=10..30,

LegendText = "No diversity", Color = RGB::Red),

plot::Function2d(Out_2, SNRTH=10..30,
LegendText = "Time diversity, N=2", Color = RGB::Blue),

plot::Function2d(Out_3, SNRTH=10..30,

LegendText = "Time diversity, N=3", Color = RGB::Green),Coordina-
teType = LinLog,LegendVisible = TRUE):
```

References

[1] Kedar, D. and Shlomi, A. “Urban optical wireless communication networks: The main challenges and possible solutions.” *IEEE Communications Magazine* 42 (2004): S2–S7.

[2] Kazaura, Kamugisha, Kazuhiko Wakamori, Mitsuji Matsumoto, Takeshi Higashino, Katsutoshi Tsukamoto, and Shozo Komaki. “RoFSO: A universal platform for convergence of fiber and free-space optical communication networks.” *IEEE Communications Magazine* 48 (2010).

[3] Majumdar, A. K. and J. C. Ricklin, eds. *Free-space laser communications: Principles and advances*. Vol. 2. Springer Science & Business Media, 2010.

[4] Majumdar, A. K. *Advanced Free Space Optics (FSO): A Systems Approach*. Vol. 186. Springer, 2014.

[5] Borah, D. K. and D. G. Voelz. “Pointing error effects on free-space optical communication links in the presence of atmospheric turbulence.” *Journal of Lightwave Technology* 27 (2009): 3965–3973.

[6] Sandalidis, Harilaos G., Theodoros A. Tsiftsis, George K. Karagiannidis, and Murat Uysal. “BER performance of FSO links over strong atmospheric turbulence channels with pointing errors.” *IEEE Communications Letters* 12 (2008): 44–46.

[7] Prabu, K., D. Sriram Kumar, and Reza Malekian. “BER analysis of BPSK-SIM-based SISO and MIMO FSO systems in strong turbulence with pointing errors.” *Optik-International Journal for Light and Electron Optics* 125 (2014): 6413–6417.

[8] Prabu, K. and Sriram Kumar, D. “Polarization shift keying based relay-assisted free space optical communication over strong turbulence with misalignment.” *Optics & Laser Technology* 76 (2016): 58–63.

[9] Prabu, P. and Sriram Kumar, D. “Bit error rate analysis of free-space optical system with spatial diversity over strong atmospheric turbulence channel with pointing errors.” *Optical Engineering* 53 (2014): 126108.

[10] Grave, J. E. and Drenker, S. "Advancing Free space Optical communication with Adaptive optics." *Lightwave online*, 19 (9) (2002): 105–113.
[11] Tyson, R. K. "Bit-error rate for free-space adaptive optics laser communications." *Journal of the Optical Society of America A* 19 (2002): 753–758.
[12] Zhu, X. and Kahn, J. M. "Free-space optical communication through atmospheric turbulence channels." *IEEE Transactions on Communications*, 50 (2002): 1293–1300.
[13] Zabidi, S. A., Al Khateeb, W., Islam, M. R., and Naji, A. W. "Investigating of rain attenuation impact on free space optics propagation in tropical region." *IEEE Proceedings of 4th International Conference on Mechatronics* (2011): 1–6.
[14] Fleagle, R. G. and Businger, J. A. *An introduction to atmospheric physics.* Academic Press, San Diego, CA, 1981.
[15] Short, N. M. and Robinson, J. The remote sensing tutorial. Goddard Space Flight Center, NASA, 1998.
[16] Zhang, Z. "Quantitative microplasma electron number density measurement by coherent microwave Rayleigh scattering." *IEEE Transactions on Plasma Science* 39 (2011): 593–595.
[17] Andrews, L. C. and Phillips, R. L. *Laser beam propagation through random media*. Bellingham: SPIE press, 2005.
[18] Killinger, D. Free space optics for laser communication through the air. Optics and Photon. News, 13 (2002): 36–42.
[19] Nistazakis, H. E., Karagianni, E. A., Tsigopoulos, A. D., Fafalios, M. E., and Tombras, G. S. "Average capacity of optical wireless communication systems over atmospheric turbulence channels." *IEEE Journal of Lightwave Technology* 27 (2009): 974–979.
[20] Nistazakis, H. E., Tsiftsis, T. A., and Tombras, G. S. "Performance analysis of free-space optical communication systems over atmospheric turbulence channels." *IET Communications* 3 (2009): 1402–1409.
[21] Ghassemlooy, Z., Popoola, W. and Rajbhandari, S. *Optical Wireless Communications System and Channel Modelling with MATLAB*, CRC Press, 2012.
[22] Ghassemlooy, Z., Tang, X. and Rajbhandari, S. "Experimental investigation of polarisation modulated free space optical communication with direct detection in a turbulence channel." *IET Communications* 6 (2012): 1489–1494.
[23] Motlagh, A. C., Ahmadi, V., Ghassemlooy, Z. and Abedi, K. "The effect of atmospheric turbulence on the performance of the free space optical communications." *Proceedings of IEEE 6th International Symposium on Communication Systems, Networks and Digital Signal Processing*, 2008, 540–543.
[24] Farid, A. A. and Hranilovic, S. "Outage capacity optimization for free-space optical links with pointing errors." *IEEE Journal of Lightwave Technology* 25 (2007): 1702–1710.
[25] Lee, I. E., Ghassemlooy, Z., Ng, W. P., and Uysal, M. "Performance analysis of free space optical links over turbulence and misalignment induced fading

channels." *Proceedings of IEEE 8th International Symposium on Communication Systems, Networks & Digital Signal Processing*, 2012, 1–6.

[26] Lee, I. E., Ghassemlooy, Z., Ng, W. P. "Effects of aperture averaging and beam width on Gaussian free space optical links in the presence of atmospheric turbulence and pointing error." *Proceedings of 14th IEEE International Conference on Transparent Optical Networks*, 2012, 1–4.

[27] Adamchik, V. S. and Marichev, O. I. The algorithm for calculating integrals of hypergeometric type functions and its realization in REDUCE system. *Proceedings of ACM International Symposium on Symbolic and Algebraic Computation*, 1990, 212–224.

[28] Wolfram, The Wolfram Functions Site Internet [Online] (2001). Available from: ⟨http://functions.wolfram.com⟩.

[29] Prabu, K., Sumanta Bose, and D. Sriram Kumar. "BPSK based subcarrier intensity modulated free space optical system in combined strong atmospheric turbulence." *Optics Communications* 305 (2013): 185–189.

[30] Fried, D. L. "Aperture averaging of scintillation." *Journal of the Optical Society of America* 57 (2) (1967): 169–172.

[31] Andrews, L. C., Phillips, R. L. and Hopen, C. Y. *Laser beam scintillation with applications*. SPIE Press, 2001.

[32] Andrews, L. C., Phillips, R. L., Hopen, C. Y and Al-Habash, M. A. "Theory of optical scintillation." *Journal of the Optical Society of America A* 16 (1999): 1417–1429.

[33] Navidpour, S. M., Uysal, M., and Kavehrad, M. "BER performance of free-space optical transmission with spatial diversity." *IEEE Transactions on Wireless Communications* 6 (2007): 2813–2819.

[34] Tsiftsis, T. A., Sandalidis, H. G., Karagiannidis, G. K., and Uysal, M. Optical wireless links with spatial diversity over strong atmospheric turbulence channels. *IEEE Transactions on Wireless Communications* 8 (2) (2009): 951–957.

[35] Nistazakis, H. E. and Tombras, G. S. "On the use of wavelength and time diversity in optical wireless communication systems over gamma–gamma turbulence channels." *Optics & Laser Technology* 44 (2012): 2088–2094.

[36] Wang, Z., Zhong, W., Fu, S., and Lin, C. "Performance comparison of different modulation formats over free-space optical (FSO) turbulence links with space diversity reception technique." *IEEE Photonics Journal* 1 (2009): 277–285.

[37] Xu, F., Khalighi, A., CaussÚ, P., and Bourennane, S. Channel coding and time-diversity for optical wireless links. *Optics Express* 17 (2009): 872–887.

[38] Chen, Z., Yu, S., Wang, T., Wu, G., Guo, H., and Gu, W. "Spatial correlation for transmitters in spatial MIMO optical wireless links with Gaussian-beam waves and aperture effects." *Optics Communications* 287 (2013): 12–18.

[39] Bayaki, E. and Schober, R. "Performance and design of coherent and differential space-time coded FSO systems." *IEEE Journal of Lightwave Technology* 30 (2012): 1569–1577.

[40] Vavoulas, A., Sandalidis, H. G., and Varoutas, D. "Weather effects on FSO network connectivity." *IEEE Journal of Optical Communications and Networking* 4 (2012): 734–740.

[41] Safari, M. and Uysal, M. "Relay-assisted free-space optical communication." *IEEE Transactions on Wireless Communications* 7 (2008): 5441–5449.

[42] Datsikas, C. K., Peppas, K. P., Sagias, N. C., and Tombras, G. S. "Serial free-space optical relaying communications over gamma-gamma atmospheric turbulence channels." *IEEE Journal of Optical Communications and Networking* 2 (2010): 576–586.

[43] Tang, X., Ghassemlooy, Z., Rajbhandari, S., Popoola, W. O., Uysal, M and Wu, D. "Experimental demonstration of polarisation shift keying in the free space optical turbulence channel." *Proceedings of IEEE International Conference on Communications in China Workshops*, 2012, 31–36.

[44] Prudnikov, A. P., Brychkov, Y. A., and Marichev, O. I. *Integral and Series, vol. 3: More Special Functions*, Gordon and Breach Science Publishers, Amsterdam (1986).

[45] Prabu, K., Shashidhar Cheepalli, and D. Sriram Kumar. "Analysis of PolSK based FSO system using wavelength and time diversity over strong atmospheric turbulence with pointing errors." *Optics Communications* 324 (2014): 318–323.

[46] Chiani, M., Dardari, D. and Simon, M. K. "New exponential bounds and approximations for the computation of error probability in fading channels." *IEEE Transactions on Wireless Communications* 2 (2003): 840–845.

[47] Majumdar, A. K. Free-space laser communication performance in the atmospheric channel. *Journal of Optical and Fiber Communications Reports* 2 (2005): 345–396.

[48] Stassinakis, A. N., Hanias, M. P., Nistazakis, H. E., and Tombras, G. S. "Evaluation of outage probability for time diversity schemes in free space optical systems over IK atmospheric turbulence channels." *Proceedings of 2nd Pan-Hellenic Conference on Electronics and Telecommunications–PACET*, Thessaloniki, Greece, 2012.

[49] Liu, X. Free-space optics optimization models for building sway and atmospheric interference using variable wavelength. *IEEE Transactions on Wireless Communications*, 57 (2009): 492–498.

[50] Bayaki, E., Schober, R., and Mallik, R. K. "Performance analysis of MIMO free-space optical systems in gamma-gamma fading." *IEEE Transactions on Communications* 57 (2009): 3415–3424.

[51] Khalighi, M., Schwartz, N., Aitamer, N., and Bourennane, S. Fading reduction by aperture averaging and spatial diversity in optical wireless systems. *IEEE Journal of Optical Communications and Networking* 1 (2009): 580–593.

[52] Molisch, A. *Diversity in Wireless Communications*. Wiley-IEEE Press, 2011.

[53] Tang, X., Xu, Z., and Ghassemlooy, Z. "Coherent polarization modulated transmission through MIMO atmospheric optical turbulence channel." *IEEE Journal of Lightwave Technology*, 31 (2013): 3221–3228.

[54] Krishnan, Prabu, and D. Sriram Kumar. "Performance analysis of free-space optical systems employing binary polarization shift keying signaling over gamma-gamma channel with pointing errors." *Optical Engineering*, 53 (2014): 076105.

[55] Prabu, K. and D. Sriram Kumar. "MIMO free-space optical communication employing coherent BPOLSK modulation in atmospheric optical turbulence channel with pointing errors." *Optics Communications* 343 (2015): 188–194.

[56] Gradshteyn, I. S. and Ryzhik, I. M. *Table of Integrals, Series, and Products*. 6th ed. New York, Academic Press, 2000.

Chapter 4

Link budget for a terrestrial FSO link and performance of space time block codes over FSO channels

Apexit Shah[1], *Krithi Katta Narasimha Moorthy*[1], *Pallavi R. Kallapur*[1], *and Udupi Shripathi Acharya*[1]

4.1 Introduction

Free space optic (FSO) communication systems are ideal for setting up high-speed short distance (a few kilometers) communication links. They have been in existence for about two decades but have not gained widespread acceptance and popularity. This is primarily due to a major drawback of the FSO systems. Changing the environmental conditions (atmospheric turbulence and fog) can drastically affect their throughput and reliability. Atmospheric turbulence can affect the FSO link in a manner similar to the impairments introduced by multi-path propagation (signal fading) in a wireless channel. It is well known that the random variation of signal strength due to fading in a communication channel can result in severe bit-error-rate (BER) performance degradation and an increase in the outage probability. A number of receiver combining schemes and space-time block codes (STBCs) have been designed to improve the performance of wireless communication systems over fading channels. FSO systems have the potential to provide high data-rate communication with the advantages of quick deployment times, high security, and no frequency regulations. These features have resulted in FSO becoming a unique technology within the domain of wireless communication. FSO systems have the potential to be applied in a wide range of applications, from short-range wireless communication links providing network access to portable computers, to last-mile links bridging gaps between the end users and the existing fiber-optic communications backbones, and even laser communications in outer-space links. FSO relies on the line-of-sight (LOS) optical communication. Typically, two specialized telescopes face each other, each having a transmitting and receiving unit, allowing bidirectional communication. On one end, a modulated optical signal is generated similar to that within a traditional fiber optic system.

[1]Department of Electronics and Communications, National Institute of Technology Karnataka, India

The optical signal is amplified through an optical amplifier, and routed to the transmitting optics in the telescope, where it is then beamed through the atmosphere. On the receiving side, the light beam is collected through the receiving telescope, optionally amplified, and then detected (converted to an electrical signal). Communication is performed identically in the other direction, and transmitters and receivers can share the same telescope.

4.1.1 Terrestrial FSO

For a typical FSO terrestrial link, the operating distance can range from a few hundred meters up to 4 km. FSO links the required LOS and thus an unobstructed point-to-point straight path. In most applications, the beam path is at high elevation from the ground level and thus, the invisible laser beam is not easily accessible. As such, FSO is inherently more "secure" than other transmission methods. FSO technology typically uses two laser photo-detector transceivers to establish the LOS duplex point to point link without a guiding medium (fiber). Typically FSO links carry information at rates ranging from 10 Mbps to 2.5 Gbps, with the data rate being expected to increase to 10 Gbps. The transmitter is usually a laser diode or vertical-cavity surface-emitting laser (VCSEL) device (optical transmit power −8 dBm). The receiver typically comprises of a Si photo diode (PD) or Avalanche photo detector (APD). The selection of appropriate wavelength in FSO technology is an important issue. The most common wavelengths used for optical communication range from 0.85 to 1.55 μm. Many FSO installations utilize 0.78, 0.85, and recently 1.55 μm beams.

Despite having been around for more than two decades, FSO has not emerged so far as a credible alternative to conventional optical fiber links over short distances. This situation should be rectified because FSO technology is well suited for deployment in high-speed-short-distance communication links because laying of optical fiber in densely populated urban areas can be a serious problem. Random fluctuations in the propagation medium, namely, free space pose the most significant obstacle to the widespread adoption of these systems. Various environmental challenges for FSO include absorption, scattering, building sway, and scintillation. Inhomogeneities in the temperature and pressure of the atmosphere lead to spatial and temporal variations in optical intensity incident on a receiver (scintillation), resulting in fading. The range is particularly limited if fog/smog and clouds are present in the atmosphere. Atmospheric effects can cause performance degradation manifested by increased bit error rate (BER) and transmission delays. Physical obstructions like birds flying across the LOS path can cause temporary outages and transmit power can be limited due to eye-safety considerations. Despite having significant benefits, these limitations have prevented this technology from gaining widespread acceptance as a worthwhile competitor to guided optical communications. However, FSO technology can provide reliable high-speed data transmission with careful link design, choice of deployment location, the use of channel coding, and MIMO techniques.

Theoretically, the most effective technique to mitigate multipath fading in a wireless channel is the transmitter power control [1]. But the transmitter might not have dynamic range as high as required. In addition, for this method to be effective, channel information has to be fed back from the receiver to the transmitter, which results in throughput degradation and added complexity at both the transmitter and the receiver. Other effective techniques are time and frequency diversity. However, time interleaving results in large delays when the channel is slowly varying. Equivalently, spread spectrum techniques are ineffective when the coherence bandwidth of the channel is larger than the spreading bandwidth.

Antenna diversity is a practical, effective and, hence, a widely applied technique for reducing the effect of multipath fading. The effects of fading in FSO systems can be considerably reduced by the use of STBCs such as the Alamouti scheme and the STBC derived from cyclic code that is described in this chapter. The use of these techniques requires the deployment of multiple lasers at the transmitter and multiple photo detectors at the receiver.

The ability of an FSO link to eliminate atmospheric effects depends on the laser power, beam divergence, receiver sensitivity, coupling losses, and receiver lens area. These factors can be summarized in the link margin. To determine the maximum distance over which an FSO link can operate for a given transmit power, it is important to look at the signal power available at the receiver, efficiency of the receiver, and reliability of the link. To quantify these factors, a detailed link margin analysis needs to be carried out. The receiver typically has a specific minimum sensitivity at a given data rate, and the task is to make sure that the launch power minus the loss factor stays above the minimum sensitivity to guarantee reliable operation of the system. These variables determine the basic link performance. Further improvement in link reliability can be achieved by the use of suitable channel codes.

In the first part of this chapter, we have computed the link budget for a terrestrial FSO link over a wide range of channel (environmental) conditions. After computing the transmit powers and link margin for various channel conditions, we focus on the use of diversity techniques adapted to optical communication to improve the reliability of the link. To accomplish this, we have adapted well-known techniques from the domain of wireless communications. MIMO techniques (Alamouti code and STBCs derived from nonbinary cyclic codes) have been employed to design FSO links that are reliable and secure over a wide range of atmospheric and environmental conditions. Our link budget computations indicate that the presence of heavy fog in the atmosphere drastically reduces the range of the FSO link. However, in the absence of fog or presence of moderate fog conditions (quantified by the visibility of the atmosphere V), a combination of careful link budget design and deployment of appropriate STBC can be used to achieve reliable information transfer. Our simulations indicate that the use of the Alamouti scheme or the STBC derived from cyclic code can substantially improve the data integrity levels over the link. The STBC derived from the cyclic code over F_{5^2} achieves BER of 10^{-6} at low SNR values (10 dB for 2×2 configuration, 16 dB for 2×1 configuration) and has a coding gain of approximately 9 dB over the corresponding

Alamouti schemes. Our study indicates that the application of suitable channel-coding techniques (especially those designed for the use with multiple transmit and receive antennas) can make the channel robust against impairments introduced by turbulent weather conditions and fog. Hence, the widespread adoption of channel-coding techniques may represent the best approach to bring about the widespread adoption of this technology.

4.2 Channel modeling

An optical wave propagating through a wireless channel undergoes fluctuations in the intensity and phase due to the time-varying inhomogeneities in the refractive index of the atmosphere. To estimate the effect of the channel on the propagating signal, we need an accurate mathematical model that can mimic the random variation (fading) of the irradiance signal (also referred to as scintillation) under various environmental conditions.

Several statistical models have been proposed in literature to understand and describe the scintillation phenomenon [2–5]. The statistical models that have been found to be most accurate are the log-normal distribution, the lognormal-Rician distribution (or Beckmann model) and the gamma–gamma distribution. These models have been found to have the following limitations:

1. Lognormal model is very simple and can be used to model the channel under weak turbulence. But it shows large deviation from experimental data under strong turbulence conditions since it does not take into consideration the multiscattering effects. It underestimates the tail probability and consequently leads to large errors in calculations of BER and fade probability [6].
2. The Beckmann model shows excellent agreement with experimental data; however, there is no known mathematical expression relating the parameters of Beckmann model to the atmospheric conditions [4].
3. The gamma–gamma distribution can be used to model the probability density function (pdf) of the irradiance fluctuations of an optical wave propagating through a turbulent medium under both strong and weak turbulence regimes [2,3]. Hence, it is well suited for use in mathematical computations of the performance of FSO link operating under different environmental conditions.

4.2.1 Gamma–gamma distribution

The gamma–gamma distribution model assumes that small-scale irradiance fluctuations are modulated by large-scale irradiance fluctuations of the propagating wave, both governed by independent gamma distributions. The two parameters of the gamma–gamma pdf (α and β) can be determined using only the values of the refractive-index structure parameter C_n^2 (or Rytov variance) and inner scale l_0 (small scale size of the eddies in turbulent air motion). These parameters can be determined by the use of a scintillometer [6]. It has been documented in literature that for a wide range of turbulence conditions (weak to strong) the fading gain I_{mn}

in FSO systems can be accurately modeled by a gamma–gamma distribution [2,6,7]. The pdf of the gamma–gamma distribution is described by,

$$f_I(I_{mn}) = \frac{2(\alpha\beta)^{\frac{\alpha+\beta}{2}}}{\Gamma(\alpha)\Gamma(\beta)} I_{mn}{}^{\left(\frac{\alpha+\beta}{2}-1\right)} K_{\alpha-\beta}\left(2\sqrt{\alpha\beta I_{mn}}\right) \tag{4.1}$$

where the parameters $\alpha > 0$ and $\beta > 0$ are linked to the scintillation index (S.I)

$$\text{S.I} = \frac{1}{\alpha} + \frac{1}{\beta} + \frac{1}{\alpha\beta} \tag{4.2}$$

where α and β are the parameters that can be adjusted to achieve a good agreement between $f_I(I_{mn})$ and measurement data. Alternatively, assuming spherical wave propagation, α and β can be directly linked to the physical parameters via

$$\alpha = \left[\exp\left(\frac{0.49\chi^2}{\left(1 + 0.18d^2 + 0.56\chi^{12/5}\right)^{7/6}}\right) - 1\right]^{-1} \tag{4.3}$$

$$\beta = \left[\exp\left(\frac{0.51\chi^2\left(1 + 0.69\chi^{12/5}\right)^{-5/6}}{\left(1 + 0.9d^2 + 0.62\chi^{12/5}\right)^{5/6}}\right) - 1\right]^{-1} \tag{4.4}$$

where,

$$\chi^2 = 0.5C_n^2 k^{7/6} L^{11/6} \tag{4.5}$$

$$d = \left[\frac{kD^2}{4L}\right]^{1/2} \tag{4.6}$$

$$k = \frac{2\pi}{\lambda} \tag{4.7}$$

Here, λ, D, C_n^2, and L are the wavelength in meters, the diameter of the receiver's aperture in meters, the index of refraction structure parameter, and the link distance in meters, respectively.

4.3 Link budget

Link budget is the quantitative assessment of the overall system performance of a link, taking care of all the gains and losses from the transmitter to the receiver due to the medium in a communication system. The amount of the received power that exceeds the minimum required received power for a successful communication (determined by receiver sensitivity) is called the link margin. Transmit power can be increased to have a larger link margin than the minimum nominal received power level so that a larger buffer (*fade margin*) is available to make the link more reliable. However, the same increase in signal power can also be used to increase the distance between the transmitter and receiver. This shows a trade-off between

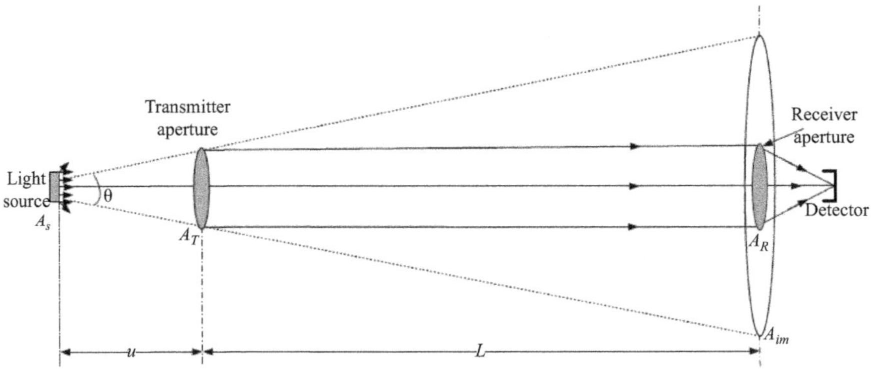

Figure 4.1 Geometric loss due to beam divergence [9]

distance and reliability. The amount of margin required can be reduced by using error control coding and space-time diversity techniques.

4.3.1 Geometric loss

The loss that occurs due to the divergence of the optical beam is referred to as geometrical loss (Figure 4.1). The beam spreads out while traversing the channel causing the size of the received beam to be greater than the receiver aperture. In FSO systems, generally beams are designed to diverge so that even when the transmitter and the receiver are not exactly in LOS, the receiver can receive some portion of the large received beam. Geometric loss dominates for large distances [8].

Geometric loss is mathematically expressed as the ratio of area of the receiver lens to the area of the beam spread over the receiver plane which is assumed to be perpendicular to the transmitted beam. Assuming conically diverging single beam, geometric loss (in dB) can be calculated using

$$L_{\text{Geom}} = -10\log\frac{A_R}{A_B} = -20\,log\left[\frac{D_R}{D_T + \theta * \text{L}}\right] \tag{4.8}$$

where

A_R is the area of the receiver aperture.
A_B is the area of the beam at the receiver plane.
D_R is the diameter of the receiving aperture in meters.
D_T is the diameter of the transmitting aperture in meters.
L is the link distance in kilometer.
θ is the divergence angle of the beam given in milli-radians.

4.3.2 Attenuation due to atmospheric turbulence

In FSO communication systems, the small variations in the index of refraction can cause the received field strength to exhibit random fluctuations. This is referred to as scintillation. The fluctuations in the index of refraction, called optical

Table 4.1 Values of C_n^2 for different turbulence conditions [7]

Turbulence	C_n^2
Weak	10^{-16}
Moderate	10^{-15}
Strong	10^{-14}

turbulence, originate from the atmospheric turbulence that creates thermal inhomogeneities along the path. The index of refraction value in the atmosphere varies with temperature, pressure, and humidity of air and on the wavelength used for the transmission.

The refractive index structure parameter is a measure of the strength of the fluctuations in the refractive index. Typical values of refractive index structure parameter are shown in Table 4.1.

Multiple scattering effect increases with increasing atmospheric turbulence or increasing link distances. It is also observed that as the turbulence becomes stronger beyond a point, the deflected beam adds back to the main beam and scintillation saturates [10].

In literature, there exist a few physical descriptions of attenuation effects due to turbulence. The Rytov approximation and Andrews' method are widely used currently in technical literature.

4.3.3 Rytov approximation

The Rytov approximation model relates relative variance of optical intensity $\left(\sigma_{I,rel}^2\right)$ to the index refraction structure parameter (C_n^2) [11].

$$\sigma_{I,rel}^2 = 0.5\, C_n^2 \,.\, k^{7/6} .\, L^{11/6} \tag{4.9}$$

This equation is valid for spherical wave and for $\sigma_{I,rel}^2 < 1$

The scintillation variance is given by

$$\sigma_\chi^2 = 23.17\, C_n^2 \,.\, k^{7/6} .\, L^{11/6} \tag{4.10}$$

The turbulence attenuation is related to the scintillation as

$$\alpha_{\text{Rytov}} = 2.\sigma_\chi \tag{4.11}$$

Thus, the turbulence attenuation α_{Rytov} is given by

$$\alpha_{\text{Rytov}} = 2.\sqrt{(23.17).\, C_n^2 \,.\, k^{7/6} .\, L^{11/6}} \tag{4.12}$$

This method does not take into consideration the receiver optical lens diameter D_R.

4.3.4 Andrews's method

This method takes aperture averaging effect into account and hence the size of the receiver optical lens D_R. The variance of optical intensity is given by [12]

$$\overline{\sigma}_I^2(D_R) \approx \exp\left[\frac{(0.49)\chi^2}{\left(1+0.18d^2+0.56\chi^{12/5}\right)^{7/5}} + \frac{0.51\chi^2\left(1+0.69\chi^{12/5}\right)^{-5/6}}{1+0.9d^2+0.62d^2\chi^{12/5}}\right] - 1 \tag{4.13}$$

The parameter d contains information about the wavelength λ of optical source, distance between the transmitter and receiver L, and the receiver optical lens diameter D_R.

$$d = \sqrt{\frac{2\pi}{4.\lambda.L}} \cdot D_R \tag{4.14}$$

The turbulence attenuation in dB is given by the following relation where $\overline{\sigma}_I^2(D_R)$ is the variance of optical intensity.

$$L_a = 10\log\sqrt{1-\overline{\sigma}_I^2(D_R)} \tag{4.15}$$

4.3.5 Atmospheric extinction loss

The optical power traversing the atmosphere is attenuated by the air molecules and aerosols. Some of the photons are absorbed by these molecular constituents (called extinction effect) whereas other photons change their direction of propagation (called scattering). Even under clear sky conditions, the transmitted laser beam gets attenuated. The Beer–Lambert Law explains the transmittance of the laser beam in terms of attenuation through the atmosphere [13]. Consider the FSO link transmitting the optical signal through the atmosphere. The transmittance T of laser radiation that has propagated over a distance L (in km) is described by Beer's law:

$$T = \frac{P_R}{P_T} = \exp(-\alpha_e(\lambda)\,.\,L) \tag{4.16}$$

The positive extinction coefficient $\alpha_e(\lambda)$, called as atmospheric attenuation coefficient, describes the extinction level of the medium. It is usually given in km^{-1} and is a function of wavelength. In addition to absorption, scattering is a major cause of attenuation and has to be taken into account. The wavelength band used in FSO communication (0.5 to 2 μm) matches with the fog particle size causing attenuation. This attenuation is quantified by the *Mie scattering* [14,15] phenomenon. The attenuation coefficient can be expressed in terms of *meteorological visibility V*, defined as a distance at which transmittance falls to a certain value (taken as 2% of the original value). For the case of wavelengths in the range from the visible to the near-infrared, the extinction coefficient

Table 4.2 Weather conditions and their visibility range values [9]

Weather condition	Visibility range, V (km)
Thick fog	0.2
Moderate fog	0.5
Light fog	0.770–1.0
Thin fog/heavy rain (25 mm/h)	1.9–2.0
Haze/medium rain (12.5 mm/h)	2.8–40.0
Clear/drizzle (0.25 mm/h)	18.0–20.0
Very clear	23.0–50.0

$\alpha_e(\lambda)$ can be calculated according to the semi-empirical Kruse formula as follows:

$$\alpha_e(\lambda) = \frac{3.91}{V}\left(\frac{\lambda}{550}\right)^{-\delta} \tag{4.17}$$

In this equation, the visibility (V) is expressed in kilometers and the exponent δ, called as *particle size distribution coefficient*, is given by [15]:

$$\delta = \begin{cases} 1.6 \text{ for } V > 50\,\text{km} \\ 1.3 \text{ for } 6\,\text{km} < V < 50\,\text{km} \\ 0.585\, V^{1/3} \text{ for } V < 6\,\text{km} \end{cases} \tag{4.18}$$

The visibility range values (V) values under different weather conditions are enumerated in Table 4.2.

Atmospheric attenuation expressed in the decibel scale is related to transmittance by

$$L_e(\text{dB}) = -10\log_{10}(T) = -10\log_{10}(\exp(-\alpha_e(\lambda)\,.\,L)) \tag{4.19}$$

Hence, the atmospheric attenuation coefficient (in decibels) is approximately:

$$L_e(\text{dB}) = 4.343\alpha_e(\lambda)L \tag{4.20}$$

4.3.6 Link budget

We calculate the losses due to the abovementioned effects that depend on beam divergence, turbulence conditions, link distance and receiver lens area, etc. and determine the transmitted power required to sustain reliable FSO communication. The required transmit power P_T (in dBm) is quantified as,

$$P_T = LM + P_R + L_a + L_e + L_{\text{Geom}} \tag{4.21}$$

where LM denotes the link margin (in dB), P_R denotes the receiver sensitivity (in dBm), L_a denotes the fading loss determined by Andrew's method (in dB), L_e and L_{Geom} denote extinction loss and geometric loss (both in dB), respectively.

Table 4.3 Magnitude of losses for links of different lengths, visibility range (V) = 2 km (heavy rain)

Link distance L (m)	L_{Geom} (dB)	L_e (dB)	L_a (dB)
1,000	22.1102	3.9563	0.2634
1,500	25.5751	5.9344	0.4986
2,000	28.0452	7.9125	0.7900
2,500	29.9662	9.8907	1.0782
3,000	31.5383	11.8688	1.1494
3,500	32.8691	13.8469	0.8492
4,000	34.0227	15.8251	0.4344

Table 4.4 Magnitude of losses for links of different lengths, visibility range (V) = 0.5 km (moderate fog)

Link distance L (m)	L_{Geom} (dB)	L_e (dB)	L_a (dB)
1,000	22.1102	20.9927	0.2634
1,500	25.5751	31.4891	0.4986
2,000	28.0452	41.9855	0.7900
2,500	29.9662	52.4819	1.0782
3,000	31.5383	62.9782	1.1494
3,500	32.8691	73.4746	0.8492
4,000	34.0227	83.9710	0.4344

4.3.6.1 Variation of parameters with link distance

For a given weather condition, different losses are calculated and compared for various link distances. Tables 4.3 and 4.4 correspond to heavy rain and moderate fog weather conditions, respectively.

In strong turbulence or along long paths, the scintillation-index saturates. The values of Andrew's loss (L_a)in Table 4.3 match this expected pattern. (L_a increases with distance initially, takes on maximum value at 3,000 m and then decreases.)

We note from this table that L_{Geom} and L_a are identical to the values in Table 4.3, but the loss due to extinction has increased significantly. The presence of fog in the atmosphere (indicated by the lower value of visibility) is the main reason for the increase in L_e.

4.3.6.2 Variation of parameters with weather conditions

The effect of weather conditions on the extinction loss is observed in Table 4.5. In Table 4.5, extinction loss is calculated under different weather conditions for link distances of 2 km and 4 km.

4.3.6.3 Required transmitted power

For practical applications, the estimation of the power to be transmitted to achieve a certain link margin is necessary. For different values of link margin, the power

Table 4.5 Variation of the magnitude of extinction loss with visibility

Visibility V (km)	L_e (dB) for $L = 2{,}000$ m	L_e (dB) for $L = 4{,}000$ m
0.2 (heavy fog)	119.1318	238.2636
0.5 (moderate fog)	41.9855	83.9710
1.0 (light fog)	18.5253	37.0506
2.0 (heavy rain)	7.9125	15.8251
20.0 (drizzle)	0.4416	0.8832
51.0 (clear sky)	0.1269	0.2538

Table 4.6 P_T required for $V = 50$ km (clear sky), $P_R = -43$ dBm

LM (dB)	P_T (mW) for $L = 1{,}000$ m	P_T (mW) for $L = 2{,}000$ m	P_T (mW) for $L = 3{,}000$ m	P_T (mW) for $L = 4{,}000$ m
10.0	0.0878	0.3947	0.9723	1.4828
6.0	0.0350	0.1571	0.3871	0.5903
3.0	0.0175	0.0787	0.1940	0.2959

Table 4.7 P_T required for $V = 2$ km (heavy rain), $P_R = -43$ dBm

LM (dB)	P_T (mW) for $L = 1{,}000$ m	P_T (mW) for $L = 2{,}000$ m	P_T (mW) for $L = 3{,}000$ m	P_T (mW) for $L = 4{,}000$ m
10.0	0.2153	2.3702	14.3105	53.4833
6.0	0.0857	0.9436	5.6971	21.2921
3.0	0.0430	0.4729	2.8553	10.6713

requirements are calculated and tabulated. Tables 4.6, 4.7, 4.8, and 4.9 show the power required for different link distances when the link margin is fixed as 10, 6, and 3 dB. The receiver is assumed to be APD, so the sensitivity is taken as −43 dBm.

For a 4 km long link, to maintain a link margin of 10 dB, the transmit power required is 1.5 mW in clear sky conditions and increases to 50 mW in heavy rain conditions. Under moderate fog conditions, the power required for a distance of 2 km and a link margin of 3 dB is 1.2 W, which is very high.

Figure 4.2 illustrates the maximum link distance for which the FSO system can maintain a positive link margin under different weather conditions. The solid and dashed lines represent transmitted power level of 50 and 10 mW, respectively.

As it is observed from Figure 4.2, for heavy rain conditions, even with a transmitted power of 10 mW, we can have sufficient positive link margin for distances up to 4 km.

In light fog conditions, the maximum range is 3 km with 50 mW transmitted power and 2.5 km for 10 mW power to have a positive link margin. When the

Table 4.8 P_T *required for* $V = 1$ *km (light fog),* $P_R = -43$ *dBm*

LM (dB)	P_T (mW) for $L = 1{,}000$ m	P_T (mW) for $L = 2{,}000$ m	P_T (mW) for $L = 3{,}000$ m	P_T (mW) for $L = 4{,}000$ m
10.0	0.7305	27.2930	559.2003	7,091.9965
6.0	0.2908	10.8655	222.6217	2,823.3747
3.0	0.1458	5.4457	111.5751	1,415.0393

Table 4.9 P_T *required for* $V = 0.5$ *km (moderate fog),* $P_R = -43$ *dBm*

LM (dB)	P_T (mW) for $L = 1{,}000$ m	P_T (mW) for $L = 2{,}000$ m	P_T (mW) for $L = 3{,}000$ m	P_T (mW) for $L = 4{,}000$ m
10.0	10.8802	6,054.4267	1.847×10^6	3.489×10^8
6.0	4.3315	2,410.3107	7.355×10^5	1.389×10^8
3.0	2.1709	1,208.0169	3.686×10^5	6.963×10^7

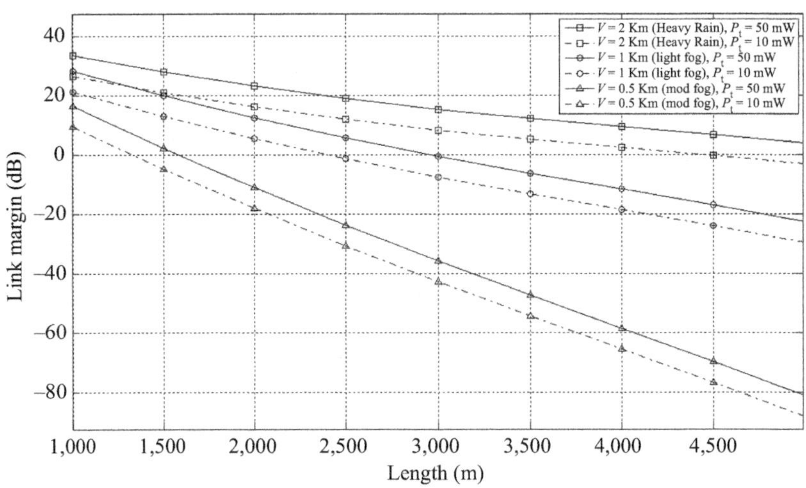

Figure 4.2 Link margin for different weather conditions with $\mathbf{P_T}$ *of 50 and 10 mW*

weather deteriorates and exhibits moderate fog conditions, the link can have a positive link margin up to the distance of about 1.6 km. From this plot, we infer that FSO systems can be effective in carrying high-speed data under favorable environmental conditions. They can be advantageously deployed in many parts of the world where conditions of fog are not encountered. However, adverse weather conditions in the form of fog can reduce their effectiveness. This problem can be addressed by the use of channel codes (specifically space-time codes). In the second part of this chapter, we have applied the techniques of channel coding (specifically STBCs like the Alamouti code and the STBC derived from non binary cyclic code) to improve the link performance. We have demonstrated by conducting simulations that it is possible to achieve BERs of the order of 10^{-6} under adverse weather conditions.

4.4 Diversity

In wireless communications, time diversity, frequency diversity, and space diversity have been employed to improve the performance of communication systems under harsh channel conditions. Time diversity techniques typically reduce the information throughput over the channel (increased latency) while frequency diversity techniques result in the wastage of spectrum. Hence, we have concentrated on space diversity techniques in our search of methods to improve the reliability of information transfer under adverse environmental conditions.

4.4.1 Space diversity

Spatial diversity in optical communication is realized by the use of multiple lasers and multiple receivers to mitigate signal strength variation due to fading in the received signal. The coherence length at wavelengths employed in optical communication is of the order of centimeters and so the multiple transmitters or receivers need to be placed only centimeters apart to realize approximately independent channel conditions.

4.4.2 Space-time diversity

When multiple copies of a signal are sent across many antennas, they will be affected by different channel conditions and hence the received copies can be optimally combined to achieve communication with higher reliability. This concept has led to the design and deployment of STCs. Space-time block codes (STBCs) and space-time Trellis codes (STTCs) have been extensively described in literature [16]. In the following sections, we have employed the Alamouti STC and STBCs derived from non binary cyclic code to improve the reliability of FSO communication.

4.4.3 Alamouti space-time code

We have adapted the Alamouti code for FSO communication by replacing the transmit antennas with laser sources and receive antennas by optical detectors. This code has the following features. This scheme requires no bandwidth expansion, as redundancy is applied in space across multiple antennas, not in time or frequency. It can improve the error performance, data rate, or capacity of wireless communications systems. A brief description of the Alamouti code is presented in this section.

4.4.3.1 Description of 2×1 Alamouti scheme

The description of Alamouti two-branch transmit diversity with one receiver is presented in Figure 4.3.

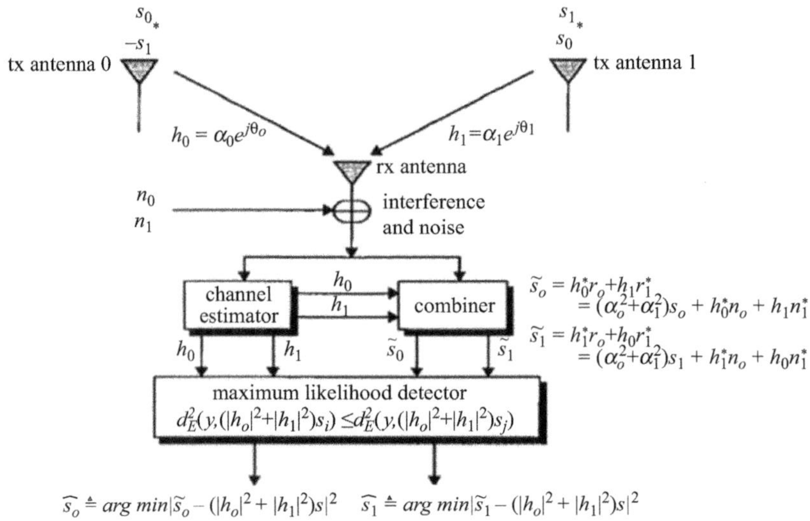

Figure 4.3 Alamouti two-branch transmit diversity with one receiver [1]

4.4.3.2 The encoding and transmission sequence

At a given symbol period, the signal transmitted from antenna-0 is denoted by s_o and from antenna-1 by s_1. During the next symbol period, signal ($-s_1^*$) is transmitted from antenna-0, and signal s_o^* is transmitted from antenna-1, where * is the complex conjugate operation.

$$r_o = r(t) = s_o h_o + s_1 h_1 + n_o \tag{4.22}$$

$$r_1 = r(t+T) = s_o^* h_1 - s_1^* h_0 + n_1 \tag{4.23}$$

where r_o and r_1 are the received signals at time t and $t+T$ and n_o and n_1 are complex random variables representing receiver noise and interference, respectively.

4.4.3.3 The combining scheme

The combiner computes the following two combined signals that are sent to the maximum likelihood detector:

$$\tilde{s_o} = h_0^* r_o + h_1 r_1^* = (\alpha_o^2 + \alpha_1^2) s_o + h_0^* n_o + h_1 n_1^* \tag{4.24}$$

$$\tilde{s_1} = h_1^* r_o - h_0 r_1^* = (\alpha_o^2 + \alpha_1^2) s_o + h_1^* n_o - h_0 n_1^* \tag{4.25}$$

4.4.3.4 The maximum likelihood decision rule

The outputs of the combiner $\tilde{s}_o$ and $\tilde{s}_1$ are processed by a maximum likelihood decoder (MLD) to determine the most likely transmitted symbols $\widehat{s}_o$ and $\widehat{s}_1$. This operation is expressed mathematically as

$$\widehat{s}_0 \triangleq \underset{s \in S}{argmin} |\tilde{s}_0 - \left(|h_o|^2 + |h_1|^2\right)s|^2 \quad (4.26)$$

$$\widehat{s}_1 \triangleq \underset{s \in S}{argmin} |\tilde{s}_1 - \left(|h_o|^2 + |h_1|^2\right)s|^2 \quad (4.27)$$

The reconstructed symbol stream at the receiver is now compared with the transmitted symbol stream to determine the BER under various turbulence conditions and signal-to-noise ratios.

4.4.4 *Description of* 2 × 2 *Alamouti scheme*

In applications where it is feasible to employ multiple receivers, we can improve the error performance of the FSO system further, by using two antennas at the receiver end also. Therefore, we have considered Alamouti 2 × 2 scheme next Figure 4.4. Employing more than two antennas at the transmitter/receiver end increases the decoding complexity significantly. Hence, we have studied STBCs

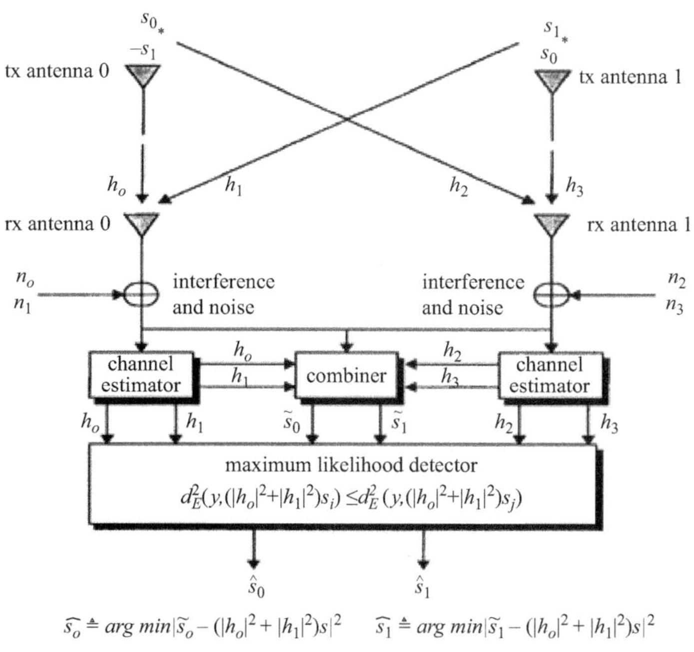

Figure 4.4 Alamouti two-branch transmit diversity with two receivers [1]

derived from non binary cyclic codes next, instead of the Alamouti schemes of higher order, to improve the error performance further.

4.4.4.1 The encoding and transmission sequence

At a given symbol period, the signal transmitted from antenna-0 is denoted by s_o and from antenna-1 by s_1. During the next symbol period, signal ($-s_1^*$) is transmitted from antenna-0, and signal s_o^* is transmitted from antenna-1, where * is the complex conjugate operation.

$$r_o = s_o h_o + s_1 h_1 + n_o \tag{4.28}$$

$$r_1 = s_o^* h_1 - s_1^* h_0 + n_1 \tag{4.29}$$

$$r_2 = s_o h_2 + s_1 h_3 + n_2 \tag{4.30}$$

$$r_3 = s_o^* h_3 - s_1^* h_2 + n_3 \tag{4.31}$$

where r_o and r_1 are the received signals at receiver-0 at time t and $t + T$, r_2 and r_3 are the received signals at receiver-1 at time t and $t + T$, respectively, and n_o, n_1, n_2, n_3 are complex random variables representing receiver thermal noise and interference.

4.4.4.2 The combining scheme

The combiner computes the following two combined signals that are sent to the maximum likelihood detector:

$$\begin{aligned}\tilde{s_o} &= h_0^* r_o + h_1 r_1^* + h_2^* r_2 + h_3 r_3^* \\ &= \left(\alpha_o^2 + \alpha_1^2 + \alpha_2^2 + \alpha_3^2\right) s_o + h_0^* n_o + h_1 n_1^* + h_2^* n_2 + h_2 n_2^* \end{aligned} \tag{4.32}$$

$$\begin{aligned}\tilde{s_1} &= h_1^* r_o - h_0 r_1^* + h_3^* r_2 - h_2 \\ &= \left(\alpha_o^2 + \alpha_1^2 + \alpha_2^2 + \alpha_3^2\right) s_o + h_1^* n_o - h_0 n_1^* + h_3^* n_2 - h_2 n_3^* \end{aligned} \tag{4.33}$$

4.4.4.3 The maximum likelihood decision rule

The decision criteria of the maximum likelihood decoder are expressed mathematically as

$$\hat{s_0} \triangleq \underset{s \epsilon S}{argmin} |\tilde{s_0} - (|h_0|^2 + |h_1|^2 + |h_2|^2 + |h_3|^2) s|^2 \tag{4.34}$$

$$\hat{s_1} \triangleq \underset{s \epsilon S}{argmin} |\tilde{s_1} - (|h_0|^2 + |h_1|^2 + |h_2|^2 + |h_3|^2) s|^2 \tag{4.35}$$

4.4.5 *Modified Alamouti code*

When the modulation scheme being used in FSO is unipolar like OOK or PPM, the complement of a signal x_i defined by $\overline{x}_i$, is the signal waveform obtained by reversing the roles of "on" and "off." For example, if $x_i = s_i$, then $\overline{x}_i = -x_i + A = -s_i + A$, where A is the amplitude of the transmitted signal corresponding to the "on" state. Also, since we deal with real signals, complex conjugate concept is absent [17].

The equations now simplify to the equations shown in Table 4.10.

Table 4.10 *Modified Alamouti's scheme equations*

	2×1 Alamouti Code equations	**2×2 Alamouti Code equations**
Received signal	$r_0 = s_0h_0 + s_1h_1 + n_0$ $r_1 = s_0h_1 + \bar{s}_1h_0 + n_1$	$r_0 = s_0h_0 + s_1h_1 + n_o$ $r_1 = s_0h_1 + \bar{s}_1h_0 + n_1$ $r_2 = s_2h_2 + s_3h_3 + n_2$ $r_3 = s_2h_3 + \bar{s}_3h_2 + n_3$
Combiner equations	$\tilde{s_o} = h_0^*r_o + h_1r_1^* - h_0^*h_1A$ $\tilde{s_1} = h_1^*r_o - h_0r_1^* + h_0h_1^*A$	$\tilde{s_0} = h_0^*r_0 + h_1r_1^* + h_2^*r_2 + h_3r_3^* - h_0^*h_1A - h_2^*h_3A$ $\tilde{s_1} = h_1^*r_0 - h_0r_1^* + h_3^*r_2 - h_2r_3^* + h_0h_1^*A + h_2h_3^*A$
ML decision criteria	$\hat{s_0} \triangleq \underset{s \epsilon S}{argmin} \lvert\tilde{s_0} - (\lvert h_0\rvert^2 + \lvert h_1\rvert^2)s\rvert^2$ $\hat{s_1} \triangleq \underset{s \epsilon S}{argmin} \lvert\tilde{s_1} - (\lvert h_0\rvert^2 + \lvert h_1\rvert^2)s\rvert^2$	$\hat{s_0} \triangleq \underset{s \epsilon S}{argmin} \lvert\tilde{s_0} - (\lvert h_o\rvert^2 + \lvert h_1\rvert^2 + \lvert h_2\rvert^2 + \lvert h_3\rvert^2)s\rvert^2$ $\hat{s_1} \triangleq \underset{s \epsilon S}{argmin} \lvert\tilde{s_1} - (\lvert h_o\rvert^2 + \lvert h_1\rvert^2 + \lvert h_2\rvert^2 + \lvert h_3\rvert^2)s\rvert^2$

4.5 STBCs derived from non binary cyclic codes

4.5.1 Cyclic code

Cyclic codes are a subclass of linear codes with additional structure. An (n,k) linear block code is said to be cyclic if every codeword after any arbitrary number of cyclic shifts gives another codeword. With every codeword $c = (c_0, c_1, c_2 \ldots c_{n-1})$ from a cyclic code, we associate the codeword polynomial $c(x) = c_0 + c_1x + c_2x^2 + \ldots + c_{n-1}x^{n-1}$, where $c_0, c_1, c_2 \ldots c_{n-1} \in F_q$. A cyclic code is an ideal of the ring $F_q[x]/(x^n - 1)$.

4.5.2 Rank distance

The Hamming distance metric has been widely used to quantify the distance properties of block codes. However, Tarokh *et al.* [16] have shown that the appropriate metric that can be used to characterize STCs designed for multiple-input multiple-output (MIMO) systems is rank distance. The rank distance for codes over finite fields is defined in the following manner: let $\boldsymbol{C}$ be a (n, k) linear code over F_{q^m} (q is a prime or a power of a prime). For any pair of codewords $\boldsymbol{c_1}$,$\boldsymbol{c_2}$ belonging to $\boldsymbol{C}$, the rank distance between them is the F_q rank of the $m \times n$ matrix corresponding to $(\boldsymbol{c_1}$-$\boldsymbol{c_2})$ obtained by expanding each entry of $(\boldsymbol{c_1}$-$\boldsymbol{c_2})$ as an m-tuple along a basis of F_{q^m} over F_q. In [16], the authors have derived criteria for the design of STCs. Let matrix **B** denote the error matrix introduced by the channel. **B** is computed by taking the difference between the transmitted codeword $\boldsymbol{C}$ and the codeword $\widehat{\boldsymbol{C}}$ estimated by processing the received matrix.

$$\mathbf{B} = \begin{bmatrix} C_1^1 - \hat{C}_1^1 & C_1^2 - \hat{C}_1^2 & \cdots & C_1^l - \hat{C}_1^l \\ C_2^1 - \hat{C}_2^1 & C_2^2 - \hat{C}_2^2 & \cdots & C_2^l - \hat{C}_2^l \\ \vdots & \vdots & \vdots & \vdots \\ C_{N_t}^1 - \hat{C}_{N_t}^1 & C_{N_t}^2 - \hat{C}_{N_t}^2 & \cdots & C_{N_t}^l - \hat{C}_{N_t}^l \end{bmatrix} \tag{4.36}$$

The matrix $\mathbf{A} = \mathbf{B}\mathbf{B}\dagger$ ($\mathbf{B}\dagger$ denotes the conjugate transpose of **B**) is Hermitian. The pairwise probability of error (probability that a ML receiver decides erroneously in favor of a codeword $\widehat{\boldsymbol{C}}$, assuming that the codeword $\boldsymbol{C}$ was transmitted) is given by

$$\Pr\left\{\mathbf{C} \rightarrow \widehat{\mathbf{C}}\right\} \leq \left(\prod_{i=1}^{N_t} \frac{1}{1 + \lambda_i \Gamma_s}\right)^{N_r} \tag{4.37}$$

where $\Gamma_s = \frac{E_s}{4N_0}$, ($E_s$ is the average energy per complex symbol). Let r be the rank of matrix **A** and $\lambda_1, \lambda_2, \cdots, \lambda_r$ be the nonzero Eigen values of **A**. Then (4.2) can be reduced to the form,

$$\Pr\left\{\mathbf{C} \rightarrow \widehat{\mathbf{C}}\right\} \leq \left(\prod_{i=1}^{r} \lambda_i\right)^{-N_r} \left(\frac{1}{\Gamma_s}\right)^{rN_r} \tag{4.38}$$

Thus, a diversity gain of rN_r and a coding gain of $(\lambda_1\lambda_2\lambda_3\cdots\lambda_r)^{\frac{1}{r}}$ are achieved. Equations (4.37) and (4.38) lead to the following design rules for the construction of STCs (both STBCs and STTCs) [16]. These may be briefly stated as follows:

- Rank criterion: In order to achieve the maximum diversity N_tN_r, the matrix $\mathbf{B}(\mathbf{C}_1, \mathbf{C}_2)$ has to be full rank for any codewords $\mathbf{C}_1$ and $\mathbf{C}_2$.
- Determinant criterion: The minimum of the product of nonzero Eigen values of the matrix $\mathbf{A}(\mathbf{C}_1, \mathbf{C}_2)$ (i.e. $\lambda_1, \lambda_2, \lambda_3, \cdots, \lambda_r$) taken over all pairs of distinct code vector sequences $\mathbf{C}_1$ and $\mathbf{C}_2$ is to be made as large as possible.

From the statement of the Rank criteria, we infer that a good STBC should possess the property that the minimum value of rank distance over all pairs of distinct codewords belonging to the code should be as large as possible. The determinant criterion has secondary importance. In practice, codes are designed so that minimum value of rank distance between code words is as large as possible and the minimum of the product of nonzero Eigen values of the matrix $\mathbf{A}(\mathbf{C}_1, \mathbf{C}_2)$ is as large as possible. The rank distance properties of non binary (cyclic codes over F_{q^m}) have been studied by using the transform domain description of these codes [18]. In Theorem 1 [18], we have described a procedure which can be used to construct a non binary cyclic code satisfying the property that all of its codewords when viewed as matrices have full rank. A brief description of the relevant transform domain description of cyclic codes is provided in the next section.

In this approach for the synthesis of STBCs for channels affected by turbulence, we start by synthesizing a cyclic code over F_{q^m} with the property that all code words when viewed as $m \times n$ matrices over F_q have full rank. In the second step, we use the rank-preserving maps specified by [19] to obtain the codewords of the STBC. Simulation studies presented in the results section of this chapter indicate that these codes are well suited for use on optical wireless channels disturbed by atmospheric turbulence and give a coding gain of approximately 8 dB over the 2 × 2 Alamouti code in this application.

4.5.3 Transform domain description of cyclic codes

Let $\boldsymbol{a} = (a_0, a_1 .. a_{n-1}) \in F_{q^m}^n$ and gcd*(n,q)* = 1. Let n be the positive integer such that $n|q^m - 1$ and $\alpha \in F_{q^m}$ be an element of order n. Then the time domain vector $\boldsymbol{a} = (a_0, a_1 .. a_{n-1})$ and the corresponding transform domain vector $A = (A_0, A_1 .. A_{n-1})$ are related by [20],

$$A_j = \sum_{i=0}^{n-1} \alpha^{ij} a_i, j = 0, 1, 2 \ldots n-1 \tag{4.39}$$

$$a_i = \frac{1}{nmodp} \sum_{j=0}^{n-1} \alpha^{-ij} A_j, i = 0, 1, 2 \ldots n-1 \tag{4.40}$$

To specify a cyclic code, it is sufficient to specify the set in which the transform domain components are zero.

4.5.4 Cyclotomic coset

Let $I_n = \{0, 1, 2, \ldots n-1\}$. For any $j \epsilon I_n$ and for any divisor d of m, the q^d-cyclotomic coset of j modulo n is defined to be the set: $\left\{[j]^d = \{i \epsilon I_n | j = iq^{dt} modulo\ n\ for\ some\ t \geq 0\right\}$

The cardinality of this set is denoted by $e_j^{(d)}$. When $d = 1$, we will denote the q-cyclotomic coset of *jmodulo n* by $[j]$ and its cardinality by e_j. Using this notation, we present the following theorem.

Theorem: *Let **C** be a cyclic code of length $n/q^m - 1$ over F_{q^m} such that the transform domain component $A_{jq^s} \in A_{[j]}, \left(|[j]| = e_j, 0 \leq s \leq e_j - 1\right)$ is free and all other transform components are constrained to zero. Then $Rank_q(C) = e_j$. In other words, for a length $n/q^m - 1$ cyclic code over F_{q^m}, with a single nonzero transform component in only one q^m cyclotomic coset, the rank of the code is equal to the size of the q-cyclotomic coset to which the free transform domain component belongs [18].*

4.5.5 Gaussian integer map [19]

Let q be a prime of the form $q = 4a + 1$ i.e. $q = 5,\ 13,\ 17, \cdots\cdots$ etc. By definition, a Gaussian integer ω is a complex number defined as $\omega = a + ib$; $a, b \in Z$, $i = \sqrt{-1}$. From number theory, it is known that every prime number q of the form $q \equiv 1 \bmod 4$ can be written as $q = (u + iv) \times (u - iv) = u^2 + v^2$. The number $\Pi = u + iv$ is known as Gaussian prime number, where $u, v \in Z$. Let $\Pi' = u - iv$. Then calculation modulo Π is defined as $\zeta = \omega \text{ modulo } \Pi = \omega - \left[\frac{\omega\Pi'}{\Pi\Pi'}\right]\Pi$, where $[\bullet]$ performs the operation of rounding to the nearest Gaussian integer. In [19], Lusina *et al.* have proved that the Gaussian integers modulo Π form a field, $G_\Pi = \left\{\zeta_0 = 0, \zeta_1 = 1, \zeta_2, \cdots, \zeta_{q-1}\right\}$ and that the map $\xi : F_q \Rightarrow G_\Pi$ given by $\zeta_i = i \bmod \Pi$, $i = 0, 1, 2, \cdots, q-1$ is an isomorphism [19]. Therefore, when we map codewords from a linear cyclic codes over F_{q^m}, $q = 5, 13, 17, \ldots$ which are basically $m \times n$ matrices over F_q to $m \times n$ matrices over the complex Gaussian field, the full-rank property of the code over F_{q^m} is preserved. The constellation map of $F_5 \rightarrow G_{1+2i}$ is tabulated in Table 4.11 and is shown in Figure 4.5.

Table 4.11 Mapping from finite field to Gaussian integer [18]

i	0	1	2	3	4
ξ_i	0	1	0 + 1i	0 − 1i	−1

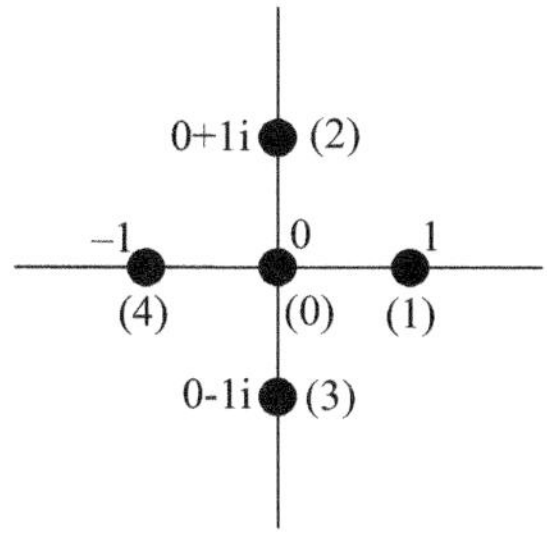

Figure 4.5 Constellation map of $F_5 \rightarrow G_{1+2i}$

4.5.5.1 Decoding of STBC derived from non binary cyclic code

The Frobenius norm, which is the matrix norm of an $m \times n$ matrix **A**, is defined as the square root of the sum of the absolute squares of its elements,

$$\|A\|_F = \sqrt{\sum_{i=1}^{m}\sum_{j=1}^{n}|a_{ij}|^2} \tag{4.41}$$

The Frobenius norm can also be considered as a vector norm. It is equal to the square root of the matrix trace of AA^H, where A^H is the conjugate transpose, i.e.,

$$\|A\|_F = \sqrt{\text{trace}(AA^H)} \tag{4.42}$$

The trace of an $n \times n$ square matrix A is defined to be the sum of the elements on the main diagonal of A. The decoding employed is the maximum likelihood decoding, given by:

$$\hat{X}_{ML} \triangleq \underset{\mathbf{X}\in C}{arg\min trace}[(\mathbf{Y} - \mathbf{HX})(Y - \mathbf{HX})^H] \tag{4.43}$$

where X is the transmitted codeword matrix, **H** is the channel matrix comprising channel coefficients, **Y** is the received codeword matrix. Let n denote the length of the codewords of the code, N_T denote the number of transmit antennas, and N_R denote the number of receive antennas. The dimensions of **X, H,** and **Y** are, respectively, N_T x n,N_Rx N_T and N_R x n.

4.5.6 Description of non binary cyclic code used

We have constructed a length-3 non-primitive cyclic code over F_{5^2} by choosing A_1 ($j = 1$) as the free transform domain component and constraining transform domain components A_0 and A_2 to zero. Since $|[j] = 2|$, by Theorem 1, all non zero codewords of this code have F_5 rank equal to 2. As all the matrices are of dimension 2×3, all nonzero matrices are full rank. The 25 codewords of this code and their respective F_5 ranks are shown in Table 4.12.

Table 4.12 Codewords with their corresponding rank

$c_1 = \begin{bmatrix} 0 & 0 & 0 \\ 0 & 0 & 0 \end{bmatrix}$ Rank = 0;	$c_2 = \begin{bmatrix} 1 & 3 & 1 \\ 0 & 2 & 3 \end{bmatrix}$ Rank = 2;
$c_3 = \begin{bmatrix} 0 & 1 & 4 \\ 1 & 1 & 3 \end{bmatrix}$ Rank = 2;	$c_4 = \begin{bmatrix} 3 & 3 & 4 \\ 4 & 0 & 1 \end{bmatrix}$ Rank = 2;
$c_5 = \begin{bmatrix} 2 & 0 & 3 \\ 4 & 3 & 3 \end{bmatrix}$ Rank = 2;	$c_6 = \begin{bmatrix} 2 & 4 & 4 \\ 3 & 2 & 0 \end{bmatrix}$ Rank = 2;
$c_7 = \begin{bmatrix} 4 & 1 & 0 \\ 4 & 2 & 4 \end{bmatrix}$ Rank = 2;	$c_8 = \begin{bmatrix} 2 & 1 & 2 \\ 0 & 4 & 1 \end{bmatrix}$ Rank = 2;
$c_9 = \begin{bmatrix} 0 & 2 & 3 \\ 2 & 2 & 1 \end{bmatrix}$ Rank = 2;	$c_{10} = \begin{bmatrix} 1 & 1 & 3 \\ 3 & 0 & 2 \end{bmatrix}$ Rank = 2;
$c_{11} = \begin{bmatrix} 4 & 0 & 1 \\ 3 & 1 & 1 \end{bmatrix}$ Rank = 2;	$c_{12} = \begin{bmatrix} 4 & 3 & 3 \\ 1 & 4 & 0 \end{bmatrix}$ Rank = 2;
$c_{13} = \begin{bmatrix} 3 & 2 & 0 \\ 3 & 4 & 3 \end{bmatrix}$ Rank = 2;	$c_{14} = \begin{bmatrix} 4 & 2 & 4 \\ 0 & 3 & 2 \end{bmatrix}$ Rank = 2;
$c_{15} = \begin{bmatrix} 0 & 4 & 1 \\ 4 & 4 & 2 \end{bmatrix}$ Rank = 2;	$c_{16} = \begin{bmatrix} 2 & 2 & 1 \\ 1 & 0 & 4 \end{bmatrix}$ Rank = 2;
$c_{17} = \begin{bmatrix} 3 & 0 & 2 \\ 1 & 2 & 2 \end{bmatrix}$ Rank = 2;	$c_{18} = \begin{bmatrix} 3 & 1 & 1 \\ 2 & 3 & 0 \end{bmatrix}$ Rank = 2;
$c_{19} = \begin{bmatrix} 1 & 4 & 0 \\ 1 & 3 & 1 \end{bmatrix}$ Rank = 2;	$c_{20} = \begin{bmatrix} 3 & 4 & 3 \\ 0 & 1 & 4 \end{bmatrix}$ Rank = 2;
$c_{21} = \begin{bmatrix} 0 & 3 & 2 \\ 3 & 3 & 4 \end{bmatrix}$ Rank = 2;	$c_{22} = \begin{bmatrix} 4 & 4 & 2 \\ 2 & 0 & 3 \end{bmatrix}$ Rank = 2;
$c_{23} = \begin{bmatrix} 1 & 0 & 4 \\ 2 & 4 & 4 \end{bmatrix}$ Rank = 2;	$c_{24} = \begin{bmatrix} 1 & 2 & 2 \\ 4 & 1 & 0 \end{bmatrix}$ Rank = 2;
$c_{25} = \begin{bmatrix} 2 & 3 & 0 \\ 2 & 1 & 2 \end{bmatrix}$ Rank = 2;	

Each codeword requires can be transmitted over three symbol durations by employing two antennas. Hence, every codeword can be represented by a 2×3 matrix. We model the channel affected by atmospheric turbulence by the gamma–gamma distribution. Noise is assumed to be additive white Gaussian with zero mean and noise variance is determined by the SNR specification. For fair comparison of the three schemes, the total energy transmitted over the symbol duration is kept equal to unity in all cases.

4.6 Results

4.6.1 Comparison of the Alamouti scheme and STBCs derived from non binary cyclic code

We have compared the error correcting performance of the Alamouti Scheme and the STBC derived from non binary cyclic code in 2×1 and 2×2 configurations, under typical and strong turbulence channel conditions, in Figures 4.6 and 4.7, respectively.

We observe that the STBCs derived from cyclic codes achieve a given BER at lower SNRs than the corresponding Alamouti codes. Hence, we can use these coding schemes to improve the error performance of an FSO system at a fixed transmit power or to reduce the transmit power required to achieve a given BER requirement.

Table 4.13 shows the transmit power required to achieve BER $= 10^{-6}$, for a 4 km long FSO link, under typical turbulence conditions, with a link margin of 10 dB, in the presence of thin fog/heavy rain and light fog.

This graph indicates that communication with acceptably low BER ($\leq 10^{-5}$) is not possible through a channel of strong turbulence without the use of error control codes. The Alamouti scheme and STBCs derived from non binary cyclic codes enable us to achieve the required low BER. We observe that BER $= 10^{-6}$ can be obtained at a SNR of 9 dB if 2×2 STBC derived from cyclic code is employed, and at a SNR of 15 dB if 2×1STBC derived from cyclic code is employed, at a SNR of 18 dB if 2×2 Alamouti scheme is used and at a SNR of 25 dB if 2×1 Alamouti scheme is employed.

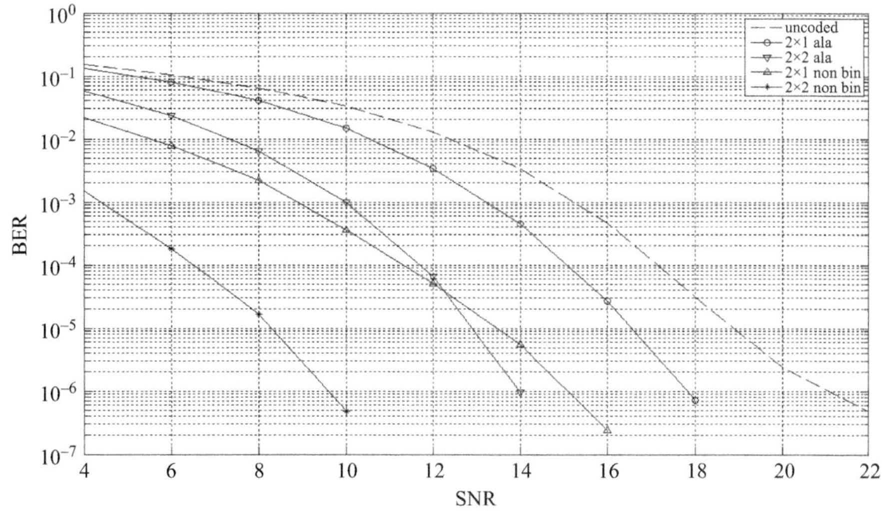

Figure 4.6 Comparison of Alamouti scheme and STBCs derived from cyclic codes under typical turbulence channel conditions

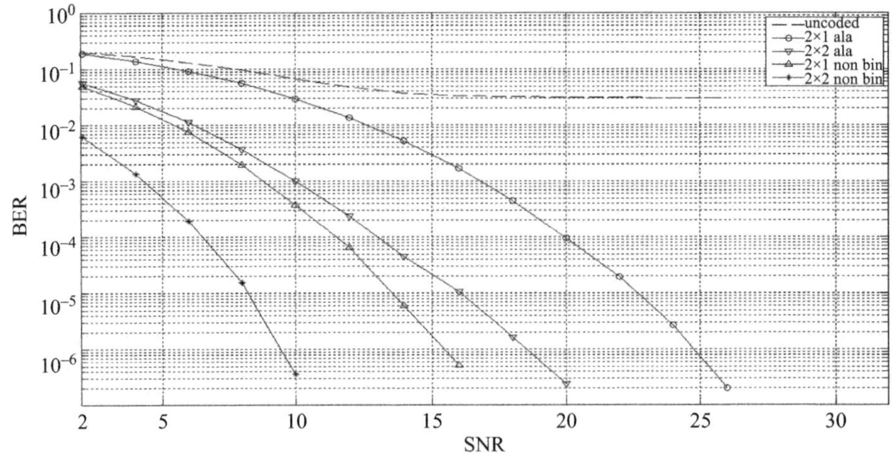

Figure 4.7 Comparison of Alamouti scheme and STBCs derived from cyclic codes under strong turbulence channel conditions

Table 4.13 Comparison of transmit power required for systems employing different coding schemes

Channel condition	P_T for uncoded transmission (mW)	P_T for 2×1 Alamouti scheme (mW)	P_T for 2×1 STBC derived from cyclic code (mW)	P_T for 2×2 Alamouti scheme (mW)	P_T for 2×2 STBC derived from cyclic code (mW)
Heavy rain	53.4833	26.8046	13.4343	10.6713	3.7863
Light fog	7,091.99	3,554.3	1,781.4	1,415.0	502.0736

Table 4.14 Coding gain of STBCs derived from cyclic code over the Alamouti code

BER	Gain (dB) for 2 × 2 STBC over 2 × 2 Alamouti	Gain (dB) for 2 × 1 STBC over 2 × 1 Alamouti
10^{-5}	8 dB	9 dB
10^{-6}	9 dB	9.5 dB

Table 4.14 shows the advantage obtained by using STBCs derived from non binary cyclic codes over the Alamouti code, for 2 × 2 and 2 × 1 configuration, in an FSO link operating in strong turbulence channel conditions.

It is apparent from this plot that an FSO system employing the STBCs derived from cyclic codes provides substantial coding gain over the corresponding

Alamouti scheme. This would translate into reduced transmit power requirement. Given that a coding gain of 8 dB implies a reduction in transmit power requirement by a factor of 0.158, it is apparent that the use of suitable STCs can make terrestrial FSO links practically deployable over a wide range of distances and applications. The use of suitable channel codes can ensure that FSO links can be used to communicate at high speeds over adverse environmental conditions while simultaneously complying with eye safety regulations [8,21].

4.7 Conclusions and scope for future work

Computation of the link budget is an important step that has to be completed before setting up any communication link. We have computed the link budget for terrestrial FSO links operating over various distances (1,000 m to 4,000 m) under a variety of weather (channel) conditions. Eye safety considerations often require that the output power of the transmitter be constrained. To facilitate error free, high-speed communication under the constraint of limited transmit power, the use of channel codes becomes necessary. In this chapter, we have suggested the use of two STCs, the widely known Alamouti scheme and a newer STBC derived from cyclic code. Our simulation results indicate that under conditions of typical turbulence, the use of an appropriate channel code can reduce transmit power requirement by a factor of ten. In addition, simulations carried out under strong turbulence conditions indicate that FSO communication over distances of the order of 4,000 m is not possible in the absence of a channel code. However, the performance of the STBCs considered by us indicate that by using a suitable channel code, reliable communications can be established under these conditions as well. In performing these simulations, we have assumed the availability of perfect channel state information at the receiver.

Considering the performance of the STBCs and their role in transforming a bad channel into one which can give accurate and substantial data throughput, it seems reasonable to explore several other STBCs that can be derived from the non binary cyclic codes. In this context, STBCs derived from non binary cyclic codes over larger finite fields (F_{13^2}, F_{17^2}) can be explored. In recent years, a large number of channel codes designed to ensure efficient information transfer over fading wireless channels have been proposed [2]. A comprehensive study of the performance of these codes can yield insight into the optimum channel code to be used in this application.

In conclusion, we feel that FSO technology aided by suitable channel-coding techniques can be used to realize reliable high-speed communication short-distance links across distances less than 4 km under good as well as adverse weather conditions with the level of data integrity being as good as is required by the application. This development can help in the widespread deployment of this attractive technology with its benefits of enabling high data-rate communication with the advantages of quick deployment time, high security, and no spectral licensing requirements.

References

[1] S. M. Alamouti, "A simple transmit diversity technique for wireless communications," *IEEE Journal on Select Areas in Communications*, vol. 16, no. 8, pp. 1451–1458,1998.

[2] F. Xu, A. Khalighi, P. Caussé, and S. Bourennane, "Channel coding and time-diversity for optical wireless links," *Optics Express*, vol. 17, no. 2, pp. 872–887, 2009.

[3] M. Uysal, J. (Tiffany) Li, and M. Yu, "Error rate performance analysis of coded free-space optical links over gamma-gamma atmospheric turbulence channels," *IEEE Transactions on Wireless Communication*, vol. 5, no. 6, pp. 1229–1233, 2006.

[4] L. C. Andrews and R. L. Phillips, "I-K distribution as a universal propagation model of laser beams in atmospheric turbulence," *Journal of the Optical Society of America A*, vol. 2, no. 2, pp. 160–163, 1985.

[5] L. C. Andrews, R. L. Phillips, C. Y. Hopen, and M. A. Al-Habash, "Theory of optical scintillation," *Journal of the Optical Society of America A*, vol. 16, no. 6, pp. 1417–1429, 1999.

[6] L. C. Andrews and R. L. Phillips, *Laser Beam Propagation through Random Media*, 2nd ed. Bellingham,WA: SPIE Press 2005.

[7] M. A. Al-Habash, L. C. Andrews, and R. L. Phillips, "Mathematical model for the irradiance probability density function of a laser beam propagating through turbulent media," *Optical Engineering*, vol. 40, no. 8, pp. 1554–1562, 2001.

[8] H. Willebrand and B. S. Ghuman, *Free-Space Optics: Enabling Optical Connectivity in Today's Networks*. Indianapolis, IN: Sams Publishing, 2002.

[9] Z. Ghassemlooy and W. O. Popoola, "Terrestrial free-space optical communications," *InTech*, pp. 355–392, 2010.

[10] H. Henniger, O. Wilfert, "An introduction to free-space optical communications," *Radioengineering*, vol. 19, no. 2, pp. 203–212, 2010.

[11] J. Vitásek, J. Látal, S. Hejduk, *et al.*, "Atmospheric turbulences in free space optics channel," in *2011 34th International Conference on Telecommunications and Signal Processing (TSP)*, pp. 104–107. IEEE, 2011.

[12] L. C. Andrews, *Field Guide to Atmospheric Optics*. Bellingham, WA: SPIE Press, 2004.

[13] C. F. Bohren and D. Huffman, *Absorption and Scattering of Light by Small Particles*. Hoboken, NJ: Wiley 1983.

[14] I. Kim, B. Mcarthur, and E. Korevaar, "Comparison of laser beam propagation at 785 and 1550 nm in fog and haze for opt. wireless communications," *Proc. SPIE* 4214, pp. 26–37, 2001.

[15] R. Srinivasan and D. Sridharan, "The climate effects on line of sight (LOS) in FSO communication," *International Conference on Computational Intelligence and Computing Research*, IEEE, 2010.

[16] A. R. Calderbank, N. Seshadri, and V. Tarokh, "Space-time codes for high data rate wireless communications: Performance criterion and code construction," *IEEE Trans. Information Theory*, vol. 45, no. 2, pp. 1456–1467, 1999.

[17] M. K. Simon and V. Vilnrotter, "Alamouti-type space-time coding for free-space optical communication with direct detection," *IEEE Transactions on Wireless Communications*, vol. 4, no. 1, pp. 35–39, 2005.
[18] U. Sripati, "Space-time-block codes for MIMO fading channels from codes over finite fields," Ph.D. dissertation, Dept. of ECE, IISc Bangalore, Oct. 2004.
[19] P. Lusina, E. Gabidulin, and M. Bossert, "Maximum-rank distance codes as space time codes," *IEEE Trans. On Information Theory,* vol. 49, no. 10, pp. 2757–2760, 2003.
[20] T. K. Moon, *Error Correction Coding: Mathematical Methods and Algorithms*. Hoboken, NJ: Wiley-Interscience, 2005.
[21] O. Bouchet, H. Sizun, C. Boisrobert, F. de Fornel, P.-N. Favennec, *Free-Space Optics Propagation and Communication*. London: ISTE Ltd., 2006.

Chapter 5

FSO channel—atmospheric attenuation and refractive index (C_n^2) modeling as the function of local weather data

Arockia Bazil Raj[1] and Julian Cheng[2]

5.1 Introduction

Free space optical (FSO) communication is an upcoming attractive alternative technology for transporting high-bandwidth data when the existing RF/fiber-optic communication is neither realistic nor viable. However, the presence of FSO channel turbulences such as fog, smoke, rain, dust, snow, and/or sand can critically degrade the quality of the FSO communication system. There is a great technical development in today's optical components such as LED, laser, optical detector, detector's sensitivity at high bandwidth, modulation techniques, power requirements, total weight, and total size. In spite of all these technological developments, the major limitation of FSO communication quality is the atmosphere [1,2]. Optical absorption and scattering due to the FSO channel's components in the atmosphere drastically reduce the transmitted optical power. Further, the arbitrary atmospheric formation due to random fluctuation of optical turbulence can alter the wavefront quality of the traveling optical signal, develop the intensity fading and thus result in random signal losses and inter symbol interference (ISI) at the receiver plane. Weather conditions thus ultimately determine the FSO communication system quality not only in ground-to-ground FSO applications but also for deep space laser satellite optical communications because a portion of the optical beam always in the atmospheric turbulence medium that causes time-varying scattering effects [1–4].

FSO communication is an age-long technology that entails the data laden with optical beam through the weather from one point to other. The performance-limiting factors in FSO communication channel depend on the local climatic and geographical conditions that cause the traveling optical beam to be deflected, absorbed and/or scattered [5–7]. The coefficient of scattering is given as the ratio of original optical power to attenuated power [1,3,8,9]. Even though these effects are locally small, the effects build up over the optical beam traveling path, i.e., FSO

[1]Department of Electronics Engineering, Defence Institute of Advanced Technology, India
[2]School of Engineering, University of British Columbia, Canada

channel and can lead to scintillation, beam wandering and wavefront aberration that vary in the order of millisecond or more. Hence, it can be concluded that in a real FSO communication environment, the optical channel exhibits a random time varying characteristics.

The atmospheric turbulence strength is observed greater near the ground and falls off exponentially with increasing altitude. Thus, the effects of turbulence strength are much less in vertical path than in horizontal path. Therefore, the optical beam severely affected in the horizontal propagation rather than the vertical. In the terrestrial FSO channel, the variations of the atmospheric turbulence strength depend on the location, time, and local weather data. Experimental study on the atmospheric attenuation and turbulence strength (C_n^2) as the function of local meteorological parameters becomes significant to understand the channel effects on the propagating optical beam to understand the maximum data rate possible at that location. The bird's-eye view of the experimentation field location showing the transmitter (in the left building) and receiver (in the right tower) built for this study is shown in Figure 5.1. More accurate and compact wireless weather stations are built as reported in [2] and deployed near the transmitter, receiver, and at a in-between point. The measured weather data, wind speed, temperature, relative humidity and pressure, are recorded in the computer along with other optical measurements. More details on the working principle and interface protocols of the sensors used in the designed weather station and their associated MATLAB®/ VHDL codes can be found in [5,6].

Since the atmospheric particles suspended in the FSO channel vary in spatial/ time domains, it becomes complicated to calculate the atmospheric attenuation and turbulence strength by using available classical/theoretical approaches [6–8]. Therefore, the predictions of atmospheric attenuation and turbulence strength using empirical models (designed based on the experimental data) are coming up. Although various models are available to estimate the atmospheric attenuation and turbulence strength, no model provides generalization [2–8]. Development of the

Figure 5.1 Satellite view of our outdoor FSO link test-bed

precise and locally suitable model to calculate the atmospheric attenuation (A_{att}) and turbulence strength (C_n^2) as the function of local weather data and validating their prediction accuracy against the existing models become attractive and significant which is the major part of our discussion in this chapter.

5.2 Design of FSO link experimental test-bed

Transmitter and receiver laboratories are built for the link range of 0.5 km at an altitude of 15.25 m for the studies on prediction of atmospheric attenuation (A_{att}) and turbulence strength (C_n^2) according to local weather data. A modulatable laser source, laser power supply and driver unit and transmitting-optics (telescope) are the opto-electronic devices at the transmitter station (building). The transmitting optics is used to expand the incident beam of diameter 3 to 9 mm to significantly reduce the beam divergence at the aperture of the receiving telescope [6]. Binary "1," i.e., +5 V is maintained at the terminal of "modulation input" throughout the experiment period to transmit the unmodulated optical beam that is required for the measurement of optical attenuation (A_{att}) at the receiver and turbulence strength (C_n^2) of the optical channel. The receiver station (tower) consists of telescope, optical filter, variable beam splitter, photodiode, trans impedance amplifier (TIA), power meter, automated data acquisition system [5], data logging in personal computer (PC) [5], position detector [5,6], and mono-pulse (MP) arithmetic circuit [5,6]. The opto-electronics equipments are assembled such as to measure power fluctuation in mW and beam displacement in V as reported in [3,4]. The schematic diagram of experimental setup is shown in Figure 5.2 and all the opto-electronic devices are mounted on the vibration-damped optical breadboards.

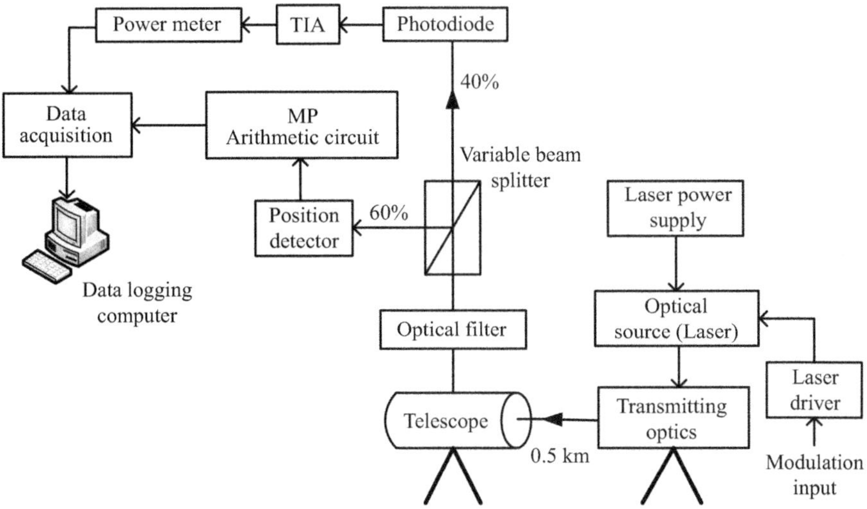

Figure 5.2 Schematic diagram of experimental set-up

The receiving telescope captures the arrived beam and concentrates on them to the optical filter that permits only the 850 nm optical beam and blocks the other wavelengths including ambient light. The variable beam splitter divides the incident beam into two directions: reflecting and propagating beam. The reflected beam falls on the position detector to measure the beam centroid displacement information using the MP arithmetic calculation [5,6]. The propagating beam falls on the photodiode and the electrical output is measured using an optical power meter. A weather station is built with commercially available sensors to continuously measure the atmospheric parameters such as wind speed (W_s), temperature (T), relative humidity (R_h), and pressure (Pr) [2,5,6].

The developed weather monitoring station is deployed at an altitude of 15.25 m at the transmitter, receiver, and at a middle point above the mechanical workshop, i.e., at a point in between the transmitter and receiver. The weather data are acquired once in every second during different local climatic conditions and the data are averaged over every 10 min interval. A graphical user interface (GUI) is designed in MATLAB environment [9] for automated data acquisition, data logging, and on/off line calculations. More details on the main specifications of the opto-electronics devices used in this experimental setup, their interface protocol, and associated MATLAB/VHDL codes can be found in [5,6]. The required parameters and their values are obtained from the measurement data and inputted to the selected models to calculate the attenuation (A_{att}) and turbulence strength (C_n^2).

5.3 Measurement of atmospheric attenuation (A_{att}) and turbulence strength (C_n^2)

A fundamental law to estimate the atmospheric attenuation of optical signal, propagating along the FSO channel, on the basis of the atmospheric visibility (V) is Koschmieder law. In this law, the visibility is defined as the distance to the optical intensity at which the transmittance drops to certain value, i.e., threshold (T_{th}) of the original visual intensity, i.e., 100% along the propagation path [3,4,6,7]. The meteorological visibility can therefore be expressed with the Beer–Lambert law as [6]

$$V = \frac{10\log_{10}(T_{th})}{\beta_\lambda} = \frac{4.343\log_{10}(T_{th})\,R}{\log_{10}(T)} \tag{5.1}$$

where T_{th} is the transmission threshold in %, β_λ is the atmospheric attenuation coefficient in dB/km, R is the link range in km, and T is the transmittance in dB. In (5.1), the smaller value of T_{th} defines a larger visibility range. The selection of T_{th} decides the exact definition of the visibility and the corresponding atmospheric optical attenuation. The transmittance is measured using appropriate power meter as received power (P_R)/transmitted power (P_T) at the peak wavelength of 550 nm. The transmittance of the 850 nm is also simultaneously measured. The atmospheric turbulent channel optical attenuation along the propagation path is directly measured using the optical power in dB/km as

$$\text{Attenuation} = \left(\frac{10\log_{10}(P_{\mathrm{T}}/P_{\mathrm{R}})}{R}\right) \tag{5.2}$$

The Rytov variance is a measure of the strength of atmospheric scintillation (σ_{I}^2). The optical beam propagating through the atmospheric turbulent medium will experience irradiating fluctuations known as scintillation [2,6,7]. The relation between atmospheric turbulence strength (C_n^2) and the relative variance of optical intensity (σ_{I}^2) as per the Rytov variance is computed as [6,7]

$$\sigma_{\mathrm{I}}^2 = KC_n^2 k^{7/6} L^{11/6} \tag{5.3}$$

where C_n^2 is the atmospheric turbulence strength parameter, k represents the optical wave number ($k = 2\pi/\lambda$), and L is the distance between the transmitter and receiver of the optical FSO link and K is a constant (K is 1.23 and 0.5 for plane and spherical waves, respectively). The scintillation data, i.e., signal intensity variations at the receiver plane are continuously recorded and the scintillation index σ_{I}^2 is estimated by the relation

$$\sigma_{\mathrm{I}}^2 = \frac{\langle I^2 \rangle - \langle I \rangle^2}{\langle I \rangle^2} \tag{5.4}$$

where I is the measured optical wave irradiance and the angle brackets $<\cdots>$ denote an ensemble average. From (5.3) and (5.4), the C_n^2 can be calculated using

$$C_n^2 = \frac{\sigma_{\mathrm{I}}^2}{1.23 k^{7/6} L^{11/6}} \tag{5.5}$$

The atmospheric scintillation (σ_{I}^2) is computed from the irradiance fluctuation of the optical signal observed by an optical detector [5,7]. The regime of weak and strong fluctuations occurs when $\sigma_R^2 < 1$ and $\sigma_R^2 < 1$, respectively, while the saturation occurs when $\sigma_R^2 \to \infty$ [2,4,6,7].

5.4 Existing attenuation and turbulence models

The optical signal traveling through the FSO channel interacts with the molecular constituents and causes some of the photons extinguish. This effect finally results in power loss, and temporal and spatial distortions that depend on the local climatic conditions/seasons [1–4,6–9]. Although various models are available, the models showing the less root mean square (RMS) error in the comparative analysis are selected for the comparative analysis with the proposed models. The considered models are listed in this section with their limitations. The detailed descriptions, sample calculations, and MATLAB codes and MATHCAD® work sheets of the selected models can be found in [1,2,6,8,10–14]. However, MATLAB code for the PAMELA model of atmospheric turbulence strength (C_n^2) estimation is given at the end of this section.

5.4.1 Atmospheric optical attenuation

Atmospheric optical attenuation due to the combined effect of absorption and dispersion of the optical field in FSO channel by the suspended turbidity depends on the local atmospheric conditions and terrain types. The molecular and aerosol behavior for the scattering and absorption process is wavelength dependent. Aerosol scattering becomes the more dominating factor along the FSO channel due to the variations of meteorological parameters. The models selected for the comprehensive analysis of atmospheric attenuation are listed in Table 5.1 and interpreted below. The attenuation is estimated/measured in dB/km. Selected atmospheric attenuation models are experimentally verified for testing their generalization and prediction accuracy when applied in real-world open atmospheric FSO link for the most common wavelength of 850 nm.

In (5.6), V is the visibility in km, λ is the source wavelength in nm, $q(t)$ is $0.1428\lambda - 0.0947$ and $0.8467\lambda - 0.5212$ for fog and smoke, respectively. The general expression for $q(t)$ is obtained for a reference wavelength of 550 nm [8]. The open atmospheric FSO channel consists of not only fog or smoke (as considered independently in this model), but also of various parameters based on the experimental spot's terrain type, altitude, geographical locations, and local weather conditions. None of the occasions smoke intervened at the channel of our FSO link, thus, the smoke function is neglected and only fog is considered for comparative analysis.

Table 5.1 Atmospheric attenuation models selected for comparative analysis

M Ijaz model [8]	$\beta_\lambda = \frac{17}{V}\left(\frac{\lambda}{\lambda_0 = 550\ \text{nm}}\right)^{-q(t)}$	(5.6)
M S Awan model [10]	$\beta_{\text{atten}} = 25.75\exp\left(-\left(\frac{R_h - 83.85}{1.026}\right)^2\right) + 93.89\exp\left(-\left(\frac{R_h - 113.8}{21.77}\right)^2\right) + 24.46\exp\left(-\left(\frac{R_h - 85.64}{0.4174}\right)^2\right)$	(5.7)
Itai Dror model [11]	$\sigma_{\text{ext}} = \frac{1}{(1 - RH/100)^4}\exp\left(-\left(a_o + \sum_{i=1}^{5}\beta_i X_i\right)\right)$	(5.8)
Bataille model [12]	$\sigma_m = -\ln\begin{pmatrix} B_1 + B_2T' + B_3R_h + B_4T'R_h + B_5T'^2 + B_6R_h^2 \\ +B_7T'R_h^2 + B_8T'^2R_h + B_9R_h^3 + B_{10}T'^3 \end{pmatrix}$	(5.9)

In (5.7), R_h is relative humidity in %. Actually in [10], two models are proposed as a function of R_h and temperature (in °C), respectively. RMS error of the temperature model appears very high, thus only R_h model is considered. The R_h model is designed based on the Gaussian fitting by employing nonlinear least square method [10].

In (5.8), a_0 is a constant equal to 10.01, β_i are regression coefficients ($\beta_1 = -13.25$, $\beta_2 = 5.64$, $\beta_3 = 0.0007678$, $\beta_4 = -0.0000281$, $\beta_5 = 0.002408$) and X_i are meteorological parameters ($X_1 = 1 - R_h/100$, $X_2 = (1 - R_h/100)^2$, $X_3 = T/(1 - R_h/100)^3$, $X_4 = T^2/(1 - R_h/100)^3$, $X_5 = (1 - R_h/100)^3\ Ws^3$) [11]. This model is designed using regressive analysis and statistical analysis system (SAS) software to predict the atmospheric extinction coefficient.

In (5.9), $T' = T/273.15$ is the reduced temperature of the air, R_h is the absolute humidity in gm^{-3}, and $B_1 = 0.9953$, $B_2 = 0.01311$, $B_3 = -0.00148$, $B_4 = 0.00137$, $B_5 = -0.012162$, $B_6 = 0.00000038$, $B_7 = 0.00000026$, $B_8 = -0.0004221$, $B_9 = 0$, and $B_{10} = 0.003763$ are the model's regression coefficients. The model accepts the inputs with the variations of 0 to 8 ms^{-1} for wind speed, 15 to 45 °C for temperature and 10% to 85% for relative humidity [12].

5.4.2 Atmospheric optical turbulence strength (C_n^2)

Most of the existing models predict the atmospheric turbulence strength (C_n^2) as a function of altitude and/or applicable only at a specific locations such as rocky terrain, HV–night time, submarine laser communication (SLC)-day, Hufnagel–Valley, Greenwood, Gurvich, and SLC–Day, etc. Prediction accuracy of these models greatly differs from the measurement data of horizontal path turbulence strength at our test field. Actually, the C_n^2 not only varies as a function of altitude, but also according to several parameters including local meteorological conditions, geographic location, terrain type, and time of day [2,6,13–15], etc. The existing models predicting the horizontal turbulence strength and showing less sum of absolute error (SAE) are only selected for the comparative analysis, listed in Table 5.2 and interpreted below.

In (5.10), φ_h is the temperature gradient, φ_m is the wind shear, ζ is the eddy dissipation rate, H is the heat flux, C_p is the specific heat, ρ is the mass density, and u_* is the friction velocity. The dynamic inputs to this model are the altitude (h), local conditions (terrain type), geographical location, cloud cover, meteorological values, latitude, longitude, number of day in year, Greenwich mean time (GMT), surface roughness length, and local time of day [2,6,13,14]. The C_n^2 computation flow graph as per this model is shown in Figure 5.3.

In (5.11), A is the nominal value of C_n^2 in m$^{-2/3}$ at the ground and v is the RMS wind speed in ms^{-1}. This model predicts the C_n^2 for inland sites and daytime viewing conditions [6,14,15].

In (5.12), x_c and y_c are the Cartesian coordinates of the center of the Gaussian light spot on the position detector, γ_c is radial displacement in mm, W is the beam waist (width) of the Gaussian beam in mm, and R is length of the optical link in km. The beam wander is derived as a function of displacement of the instantaneous

Table 5.2 Atmospheric turbulence (C_n^2) models selected for comparative analysis

PAMELA model [2,5,6,13,14]	$C_n^2 = 5.152\varphi_h \left(\frac{1}{\varphi_m - \xi}\right)^{0.33} \left(\frac{77.6 \times 10^{-6}\mathrm{Pr}}{T^2}\right)^2 h^{-0.667} \left(\frac{-H}{C_p \rho u_*}\right)^2$	(5.10)
Hufnagel-Valley model [6,14,15]	$C_n^2(\mathrm{h}) = \mathrm{A}\exp(-h/100) + 5.94 \times 10^{-53}(v/27)^2 h^{10}\exp(-h/1,000) + 2.7 \times 10^{-16}\exp(-h/1,500)$	(5.11)
Beam-wandering model [6,16,17]	$\gamma_c = \sqrt{x_c^2 + y_c^2 C_n^2} = \frac{\langle \gamma_c{}^2 \rangle}{2.42\, R^3 W^{-1/3}}$	(5.12)
Polynomial regression [6,18–20]	$\mathrm{TCSA} = 9.96 \times 10^{-4} R_h - 2.75 \times 10^{-5} R_h^2 + 4.86 \times 10^{-7} R_h^3 - 4.48 \times 10^{-9} R_h^4 + 1.66 \times 10^{-11} R_h^5 - 6.26 \times 10^{-3} \ln R_h - 1.37 \times 10^{-5} \mathrm{SF}^4 + 7.30 \times 10^{-3}$ $C_n^2 = 5.9 \times 10^{-15}\mathrm{W} + 1.6 \times 10^{-15} T - 3.7 \times 10^{-15} R_h + 6.7 \times 10^{-17} R_h^{2'} - 3.9 \times 10^{-19} R_h^3 - 3.7 \times 10^{-15} W_s + 1.3 \times 10^{-15} W_s^2 - 8.2 \times 10^{-17} W_s^3 + 2.8 \times 10^{-14}\mathrm{SF} - 1.8 \times 10^{-14}\mathrm{TCSA} + 1.4 \times 10^{-14}\mathrm{TCSA}^2 - 3.9 \times 10^{-13}$	(5.13)

center of the received beam on the position detector. The stochastic polar variable of the beam center is computed using the measurement values of Cartesian (x,y) or polar (r,θ) coordinates by coordination transformation techniques presented in [16,17].

In (5.13), W is the temporal-hour weight values taken from [19] for computations, SF is the solar flex (kWm^{-2}), and TCSA is the total cross-sectional area ($\mathrm{cm}^2/\mathrm{m}^3$). The inputs to this model are macroscale meteorological data which can be measured directly by a suitable weather station. The concept of temporal or solar hour is introduced. Temporal-hour at sunrise is 00:00, at noon is 06:00 and at sunset is 12:00 in any day [6,18–20]. The allowed input variations of this model are 0 to 10 ms^{-1} for wind speed, 9 to 35 °C for temperature and 14% to 92% for relative humidity [18,20].

MATLAB version of PAMELA model for C_n^2 estimation is as follows: the assumed input values are (i) number of the day in the year (N_d) is 151, (ii) latitude in degrees (lat) is 32, (iii) longitude in degree (long) is 106, (iv) define start of 24 h period for diurnal calculations in terms of GMT (t_{start}) is 6, (v) percent cloud cover (cc) is 3, (vi) average wind speed in m/s (v) is 2, (vii) surface roughness length in m (h) is 0.1, (viii) height above ground in m (h) is 4, (ix) atmospheric pressure in mb (Pa) is 1,000, and (x) atmospheric temperature in °F (T_f) is 85. The mathematics associated to PAMELA model can be found in the given references mainly in [14]. The MATLAB code given below follows the computation flow graph shown in Figure 5.3.

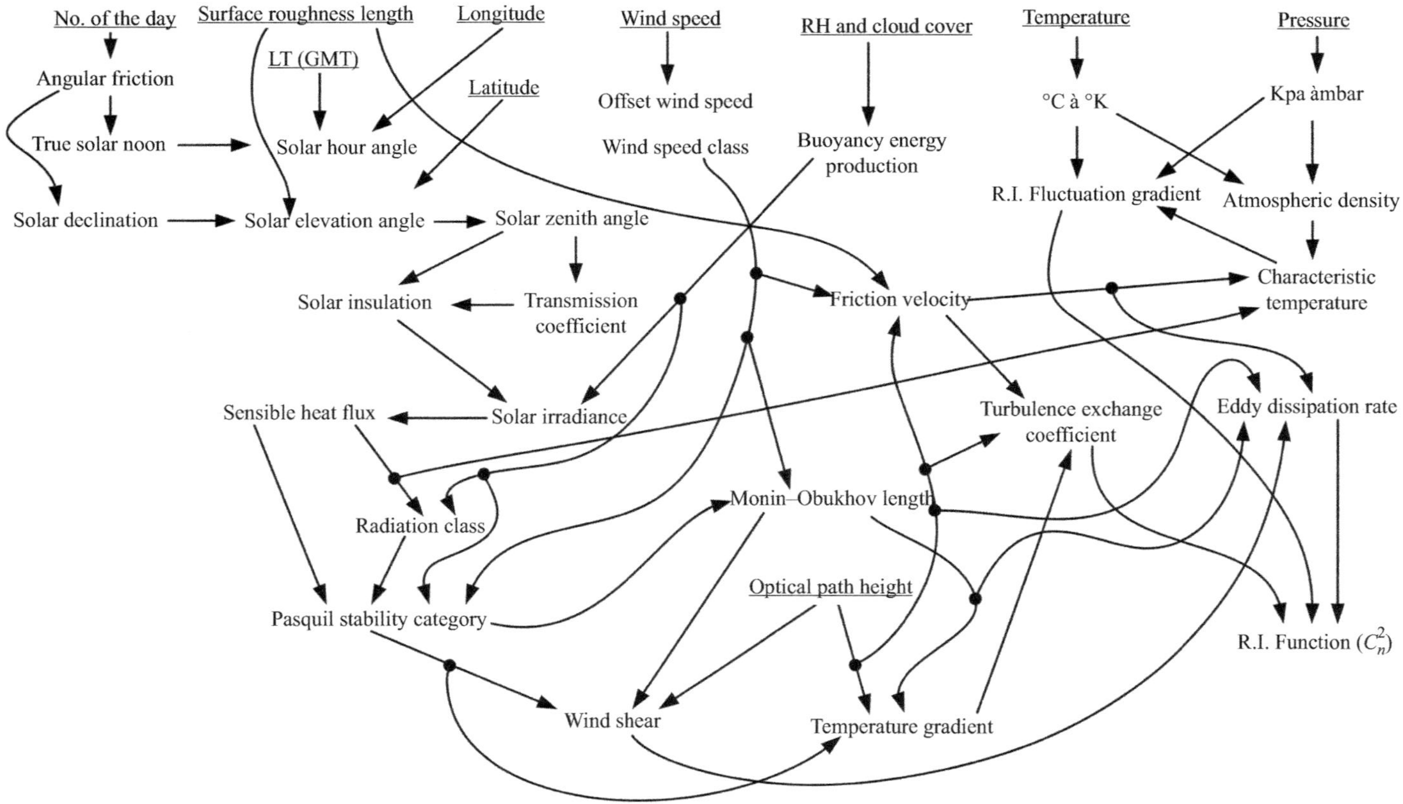

Figure 5.3 Experimental modeling of C_n^2 computation flow graph

MATLAB code:

```
clc;
clear all;
clear memory;
clear screen;
Nd=151;
lat=32;
long=106;
tstart=6;
cc=3;
v=2;
hr=0.1;
h=4;
Pa=1000;
Tf=85;
if v<3,
    v0=3;
else
    v0=v;
end;
T=(5/9)*(Tf-32)+273.15
gmt1=tstart:0.1:tstart+24;
d=(Nd-1)*(360/365.24);
```

```
r=279.93+d;

m=(12+ 0.12357*sind(d)- 0.004289*cosd(d) +.....

 0.1538*sind(2*d)+0.06078*cosd(2*d));

% for i=1:1:241,

% gmt=gmt1(i);

gmt=18;

ht=(15*(gmt-m)-long);

beta=(r+ 0.4087*sind(r) +  1.8724*cosd(r) - 0.0182*sind(2*r)
.....

+ 0.0083*cosd(2*r));

delta=asin(0.39785*sind(beta));

alfa=asin((sind(lat)*sin(delta))+....

    cosd(lat)*cos(delta)*cosd(ht));

slz=90.57- alfa*57.29;

Tc=0.57+0.0045*slz;

Itemp=1353*cosd(slz)* Tc^(secd(slz));

if Itemp<0,

 I=0;

else

 I=Itemp

end

B=[1.07 0.89 0.81 0.76 0.72 0.67 0.59 0.45 0.23];
```

```
b=B(cc+1)

R=b*I;

if I > 0,

    H=0.4*(R-100)

else

    H=-40;

end

if H > 0,

    crtemp=R/300

elseif (H <= 0 && cc >= 4);

    crtemp=-1

elseif (H <= 0 && cc < 4);

    crtemp=-2

end

if crtemp <= 0

   cr=3

else

 cr=crtemp

end

if v0<=8,

    cw=v0/2;
```

```
else

    cw=4;

end

if cc==8

    Pt=0

else

  if H<=0

     Pt=0.5*(4-cw+abs(cr))

  elseif H>0

     Pt=-0.5*(4-cw+abs(cr))

  end

end

L=((0.004349*Pt+0.003724*Pt^3)*hr^-(0.5034-
0.231*abs(Pt)+0.0325*Pt^2))^-1

E=(h/L);

if Pt<=0

    pim=(1-16*E)^-0.25

else

    pim=1+5*E

end

if Pt<=0

    y=(1-16*E)^0.25;

    sim=log(((1+y^2)/2)*((1+y)/2)^2)-2*atan(y)+pi/2
```

```
else
    sim=-5*E
end;
if Pt<=0
    fih=0.74*(1-9*E)^-0.5
else
    fih=0.74+4.7*E
end;
Ustar=((0.4*v0)/(log(h/hr)-sim))
row=(Pa/(2.87*T));
Tstar=(-H/(1004*row*Ustar))
Kh=(0.4*Ustar*h)/fih
dndz=-(((77.6*10^-6)*Pa*Tstar*fih)/(0.4*h*T^2))
epsalan=(pim-E)*((Ustar^3)/(0.4*h))
Cn2=((2.8*Kh*dndz^2)/(epsalan^0.33333))
% Cn2(i)=((2.8*Kh*dndz^2)/(epsalan^0.33333));
% %end
% figure(1);
% subplot(2,2,3);
% time=1:length(Cn2);
% p1=semilogy(time,Cn2);
% set(gca,'fontsize',12);
```

```
% hold on;

% set(p1,'Color','k','LineWidth',1.5);

% xlabel('IST of the day (Hrs)');

% ylabel('C_n^{2} (m^{-2/3})');

% vertical_line_gap=10;

% vertical_lines=1:vertical_line_gap:24*vertical_line_gap;

% vertical_line_label= ['  MN   ';'        ';'  2   ';'
';'  4   ';'        ';........

%                          '  6   ';'        ';'  8   ';'
';'  10  ';'        ';......

%                          '  Nn  ';'        ';'  2   ';'
';'  4   ';'        ';.........

%                          '  6   ';'        ';'  8   ';'
';'  10  ';'        '];

%
set(gca,'Xlim',[1,241],'XTick',vertical_lines,'XTicklabel',ver
tical_line_label);

% grid on;

% x=input('How many figures:=  ');

% if x==1,

% set(gcf,'color','w');

% print('Fig4', '-dtiff', '-r300');

% hold on;

% end
```

Table 5.3 PAMELA model simulation results when GMT = 18

Parameters	Used notation	MATLAB Result
Greenwich mean time	GMT	18
Temperature in Kelvin	T	302.5944
Solar insolation in W/m^2	I	819.6542
Rate of energy production due to buoyancy	b	0.7600
Sensible heat flux in W/m^2	H	209.1749
Radiation class	cr	2.0765
Pasquill stability category	Pt	−2.2882
Monin–Obukhov length in m	L	−13.1239
Dimensionless wind shear	Φ_m	0.6423
Diabetic influence function for momentum	Ψ_m	0.6001
Dimensionless temperature gradient	Φ_h	0.3825
Friction velocity	U_{star}	0.3885
Characteristic temperature	T_{star}	−0.4657
Turbulent exchange coefficient for heat	K_h	1.6252
Gradient for refractive index fluctuations	d$n$dz	9.4354e-08
Eddy dissipation rate	ϵ	0.0347
Refractive index structure parameter in m$^{-2/3}$	C_n^2	1.2419e-13

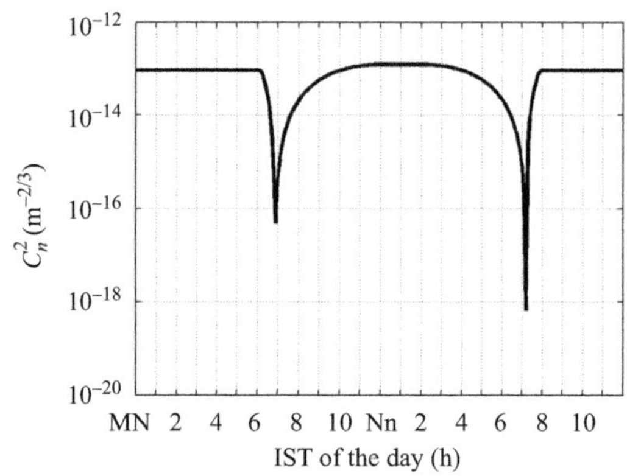

Figure 5.4 PAMELA model simulated profile of C_n^2

The results corresponding to the above MATLAB code of PAMELA model for the given input values are given in Table 5.3. Upon removing the comments, i.e., % in the above MATLAB code, commenting the lines "`gmt=18;`" and "`Cn2=((2.8*Kh*dndz^2)/(epsalan^0.33333))`" and running the same code, we will get the diurnal C_n^2 profile for the selected day as shown in Figure 5.4, whose interpretations can be found in [14].

5.5 Design of regressive model for attenuation and C_n^2 estimation

Regressive models, based on the outcomes of the experiments, can be built using the MINITAB software. Prior to model building, design of experiment (DOE) has to be planned in such a way to collect all the possible combinations of input variables/values and their response variables/values. The DOE is an active method to model a process as opposed to passively observing a process [21]. The DOE schedule enables the operators to evaluate the changes occurring in the output i.e., response of a process while varying one or multiple input(s) variables. The DOE and TAGUCHI technique provide a platform for getting predictions about the outcomes of experiments and/or measurements to design more suitable regressive models with most possible prediction accuracy. The combinations of input variables/values and corresponding output variables/values have to get entered in the MINITAB Worksheet. The statistical details behind the analysis of inputs, outputs, and response surface can be done by the following steps:

(i) open the sample data work sheet
(ii) select stat-> DOE -> response surface (SR) -> analyze SR design
(iii) input: strength
(iv) choose graphs
(v) select 4 in 1 residual graphs
(vi) Click ok in all popup windows.

The dialog boxes show the statistics of the input, output values, and response surface. A general structure of a regression equation describing the relation between the response variables and input variables with more than one term is given in (5.14). The algebraic representation of a regression equation of the model is

$$y = b_0 + b_1X_1 + b_2X_2 + \cdots + b_kX_k \tag{5.14}$$

where y is the output factors, b_0 is the constant, b_1, b_2, ..., b_k are the equation coefficients and X_1, X_2, ..., X_k are the values of the input factors. In the regression equation, the uncoded terms interpret the coefficients in natural units of the factors. Typical natural units of the factors are -1 and $+1$ to indicate low- and high-level terms, respectively. The center level term is denoted by the midpoint 0 [21]. Because the model is averaged over blocks, no coefficients for any blocks are left in the model. The general steps to be followed for designing the regressive model is as follows:

(i) creation of factorial design
(ii) select 2-level default factorial generator
(iii) input: numbers No
(iv) designs highlight
(v) input: corner points No
(vi) select ok

(vii) select worksheet
(viii) select ok
(ix) input: factor names
(x) MINITAB creates a worksheet DOE array
(xi) analyze factorial design
(xii) enter response open window
(xiii) select on
(xiv) select the terms you want in the model (in general all the input factors)
(xv) select the term
(xvi) select ok
(xvii) select plots
(xviii) chose report
(xix) click ok.

Now, MINITAB will show the designed regressive model equation. The MINITAB has provision to effectively exhibit the prediction results of the model and experimental outcomes using the analysis of variance (ANOVA) tools. An ANOVA factor "S," a measure in the units of the response, represents the variation of how far the data deviate from the actual data, i.e., to accesses how well the model describes the actual response/system. Lower the value of *S*, better the model describes the response. Few more important factors can be found graphically in the residual plots [21]. Another ANOVA factor "coefficient of determination (R^2)" shows the accuracy of model fitness, i.e., variance strength to the DOE data set in %. The R^2 can be raised to ≈99% by adequately scheduling the DOE, performing the experiment and selecting the more significant terms in the model. The model designer can consider reducing the model to remove the terms that are not significant. The predicted R^2 of 0 suggests that this model is over fit or not at all fit which supports redesigning the model. The response indicates that the designed model is in natural units when the model is at a hierarchical.

Most of the existing atmospheric attenuation and turbulence strength models are derived from the data corresponding to the local atmospheric conditions, and thus, all these models failed to attain the generalization on predicting the optical attenuation (A_{att}) and atmospheric turbulence strength (C_n^2) [6,13–20]. These models also do not offer any suitable means to statistically alter their parameters to fit to new location. Therefore, designing a locally more suitable atmospheric attenuation and turbulence strength prediction models at the location of our experimentation becomes significant to have accurate prediction of those unknowns. The formulation of these two proposed models using the MINITAB software is defined and discussed in the following section.

5.5.1 Atmospheric attenuation (A_{att}) model

Prior to model building, appropriate pilot survey experiments are conducted as per the DOE schedule using the developed experimental setup shown in Figure 5.2 in various open climatic conditions. The time average of measured weather (W_s, T, R_h, Pr), visibility (V), transmittance (Tr), scintillation (σ_I^2), and attenuation (A_{att}) data for the day-long period are recorded in the data logging PC for a longer period. Model design data of size [2,000 ×4] with three input variables W_s in ms^{-1}, T in °C, and R_h in % and one output variable: A_{att} in dB/km is applied to the MINITAB software [21]. The optimal regression equations of different grouping of the input factors and significant terms at the confidence accuracy of 98% are formulated. Different possible regressive models such as linear equation, quadratic equation, interaction equation, full model equation, and cubic equation are developed for the prediction of optical attenuation (A_{att}) as a function of locally measured weather data. The individual and combined effects of the input parameters on the performance measures, i.e., prediction accuracy, are analyzed and subsequently mathematical models are redesigned. The ANOVA tools are used to analyze the individual and combined effects of atmospheric parameters (Rs, T, R_h) on the optical attenuation. The final ANOVA results are used to determine the most significant input variables on the output plane and those groups are only considered while optimizing the model.

$$
\begin{aligned}
\text{Atten} = &-1.09734 - 3.18118\,W_s + 2.49896\,T + 0.277543\,R_h \\
&+ 0.0667605\,W_s^2 + 0.0381112\,W_sT + 0.0264432W_sR_h \\
&- 0.106594\,T^2 + 0.00641653\,TR_h - 0.0148677\,R_h^2 \\
&- 0.0060649\,W_s^3 + 0.00118457\,T^3 + 9.56075\text{e} - 005\,R_h^3
\end{aligned}
\tag{5.15}
$$

Utmost prediction accuracy, i.e., model fitness to the experimental data, is only considered while ignoring the less significant groups. All the possible combination models are examined in order to find a more suitable model for predicting the atmospheric attenuation as a function of weather parameters. The R^2 values are computed using all the models with new test data sets. The results proved that the cubic model exhibits the good correlation with the measurement values, thus, the cubic model (5.15) alone selected for practical implementation to predict the optical attenuation (A_{att}) as a function of the measured local metrological data (W_s, T, R_h).

5.5.2 *Atmospheric turbulence strength (C_n^2) model*

Atmospheric turbulence strength (C_n^2) values measured in different local weather conditions are related to the measured weather parameters during the same experimentation day using the MINITAB ANOVA tools and the regression model analyses are performed. The data sets of size [2,000 × 4], i.e., 3 input factors: (W_s, T, and R_h) and 1 response factor: (C_n^2) is used to develop the regression models as model equation-I through *V*. The individual and combined effects of the input variables on the prediction accuracy measures are analyzed and subsequently mathematical models are formulated. The sensitivity of the atmospheric parameters on the turbulence strength is also analyzed using the ANOVA tools and statistical graphs. The ANOVA method is used to analyze the individual and combined effects of atmospheric parameters on the turbulence strength (C_n^2). Different possible combined effects of W_s, T, and R_h on the turbulence strength are analyzed and examined. The calculated ANOVA results are used to determine the most dominating input variables on the output surface and those groups are only selected while optimizing the equation. Other regressors are neglected. Utmost prediction accuracy to have more fitness of the model response to the DOE is only considered while ignoring the less significant groups. All models are examined in order to find a model more suitable for predicting the turbulence strength (C_n^2) as a function of meteorological parameters.

$$\begin{aligned} C_n^2 = 1 \times 10^{-14}(&53.63 + 21.04\, W_s - 281.7\, T - 63.5\, R_h - 0.043\, W_s^2 \\ &- 0.1015\, W_s T - 0.271\, W_s R_h + 2.195\, T R_h - 0.264\, W_s^3 + 0.1992\, T^3 \\ &+ 0.0168798\, T R_h^2 + 0.000579369\, R_h^3 - 0.001449\, W_s^4 + 0.0101365\, W_s^3 T \\ &+ 0.00092494\, W_s^3 R_h - 0.00159\, W_s^2 T^2 + 0.000118693\, W_s^2 R_h^2 \\ &- 0.00265882\, T^4 - 0.000436822\, T^3 R_h - 0.000335\, T^2 R_h^2 \\ &+ 7.60425 \times 10^{-6} R_h^3 W_s - 6.82 \times 10^{-5} R_h^3 T + 1.65979 \times 10^{-6} R_h^4) \end{aligned} \tag{5.16}$$

The R^2 values of proposed turbulence strength prediction models are computed and the results proved that the model equation-*V* (5.16) shows the good correlation with the measurement values and hence, this model alone is selected for the practical implementation to predict the C_n^2 according as the function of measured local weather data. Accuracy of the proposed models is validated using the repetitive confirmatory test experiments [6,21].

5.6 Experimental validation of prediction accuracy of proposed models

The measured meteorological and visibility data are applied as the inputs to all the model equations (existing and proposed) discussed in the above sections and the prediction accuracies of all of them are intensively analyzed. A set of experimentally acquired input data and their corresponding model/measurement values are presented in this section. These results are used to test the potential and feasibility of all the selected and proposed models. The day-long period profile of weather parameters is analyzed in terms of the min–max values and standard deviation (Std). The more suitable models predicting the atmospheric attenuation (A_{att}) and turbulence strength (C_n^2) are studied. Investigations on the validation of estimated A_{att} and C_n^2 with the measurement readings are described and the accuracy of the estimation is analyzed. The meteorological profile data are acquired using designed weather monitoring station [1–4]. The weather sensor interface architectures transfer the measured weather data, in packed form/frame (header-stop bit-data-start bit-Footer), to the RS232 port of the data logging PC where the data reading MATLAB program collect the frames of weather data every second. The data-reading code also unpacks the data packet and stores only the data values. Finally, the outputs of the weather sensors are logged in the computer in radix-2 format.

Table 5.4 Measured meteorological data recording Excel work sheet

Data ID	Sample data and time	W_s (ms^{-1})	T (°C)	R_h(%)	V (km)	Pr (kPa)
⋮	⋮	⋮	⋮	⋮	⋮	⋮
41406	06-Sep-2018 10:30:06	2.20	37.0	38.0	6	100.6
41407	06-Sep-2018 10:30:07	2.40	36.0	37.0	6	100.6
41408	06-Sep-2018 10:30:08	2.30	37.4	38.0	6	100.6
41409	06-Sep-2018 10:30:09	2.50	37.0	38.0	6	100.6
41410	06-Sep-2018 10:30:10	2.30	36.0	37.6	6	100.6
⋮	⋮	⋮	⋮	⋮	⋮	⋮

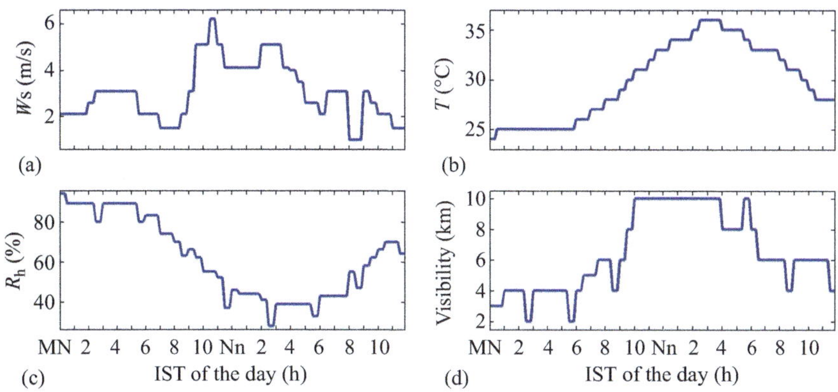

Figure 5.5 Diurnal time series meteorological measurement values

The data frames are manipulated (frame concatenation) to obtain the corresponding real time changes of the true values of weather parameters. The corresponding plots and data recording file (.xls or .doc) in real time are updated every second. A sample data logging .xls file work sheet generated on 6.09.2018 (Thursday) is given in Table 5.4. Investigation on the validation of estimated A_{att} and C_n^2 with the scintillometer readings is carried out and the accuracy of the estimation in terms of correlation is analyzed.

5.6.1 Comparison of predicted and measured A_{att} data

The local meteorological (W_s, T, R_h) and visibility (V) (using (5.1)) data are measured for every second in the diurnal period and averaged over 10 min data are recorded as shown in Figure 5.5 and the environmental changes are studied. The W_s, T, R_h, and V data are substituted in (5.6–5.9) and (5.15) as required and prediction results are compared against the measured attenuation (A_{att}) using (5.2). The selected (5.6)–(5.9) and proposed (5.15) models are implemented in the MATLAB environment. The accuracies of the prediction of the models are

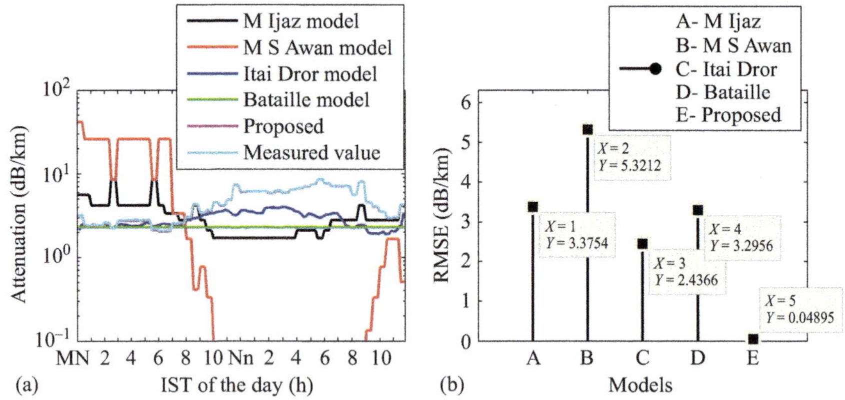

Figure 5.6 Comparison of time series atmospheric attenuation predicted and measured: (a) and RMS error values of prediction results: (b)

carefully analyzed in terms of RMS error. The profile of recorded real-time W_s, T, R_h, and V data corresponding to September 1, 2018 and prediction/measurement values of optical attenuation are described below. The weather parameter variations from 0.2 to 6.2 ms^{-1} with the Std of 1.246 ms^{-1} for W_s, 24 °C to 36 °C with the Std of 3.88 °C for T, 28% to 94% with the Std of 19.7% for R_h and 2 km to 10 km with the Std of 2.58 km for V are observed as in Figure 5.5. The Std on the R_h, T, and V are high while W_s is low. The different climatic conditions of the atmosphere observed on the day of experiment are scattered cloudy, mostly cloudy, drizzly, light drizzly, mist, and hazy.

Figure 5.6(a) shows the experimental time series plot of attenuation predicted from local meteorological (W_s, T, R_h) and visibility (V) data and measured values. The min–max values of measured attenuation are ≈2.07 dB/km and 8.54 dB/km about 02.30 AM and 5.15 PM, respectively, with the Std of 1.15 dB/km. An average of ≈7.44 dB/km attenuation is seen in the time span of Noon (Nn) to 9.00 PM due to the more fluctuating wind speed and high temperature variations and low relative humidity. In a comparison of the selected models, the prediction values of attenuation obtained using M Ijaz and M S Awan models greatly differ from the measured values with the maximum deviations of 20.32 dB/km and 12.24 dB/km, respectively, as in Figure 5.6(a) and are considered to be not accurate at the our test field. This is because these models are developed to specific situations with small data sets, therefore, they failed when applied in real-world open atmosphere. M Ijaz model gives low and high attenuation of ≈1.68 dB/km and ≈8.42 dB/km with the Std of ≈1.55 dB/km when the visibility is high: 10 km and low: 3 km in the time span of around Nn and mid-night (MN) to 7.30 AM, respectively. Reasonable correlation is seen at an average value of 2.99 dB/km about 8.00 AM to 9.00 AM. However, the pattern of exhibiting less correlation and disagreement with the measurement data over the longer periods is observed. The min–max values of attenuation predicted by M S Awan model are 1.68×10^{-5} dB/km and 41.06 dB/km with the std of 11.65 dB/km about 8.00 AM to 10.00 PM and MN to 7.45 AM, respectively, when the R_h is very low: 28% and high: 89%. The statistics are inappropriate and the model exhibits adequate deviations from the other models as well as measurement values.

The min–max values of attenuation predicted by the Itai Dror model, as in (5.8), are 1.88 and 3.97 dB/km with the std of 0.59 dB/km about 10.00 PM and 3.00 PM, respectively, when the W_s, T are normal: $\approx$2.1 ms^{-1}, $\approx$29 °C; R_h is low: 47% and W_s, T are high: 5.1 ms^{-1}, 36 °C; R_h is low: 44%. These statistics significantly differ from the measured values. At few instants of time: MN to 8.00 AM and after 11.00 PM, the prediction results hold good agreement with the measured data since the W_s is low: $\approx$1.5 ms^{-1}, T is low: $\approx$28 °C and R_h is high: $\approx$70%. In particular, the results obtained during the daytime do not match with the measured values. The min–max values of the Bataille model are 2.25 and 2.29 dB/km with the Std of 0.008 dB/km. These results repeat themselves and exhibit almost constant on all the time of the day with an average value of 2.25 dB/km. The min–max values of the proposed model are 2.02 dB/km and 8.51 dB/km with the Std of 1.95 dB/km around 7.00 AM and 9.00 AM to 7.00 PM, respectively, when the W_s and T are low: 2 ms^{-1}, 25 °C; R_h is high: 89% and W_s, T are high: 6.2 ms^{-1}, 36 °C; R_h is low: 39%. The values predicted using the proposed model, as in (5.15), could yield good correlation to the measured values at our test field and the results reveals that the proposed model exhibits the best prediction with the deviation error $\approx$0 dB/km throughout the diurnal periods of experimentation. Accuracy in prediction is validated by computing the RMS error using the predicted and measured values. The results show, as in Figure 5.6(b), that the proposed model gives very less RMS error value of 0.048 dB/km whereas other models are 3.37, 5.32, 2.43, and 3.29 dB/km.

5.6.2 Comparison of predicted and measured C_n^2 *data*

The necessary data and measured values of meteorological (W_s, T, R_h, Pr), geographical, geometrical, and received beam parameters are substituted in (5.4) and (5.5), (5.10)–(5.13) and (5.16) as required. The prediction results of C_n^2 are recorded and compared against the measured C_n^2 values. The accuracy of prediction of the models is carefully analyzed in terms of the sum of absolute (SA) error. The analysis on the profile of real-time meteorological data, predicted, and measured C_n^2 values corresponding to September 6, 2018 are presented below. The weather parameters variations from 1 to 6.2 ms^{-1} with the Std of 1.60 ms^{-1} for W_s, 26 °C to 37 °C with the Std of 3.68 °C for T, 21% to 69% with the Std of 14.83% for R_h and 100.3 kPa to 100.8 kPa with the Std of 0.16 kPa for Pr are observed from Figure 5.7. The Std on the T and R_h is high; W_s and Pr are relatively low. The different climatic conditions observed on the day of experimentation are partially cloudy, hazy, and scattered cloudy. Uneven W_s and R_h variations are observed as in Figure 5.7(a) and (c).

Figure 5.8(a) shows the experimental time series plot of C_n^2 predicted from local meteorological data and measured values. The min–max values of measured C_n^2 are $\approx 3.07\times10^{-14} m^{-2/3}$ and 3.81×10^{-13} $m^{-2/3}$ about MN to morning and

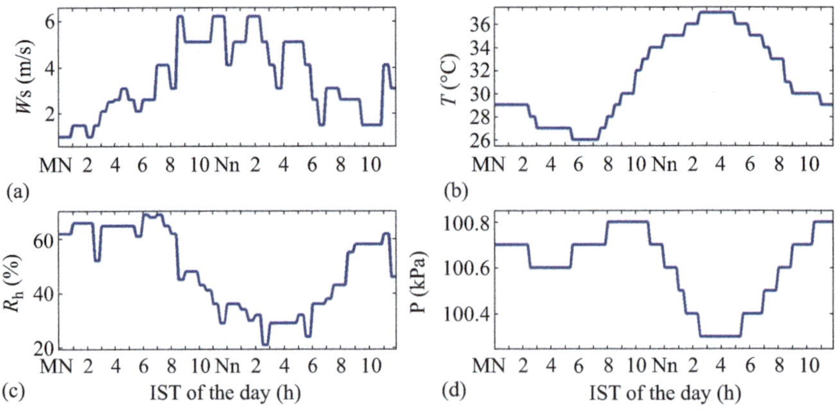

Figure 5.7 Diurnal time series meteorological measurement values

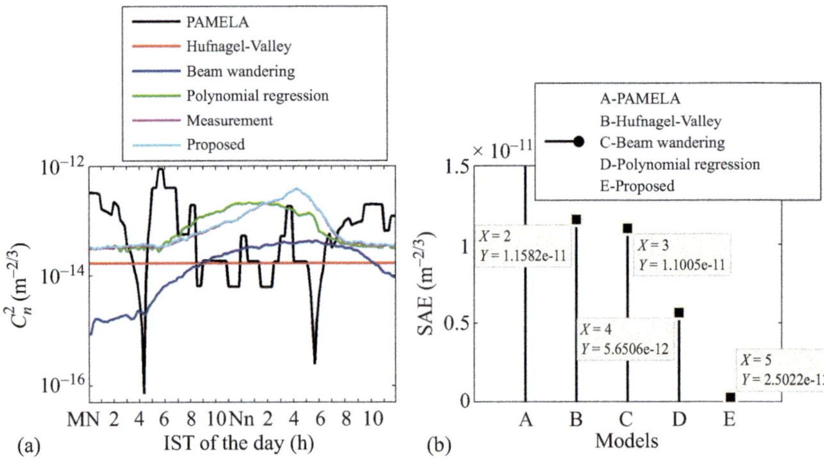

Figure 5.8 Comparison of time series atmospheric turbulence strength (C_n^2) predicted and measured: (a) and SAE value of prediction results: (b)

4.30 PM, respectively, with the Std of 9.16×10^{-14} m$^{-2/3}$ when W_s, T are low: 1 ms^{-1}, 29 °C; R_h is high: ≈62% and W_s, T are high: around 6 ms^{-1}, around 36 °C; R_h is low: 21%. The min–max values of C_n^2 predicted by the PAMELA model are 6.32×10^{-15} m$^{-2/3}$ and 8.85×10^{-13} m$^{-2/3}$ about 6.00 AM and 8.30 AM to 8.00 PM when W_s, T are normal: ≈3.1 ms^{-1}, ≈27 °C, R_h, Pr are high: 69%, 100.7 kPa and W_s, T are high: 6.2 ms^{-1}, 37 °C, R_h and Pr are low: R_h: 24%, 100.3 kPa, respectively. Local minima are seen in the early morning and late evening at 7.24×10^{-17} m$^{-2/3}$ and 2.48×10^{-26} about 4.20 AM and 5.50 PM, respectively. A greatly fluctuating turbulence pattern is predicted by the PAMELA model during 6.00 AM to 5.45 PM, since the model keeps high sensitivity to W_s. The results yield less correlation with the measurements almost all the time of the day as shown in Figure 5.8(a). The min–max value of C_n^2 predicted by the Hufnagel-Valley model is $\approx 1.65\times 10^{-14}$ m$^{-2/3}$ with the std of 1.9×10^{-29} m$^{-2/3}$, i.e., this model yields a constant C_n^2 throughout the day as in Figure 5.8(a) with the maximum deviation of 3.56×10^{-13} m$^{-2/3}$ from the measurement values.

The min–max values of C_n^2 predicted based on beam wandering are 8.69×10^{-16} m$^{-2/3}$ and 4.29×10^{-14} m$^{-2/3}$, respectively, with the std of 1.44×10^{-14} m$^{-2/3}$ about MN and 5.00 PM when W_s, T are low: 1 ms^{-1}, 29 °C; R_h, Pr are high: 62%, 100.7 kPa and W_s, T are high: 5.1 ms^{-1}, 36 °C; R_h, Pr are low: 24%, 100.3 kPa, respectively. The model predicts a smooth turbulence pattern roughly similar to measurement with a maximum deviation of 3.24×10^{-14} m$^{-2/3}$ as in Figure 5.8(a). The results became very close to the measurements at 4.03×10^{-14} m$^{-2/3}$ about 7.50 PM when W_s, T, R_h, Pr are normal: 1.5 ms^{-1}, 31 °C, 24%, 100.4 kPa. The min–max values of polynomial regression model are 2.72×10^{-14} m$^{-2/3}$ and 2.10×10^{-13} m$^{-2/3}$, respectively, with the Std of 6.60×10^{-14} about MN to 6.00 AM, 7.00 PM to MN and Nn when W_s, T are low: ≈2.6ms^{-1}, ≈28 °C; R_h is high: 78% and W_s, T are high: 6.2 ms^{-1}, 35 °C; R_h is low: 36%, respectively, as in Figure 5.7(a–c). This model exhibits a significant correlation with the measurement with the maximum deviation of 2.52×10^{-13} m$^{-2/3}$. The min–max values of the proposed model are 3.17×10^{-14} m$^{-2/3}$ and 3.82×10^{-13} m$^{-2/3}$, respectively, with the Std of 9.15×10^{-14} m$^{-2/3}$ about MN to 7.00 AM, 7.00 PM to MN and 3.00 PM when the W_s, T are low: 2.1 ms^{-1}, 28 °C; R_h is high: 66% and W_s and T keep increasing while R_h decreases as in Figure 5.7(a–c) and Figure 5.8(a). The proposed model yields a good agreement with the measurements all the time of the day. The computed SA errors using the measured and predicted values are shown in Figure 5.8(b) against the respective models and prove that the proposed model gives very less SA error of 2.5×10^{-13} m$^{-2/3}$ whereas the other models are 2.27×10^{-10} m$^{-2/3}$, 1.15×10^{-11} m$^{-2/3}$, 1.10×10^{-11} m$^{-2/3}$, and 5.65×10^{-12} m$^{-2/3}$.

These experimental studies strongly conclude that designing a more generic and location-independent models to predict the atmospheric attenuation and turbulence strength is really difficult. Therefore, building locally more suitable models is the right way to have the accurate prediction of the A_{att} and C_n^2. The prediction results of designed A_{att} (5.15) and C_n^2 (5.16) models can be used to tune the parameters of a real FSO communication system to adjust the data rate, transmitting power, wavelength selection, decide the coding technique, and/or enable the multiple transmissions/receptions etc., to establish a data link with the acceptable bit error rate during all the weather conditions.

References

[1] Oliver Bouchet, Herve Sizun, Christian Boilsrobert and Frédérique de Fornel, 'Free Space Optics: Propagation and Communication', ISTE, UK, 2016.

[2] Arockia Bazil Raj, A, Arputha Vijaya Selvi, J and Raghavan, S, 'Real-time Measurement of Meteorological Parameters for Estimating Low Altitude

Atmospheric Turbulence Strength (C_n^2)', IET- Science Measurement & Technology, vol. 8, no. 6, pp. 459–469, 2014.

[3] Sudhir Kumar Gupta, Shibu K. Mathew and Venkatakrishnan, P, 'Development of Solar Scintillometer', Journal of Astrophysics, vol. 27, no. 2–3, pp. 315–320, 2006.

[4] Larry C Andrews, Ronald L Phillips and Cynthia Y Hopen, 'Laser Beam Scintillation with Applications', SPIE, USA, 2001.

[5] Arockia Bazil Raj, 'FPGA-Based Embedded System Developer's Guide', CRC Press, USA, 2017.

[6] Arockia Bazil Raj, 'Free Space Optical Communication: System Design, Modeling, Characterization and Dealing with Turbulence', de Gruyter, Gmbh, 2015.

[7] A Arockia Bazil Raj, J Arputha Vijaya Selvi, and S Durairaj, 'Comparison of Different Models for Ground-level Atmospheric Turbulence Strength (C_n^2) Prediction with a New Model According to Local Weather Data for FSO Applications', Applied Optics, vol. 54, no. 4, pp. 802–815, 2015.

[8] Muhammad Ijaz, Zabih Ghassemlooy, Jiri Pesek, Ondrej Fiser, Hoa Le Minh and Edward Bentley, 'Modeling of Fog and Smoke Attenuation in Free Space Optical Communications Link Under Controlled Laboratory Conditions', Journal of Lightwave Technology, vol. 31, no. 11, pp. 1720–1726, 2013.

[9] Scott T Smith, 'MATLAB Advanced GUI Development', Dog Ear Publishing, USA, 2006.

[10] Muhammad Saleem Awan, Laszlo Csurgai Horwath, Sajid Sheikh Muhammad, Erich Leitgeb, Farukh Nadeem and Muhammad Saeed Khan, 'Characterization of Fog and Snow Attenuations for Free Space Optical Propagation', Journal of Communications, vol. 4, no. 8, pp. 533–545, 2009.

[11] Itai Dror, Atar, S, Shlomit Grossman and Norman S Kopeika, 'Accurate Method for Prediction of Atmospheric Transmission According to Weather', Optical Engineering, vol. 35, no. 9, pp. 2548–2555, 1996.

[12] Bataille, P, 'Analysis of the Behavior of a Telecommunications System Operated Optical 0.83 microns in the Lower Atmosphere', Ph.D. Thesis, University of Rennes, France, 1992.

[13] Steve Doss-Hammel, Eun Oh, Jennifer Ricklin, Frank Eaton, Charmaine Gilbreath and Dimitri Tsintikidis, 'A Comparison of Optical Turbulence Model', Proceedings of SPIE, Free-Space Laser Communications IV, vol. 5550, pp. 236–245, 2004.

[14] Arun K Majumdar and Jennifer C Ricklin, 'Free-Space Laser Communications: Principles and Advances: Optical and Fiber Communications Reports', Springer, USA, 2008.

[15] Carlos O Font, Mark, PJL, Chang, Eun Oh and Charmaine Gilbreath, 'Humidity Contribution to the Refractive Index Structure Function C_n^2', Proceedings of SPIE 6215, Atmospheric Propagation III, Orlando (Kissimmee), FL, pp. 1–9, 2006.

[16] Vinicius NH Silva, Andres, PL, Barbero and Ricardo M Ribeiro, 'A New Triangulation-like Technique for the Evaluation of the Refractive Index

Structure Constant (C_n^2) in Free-space Optical Links', Journal of Lightwave Technology, vol. 29, no. 24, pp. 3603–3610, 2011.

[17] Jaume Recolons, Larry C Andrews and Ronald L Philips, 'Analysis of Beam Wander Effects for a Horizontal-Path Propagating Gaussian-Bam Wave: Focused Beam Case', Optical Engineering, vol. 46, no. 8, pp. 1–11, 2007.

[18] Narottam Das, 'Optical Communications Systems', INTECH, Croatia. Available from INTECH Books [3 October 2012], 2012.

[19] Troy T Leclerc, Ronald L Philips, Larry C Andrews, David T Wayne, Paul Saucer and Robert Crabbs, 'Prediction of the Ground-level Refractive Index Structure Parameter from the Measurement of Atmospheric Conditions', Proceedings of SPIE 7685, Atmospheric Propagation VII, Orlando, Florida, USA, pp. 76850A–76858, 2010.

[20] Sergey Bendersky, Norman S Kopeika and Natan Blaunstein, 'Atmospheric Optical Turbulence over Land in Middle East Coastal Environments: Prediction Modeling and Measurements', Applied Optics, vol. 43, no. 20, pp. 4070–4079, 2004.

[21] Gary W Oehlert, A First Course in Design and Analysis of Experiments, W. H. Freeman, Minneapolis, 2000.

Chapter 6

Spectral analysis and mitigation of beam wandering using optical spatial filtering technique in FSO communication

Ucuk Darusalam[1], *Fitri Yuli Zulkifli*[2], *Purnomo Sidi Priambodo*[2], *and Eko Tjipto Rahardjo*[2]

6.1 Introduction

Free-space optical (FSO) communications are very promising and competitive technology to be applied in the future for deep-space and terrestrial communications [1,2]. The benefits of FSO are very attractive due to its large bandwidth, close to fiber-optics system, and flexibility of point-to-point communication such as microwave system. Low-cost deployment and offering high rate transmission capacity are the advantages for FSO where bit-rate transmission capacity is nearly as the optical fiber communication system. The rate transmission of 2.5 Gbps has been investigated by Nykolak *et al.* and has successfully reached the optical link distance at 4.4 km [3]. FSO also has implemented wavelength division multiplexing (WDM) which carried rate transmission on the scale of Gbps [4–6]. In [18], Song *et al.* have expanded FSO capacity using 4 WDM channels of 4×10 Gbps and has reached an optical link at 1.2 km. In [5], Jeong *et al.* have expanded the optical link using EDFA as an optical repeater, by multiplexing 8 channels of WDM with each capacity is 10 Gbps, and has successfully reached at 3.4 km. In [6], Ciaramella has made the tremendous improvement capacity by employing 40 WDM channels and has successfully transmitted 1.28 Tbps over an optical link distance at 210 m.

Besides these promising points, however, FSO technology has major weaknesses as a challenge that needs to be solved intensively to support its wide applications into many platforms of telecommunications technology. The major weaknesses are caused by the existence of static attenuation and temporal or dynamic intensity fluctuation along the propagation path. Static attenuation is caused by absorption and light scattering due to various atmospheric gases and particles in atmospheric propagation medium, while the dynamic intensity

[1]Department of Informatics Engineering, Universitas Nasional, Indonesia
[2]Department of Electrical Engineering, Universitas Indonesia, Indonesia

fluctuation is caused by atmospheric turbulence, which is due to temporal and spatial temperature gradient in atmosphere. The atmospheric turbulence produces various distributions of pockets of non-homogeneity optical medium as refraction zones. These refraction zones act as multiple lens effects. If the size of turbulences is only about a several wavelengths of the light, then it is called as micro turbulence that causes optical scattering and attenuation, where the optical axis path is considered remains the same. The temporal micro turbulence produces high frequency of noise that goes fluctuation and random at the receiver. If the size of turbulence is sufficiently large or called as macro turbulence, these causes beam wandering effect, which leads to shift optical axis from transmitter (Tx) to receiver (Rx). The beam wandering causes slow intensity fluctuation at Rx, which is called as turbulence-induced scintillation. The overall effects, spatial noise, beam wander and scintillation limit propagation distance and bandwidth of FSO [7–10].

Based on the aforementioned results, there is a clear trend line in FSO which high rate transmission through atmospheric turbulence, tends to shortening propagation path length. High rate transmission over the atmosphere is very vulnerable to be induced by turbulence maximally. An optical propagation which carry n-channels of WDM system interacts independently with atmospheric turbulence. Hence, each channel degrades maximally where the average value of signal-to-noise ratio (SNR) decreases significantly as well. Long distance for optical propagation path also contributes to higher accumulation for each channels to interact with atmospheric turbulence in random and fluctuative as well. Each WDM channel has specific center frequency or signal spectrum, which is 100 GHz spacing (ITU-T G.694.1). For n-channels of WDM that are transmitted into atmospheric turbulence, hence based on the fourth order of strong fluctuation theory, each channel of signal spectral modulates noise [11,12]. Thus, maximum noise is accumulated with the scale of turbulence and optical link distance on WDM of FSO. The strong level of turbulence and longer of optical link means maximum noise is modulated in signal spectrum. It means that the signal spectrum is deteriorated caused by this random process. Thus, noise modulation in signal spectrum exhibits the lower value of SNR for each channel of WDM [13–15]. Noise modulation in signal spectrum under the influence of atmospheric turbulence also degrades FSO performance [16,17]. Furthermore, these may shorten the optical link distance in FSO.

Turbulence-induced noise and scintillation at Rx has been studied and investigated extensively. Several techniques have been developed to model turbulence-induced noise and scintillation [18–20]. There are at least two important statistical parameters related to turbulence-induced noise and scintillation, i.e. (1) d_0, the correlation length of random intensity fluctuation and (2) τ_0, the correlation time of random intensity fluctuation [21]. Several improvements on detection techniques have been developed based on the statistical properties of turbulence-induced noise and scintillation by temporal or spatial-domain technique approaches. In temporal domain technique approach, Rx should be designed in such a way that Rx observation interval satisfies the condition $T_0 > \tau_0$, where turbulence-induced noise can be much reduced by temporal-averaging method at the Rx. In the same way, for spatial domain technique approach, Rx aperture diameter $D_0 > d_0$, such that turbulence-induced scintillation can be much reduced by spatial-aperture averaging method at Rx. Another improvement on spatial-aperture averaging method is in the form of spatial diversity method that is applying multi-Tx and multi-Rx, which is called multi-input multi-output (MIMO) and combined with Q-ary pulse-position modulation (QPPM) [22]. Several research groups have achieved developing FSO for middle distance up to 1 km with 10-Gbps bit rate [23].

The objective of this chapter is to describe the effects of absorption and scattering on the optical propagation due to the interaction with random media of atmospheric turbulence. The optical propagation phenomena which leads to random attenuation cause major problem of signal intensity fluctuation or the so-called scintillation. Beside this, random scattering of optical propagation leads to spatial and temporal variations of the signal spectrum which degrades the quality of signal modulation. By considering those deterioration effects, an optical spatial filter (OSF) is proposed in order to suppress the scintillation and temporal variation of signal spectrum. The model of the OSF that is implemented on the receiver (Rx) of FSO uses pinhole, cone reflector, and multi-mode fiber (MMF). Pinhole governs Fresnel diffraction for the incident of optical propagation from the receiver lens in order to suppressing the variance of spatial noise. Meanwhile, the deflection of optical propagation that is caused by beam wandering is collected by cone reflector radially to be guided into pinhole. Regarding the reception of signal in FSO uses direct-detection method or the so-called DD, where the signal is received directly without any optical treatment beforehand to filter the turbulence effects, the OSF which is integrated with MMF can be used as detection method as well. In this section, the OSF that is integrated with MMF has some benefits, which are noise suppression in the received of signal spectrum can be achieved optimally and the signal detection can be integrated with an optical amplifier in order to enhance the optical signal amplification in the scheme of forward-relaying FSO. The benefit of implementing the pinhole and cone reflector as the OSF is suppression for the spatial noise and temporal variance of signal spectrum goes on simultaneously within small diameter of pinhole before photodetector (PD) or MMF. This means, optical spatial filtering minimizes the effects of beam wander and spatial noise modulation in order to reduce scintillation. In fact, random scattering also causes the major problem for the performance degradation in bit-error-rate (BER). Scattering that goes random along the optical propagation path in turbulence media leads to spatial and temporal variance in the signal spectrum as well. The modulation of temporal noise in the signal spectrum can degrade the signal modulation, specifically in the WDM on FSO which carry multi-channel of wavelength. This phenomena can be minimized by the optical spatial filtering technique where receives the optical propagation within small area of pinhole can suppress the temporal variance in the signal spectrum. Thus, by implementing the OSF on the receiver of FSO, the scintillation and temporal variation in the signal spectrum can be reduced significantly prior to the signal intensity degradation which is caused by random absorption and random scattering in the turbulence media of atmosphere. The objective of this chapter is to show the effect of atmospheric turbulence that leads to image dancing or random movement of hot spot on the focus spot or focal-spot. This phenomena is very well known as beam wandering where random incident angle of the beam in receiver lens leads to random movement of hot spot at the focus spot. Meanwhile, those phenomena gather with spatial noise modulation that occurs simultaneously in the hot spot as well. Those effects caused scintillations as well. By considering those problems, the OSF is installed in the focus spot before signal intensity is received by PD or received by multi-mode fiber where the optical signal can be delivered into an optical amplifier scheme in order to suppress the effects of beam wander and spatial noise modulation.

In this chapter, the main contribution by proposing the OSF as the detection method in FSO is producing an optical method that capable to filtering noise and minimizing scintillation in real time where these advantages are accordance to the benefit of FSO as the low cost platform of telecommunication. The OSF-detection method capable to solve the weaknesses of DD method where spatial noise, beam wander and scintillation as the results of turbulence effects can be suppressed optimally before the signal detected by PD. The OSF detection method also can solve the major problem in optical amplification using Erbium-doped fiber amplifier (EDFA) where beam wandering can cause the loss signal that should be coupled into the stage of amplification process. Thus, implementing the OSF as a detection method in FSO brings significance for the improvement of performance.

6.2 Pinhole as the optical spatial filter

In this section, we propose optical spatial filter to overcome the turbulence-induced noise on FSO by using a different way. Instead of suppressing the turbulence-induced noise by using temporal and spatial-domain statistical technique approaches as explained above, we propose to use Fourier optics filter concept. Atmospheric turbulence along the propagation medium produces a high-frequency temporal and spatial fluctuated wavefront right in front of the Rx lens. The fluctuated wavefront based on Fourier optics analysis concept consists of fundamental and higher-order spatial harmonics planar wavefronts. The temporal noise mostly rides on the spatial higher-order ones. Pinhole as a low-pass Fourier optics filter and commonly used in interferometric process [24] is used to suppress the higher-order spatial harmonics (consisting noisy signal parts) of the receiving turbulence-induced noise. After pinhole filter, the detector receives clean signal, which is the fundamental order of optical carrier modulated by the information signal. In this section, suppression is limited on high-frequency fluctuation noise and not the scintillation, due to the limitation of our laboratory experiment.

The noisy wavefront as information carrier is received by the receiving lens aperture then focused onto the detector. The light-focusing process is an equal gain-combining detection process. An ideal condition is the receiver diameter $D_0 \gg d_0$, the equal gain combining detection averages the received noisy signal and can vanish the addition noise on the signal. However, mostly the condition is $D_0 < d_0$. The limitation of receiving aperture diameter causes Rx lens focusing system cannot perfectly average the received noisy wavefront in order to clean the noise. At the focal point, the light beam looks like a waist. For a detector that has an area larger than the beam waist area and located right behind the focal point, then the overall received signal that consists of temporal noises and the original signal will be detected by PD. This decreases the SNR and increases the BER.

In this section, we consider that the received noisy wavefront $U(x,y)$ at Rx is constructed by superposition of fundamental clean signal and noisy harmonic planar wavefronts, which propagates in atmospheric turbulence along the optical axis of k_z, as shown in Figure 6.1 [24]. Then it can be stated as follows,

$$U(x,y) = \iint F(\nu_x, \nu_y) \exp\left[-j2\pi(\nu_x x + \nu_y y)\right] d\nu_x \, d\nu_y, \tag{6.1}$$

$$P_r = \iint |U(x,y)|^2 dx \, dy, \tag{6.2}$$

where ν_x and ν_y are the spatial frequencies in the x- and y-directions in the X–Y plane, respectively, and $F(\nu_x, \nu_y)$ is the amplitude of the mixing spatial frequencies and $U(x,y)$ is the total electromagnetic field at point (x,y). The total received power at receiving lens is P_r and the fundamental or clean signal power is P_0. The SNR without filtering can be expressed as

$$SNR = \frac{P_0}{P_r - P_0}. \tag{6.3}$$

In this filtering process, pinhole works to block the higher-order spatial harmonics focal points. Only fundamental order wavefront can pass the pinhole to reach the receiving detector. In this case, pinhole acts as a Fourier optics low-pass spatial filter. It blocks higher-order spatial harmonics, which are overridden by temporal noise at the focal points.

Figure 6.1 illustrates the wavefront that is induced by atmospheric turbulence which the results are superposition of spatial harmonics, where each wave vectors

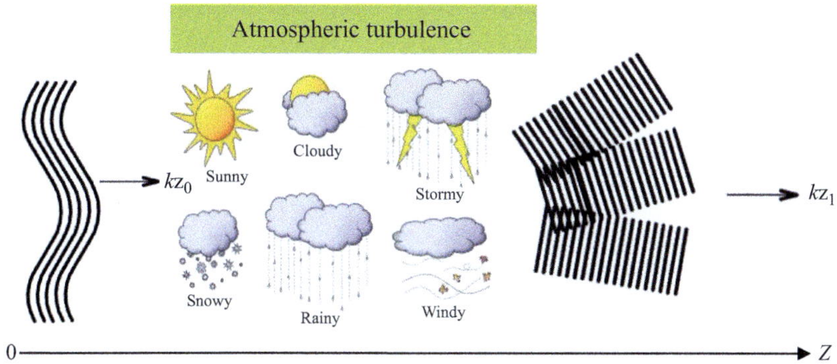

Figure 6.1 Fourier optics principles: turbulence-induced wavefront in free space is analyzed as a superposition of wavefronts, where each wavefront representing resultant vector of spatial harmonic frequencies (k_x and k_y) in the X–Y plane and propagation vector k_z

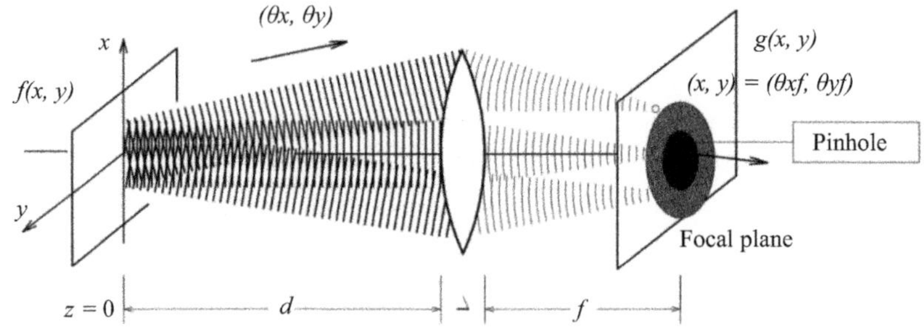

Figure 6.2 The receiving lens functions as a spatial Fourier transformer. It transforms spatially noisy wavefront becomes spatially harmonic focal points on focal point plane. Illustration of pinhole blocks high-order spatial noise overridden by temporal noise

propagates in the *Z*-axis. The higher spatial harmonic frequencies cause larger angle of propagation with refer to the optical *Z*-axis. Referring to optical propagation in FSO, the wavefront that is incident on the receiver lens has variations in temporal and spatial functions. The temporal function overrides on the amplitude or intensity of harmonic spatial cross-section function.

As shown in Figure 6.2, the receiver lens transforms the wavefront becomes spatially harmonic on focal point plane. In this filtering process, pinhole functions to block the high-order harmonic focal points. Hence, only low-order harmonics and fundamental wavefront can pass the pinhole to reach the receiving detector. In this case, pinhole acts as a Fourier optics low-pass spatial filter. Blocking the spatial higher-order harmonics or noise, which is overridden by temporal noise, means that pinhole suppresses the noise and improving SNR. Only fundamental spatial order that brings clean information signal is expected to be detected by the receiving detector. The SNR by using pinhole will drastically increase, even though the total receiving power decreases.

In order to detect the original information, signal on the optical propagation from Tx to Rx is conducted by focusing the receiving turbulence-induced wavefront with the receiving lens to the detector. We can assume that the light-focusing process is an equal gain combining detection process. Ideally, if $D_0 \gg d_0$, then the average of signal can vanish the superposition of noise. However, mostly the condition is $D_0 < d_0$. Because of aperture limitation, the receiver lens cannot perfectly averaging the statistical of spatial and temporal wavefront. Then, both of spatial and temporal wavefront is focused by the receiver lens into area of detection. The temporal and spatial fluctuated input wavefront is then focused by the receiving lens. In the focal point, the light beam will look like a waist. The spatial higher-order harmonics will be splitted from the fundamental order signal, right after focusing lens. However, the lower-order spatial frequency noises, due to paraxial to the fundamental order, are then splitted relatively close to the fundamental waist. For a detector that has area more than the beam waist area and located right behind the focal point, all of temporal noises riding on various spatial

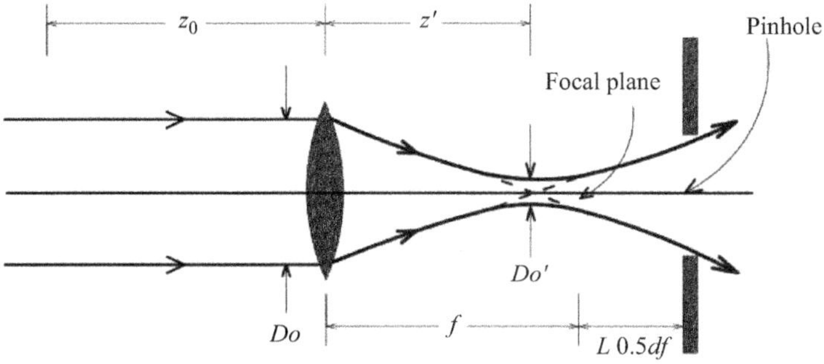

Figure 6.3 Receiving lens and pinhole alignment works as a Fourier optics transformer and low-pass filter. Pinhole is located behind the focal point to block the spatial higher-order harmonics components overridden by the temporal noise

harmonic orders along with the original signal will be detected by the detector. This will decrease the SNR. In this section, we propose a temporal and spatial noise suppression method by blocking all the spatial higher-order harmonics, which are overridden by temporal noise by using low-pass Fourier spatial filter in the form of pinhole as illustrated on Figure 6.3.

Pinhole as the OSF is located behind the focal point. The pinhole diameter has to be equal or slightly larger than the beam waist, in order that the whole clean signal can be detected, while the spatial higher-order harmonics or noisy signal is blocked. The relationship between receiving lens and beam waist radius at focal-point is formulated in the following equation (6.4),

$$D_0' \approx \frac{4\lambda}{\pi D_0} f, \tag{6.4}$$

where D_0 is the receiving beam or wavefront diameter right in front of the receiving lens. D_0' is the beam waist diameter at the focal point. Equation (6.3) shows that if the receiving beam diameter gets larger with constant focal distance f, then proportionally decreases the beam waist diameter at focal point. Moreover, longer focal point f with constant receiving beam diameter causes enlarging beam waist diameter at the focal point. For instance, if the receiver lens diameter is 5 cm and the distance of focal point is 15 cm, then the beam waist diameter is 3.82λ, where $\lambda = 1{,}550$ nm, Based on the experiment, the pinhole diameter of 3.82λ is very difficult for optical alignment. This very small diameter of pinhole faces problem when beam wandering also rises as well. Regarding the small diameter of pinhole, to anticipate the difficulty in optical alignment and beam wander effect that sometimes rises in FSO, it is recommended to locate the pinhole not exactly right on the focal point, however, a little shifted in a 0.5 of depth of focus spot ($L_{0.5df}$), and pinhole diameter is enlarged at minimum of $\sqrt{2}$ times, as stated in (6.5),

$$L_{0.5df} \approx \frac{\pi D_0'^2}{4\lambda}. \tag{6.5}$$

For the above typical example, the pinhole is suggested to be located at a distance of about 11.5λ from the focal point with pinhole diameter at least 5.4λ. Based on our experience, pinhole diameter ranging from 6 to 30λ is an ideal condition to be exercised.

To estimate the SNR gain due to the pinhole effect and to simplify derivation, it we should assume that the noise power is distributed equally from low to high spatial frequency noise. SNR gain is defined as an SNR comparison between different pinhole diameters. Assuming that the fundamental clean signal waist radius is $R_0 = 0.5\,D_0$ at the focal plane of the receiving lens and R_1 is small and R_2 is large pinhole radii, by recalling (6.3), then the SNR gain can be formulated as (6.6),

$$SNR_{\text{Gain}} = \frac{R_2^2 - R_0^2}{R_1^2 - R_0^2}. \tag{6.6}$$

In order to estimate BEP, we assume that the turbulence channel is a log-normal distribution model [25,26] as a representation of weak-turbulence inside the propagation simulator box. The spatial spectrum distribution as a result of Fourier optical transformation by the receiving lens should represent the probability density distribution of the receiving signal. The modulation is on–off keying (OOK). By assuming that the noise due to turbulence is a Gaussian distribution and larger than the receiver noise figure, the BER improvement or gain formula can be derived from [27]. The probability of error when receiving a transmitted "1" is,

$$\begin{aligned} P(r|\text{“1”}) &= \int_{-\infty}^{\frac{1}{2}\sqrt{E_b}} p(r|\text{“1”})dr = \frac{1}{\sqrt{\pi N_0}}\int_{-\infty}^{\frac{1}{2}\sqrt{E_b}} e^{-\left(r-\sqrt{E_b}\right)^2/N_0}\,dr \\ &= \frac{1}{\sqrt{2\pi}}\int_{\sqrt{\frac{E_b}{2N_0}}}^{\infty} e^{-x^2/2}\,dx = Q\left[\sqrt{\frac{E_b}{2N_0}}\right] \end{aligned} \tag{6.7}$$

The probability of error when receiving a transmitted "0" is

$$\begin{aligned} P(r|\text{“0”}) &= 2\int_{\frac{1}{2}\sqrt{E_b}}^{\infty} p(r|\text{“0”})\,dr = \frac{2}{\sqrt{\pi N_0}}\int_{\frac{1}{2}\sqrt{E_b}}^{\infty} e^{-(r)^2/N_0}\,dr \\ &= \frac{2}{\sqrt{2\pi}}\int_{\sqrt{\frac{E_b}{2N_0}}}^{\infty} e^{-x^2/2}\,dx = 2Q\left[\sqrt{\frac{E_b}{2N_0}}\right], \end{aligned} \tag{6.8}$$

where $P(r|\text{“1”})$ and $\text{p}(r|\text{“1”})$ are probability amplitude error distribution functions for transmission "1" and "0," respectively. Q or complementary error function is the tail probability in a Gaussian random variable. E_b is the received bit energy and N_0 is the noise intensity due to turbulence or signal covariance at the focal plane of the receiving lens as stated in (6.9),

$$N_0 = \sigma_N^2 = \frac{P_r - P_0}{\pi\left(R^2 - R_0^2\right)}, \tag{6.9}$$

where P_r is the total received power and P_0 is the fundamental clean signal and R and R_0 are radii of beam waist of total signal and fundamental signal, respectively, at the focal plane of the receiving lens.

Since the signal "1" and "0" are equally likely to be transmitted, the overall average probability of error is computed as

$$P_e = \frac{1}{2}P(e|\text{“}1\text{”}) + \frac{1}{2}P(e|\text{“}0\text{”}) = \frac{3}{2}Q\left[\sqrt{\frac{E_b}{2N_0}}\right]. \tag{6.10}$$

The probability of error P_e or sometimes is called as bit error probability (BEP) is the expectation value of the BER. The value of E_b/N_0 is in agreement with the SNR value. Then (6.10) can be written as

$$BEP = \frac{3}{2}Q\left[\sqrt{\frac{1}{2}SNR}\right]. \tag{6.11}$$

The BEP improvement (BEP_{Gain}) of the small radius pinhole R_1 in comparison to the large radius pinhole R_2 is written as stated in (6.12),

$$BEP_{\text{Gain}} = \frac{\frac{3}{2}Q\left(\sqrt{\frac{1}{2}SNR_1}\right)}{\frac{3}{2}Q\left(\sqrt{\frac{1}{2}SNR_2}\right)} = \frac{Q\left(\sqrt{\frac{E_b}{2N_o\pi\left(R_1^2-R_0^2\right)}}\right)}{Q\left(\sqrt{\frac{E_b}{2N_o\pi\left(R_2^2-R_0^2\right)}}\right)}. \tag{6.12}$$

In order to prove the concept of pinhole as spatial filtering, the experiment is performed using laser He–Ne as the source of optical propagation. The purpose of this experiment is to show the noise suppression effect by using pinhole visually. The reason we use the wavelength of 632.8 nm and not 1,550 nm is because of visuality. In this experiment, we use He–Ne laser of 632.8 nm wavelength with output power 10 mW. The total length of the experiment setup is about 150 cm on top of the optical table. The size of turbulence simulator box is 50 cm × 30 cm × 30 cm and sufficient to simulate the fog and local turbulence and supported by programmable speed exhaust fan. Figure 6.4 shows the experiment setup for turbulence-induced noise suppression by pinhole.

We enlarge laser beam diameter from 2 to be 3 cm such as in real FSO. The enlarged collimated beam is treated with hot water vapor and shocking air pressure to resemble such as a long-distance turbulence-induced beam propagation. The receiving lens focuses the incoming turbulence-induced wavefront onto the focal plane. In order to prove the suppression effect on the higher spatial orders of the turbulence-induced wavefront, we set two different measurements, first without pinhole (no spatial noise suppression) and second incorporated with pinhole of 15 μm diameter. The observation screen is extended to be located 30 cm from the focal point to assure for better picture quality. The pinhole is located about 20–25 μm behind the focal point. Figure 6.5(a) shows the image on observation screen as a result beam propagation through the turbulence medium and treated without pinhole. As shown on the Figure 6.5, the fundamental order is superposed by the spatial

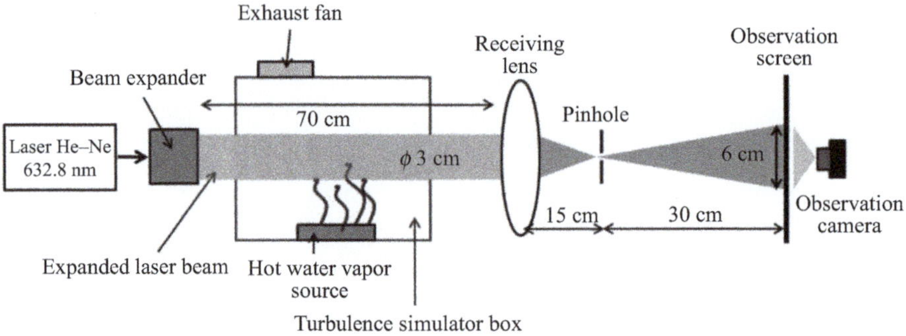

Figure 6.4 Experiment setup for illustrating the turbulence-induced noise suppression by pinhole

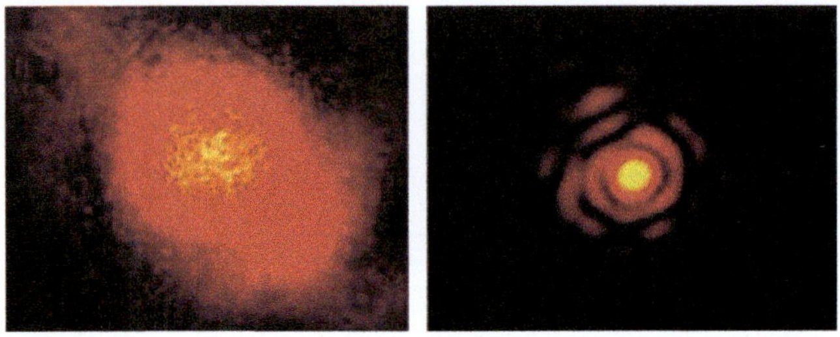

Figure 6.5 The observation images as the results of optical propagation that captured at the observation screen where shows the two conditions of experiments

Table 6.1 Average power measurement in reference and turbulence chambers, where Tx power (after EDFA) is 23.5 dBm and P_i is the received signal power [28]

Pinhole (μm)	**Turbulence**	**P_i (dBm)**
No pinhole	No	15.0
50	No	12.5
50	Yes	−2.1
30	No	12.0
30	Yes	−3.5
20	No	11.5
20	Yes	−5.0
10	No	11.0
10	Yes	−6.0
5	No	2.3
5	Yes	−12.0

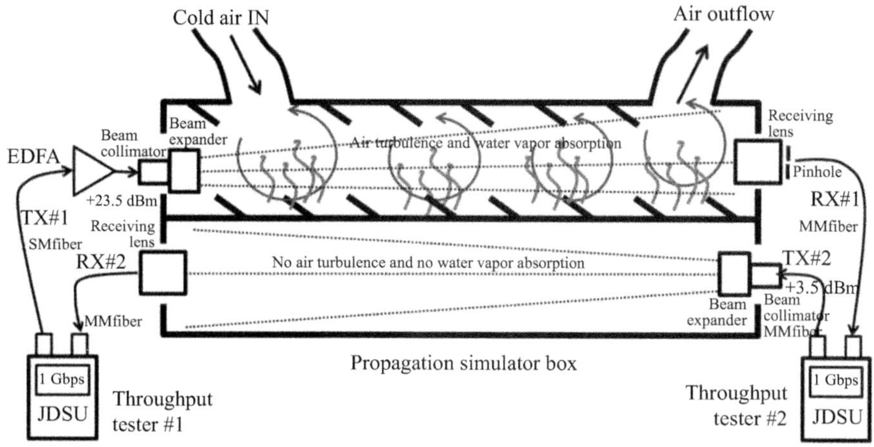

Figure 6.6 Experiment setup for turbulence-induced noise by pinhole, a Fourier optic low-pass filter. The quality of BER is measured by a pair of throughput testers, JDSU type: smart class ethernet

higher-order harmonics, which overridden by temporal noise. Figure 6.5(b) shows the image on observation screen, where the spatial higher order harmonics are blocked by 25 μm pinhole. The clean bright circle in the center of Figure 6.5(b) is the fundamental order without noise. The dark-bright pattern surrounding the bright fundamental circle is airy effect due to pinhole usage.

This visual inspection experiments show that pinhole is very effective to block the spatial higher-order harmonics that are overridden by temporal noise. It is a fact that the total power of the receiving signal is very less for the system incorporating pinhole, however it will have a lower BER in comparison to the system without pinhole. In the following, the second experiment will prove this fact.

This second setup is designed for wavelength 1,550 nm, which is an ideal wavelength for terrestrial FSO. This second experiment is intended to prove the improvement of BER performance for FSO system using pinhole filter. Figure 6.6 is the setup of the experiment. The propagation simulator box as shown on Figure 6.6 consists of two types of propagation channels, one is affected by air turbulence due to air flow through obstacles and water vapor (as shown as upper part). The second one is not affected by turbulence unless beam divergence less than 5° (the lower part). The overall dimension of the box is 4 m × 1 m × 0.5 m. The inflow air has temperature of 14 °C, while the hot water vapor has temperature of nearly 100 °C, sufficient to create multiple lens scattering and absorption effects. For this experiment, we assume it is sufficiently to model a medium turbulence condition that may affect the FSO.

In order to prove the enhancement on detection quality by using pinhole, in this experiment, the comparison between non-pinhole detection and various diameter of pinhole detections is investigated. Various diameters of pinhole are used, i.e., 50, 40, 30, and 20 μm and located about 20 μm behind the focal point.

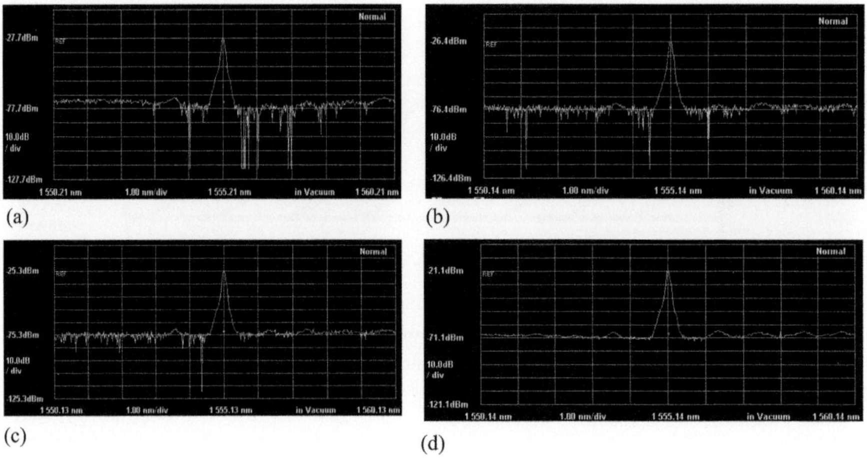

Figure 6.7 Signal spectrum measurements for FSO using pinhole as spatial filtering: (a) $D_{P1} = 50.0\,\mu m$; (b) $D_{P2} = 40.0\,\mu m$; (c) $D_{P3} = 30.0\,\mu m$; and (d) $D_{P4} = 20.0\,\mu m$

The resulting signal spectrum measurements for the pinhole diameters D_{P1}, D_{P2}, D_{P3}, and D_{P4} are shown in Figure 6.7.

As shown in Figure 6.7, signal spectrum that is received by PD is recovered by the pinhole. Thus, the temporal variations that are modulated in signal spectrum of optical propagation caused by turbulence effects is filtered out optimally. It is also shown that signal spectrum tends to be recovered optimally as the pinhole diameter goes smaller where noise suppression achieve optimum for pinhole diameter, $D_{P4} = 20\,\mu m$. The idea of using pinhole as the optical spatial filter has shown good results to filtering noise as shown by Figure 6.7 as well. Small diameter of pinhole has successfully blocked the spatial noise in the focus spot. By this blocking mechanism, noise modulation in the signal spectrum of FSO can be achieved in minimum where it is shown clearly on the result of measurement on signal spectrum at $D_{P4} = 20\,\mu m$.

Figure 6.8 shows the experiment results, when various sizes of pinhole are applied. It shows that when pinhole diameter getting smaller, then the performances of BER are improved (to a lower value). In comparison to the larger diameter of pinhole, the smaller diameter of pinhole filtered out spatial noise optimally, it proves our proposed hypothesis that the diameter of pinhole should consider (6.4) and (6.5).

Based on the results of BER measurement (Figure 6.8), it can be predicted that that BER still improves until it reaches an optimal diameter of 9 μm as shown in the dashed line in Figure 6.8. If pinhole diameter is less than the optimal diameter, then BER degrades to a higher value. The reason for this condition is caused by signal power that is resulted from the larger diameter of pinhole is lower. Then the lower signal that is detected by PD will compete with noise figure of PD. Figure 6.8 also shows the experiment measurement and theoretical results, when various sizes of pinhole are applied. Figure 6.5 shows both measurement results and theoretical

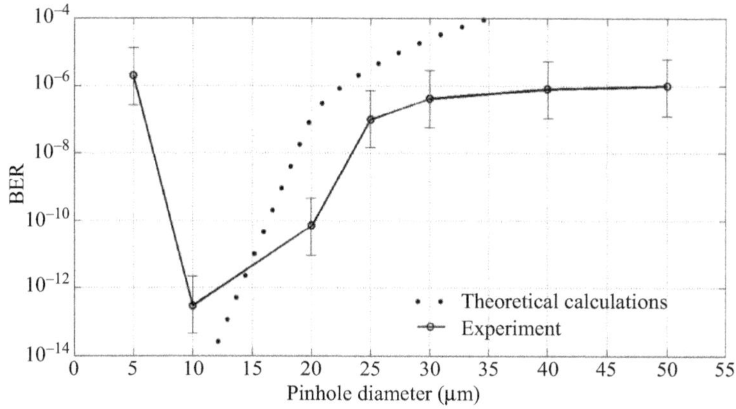

Figure 6.8 BER measurement and theoretical calculations vs. various pinhole diameters. The theoretical calculations use the following assumptions: $E_b = 8.1 \times 10^{-12}\ W^{-s}$; $N_o = 7.0 \times 10^{-9}\ W/mm^2$ *and 1 Gbps bit rate*

calculations based on (6.11). It also emphasizes our proposed hypothesis that small diameter of pinhole filtered out spatial noise effectively in comparison to the larger diameters of pinhole. It proves our proposed hypothesis that smaller pinhole diameter will block more the spatial higher-order harmonics, which are overridden by temporal noise, in comparison to the larger diameters. It agrees with the theoretical derivation as well. However, there is a situation that contradictive with our hypothesis, when pinhole diameter is less than the optimal diameter (in this case, the optimal diameter is 6 μm), then the BER degrades to higher value.

The weakness of pinhole implementation for optical spatial filtering is in the condition when beam wander occurs where cannot anticipate high deviation of focus spot around the optical axis. Pinhole filtering system is not immune to the beam-wandering effect; the system will be temporally misaligned and produce the scintillation effect. In order to take the benefits of pinhole advantages in suppressing high-order spatial noises and to avoid the weakness of pinhole, the OSF needs to be improved with an optical element before the pinhole could anticipate the beam-wandering effects. However, a new idea of propagation noise suppression in FSO, by using a pinhole as the OSF, has been presented. By experiment, it has been proven to improve the BER performance in medium turbulence-induced scintillation propagation from 10^{-7} down to 10^{-11} (up to 10^4 BER improvement), and even more.

6.3 Pinhole and cone reflector as the optical spatial filter

In this section, the OSF that is composed of pinhole and cone reflector has been proposed for suppression in the signal intensity fluctuation and noise reception before PD or simply as detection method to complement the DD [28–30]. The OSF

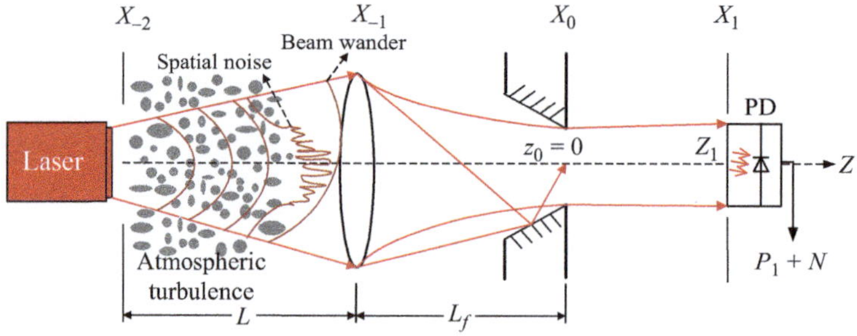

Figure 6.9 The optical propagation induced by atmospheric turbulence which modulates noise in the signal spectrum is filtered by the OSF at X_0

is designed to minimize the noise bandwidth and fluctuation of signal intensity through suppression of beam wander and spatial noise, hence fluctuation of temporal noise can be limited to a small area of reception which is the pinhole. In order to achieve this process, pinhole governs Fresnel diffraction in a narrow region. It is designed to minimize the noise that is modulated in the focus spot. Cone reflector governs directed reflectance for beam wandering into pinhole where incident angle goes random. Through this, the OSF performs simultaneous suppression of beam wander and spatial noise in order to produce the signal spectrum with minimum noise modulation. Furthermore, in order to investigate the results of filtering, we compare the signal spectrum from the OSF and DD method.

The optical propagation that is induced by atmospheric turbulence is illustrated in Figure 6.9. It is transmitted out from laser sources, experiences beam spreading, diffraction, absorption, and scattering that undergoes random. This process leads to the scintillation effect, spatial noise, and beam wander [28–30]. The optical propagation interacts stochastically with random media of atmospheric turbulence which is composed of random outer- and inner-scale dimension of the refraction index. Through this process, when optical propagation reaches on the receiver lens, it is no longer originally as it was transmitted out by laser source or transmitter. Spatial noise that is caused by random scattering through atmospheric turbulence along the propagation path is modulated. Higher accumulation of spatial noise along the propagation path leads to noise modulation in signal spectrum. Thus, focus spot from receiver lens should be recovered in order to maintain the signal spectrum as it was transmitted originally from the laser source at the transmitter.

Based on the fourth order of strong fluctuation theory, the signal spectrum of optical propagation for path length L and coincidence on X_0 is [31],

$$S_I(\omega) = \frac{4}{\omega_t}\int_0^{\infty} B_I\left(\frac{s}{\omega_t}, -L - L_f\right)\cos\left(\frac{\omega s}{\omega_t}\right)ds, \tag{6.13}$$

where ω_t represents the transition frequency at which the signal spectrum begins to decay under weak fluctuations within the Fresnel zone and B_I is the temporal covariance function of Gaussian beam that represents spatial noise modulation by atmospheric turbulence,

$$B_I\left(\frac{s}{\omega_t}, -L - L_f\right) = exp\left[\frac{0.49\sigma_R^2}{\left(1 + 1.11\sigma_R^{12/5}\right)^{7/6}} F_1\left(\frac{7}{6};\ 1;\ -\frac{k\rho^2\eta_X}{4L}\right)\right.$$
$$\left. + \frac{0.50\sigma_R^2}{\left(1 + 0.69\sigma_R^{12/5}\right)^{5/6}} \left(\frac{k\rho^2\eta_Y}{L}\right)^{5/12} K_{5/6}\left(\sqrt{\frac{k\rho^2\eta_Y}{L}}\right)\right] - 1, \tag{6.14}$$

where $\rho = V_\perp \tau$ is the spatial frequency that is induced by the mean of wind speed at the period of temporal frequency, σ_R^2 is the Rytov variance, and F_1 is the hypergeometric function where η_X and η_Y are the log-irradiance variance for outer- and inner-scale for Gaussian-beam wave, respectively.

The OSF is designed to suppress the beam wander and spatial noise effects, where fluctuation of noise in signal spectrum is limited to the pinhole diameter. Through localization of random spatial and temporal noise in pinhole diameter, noise modulation in signal spectrum can be minimized. Term of noise bandwidth B_I as shown in (6.13) is collected by direct reflectance of cone reflector into pinhole diameter. The mean irradiance for signal spectrum, as the output of OSF, is computed as [29],

$$\langle I_1(r_1, z_1 \approx 0, \phi_i)\rangle = \frac{1}{4} I_{-1}^0(0, -L_f)\frac{W_G^2}{W_0^2} SR \exp\left(-SR\frac{2(\Delta r_0)^2}{W_0^2}\right)\cos(\phi_i)$$
$$\times\left[1 - 2\cos(\nu)J_0(\nu) + J_0^2(\nu)\right], \tag{6.15}$$

where brackets of $\langle\cdot\rangle$ denotes mean value, r_0 is the radius coordinate at X_0, W_G is the effective aperture radius of receiver lens, W_0 is the focus spot radius, ϕ_i is the reflectance angle for beam wander from cone reflector into pinhole diameter with respect to the optical axis Z as shown in Figure 6.10, J_0 is the Bessel function of the

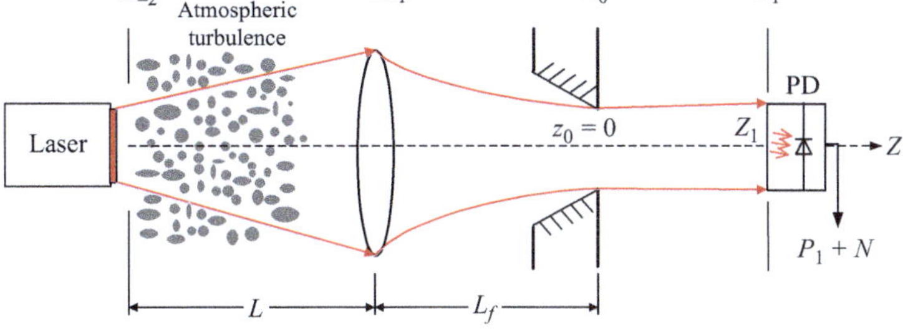

Figure 6.10 Optical system configuration of the OSF for suppression of beam wander and spatial noise effects on optical propagation that is induced by atmospheric turbulence

first kind, $I^0_{-1}(0, -L_f) = W^2_{-2}/W^2_{-1}$ is the free-space irradiance of optical propagation that incident on the receiver lens of X_{-1} where W_{-2} is the transmitted light radius from laser source on X_{-2} and W_{-1} is the incident light radius on receiver lens, $SR = 1/\left(1 + 1.63\sigma_R^{12/5}(L)\Lambda_{-1}\right)$ is Strehl ratio, $\Lambda_{-1} = 2L/kW^2_{-1}$ is the effective beam parameter of incidence light, $\sigma_R^{12/5}(L) = (1.23C_n^2k^{7/6}L^{11/6})^{6/5}$ is the Rytov variance for propagation path length L, $v = [2\pi\Delta r_0/\lambda z_1]r_1$ is the spatial frequency at radius r_1 on X_{-1} as the function of spacing distance z_1, and $k = 2\pi/\lambda$ is wave number, respectively. Based on (6.13), ω_t which represents noise bandwidth in signal spectrum is minimized by cone reflector and pinhole. Thus, signal spectrum is minimum for noise modulation.

The mechanism of noise suppression in the signal spectrum can be described as shown in Figure 6.10. The configuration of optical system for the OSF is shown in Figure 6.11. It is composed of four planes that are transmitter plane X_{-2}, receiver lens plane X_{-1}, the OSF plane X_0, and receiver plane X_1. The origin $z_0 = 0$ is located on X_0. The optical propagation from X_{-2} into X_{-1} that directs into Z experiences modulation of atmospheric turbulence. It reaches on X_{-1} where fluctuation of signal intensity is mainly caused by beam wander and spatial noise effects. Beam wandering is fluctuation of angle-of-arrival on incident beam at receiver lens of X_{-1} that leads random displacement of focus spot around the optical axis. Spatial noise is fluctuation of phase distribution on wavefront that is incident on receiver lens. The incidence of optical propagation on X_{-1} is focused by the receiver lens onto pinhole of the OSF. The mean irradiance of focus spot that is coincident on the pinhole under the atmospheric turbulence [29] is given as

$$\langle I_0(r_0, 0)\rangle = \frac{W_G^2}{W_0^2} SR\, I^0_{-1}(0, -L_f)\exp\left(-SR\frac{2r_0^2}{W_0^2}\right), \tag{6.16}$$

where the brackets of $\langle\cdot\rangle$ denotes mean value. r_0, $I^0_{-1}(0, -L_f)$, W_G, W_0, and SR are the radius coordinates of pinhole on X_0, on-axis irradiance for free-space on X_{-1}, the effective aperture radius of receiver lens, focus spot radius, and Strehl ratio,

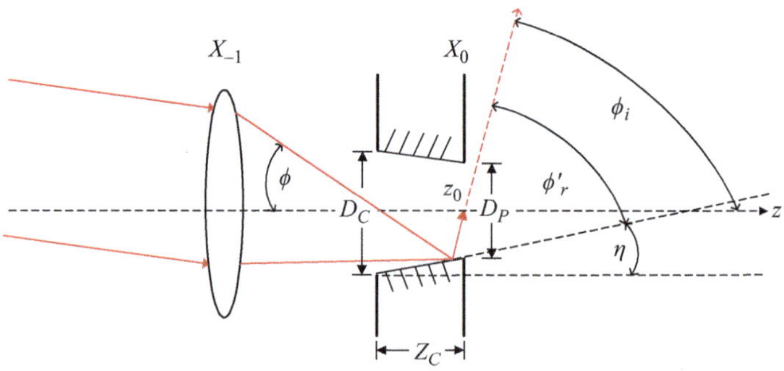

Figure 6.11 Cone reflector of the OSF for suppression angle of beam wander ϕ to produce incident angle ϕ_i at the output of pinhole diameter

respectively. Pinhole radius is $0 \leq r_0 \leq D_P/2$, where D_P is the pinhole diameter, as shown in Figure 6.11. In (6.16), $\langle I_0(r_0, 0)\rangle$ is the mean of signal intensity that modulates beam wander and spatial noise effects.

By the presence of beam wander and spatial noise effects in optical propagation, signal intensity $\langle I_1\rangle$ degrades in maximum. Thus, PD produces minimum signal power P_1 and maximum noise. In order to suppress spatial noise, the OSF that consists of pinhole governs diffraction on X_0. It produces near-field distribution of Fresnel diffraction at z_1 on X_1. The mean of signal intensity that is fundamental component of diffraction [29] is given as

$$\langle I_1(\mathbf{r_1}, z_1 \approx 0)\rangle = \frac{1}{4}\frac{W_G^2}{W_0^2} SR\, I_{-1}^0(0, -L_f)\exp\left(-SR\frac{2r_0^2}{W_0^2}\right)\left[1 - 2\cos(v)J_0(v) + J_0^2(v)\right] \tag{6.17}$$

where J_0 is the Bessel function of the first kind and $v = [2\pi(D_P/2)/\lambda z_1]r_1$. In (6.17), spatial noise on focus spot is suppressed by last term $\left[1 - 2\cos(v)J_0(v) + J_0^2(v)\right]$.

Cone reflector reflects angle of beam wander ϕ as shown in Figure 6.11. It has a radial mirror with diameter D_C, width Z_C, and the tilt angle η. ϕ under atmospheric turbulence is random. If focus spot is deflected at ϕ, then the radial mirror reflects into complementary angle ϕ_r' on pinhole radially. It produces the incident angle ϕ_i with respect to the optical axis Z at the output of pinhole between X_0 and X_1. ϕ_i is given by

$$\phi_i = \phi_r' + \eta = \phi + 2\eta = \phi + 2\tan^{-1}\left(\frac{D_C - D_P}{2Z_C}\right), \tag{6.18}$$

where η that is tilt angle of cone reflector is constituted by pinhole- and receiver lens-diameter with respect to the optical axis Z. ϕ in (6.10) ranges from $\phi_{\min}$ to $\phi_{\max}$ as given by

$$\phi_{\min} = \phi^0 + \tan^{-1}\left(\frac{\frac{1}{2}(D_G - D_P)}{L_f}\right), \tag{6.19}$$

$$\phi_{\max} = \phi_{\min} + \tan^{-1}\left(\frac{\frac{1}{2}(D_G - D_C)}{L_f - Z_C}\right), \tag{6.20}$$

where ϕ^0, D_G, and L_f are the maximum angle of focus spot at nonturbulent atmosphere, hard diameter of receiver lens, and length of focus spot, respectively. Thus, cone reflector with range of ϕ collects random displacement of focus spot to be reflected onto pinhole diameter. By considering ϕ_i in (6.18), $I_1(r_1, z_1 \approx 0)$ in (6.17) is factorized by $\cos(\phi_i)$. It is stated in (6.21) as

$$\langle I_1(r_1, z_1 \approx 0, \phi_i)\rangle = \frac{1}{4}\frac{W_G^2}{W_0^2} SR\, I_{-1}^0(0, -L_f)\exp\left(-SR\frac{2r_0^2}{W_0^2}\right) \times \cos(\phi_i)\left[1 - 2\cos(v)J_0(v) + J_0^2(v)\right]. \tag{6.21}$$

By assuming that X_0 and X_1 are almost coincident such that $z_0 \approx z_1$ and $r_0 \approx r_1$, PD receives instantaneous signal intensity $I_1(r_1, z_1 \approx 0, \phi_i)$ from the pinhole. Regarding this, PD receives the fundamental component of diffraction to produce optimum signal power P_1. The mean of the signal power $\langle P_1 \rangle$ is stated as

$$\langle P_1(r_1, z_1 \approx 0, \phi_i) \rangle = \frac{\pi}{2} \frac{W_G^2}{W_0^2} SR\, I_{-1}^0(0, -L_f) \exp\left(-SR\frac{2r_0^2}{W_0^2}\right) \cos(\phi_i) B \quad (6.22)$$

where B is the circular aperture function of the pinhole that is stated as

$$B = \frac{r_1^2}{2} - \left[\frac{2A^2 r_1^2}{3} \cos(Ar_1) J_0(Ar_1) + \frac{A^2 r_1^2 \sin(Ar_1) - Ar_1 \cos(Ar_1)}{3} J_1(Ar_1)\right] + \left[\frac{A^2 r_1^2}{2}\left(J_0^2(Ar_1) + J_1^2(Ar_1)\right)\right], \quad (6.23)$$

where $A = [2\pi(D_P/2)/\lambda z_1]$ is the spatial frequency of the focus spot that is localized in pinhole diameter and J_1 is the Bessel function of the second kind.

In comparison to the DD method [29], $\langle P_1 \rangle$ in (6.22) has minimum fluctuation of signal intensity. This is shown by the terms B and cos (ϕ_i) in (6.22). B acts as the modulator function to envelope spatial noise on the area of pinhole. Thus, spatial noise is localized in a narrow area. cos (ϕ_i) acts as the compensator function for fluctuation of signal intensity that is caused by random displacement of focus spot. Based on those operations, signal intensity $\langle I_1 \rangle$ that is received by PD goes higher. It means that pinhole operates as low-pass filter to spatial noise. Meanwhile, cone reflector compensates signal intensity to be higher through reflection of focus spot that undergoes random radially. The reflection from cone reflector guides random displacement of focus onto pinhole. Through this, the OSF suppresses beam wander and spatial noise effects simultaneously in order to minimize fluctuation of signal intensity.

Regarding the enhancement of the received signal power $\langle P_1 \rangle$ by PD in (6.21) as the compensation for confinement of signal intensity fluctuation in narrow region of pinhole through reflectance of cone reflector, $\langle SNR \rangle$ degradation can be minimized by the OSF. $\langle SNR \rangle$ is increased by suppression of signal power ratio, $P_1^0/\langle P_1 \rangle$. The OSF minimizes those ratio in order to enhance $\langle SNR \rangle$ as given below [30],

$$\langle SNR \rangle = \frac{SNR^0}{\left(\frac{P_1^0}{\langle P_1 \rangle} + \sigma_I^2(D_G)\left(SNR^0\right)^2\right)^{\frac{1}{2}}} \quad (6.24)$$

where SNR^0 and P_1^0 are the optimum values of SNR and signal power P_1 in the absence of atmospheric turbulence, respectively. In (6.24), the irradiance flux variance $\sigma_I^2(D_G)$ has a range value of 0–1 for weak to strong turbulence level. Frequently, signal power P_1 falls below P_{Th}. The OSF enhances value of

$\langle I_1(r_1, z_1 \approx 0, \phi_i)\rangle$. Thus PD produces signal power P_1 beyond P_{Th}. Signal power ratio $P_1^0/\langle P_1\rangle$ is decreased by the OSF as

$$\frac{P_1^0(\boldsymbol{r}_1, z_1 \approx 0)}{\langle P_1(\boldsymbol{r}_1, z_1 \approx 0, \phi_i)\rangle} = \frac{W_0^2}{SR\exp\left(-SR\frac{2(\Delta r_0)^2}{W_0^2}\right)\cos(\phi_i)\,B}. \tag{6.25}$$

The optimum suppression of $P_1^0/\langle P_1\rangle$ produces $\langle SNR\rangle$ that is approximated to SNR^0. In comparison to DD that $P_1^0/\langle P_1\rangle \cong 1/SR$, this ratio is decreased to a minimum value by the OSF as shown in (6.24) [29]. Thus, the OSF gives an advantage to minimize the signal power ratio $P_1^0/\langle P_1\rangle$ by suppressing the beam wander and spatial noise effects simultaneously.

Generally, probability density function uses the gamma–gamma distribution $p_I(u)$ as the channel model for FSO at atmospheric turbulence. $p_I(u)$ is the probability of signal unit $u = S/i_S$ that is influenced by α and β as the representations of small- and large-scale of atmospheric turbulence, respectively. $\langle SNR\rangle$ enhancement by the OSF minimizes the probability of $\langle BER\rangle$. In [29], $\langle BER\rangle$ can be suppressed into lower order through $\langle SNR\rangle$ enhancement. Furthermore, $\langle BER\rangle$ of OOK modulation method is derived with regard to the integral solution in [32]. $\langle BER\rangle$ is stated below,

$$\begin{aligned}\langle BER\rangle &= \frac{1}{4}p_I^2(u)erfc\left(\frac{\langle SNR\rangle u}{2\sqrt{2}}\right)\\ &\quad + \frac{1}{\langle SNR\rangle^2}erf\left(\frac{\langle SNR\rangle u}{2\sqrt{2}}\right) - \frac{u\sqrt{2}}{\langle SNR\rangle 2\sqrt{\pi}}exp\left(-\frac{\langle SNR\rangle^2 u^2}{8}\right),\end{aligned} \tag{6.26}$$

$\langle BER\rangle$ in (6.26) is determined by $\langle SNR\rangle$ and $p_I(u)$. In order to achieve lower order of $\langle BER\rangle$ under turbulence effects, $\langle SNR\rangle$ must be increased higher. Since the OSF suppresses noise, signal intensity $\langle I_1(r_1, z_1 \approx 0, \phi_i)\rangle$ increases higher than in DD. Hence, PD produces optimum of signal power $\langle P_1\rangle$ rather than noise. It means the OSF decreases the signal power ratio of $P_1^0/\langle P_1\rangle$ in order to enhance $\langle SNR\rangle$. Furthermore, by the improvement of $\langle SNR\rangle$, $\langle BER\rangle$ also decreases into lower order as well. In comparison to DD, $\langle BER\rangle$ is produced lower by the OSF, since $\langle SNR\rangle$ is increased by suppression of $P_1^0/\langle P_1\rangle$ as stated in (6.24)–(6.25).

In this experiment [29], FSO of full-duplex transmission uses wavelength $\lambda = 1{,}550$ nm as shown in Figure 6.12. It transmits 1 Gbps of bit rate with on–off keying modulation. In order to provide media of optical propagation for FSO, BTS is developed as an atmospheric chamber as shown in Figure 6.13. It has volume dimension that is 4.0 m × 0.5 m × 0.5 m. It is designed as a mixing flow of high-temperature gradient. Steam at $T_1 = 100\,°C$ with low-speed and cold air at $T_2 = 16\,°C$ with wind speed of 6.0–8.0 m/s are allowed to flow through it. The flows are broken up by ailerons inside of BTS. Furthermore, these processes produce turbulence media to induce forward direction of optical propagation along the path in BTS. In order to avoid the interference of optical propagation, backward direction that is conditioned at nonturbulent media is separated from the forward direction.

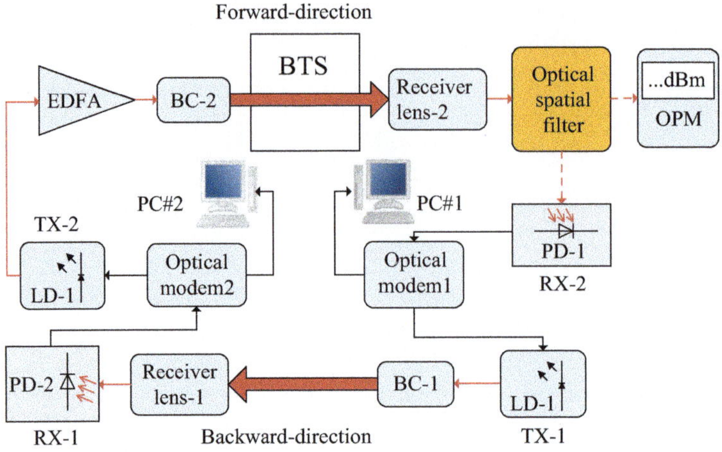

Figure 6.12 FSO of full-duplex transmission using wavelength of 1,550 nm *that transmit 1 Gbps of data rate on the turbulence media of BTS*

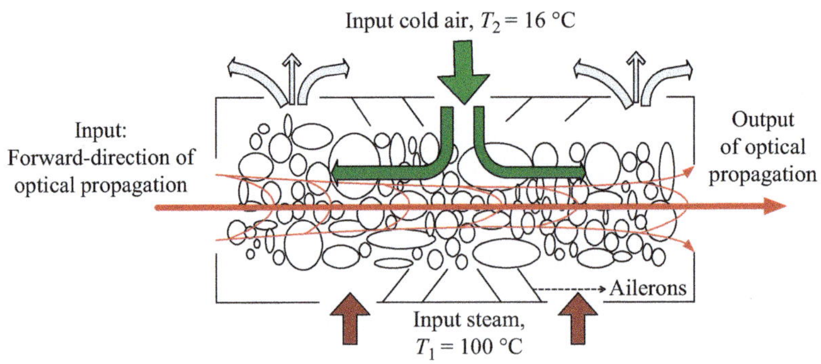

Figure 6.13 BTS providing turbulence media to induce the forward direction of optical propagation

The properties of optical propagation that are designed in the experiment are $W_{-2} = 0.015$ m, $W_{-1} \cong 0.15$ m, and $W_0 = 500$ µm. The signal power P^0_{-2} of +16.5 dBm is transmitted from beam collimator BC-2 to receiver lens-2, as shown in Figure 6.12. The properties of receiver lens are $D_G = 0.05$ m and $L_f = 0.1$ m. P_{Th} of −25.0 dBm is the threshold level P_{Th} of PD. The optical power meter, OPM is used for the measurement of signal power P_1.

The OSF is designed with constant parameters of $D_C = 1.5$ mm and $Z_C = 2.0$ mm. It is also designed with various pinhole diameters. They are 50.0, 40, 30, and 20 µm for D_{P1}, D_{P2}, D_{P3}, and D_{P4}, respectively. Based on (6.18), tilt angles η are 19.9°, 20.0°, 20.2°, and 20.3° for D_{P1}, D_{P2}, D_{P3}, and D_{P4}, respectively. Since the different values of η are not high for all D_{PS}, also based on (6.19) and (6.20), the range of beam wander angle that can be received by cone reflector is 14.0° to 38.0°.

Cone reflector guides random of ϕ onto pinhole diameter in those ranges. The OSF is made from silver in order to produce optimum reflection at 1,550 nm [33]. The OSF is installed at $z_0 = 0$ on X_0. PD is also installed on X_1 at the output of the OSF where $z_1 \approx 0$. Hence, PD receives fundamental component of diffraction from pinhole of the OSF.

The calculations of signal power $\langle P_1 \rangle$ for the OSF and DD method are shown in Figure 6.14. Based on the experiment, signal power from the laser source on X_{-2} in the absence of turbulence media on BTS, $P^0_{-2} = +16.5$ dBm is chosen as reference for the analysis. The calculations are based on the value of index structure for strong turbulence level, $C_n^2 = 5 \times 10^{-13} \text{m}^{-2/3}$. By considering (6.18) to (6.20) and parameters of the OSF in the experiment such as D_C, Z_C, and D_P, beam wanders that are 18°, 28°, and 38° for ϕ_1, ϕ_2, and ϕ_3, respectively, are chosen for the calculations and experiment as well. The properties of optical propagation and optical system are set based on the experiment.

As shown in Figure 6.14, signal power $\langle P_1 \rangle$ quite degrades in maximum for DD method, where it falls below P_{Th}. Regarding $P_1^0 = -14$ dBm, signal power $\langle P_1 \rangle$ of DD method is not produced maximum by PD since signal intensity I_0 modulates beam wander and spatial noise. But by the OSF, signal power $\langle P_1 \rangle$ is improved into the upper of P_{Th}. Figure 6.15 shows trend that signal power $\langle P_1 \rangle$ from the OSF goes higher for all D_P than DD method. It also shows trend that signal power $\langle P_1 \rangle$ is produced higher by the OSF at the presence of beam wander ϕ_1, ϕ_2, and ϕ_3. By considering values of ϕ^0, $\phi_{\min}$, and $\phi_{\max}$ in the experiment, it can be seen that ϕ of 28.0° to 38.0° is the range of beam wander angle that can be received by cone reflector in order to be reflected onto pinhole diameter. For ϕ of 14.0° to 27.0° is the range of beam wander angle that goes directly onto pinhole without reflection

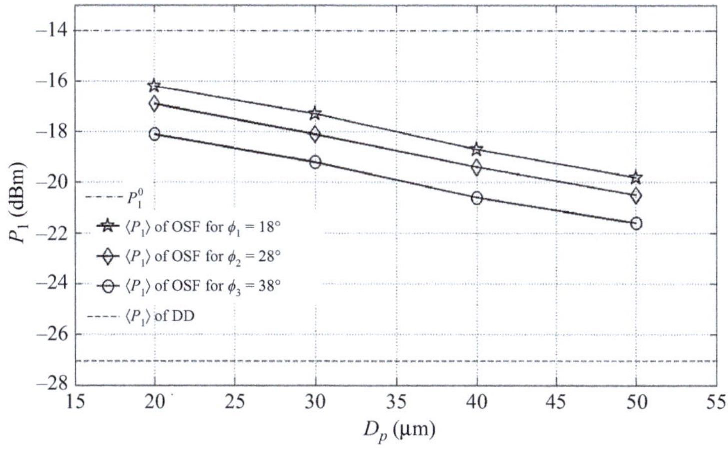

Figure 6.14 Signal power $\langle P_1 \rangle$ vs. pinhole diameter D_P from the OSF for beam wander $\phi_1 = 18.0°$, $\phi_2 = 28.0°$, and $\phi_3 = 38.0°$, where $\langle P_1 \rangle = -27.0$ dBm is from DD method and $P_1^0 = -14.0$ dBm is the signal power in the absence of atmospheric turbulence

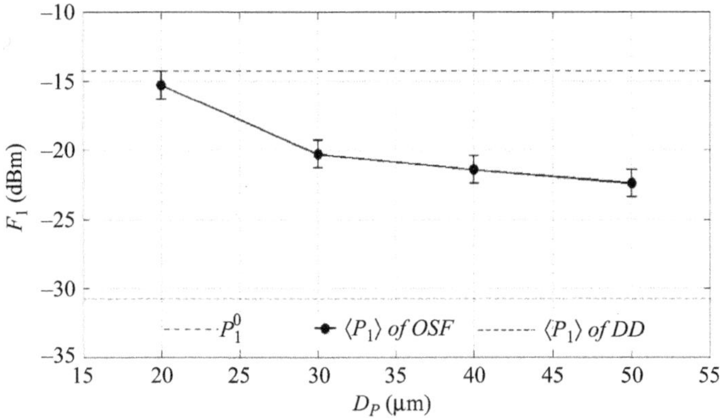

Figure 6.15 Signal power $\langle P_1 \rangle$ vs. pinhole diameter D_P for DD- and the OSF-method by experiment where $\langle P_1 \rangle = -31.0$ dBm is from DD method and $P_1^0 = -14.0$ dBm is signal power in the absence of atmospheric turbulence

from cone reflector. Thus, the OSF receives the angle of beam wander in the range of 14.0° to 38.0°. It means, the term of $\cos(\phi_i)$ as stated in (6.22) compensates signal intensity $\langle I_1 \rangle$ to be higher at the presence of beam wander. Signal power $\langle P_1 \rangle$ is produced higher by the OSF at ϕ_2 and ϕ_3 than DD method. Moreover, signal power is produced highest by the OSF at ϕ_1. Figure 6.5 also shows trend that the OSF works to suppress fluctuation of signal intensity in optimum since D_P goes smaller. Based on (6.23), the circular aperture function of pinhole B filters spatial noise through localization in narrow area. It means, the circular aperture function of B works in optimum for smaller pinhole diameter where in mathematically it performs by Bessel function as can be seen in (6.23).

As shown in Figure 6.15, DD method produces lower signal power on turbulence media of BTS where $\langle P_1 \rangle = -31.0$ dBm. $\langle P_1 \rangle$ of DD method falls below P_{Th} as well. Turbulence media of BTS induces the forward direction of optical propagation modulates beam wander and spatial noise effects. This random process leads to fluctuation of signal intensity I_0. Hence, the signal intensity $\langle I_0 \rangle$ goes in minimum. Moreover, this condition leads to performance degradation that PD produces minimum signal power and maximum noise. The OSF suppresses beam wander and spatial noise effects as shown in Figure 6.15. Signal power $\langle P_1 \rangle$ increases higher beyond P_{Th}. Moreover, signal power is produced higher by the OSF for all D_P than DD method. The signal power from all D_P almost approximates the signal power in the absence of turbulence media in BTS, P_1^0. Signal power $\langle P_1 \rangle$ from the OSF are −22.4 dBm, −21.4 dBm, −20.3 dBm, and −15.3 dBm for D_{P1}, D_{P2}, D_{P3}, and D_{P4}, respectively. Beam wander and spatial noise effects are optimally suppressed by cone reflector and pinhole in order to enhance $\langle I_1 \rangle$ as stated in (6.21). Spatial noise that is modulated in optical propagation is filtered out by pinhole in order to minimize random movement of hot spot. Therefore within narrow area of the pinhole, spatial noise is localized optimally. Meanwhile, random displacement of focus spot is minimized by cone reflector. Thus, fluctuation of signal intensity is collected within narrow area of pinhole as well. Those operations between pinhole and cone reflector bring an advantage in suppressing fluctuation of signal intensity in optimum. Since turbulence effects are stochastic and independent process that causes random angle, ϕ of beam wander, cone reflector operates to anticipate beam wander effect in order to minimize misalignment of detection into PD.

By experiment as shown in Figure 6.15, different values of D_P tend to produce the same value of signal power $\langle P_1 \rangle$. It can be seen for D_{P1}, D_{P2}, and D_{P3}. In the experiment, turbulence effects are provided by turbulence media of BTS with scale or level that has random variation. It can be assumed that for D_{P4} turbulence level is not produced greater by BTS hence turbulence media induces the forward direction of optical propagation in minimum. Thus, D_{P4} produces highest signal power $\langle P_1 \rangle$. However, the results of calculation and experiment show the similar trend that suppression of signal intensity fluctuation works in optimum for the OSF since pinhole diameter goes smaller, as shown in Figures 6.14 and 6.15. Moreover, the OSF for all pinhole diameter produces minimum fluctuation of signal intensity $\langle I_1 \rangle$ than DD method where signal power $\langle P_1 \rangle$ is produced higher.

Based on the results of calculations and experiment as shown in Figures 6.14 and 6.15, the OSF enhances signal intensity $\langle I_1 \rangle$ that has degraded by the turbulence effects in DD method. Beam wander and spatial noise effects are suppressed by cone reflector and pinhole. Through this suppression, signal intensity I_1 is received optimally by PD. As shown in (6.21), fluctuation of signal intensity that is caused by beam wander and spatial noise effects can be minimized as well. Regarding signal power in the absence of turbulence media in the experiment $P_1^0 = -14.0$ dBm, the OSF improves DD method in order to produce higher signal power $\langle P_1 \rangle$. It means, the OSF operates to suppress fluctuation of signal intensity $\langle I_0 \rangle$ optimally.

By experiment, the OSF increases signal power $\langle P_1 \rangle$ of DD method in the level of 8.6 dB, 9.6 dB, 10.7 dB, and 15.7 dB for D_{P1}, D_{P2}, D_{P3}, and D_{P4}, respectively. The OSF enhances signal intensity $\langle I_1 \rangle$ in order to produce higher of signal power $\langle P_1 \rangle$. It works to suppress beam wander and spatial noise effects in optical propagation that is induced by turbulence media. It keeps signal power $\langle P_1 \rangle$ higher beyond P_{Th}. The OSF localizes acquisition of focus spot in a narrow area of pinhole in order to filter spatial noise. Furthermore, fluctuation of signal intensity that is caused by spatial noise is suppressed in optimum within a narrow area of pinhole as well. This suppression is performed by pinhole through governing Fresnel diffraction in narrow area as shown by the circular aperture function of pinhole B in (6.23). As shown in (6.17) and (6.21), pinhole performs filtering to spatial noise in optimum. In the same time, cone reflector enhances signal intensity to be higher as well through collection of focus spot displacement around the optical axis to be reflected onto pinhole diameter. Thus, fluctuation of signal intensity $\langle I_1 \rangle$ is minimized within pinhole diameter. In order to optimize pinhole operation for suppression of spatial noise, the ratio of pinhole diameter D_P and space between the OSF and PD, z_1 should satisfy $(D_P/2) = \lambda z_1/4r_1$.

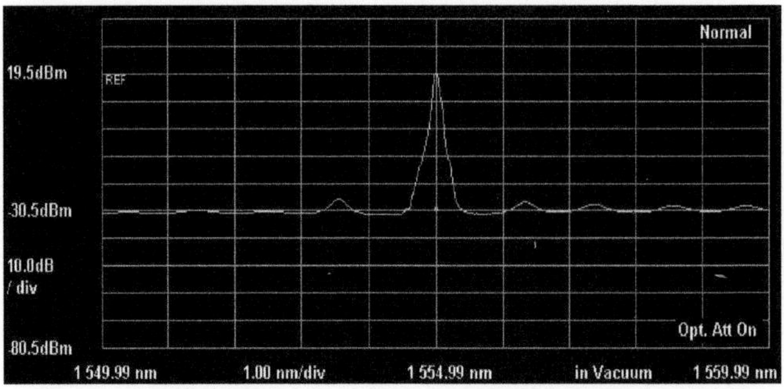

Figure 6.16 Original signal spectrum of FSO in the experiment as the output of EDFA before transmitted out into BTS also in the absence of turbulence effects

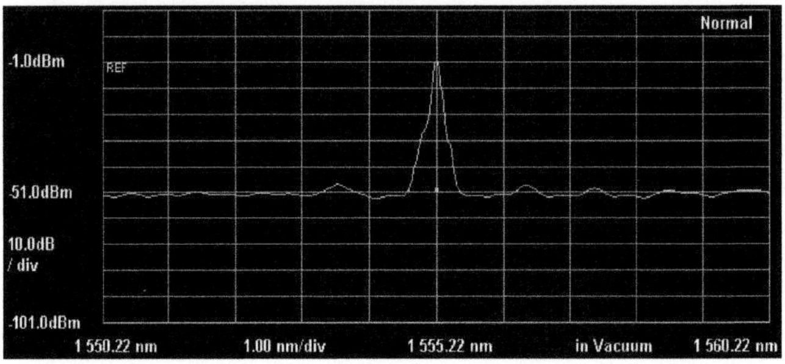

Figure 6.17 Transmitted of signal spectrum into BTS by EDFA at nonturbulent condition in BTS

The tilt angle of cone reflector, η plays an important role to reflect the random displacement of focus spot around the optical axis radially. Based on (6.18), dimension of cone reflector must consider term of $2\tan^{-1}(D_C - D_P/2Z_C)$ in order to produce minimum of ϕ_i. Hence, maximum of η is $\tan^{-1}(D_C - D_P/2Z_C) < \phi$, since ϕ is quite impossible at $\pi/2$ in FSO. Also based on (6.19) and (6.20), there is an opportunity to enlarge the range of $\phi_{\min}$ to $\phi_{\max}$ by designing optimum parameter of D_C, D_P, Z_C, and L_f to receive larger of ϕ. Through this, larger of ϕ at strong turbulence level can be collected onto pinhole diameter D_P radially. It will bring an advantage to enhance the signal intensity higher. Thus, random displacement of focus spot that exists in DD method under turbulence effects can be minimized as well.

The analysis for four signal spectrum measurements (SNR) is shown in Figures 6.16–6.19. First is the original signal spectrum measurement or direct from the output of EDFA. Second is the signal spectrum measurement after propagation on BTS at nonturbulent. Third is the signal spectrum measurement for DD method at strong turbulence. Fourth is for the OSF at strong turbulence as well. The original signal spectrum that is coupled into beam collimator is shown in Figure 6.16. The

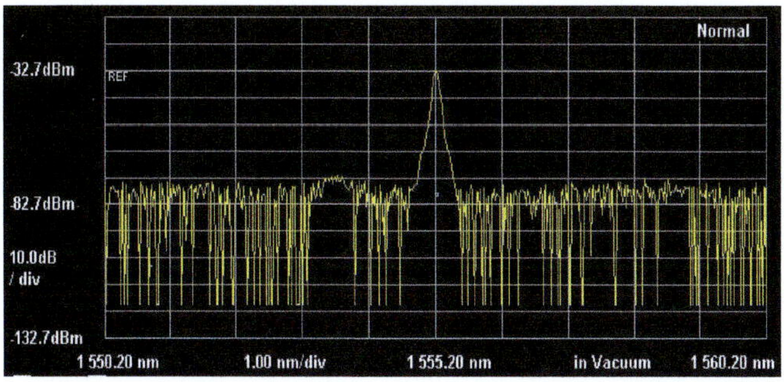

Figure 6.18 The received of signal spectrum by DD method at strong turbulence of BTS

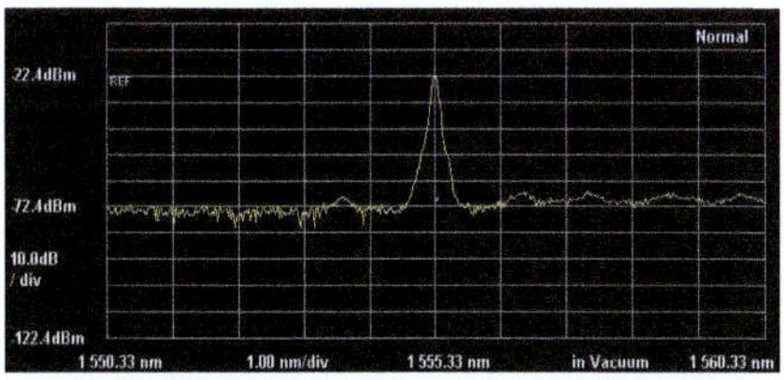

(a)

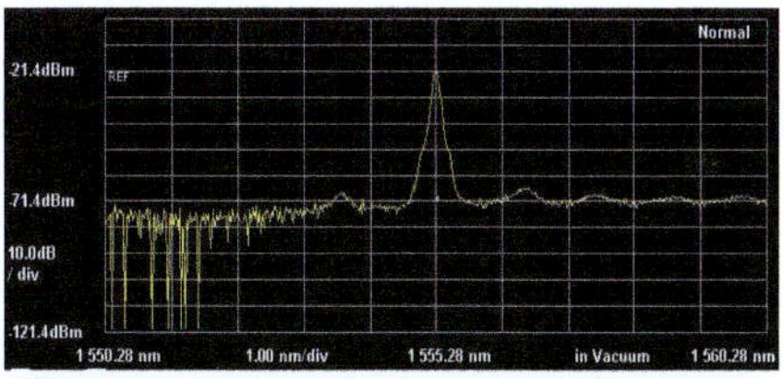

(b)

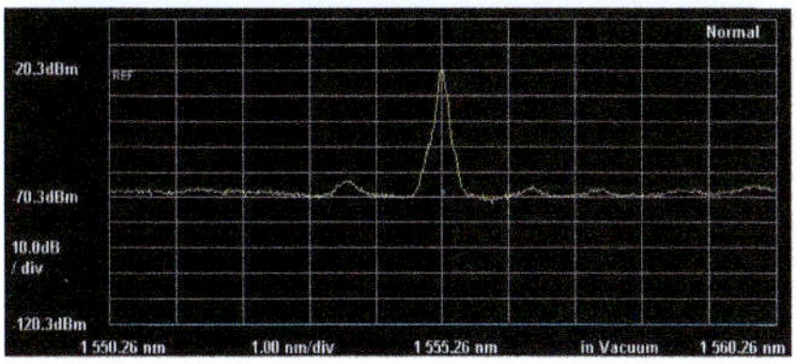

(c)

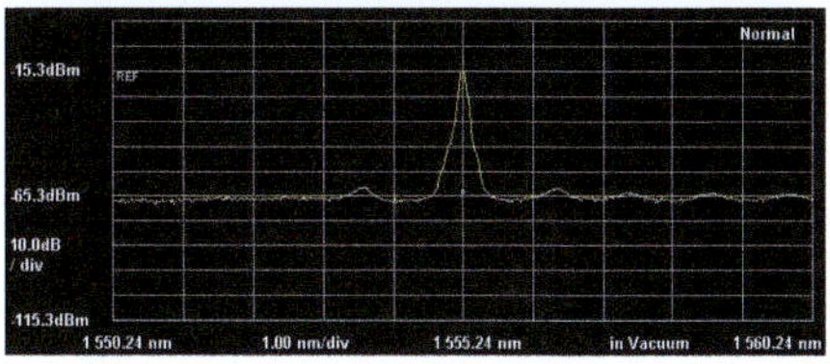

(d)

Figure 6.19 Signal spectrum measurement by the OSF: (a) $D_{P1} = 50\,\mu m$*; (b)* $D_{P2} = 40\,\mu m$*; (c)* $D_{P3} = 30\,\mu m$*; and (d)* $D_{P4} = 20\,\mu m$

peak of signal spectrum is 1,554.99 nm and its power is +19.5 dBm. It is clear that there is no spectrum noise because the signals have not yet propagated in turbulent media. The value of SNR is quite high and is constant at 39.23 dB. The second peak at 1,553.55 nm is very low and side-mode suppression ratio or SMSR is 45.79 dB.

The original signal spectrum as shown in Figure 6.16 is transmitted by beam collimator into BTS with condition of nonturbulent. The received of signal spectrum is attenuated by attenuator of 6 dB which is used in OSA. The signal power decreases at −1 dBm and the center changes to 1,555.22 nm as shown in Figure 6.17. Based on the signal spectrum in Figure 6.17, the value of SNR in the absence of turbulent media in BTS is 38.42 dB. The SMSR parameter is also high at 48.6 dB because the second peak is attenuated. The result also shows that there is no high oscillation along the width of signal spectrum, 1 nm. It means that the signal spectrum does not exhibit noise modulation in signal spectrum as a consequence for optical propagation at nonturbulent media. Thus, optical propagation is transmitted along the propagation path without any disturbance. Hence, the result for the signal spectrum is quite the same as the original signal from EDFA.

DD method is used to measure the signal spectrum at strong turbulence in BTS. Noise modulation in signal spectrum presents for a width of 1 nm or in the region of full-width at half-maximum (FWHM), as shown in Figure 6.18. The received signal power decreases extremely at −32.67 dBm and SNR goes lower to 32.6 dB. The signal spectrum is very deteriorated caused by the turbulence effects, where signal power falls below P_{Th}. Moreover, the center frequency of signal spectrum at FWHM modulates noise maximally. The signal power is almost flat with the high order of temporal noise. It means PD receives spatial noise in optimum. Those degradation is caused by the randomness of optical phenomena that occurs along the propagation path in the turbulence media if BTS. Under turbulence effects, the signal spectrum that modulates maximum noise causes degradation of the performance where the SNR value falls into high scintillation. By all means, the deteriorated signal spectrum leads to the degradation in FSO performance maximally.

In order to suppress noise modulation in signal spectrum that exists in DD method, the OSF is installed at focus spot before PD, where the front diameter for cone reflector is 1.5 mm and the width is 2 mm. The parameters for cone reflector are designed constant for all pinhole diameter. It is also made of silver that is capable of reflecting in long-wavelength optimally. The trend as shown in Figure 6.19(a)–(d) shows significant improvement to recover the signal spectrum which is deteriorated maximally by turbulence effects in BTS. We also use lower pinhole which diameter is 20 μm. The results of signal power for pinhole diameter (D_P) of 50, 40, 30, and 20 μm are −22.4, −21.42, −20.3, and −15.31 dBm, respectively. The SNR values are 37.5, 38.5, 38.7, and 39.2 dB for pinhole diameters 50, 40, 30, and 20 μm, respectively, also. The trend shows that signal spectrum modulates noise in minimum and fully recovered as the original signal spectrum from EDFA as shown in Figure 6.18.

The improvement of signal spectrum is successfully achieved by using the OSF before signal reception into PD. The intensity of signal spectrum rises-up beyond P_{Th}. It is also optimized by the operation of pinhole to sweep out the spatial noise. By implementing the OSF, noise modulation in signal spectrum can be suppressed optimally while the signal power is still received in higher value than DD method does. It brings benefit for PD to minimize the scintillation effects when strong turbulence induces optical propagation of FSO. Hence, the performances of SNR can be increased also under influences of strong turbulence as well.

The ultimate contribution by installing the OSF for suppressing beam-wandering effect in strong turbulence is the noise suppression in signal spectrum can be achieved optimally where random spatial and temporal noise that exist in the signal spectrum can be minimized through spatial localization in narrow region which is in pinhole diameter. Beam wander fluctuates randomly, hence the focus spot experiences random displacement around the optical axis of pinhole. Meanwhile, spatial noise that is caused by random scattering in turbulent media leads signal spectrum to degrade as well. Thus, signal spectrum modulates maximum noise in the presence of beam wander and spatial noise. Henceforth, by designing the optimum angle to reflect the displacement of focus spot, cone reflector can guide larger angle of beam wander into pinhole diameter optimally. Thus, signal spectrum is minimum for noise modulation. It means, minimum noise bandwidth in signal spectrum contributes to higher SNR value.

The OSF which is composed of cone reflector and pinhole works optimally to suppress noise modulation in signal spectrum that is modulated by beam wander and spatial noise. The signal spectrum which is deteriorated maximally by strong turbulence as shown in Figure 6.18 can be fully recovered with benefit of enhancement in the signal intensity and minimum noise modulation, as shown in Figure 6.19. In comparison to the DD method, the signal spectrum by the OSF is fully recovered where the characteristics are nearly to the original signal spectrum as it was produced by EDFA, as shown Figure 6.16.

Cone reflector and pinhole as the component of the OSF has successful suppressed beam wander and spatial noise effects in order to recover the signal spectrum that has been deteriorated by spatial noise modulation. The implementation of OSF in FSO also gives benefit for higher of signal power and lower noise bandwidth in comparison to DD method. Moreover, signal power is successfully enhanced beyond the threshold level of PD, where −22.4 dBm, −21.42 dBm, −20.3 dBm, and −15.31 dBm are for pinhole diameter 50, 40, 30, and 20 μm, respectively. The values of SNR increase which are 37.5, 38.5, 38.7, and 39.2 dB for pinhole diameter 50, 40, 30, and 20 μm, respectively, as well.

The calculations of $\langle SNR \rangle$ and $\langle BER \rangle$ for DD and the OSF are shown in Figures 6.21 and 6.22. $SNR^0 = 40.0$ dB is chosen as the reference for analysis. The calculations are based on the values of index structure, $C_n^2 = 5 \times 10^{-13}\text{m}^{-2/3}$. The properties of optical propagation are set based on the experiment. Based on the ϕ^0, ϕ_{min}, and ϕ_{max} that are stated in (6.3) and (6.4), $\phi_1 = 18°$, $\phi_2 = 28°$, and $\phi_3 = 38°$ are chosen as the beam-wandering angles. $\alpha = 3$ and $\beta = 2$ are set constant for calculations. α is chosen higher than β, since it is dominant factor for SNR and BER performance degradation under atmospheric turbulence.

In Figures 6.20 and 6.21, DD method produces degradation of performance by the presence of beam wander and spatial noise where $\langle SNR \rangle = 34.4$ dB and $\langle BER \rangle = 10^{-4}$. The OSF improves $\langle SNR \rangle$ degradation that is presented in DD. $\langle SNR \rangle$ increases linear as D_P of the OSF goes lower for beam wander angles of ϕ_1, ϕ_2, and ϕ_3. The performance of FSO for ϕ_1 is better than that at ϕ_2 and ϕ_3. Beam-wandering angle ϕ_1 is at the range that is received by pinhole directly without reflectance of cone reflector. The range of beam wander angle for direct reception by pinhole is 14°–27°. While the range of beam wander angle for reflectance by cone reflector is 28°–38°. However, for ϕ_2 and ϕ_3 SNR and BER performance is achieved better than DD. By recalling (6.17), $P_1^0/\langle P_1 \rangle$ determines value of $\langle SNR \rangle$. Thus, the OSF produces a lower value of $P_1^0/\langle P_1 \rangle$ for lower D_P. $\langle SNR \rangle$ performance for the OSF is better than DD since D_P goes lower as well. The OSF improves $\langle SNR \rangle$ through the suppression of beam wander and spatial noise effects simultaneously. It governs Fresnel diffraction as can be seen by the circular aperture function of pinhole B that is stated in (6.15). Hence, the OSF minimizes $P_1^0/\langle P_1 \rangle$

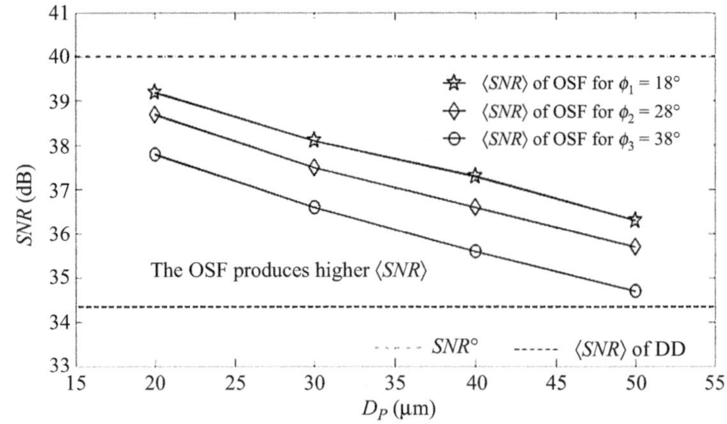

Figure 6.20 The calculation results where $\langle SNR\rangle$ *vs.* D_P *is from the OSF for* $\phi_1 = 18°$, $\phi_2 = 28°$, *and* $\phi_3 = 38°$, $\langle SNR\rangle = 34.4$ *dB is from DD and* $SNR^0 = 40.0$ *dB*

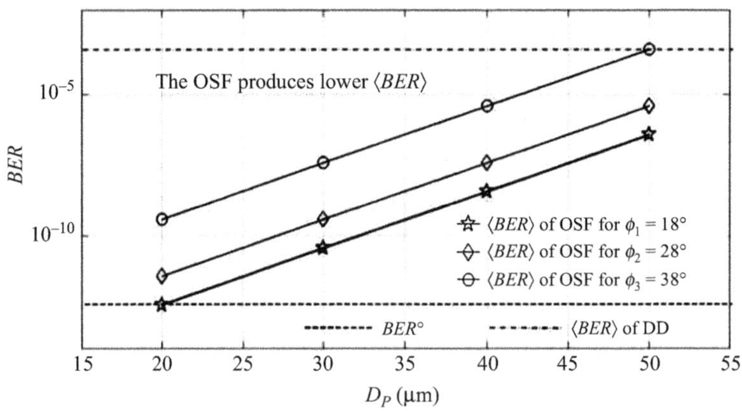

Figure 6.21 The calculation results where $\langle BER\rangle$ *vs.* D_P *is from the OSF method for* $\phi_1 = 18°$, $\phi_2 = 28°$, *and* $\phi_3 = 38°$, $\langle BER\rangle = 10^{-4}$ *is from DD, and* $BER^0 = 10^{-13}$

since D_P goes lower. Even for larger pinhole diameter D_{P1}, $\langle SNR\rangle$ is produced higher by the OSF than DD for ϕ_1, ϕ_2, and ϕ_3. The OSF also improves $\langle BER\rangle$ degradation as D_P goes lower as well. It produces higher $\langle SNR\rangle$ in order to achieve lower order of $\langle BER\rangle$ for ϕ_1, ϕ_2, and ϕ_3 as well. Suppression of beam wander and spatial noise effects by the OSF is achieved through minimizing signal power ratio $P_1^0/\langle P_1\rangle$ as shown by (6.17).

In Figure 6.22, the results of experiment for $\langle SNR\rangle$ vs. D_P that are produced from DD and OSF are shown. The value of $\langle SNR\rangle$ from DD method is 34.9 dB. Regarding $SNR^0 = 40.0$ dB, this performance quite degrades by the presence of beam wander and spatial noise modulation that arise on optical propagation. The OSF improves those

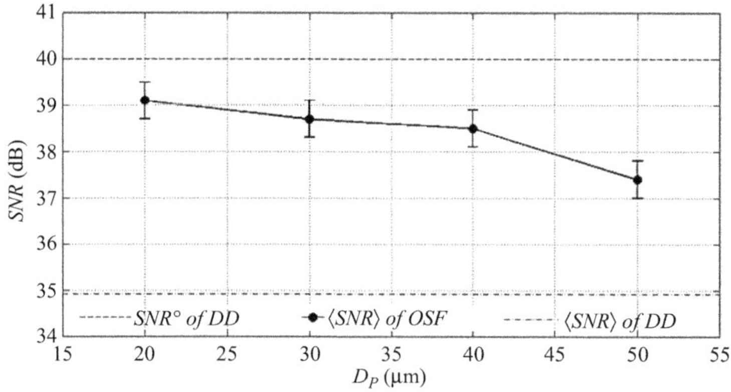

Figure 6.22 The experiment results where ⟨SNR⟩ vs. D_P from the OSF, ⟨SNR⟩ = 34.9 dB is from DD, and $SNR^0 = 40.0$ dB

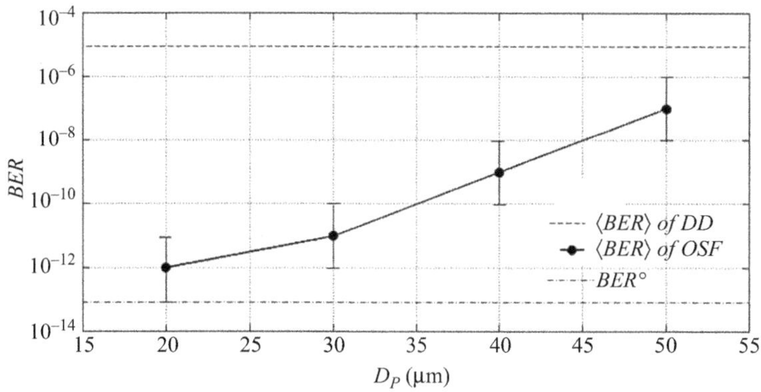

Figure 6.23 The experiment results where ⟨BER⟩ vs. D_P is from the OSF, $\langle BER\rangle = 10^{-5}$ is from DD, and $BER^0 = 10^{-13}$

degradation in scale of 2.5 dB, 3.5 dB, 3.8 dB, and 4.2 dB for D_{P1}, D_{P2}, D_{P3}, and D_{P4}, respectively. ⟨*SNR*⟩ improvement by the OSF is better than DD. But ⟨*SNR*⟩ improvement by different of D_P do not contribute to higher value. The OSF for each of D_P tends to produce the same value of ⟨*SNR*⟩. For example, D_{P2} and D_{P3} do not produce high different value of ⟨*SNR*⟩. Only D_{P4} produces highest ⟨*SNR*⟩ that approximates SNR^0. However, the results of experiment as shown in Figure 6.22 shows the same trend as the calculation in Figure 6.20, that ⟨*SNR*⟩ is increased by the OSF since D_P goes smaller.

In Figure 6.23, experiment results of ⟨*BER*⟩ vs. D_P for DD and the OSF method are shown. ⟨*BER*⟩ of DD is produced at 1.9×10^{-5}. Regarding $BER = 1.0 \times 10^{-13}$ at SNR^0, those performance quite degrades also. The OSF improves ⟨*BER*⟩ degradation in scale of 10^{-2}, 10^{-4}, 10^{-6}, and 10^{-7} for D_{P1}, D_{P2}, D_{P3}, and D_{P4}, respectively. ⟨*BER*⟩ improvements by different of D_P are significant. The OSF for all diameter of pinhole achieve improvement of ⟨*BER*⟩ in scale range of 10^2 to 10^7 for D_{P1} to D_{P4}.

The smallest diameter of pinhole, D_{P4} produces $\langle BER \rangle$ that approximates to the value of *BER* at SNR^0. Moreover, Figure 6.8 shows the same trend as Figure 6.23, that $\langle BER \rangle$ is decreased to lower order by the OSF since D_P goes lower as well.

The OSF is a competitive method to be implemented in optical amplification method. For example EDFA on FSO, noise has an opportunity to degrade the performance of SNR and BER [34,35]. Since focus spot is coupled into EDFA directly on the scheme of forward-relaying, beam wander and spatial noise effects are also modulated in maximum into the stage of amplification process. Thus, the gain of signal by EDFA is not achieved in optimum. The performance of SNR from the implementation of EDFA on FSO could degrade in maximum since noise is amplified even the signal amplification process is in the saturated region. By considering this reason, the OSF can be implemented on EDFA in order to suppress beam wander and spatial noise effects. Thus, by integration of the OSF and EDFA, maximum fluctuation of signal intensity can be minimized as well. Furthermore, by implementing the OSF to deliver signal input into EDFA, signal power can be produced by EDFA on FSO in higher with minimum of noise. By this scheme, the gain of signal intensity will be produced in optimum as well. Hence, the performance of SNR and BER can increase as well.

The suppression of beam wander and spatial noise effects by the OSF has been carried out by calculations and experiment. Both calculations and experiment show the similar trend that the OSF produces higher signal power than DD method. The suppression of beam wander and spatial noise effects is achieved in optimum by the OSF since pinhole diameter goes smaller. The OSF suppresses beam wander and spatial noise effects simultaneously in order to minimize fluctuation of signal intensity and noise modulation in the signal spectrum as well. Thus, it produces higher signal intensity $\langle I_1 \rangle$ and clean signal or signal spectrum with minimum noise modulation than DD method does. It means the OSF has successfully suppressed the turbulence effects that affects the incoming of optical signal into detection by PD. The OSF that composed of pinhole and cone reflector as a detection method capable suppressing the fluctuation of signal intensity in minimum. Thus, the performance of FSO under the turbulence effects can be improved.

6.4 Pinhole, cone reflector, and multi-mode fiber as the optical spatial filter

In addressing the problem in WDM of FSO where signal spectrum tends to degrade where each channel is induced by random media independently, the OSF that is composed of pinhole and cone reflector is integrated with multi-mode fiber. This integration brings an optical treatment of the incoming signal spectrum to be filtered out cleanly before received by PD. In [36], the OSF that is composed of pinhole, cone reflector, and multi-mode fiber has been developed successfully to suppress scintillation, beam wander, and spatial noise. The configuration of the OSF as a detection method is shown in Figure 6.24. The OSF is installed in rear focus spot before PD. Those suppression is achieved by governing diffraction process of focus spot in pinhole while beam wandering is collected by cone reflector.

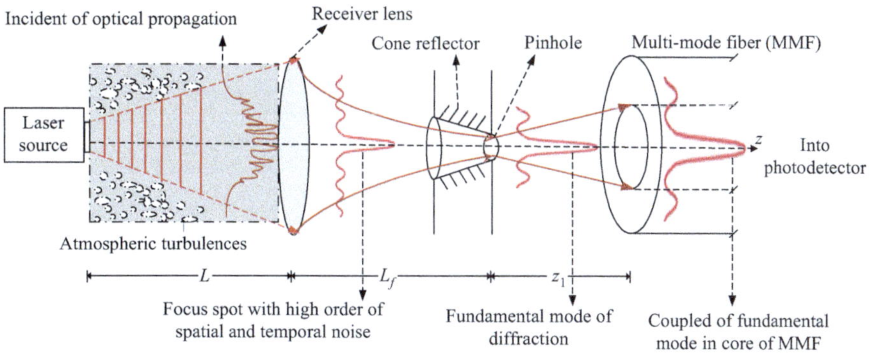

Figure 6.24 The optical propagation of FSO communications on atmospheric turbulence filtered by the OSF detection method that composed of cone reflector, pinhole, and multi-mode fiber

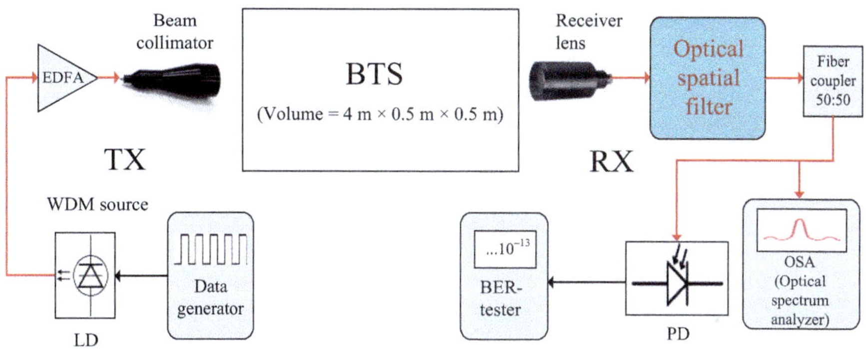

Figure 6.25 Setup of experiment for FSO communications implementing single channel of WDM where the turbulence media using BTS

In [35], single channel of 1,554.9 nm (ITU-G.694.1) is used on FSO system in the setup of experiment. FSO system implements full duplex transmission of Tx and Rx (Figure 6.25). Bit rate capacity at 1 Gbps with OOK modulation is transmitted via data generator. It is transmitted via laser diode (LD). The signal power of −2.6 dBm of LD is amplified by EDFA at the output of +23.5 dBm as shown in Table 6.2. The amplified optical signal is coupled into beam collimator. Then optical propagation with diameter of 4.0 cm is transmitted by beam collimator into turbulence media in box of turbulence simulator (BTS), Figure 6.13). BTS is the atmospheric chamber as the propagation media for the optical signal [36]. Henceforth the optical propagation is received by receiver lens and focused into the OSF.

The optical signal from OSF is coupled into fiber coupler of 50:50 and splitted out into optical spectrum analyzer (OSA) and photodiode to be measured by *BER* Tester. The measurement of *SNR* used OSA (Anritsu MS9740A). *BER* measurement used JDSU Throughput tester.

Table 6.2 The optical power budget of the FSO system

No.	Parameter of device	Value
1	Tx power of LD 2 (DFB laser)	−2.6 dBm
2	Output power of EDFA	+23.5 dBm
3	Sensitivity of photodiode (Si-PIN), P_{th}	−30.0 dBm
4	Coupling loss of beam collimator	3.0 dB
5	Fiber coupler loss (50:50)	9.0 dB
6	Receiver lens loss	3.0 dB
7	Attenuator	6.0 dB

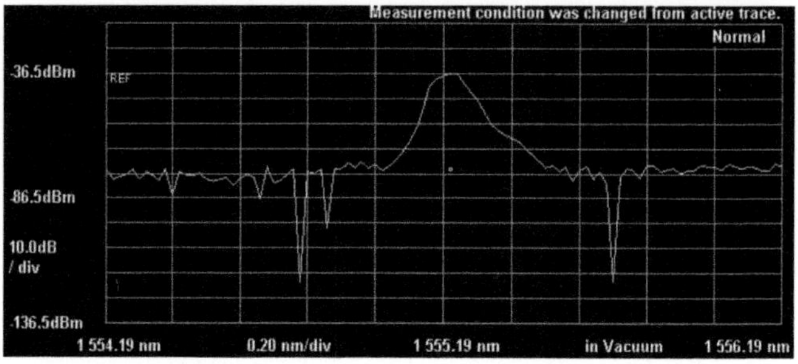

Figure 6.26 The signal spectrum from fiber-detection method at strong turbulence of BTS

The signal power of +23.5 dBm from EDFA is the original signal spectrum that is transmitted via beam collimator into BTS (Table 6.2). The original signal is induced by turbulent media of BTS. It is designed to interact with turbulent media in BTS optimally (Figure 6.13). The strength of signal power is calculated by considering the loss transmission which is caused beam collimator, thus the signal power P_S is about +19.5 dBm that propagates on the BTS.

The signal spectrum is compared for two method of detection that are fiber- and OSF-detection. The OSF-detection is categorized into two kind that are pinhole-MMF and cone reflector-pinhole-MMF. Those measurements of signal spectrum are at strong turbulence level of BTS. The results of measurement are shown in Figures 6.26–6.28.

As shown in Figure 6.26, turbulence effects deteriorate the signal spectrum. The signal power at +19.5 dBm that propagates in BTS is induced by turbulent media. Regarding turbulence effects modulates scintillation and beam wandering, signal spectrum fluctuates temporally and spatially. Thus it behaves random of the signal spectrum at the PD. The fiber detection that has a higher numerical aperture of MMF receives the signal spectrum that modulates random noise as shown in Figure 6.26.

The noise by temporal fluctuation is seen around the center of frequency where signal spectrum no longer Gaussian as come from the source of EDFA. Based on the measurements, the noise by temporal fluctuations has a trend to be modulated near at FWHM region. This causes the signal-to-noise ratio $\langle SNR \rangle$ degrades

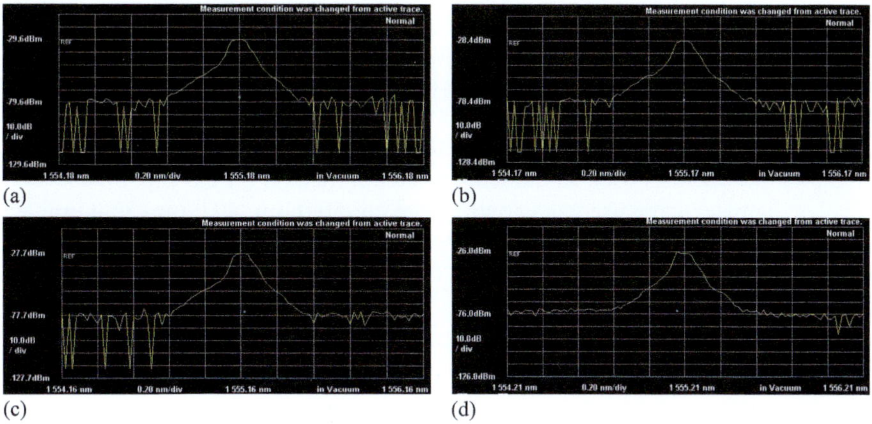

Figure 6.27 The signal spectrum measurement by OSF-detection that composed of pinhole-MMF: (a) $D_{P1} = 50.0$ μm; (b) $D_{P2} = 40.0$ μm; (c) $D_{P3} = 30.0$ μm; and (d) $D_{P4} = 20.0$ μm

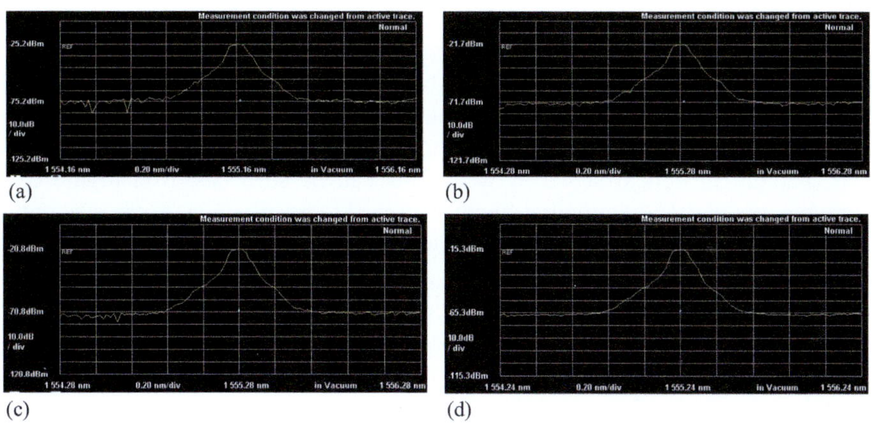

Figure 6.28 The signal spectrum measurement by the OSF-detection that composed of cone reflector-pinhole-MMF: (a) $D_{P1} = 50.0$ μm; (b) $D_{P2} = 40.0$ μm; (c) $D_{P3} = 30.0$ μm; and (d) $D_{P4} = 20.0$ μm

significantly where −36.5 dB is measured with fiber-detection method. The signal spectrum that modulates this noise is received by PD thus it produces shot noise and current fluctuation in the electronical processing. This leads to performance of $\langle BER \rangle$ goes higher. Regarding the $\langle BER \rangle$ at nonturbulent condition at BTS is 10^{-13}, strong turbulence level deteriorates the signal spectrum optimally.

The fiber-detection method received directly the spatial and temporal noise that goes random from the receiver lens. This noise should be filtered out in order to recover the signal spectrum as clean as from the source. Nonetheless, the NA of MMF is quite higher but noise modulation by temporal fluctuation still received in optimum. Also MMF cannot receive the signal spectrum for beam wandering where the focus spot goes random along the optical axis, thus this condition leads to misdetection as well. Based on this fact, fiber-detection that uses MMF is not capable of filtering random noise in the signal spectrum.

In order to recover signal spectrum that is deteriorated by turbulent media of BTS, pinhole-MMF is used as the OSF-detection method on the experiment. Four pinholes diameter were used as variation that are 50.0 μm, 40.0 μm, 30.0 μm, and 20.0 μm for D_{P1}, D_{P2}, D_{P3}, and D_{P4}, respectively. Each pinhole was coupled into MMF at the spacing distance of 2.0 mm. The results of filtering the signal spectrum are shown in Figure 6.27(a)–(d).

Noise modulation decreases drastically. Pinhole-MMF tends to suppresses spatial noise as the size diameter of pinhole goes more lower. The results of $\langle P_S \rangle$ are −29.6 dBm, −28.4 dBm, −27.7 dBm, and −26.0 dBm for D_{P1}, D_{P2}, D_{P3}, and D_{P4}, respectively. The results of $\langle SNR \rangle$ are 35.0 dB, 35.6 dB, 36.4 dB, and 36.8 dB, for D_{P1}, D_{P2}, D_{P3}, and D_{P4}, respectively. Even though pinhole-MMF detection achieves improvement of $\langle SNR \rangle$ compared to fiber-detection method. But P_S is almost and near to fall threshold level of photodiode, P_{th} that is −30 dBm for all D_P. However pinhole-MMF detection has worked to recover the signal spectrum by suppressing temporal fluctuation. The signal spectrum of fiber-detection is quite different when pinhole-MMF is implemented. The noise modulation by temporal fluctuation on the signal spectrum can be minimized in quite optimum. Noise modulation is suppressed by pinhole through Fresnel diffraction process. Also the size pinhole diameter capable to localizing the temporal fluctuation that is modulated spatially in optical signal as well. This can be seen based on Figure 6.28 where the modulation of temporal fluctuation tend to be minimum as the size of the pinhole diameter goes lower. Lower diameter of pinhole means localization of the temporal fluctuation from the optical signal occurs on a very small area of reception. This brings benefit for minimizing the spatial noise. Thus, the signal spectrum is received by PD with minimum of noise modulation as well.

In order to increase P_S by pinhole-MMF detection, cone reflector was installed in front of it. The aperture diameter, D_C is 1.5 mm and the width, Z_C is 2.00 mm where the dimension was designed constant for D_{P1} to D_{P4}. It also made from pure silver (80%) that capable to reflects in long wavelength optimally [7–9]. Signal spectrum measurement as shown in Figure 6.29(a)–(d) shows significant improvement to recover the signal spectrum that is deteriorated by turbulent media optimally. The results of $\langle P_S \rangle$ are −25.2 dBm, −21.7 dBm, −20.8 dBm, and −15.3 dBm for D_{P1}, D_{P2}, D_{P3}, and D_{P4}, respectively. The results of $\langle SNR \rangle$ are 36.5 dB, 37.5 dB, 37.7 dB, and 38.2 dBm for D_{P1}, D_{P2}, D_{P3}, and D_{P4}, respectively. OSF produces $\langle P_S \rangle$ higher beyond threshold level of PD for D_{P1} to D_{P4} and also achieve higher than pinhole-MMF detection. The trend of results show that the signal spectrum has characteristic very low noise modulation. It means signal spectrum is recovered as the original signal from the source (EDFA).

The improvement of signal spectrum is achieved by integration cone reflector, pinhole and fiber as the OSF-detection. The signal power increases beyond the threshold level of PD. The OSF suppresses bandwidth of spatial noise modulation

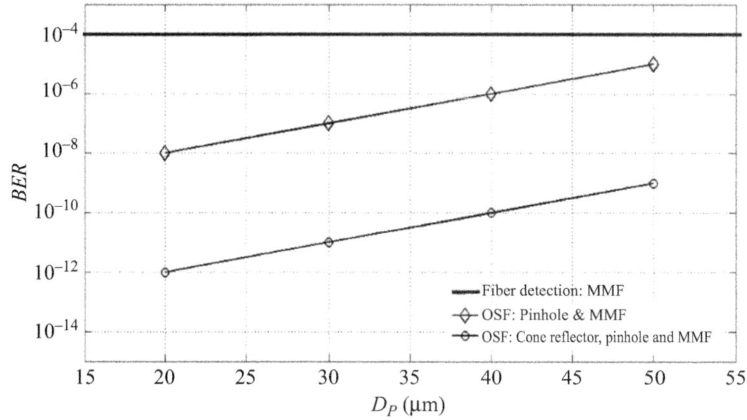

Figure 6.29 ⟨BER⟩ performances from three scheme of signal reception methods which are fiber-, pinhole-MMF-, and the OSF-detection

optimally in the signal spectrum. It produced higher intensity of signal spectrum with minimum noise modulation around the region of FWHM. It means pinhole works to suppress spatial noise caused by turbulence effects modulation on the optical signal.

It can be seen from Figure 6.28, as the results from pinhole-MMF detection, small diameter of pinhole tend to localize the spatial noise optimally. Hence the signal spectrum that modulates spatial noise is received by PD with minimum noise modulation in the signal spectrum. Then the performance of $\langle SNR \rangle$ increases higher under the influence of turbulent media from BTS to optical propagation.

The important contribution of cone reflector in the OSF is suppression for beam wandering under turbulent media of BTS can be achieved optimally. Beam wandering is random, thus focus spot dances or moves random around the optical axis. By designing proper angle of η, displacement or dancing of focus spot can be reflected into pinhole diameter to be filtered out through diffraction process. Beside, guiding random focus spot into pinhole increases the coupling efficiency into MMF. Through this, high intensity of signal spectrum with minimum noise is achieved. Indeed, the benefit for installing the cone reflector is collection of random focus spot around the optical axis can amplifies the signal intensity within small diameter of pinhole. Thus the signal power from the OSF with cone reflector is achieved higher than fiber-detection does only.

$\langle BER \rangle$ comparison by fiber- and the OSF-detection method is shown in Figure 6.29. $\langle BER \rangle$ from fiber-detection increases more higher where it is caused by temporal fluctuation in the signal spectrum where measured at 10^{-4}. The OSF shows high performances in suppressing temporal fluctuation in the signal spectrum than fiber-detection and OSF with pinhole-MMF. $\langle BER \rangle$ values by the OSF is lower in the range of 10^{-9} to 10^{-12} while for pinhole-fiber detection, $\langle BER \rangle$ values is in the range of 10^{-5} to 10^{-8}. The trend of results also show that lower $\langle BER \rangle$ performances is achieved as the size of pinhole diameter goes narrower. It means small diameter of pinhole works optimally to sweeping out the spatial noise as the result of temporal fluctuation in the signal spectrum. Hence the increment of intensity in the signal spectrum is achieved more higher with minimum of noise. This brings benefit for PD to detect signal intensity dominantly and produces current with minimum fluctuation than fiber-detection does.

The cone reflector that performs radial reflection for random focus spot around the optical axis guides the light into pinhole diameter then the optical signal is coupled into MMF. This mechanism solves the problem for beam wandering that caused temporal fluctuation goes optimum. Beam wandering modulates optimum noise for the case of strong turbulence level in the propagation media or atmosphere. This leads random scattering and multipath of optical propagation before receiver lens. Hence the phase of optical signal behaves random as well. Thus, the signal spectrum modulates optimum noise and lower intensity. By implementing cone reflector, this random phase of optical signal is suppressed by collecting temporal fluctuation into narrower area of pinhole than direct and fiber detection. This brings an impact for increasing signal rather than noise in the signal spectrum. Hence, the signal spectrum is received by PD with minimum noise modulation. Following this fact, the performance of $\langle BER \rangle$ from the OSF with cone reflector is achieved lower than fiber detection or the OSF that composed of pinhole-MMF.

The OSF has the ability to suppress temporal fluctuation on optical propagation. The deteriorated signal spectrum that is caused by turbulence effects can be recovered as the signal spectrum at nonturbulent of BTS. The recovery is achieved by two mechanisms that are localization of optical signal into very narrow area which is pinhole diameter and collection of random temporal fluctuation into this area which is performed by cone reflector.

Thus, the OSF increases signal power in higher than fiber-detection does. It also produces the signal spectrum with minimum of noise modulation compared to fiber-detection method. The signal spectrum (Figure 6.28(d)) is recovered as nearly the original signal as it was produced by EDFA. For this reason, the OSF can be implemented as the coupling method to integrate FSO and optical fiber communication system where misdetection caused by beam wandering that become one of major problem can be solved with ultimately and also with offering benefit for low-cost solution.

The OSF which is composed of pinhole, cone reflector and MMF as detection method can be implemented in FSO to recover the deterioration of the signal spectrum on optical propagation that is induced by atmospheric turbulence. In [37], focal-spot wandering is the major problem in the PD since beam wandering modulated in optical propagation. Thus, focal-spot goes random in the area of signal reception in PD. Regarding this condition, in [37], a robust controller algorithm is developed in order to track the movement of focal-spot wandering. The result by those method is focal-spot can be detected effectively without misdetection. However, those method is developed in electronically. Since beam wandering moves in very fast event, also in random fashion, the controller that is used in those method rather difficult to track the movement of random focal-spot. Thus, for the development of FSO in the scheme of forward relaying, that optical signal is forwarded by an optical amplifier, those method needs an improvement. Integration the controller algorithm and the OSF is potential solution in order to optimize the tracking of focal-spot wandering. Hence, by integrating the controller algorithm with the OSF, focal-spot can be delivered into optical fiber in order to be guided into an optical amplifier such as EDFA.

Noise modulation within the signal spectrum can be swept out optimally by using OSF that is composed of pinhole, cone reflector, and MMF. The first key operation of the OSF is pinhole that governs diffraction process to sweep out temporal fluctuation. Second key is cone reflector that is designed to suppress random displacement of focus spot caused by beam wandering on optical propagation. The dancing or displacement of focus spot that goes random around the optical axis is collected and reflected into pinhole in order to be coupled into MMF. Through this mechanism, scintillation can be minimized as well. Third key is the MMF that has a core diameter of 62.5 μm, larger than pinhole diameter. It is designed to receive fundamental component of diffraction from pinhole, that contains of the signal spectrum dominantly with minimum of temporal fluctuation in the signal spectrum. Finally, by implementing the OSF method that is composed of cone reflector, pinhole, and multi-mode fiber for minimizing noise modulation in the signal spectrum that has been deteriorated by atmospheric turbulence can be recovered optimally. The OSF also potential method of detection that can be integrated with an optical amplifier directly where focal-spot wandering can be guided via an optical fiber. Other potential implementation is the OSF can be gathered with the controller algorithm to detect the focal-spot wandering in an effectively way [37]. Thus, SNR and BER performance can be improved higher by using the OSF than other scheme of signal detection method in FSO.

References

[1] H. Hemmati, A. Biswas, and I. B. Djordjevic (2011), "Deep-space optical communications: future perspectives and applications," Proceedings of the IEEE, vol. 99, pp. 2020–2039.

[2] C. C. Davis, I. I. Smolyaninov, and S. D. Milner (2003), "Flexible optical wireless links and networks," IEEE Communications Magazine, vol. 41, pp. 51–57.

[3] G. Nykolak, P. F. Szajowski, G. Tourgee, and H. Presby (1999), "2.5 Gbit/s free space optical link over 4.4 km," Electronics Letters, vol. 35, pp. 578–579.

[4] D.-Y. Song, Y.-S. Hurh, J.-W. Cho, *et al.* (2000), "4 × 10 Gb/s terrestrial optical free space transmission over 1.2 km using an EDFA preamplifier with 100 GHz channel spacing," Optics Express, vol. 7, pp. 280–284.

[5] M.-C. Jeong, J.-S. Lee, S.-Y. Kim, *et al.* (2003), "8 × 10-Gb/s terrestrial optical free-space transmission over 3.4 km using an optical repeater," IEEE Photonics Technology Letters, vol. 15, pp. 171–173.

[6] E. Ciaramella, Y. Arimoto, G. Contestabile, *et al.* (2009), "1.28 terabit/s (32 × 40 Gbit/s) WDM transmission system for free-space optical communications," IEEE Journal on Selected Areas in Communication, vol. 27, pp. 1639–1645.

[7] S. M. Navidpour, M. Uysal, and M. Kavehrad (2007), "BER performance of free-space optical transmission with spatial diversity," IEEE Transactions on Wireless Communications, vol. 6, pp. 2813–2819.
[8] M. Ijaz, Z. Ghassemlooy, J. Perez, V. Brazda, and O. Fiser (2013), "Enhancing the atmospheric visibility and fog attenuation using a controlled FSO channel," IEEE Photonics Technology Letters, vol. 25, pp. 1262–1265.
[9] H. Kaushal, V. Kumar, A. Dutta, *et al.* (2011), "Experimental study on beam wander under varying atmospheric turbulence conditions," IEEE Photonics Technology Letters, vol. 23, pp. 1691–1693.
[10] H. Henniger (2012), "Transmission performance analysis of free-space optical communications using Gilbert-Erasure channel," IEEE Transactions on Wireless Communications, vol. 60, pp. 55–61.
[11] N. S. Prasad (2005), "Optical communications in the mid-wave IR spectrum band," Journal of Optical Fiber Communications Report, vol. 2, pp. 558–602.
[12] T. Weyrauch and M. A. Vorontsof (2004), "Free-space laser communications with adaptive optics: Atmospheric compensations," Journal of Optical Fiber Communications Report, vol. 1, pp. 355–379.
[13] D. O. Caplan (2007), Laser communication transmitter and receiver design, Journal of Optical Fiber Communications Report, vol. 4, pp. 225–362.
[14] A. K. Majumdar (2005), "Free-space laser communication performance in the atmospheric channel, Journal of Optical Fiber Communications Report," vol. 2, pp. 345–396.
[15] J. C. Ricklin, S. M. Hammel, F. D. Eaton, and L. Svetlana (2006), "Atmospheric channel effects on free-space laser communication," Journal of Optical Fiber Communications Report, vol. 3, pp. 111–158.
[16] L. B. Pedireddi and B. Srinivasan (2010), "Characterization of atmospheric turbulence effects and their mitigation using wavelet-based signal processing," IEEE Transactions on Communications, vol. 58, pp. 1795–1802.
[17] I. Toselli, L. C. Andrews, R. L. Phillips, and V. Ferrero (2009), "Free-space optical system performance for a Gaussian beam propagating through non-Kolmogorov weak turbulence," IEEE Transactions on Communications, vol. 57, pp. 1783–1788.
[18] M. A. Khalighi, N. Schwartz, N. Aitamer, and S. Bouren-nane (2009), "Fading reduction by aperture averaging and spatial diversity in optical wireless systems," Journal of Optical Communications and Networking, vol. 1, pp. 580–593.
[19] J. Cheng, Y. Ai, and Y. Tan (2008), "Improved free space optical communications performance by using time diversity," Chinese Optics Letters, vol. 6, pp. 797–799.
[20] C. A. Rjeily and A. Slim (2011), "Cooperative diversity for free-space optical communications: transceiver design and performance analysis," IEEE Transactions on Communications, vol. 59, pp. 658–663.
[21] X. Zhu and J. M. Kahn (2002), "Free-space optical communication through atmospheric turbulence channels," IEEE Transactions on Communications, vol. 50, pp. 1293–1300.

[22] S. G. Wilson, M. Brandt-Pearce, Q. Cao, and J. H. Leveque (2005), "Free-space optical MIMO transmission with Q-ary PPM," IEEE Transactions on Communications, vol. 53, pp. 1402–1412.

[23] K. Wakamori, K. Kazaura, and I. Oka (2007), "Experiment on regional broadband network using free-space-optical communication systems," Journal of Lightwave Technology, vol. 25, pp. 3265–3273.

[24] B. E. A. Saleh and M. C. Teich (1991), "Fundamentals of Photonics," A Wiley-Interscience Publication, John Wiley and Sons, Inc. Ch. 4. Fourier Optics, pp. 108–156.

[25] L. C. Andrews and R. L. Phillips (2005), "Laser beam propagation through random media", SPIE Press, 2nd Ed, Bellingham, Washington, USA.

[26] S. Arnon (2003), "Effects of atmospheric turbulence and building sway on optical wireless-communication systems," Optics Letters, vol. 28, pp. 129–131.

[27] J. G. Proakis and M. Salehi (1994), "Communication Systems Engineering," Prentice Hall, Englewood Cliffs, New Jersey, USA.

[28] P. S. Priambodo, U. Darusalam, and E. T. Rahardjo (2015), "Free-space optical propagation noise suppression by Fourier optics filter pinhole," International Journal of Optics and Applications, vol. 5, pp. 27–32.

[29] U. Darusalam, P. S. Priambodo, and E. T. Rahardjo (2015), "Optical spatial filter to suppress beam wander and spatial noise induced by atmospheric turbulence in free-space optical communications," Advances in Optical Technologies.

[30] U. Darusalam, P. S. Priambodo, and E. T. Rahardjo (2015), "SNR and BER performance enhancement on FSO induced by atmospheric turbulence using optical spatial filter," International Journal of Optics and Applications, vol. 5, pp. 51–57.

[31] U. Darusalam, P. S. Priambodo, and E. T. Rahardjo (2015), "Noise suppression in the signal spectrum induced by atmospheric turbulence on the FSO (free-space optical communications)," International Journal of Technology, vol. 4, pp. 336–344.

[32] E. W. Ng and M. Geller (1969), "A table of integrals of the error functions," Journal of Research of the National Bureau Standard – B. Mathematical Sciences, 73B, pp. 1–20. C. Si, Y. Zhang, Y. Wang, J. Wang, and J. Jia, (2012), Average capacity for non-Kolmogorov turbulent slant optical links with beam wander corrected and pointing errors, Elsevier Journal of Optik, vol. 123, pp. 1–5.

[33] J. M. Bennett and E. J. Ashley (1965), "Infrared reflectance and emittance of silver and gold evaporated in ultrahigh vacuum," Applied Optics, vol. 4, pp. 221–224.

[34] Y.-S. Hurh, K.-W. Shin, S.-H. Lee, and J.-S. Lee (2005), "Weather-insensitive optical free-space communication using gain-saturated optical fiber amplifiers," Journal of Lightwave Technology, vol. 25, pp. 4022–4025.

[35] N. Kenari, M. Karimi, and Masoumeh (2011), "Free space optical communications via optical amplify-and-forward relaying," Journal of Lightwave Technology, vol. 29, pp. 242–248.

[36] U. Darusalam, P. S. Priambodo, and E. T. Rahardjo (2017), "Minimizing the temporal fluctuation in the signal spectrum of FSO communications to improve BER performance under the influence of atmospheric turbulence," In Broadband Communication, Wireless Sensors and Powering (BCWSP), 2017 International Conference on 2017 Nov. 21 (pp. 1–5). IEEE.
[37] A. B. Raj and U. Darusalam (2016), "Performance improvement of terrestrial free-space optical communications by mitigating the focal-spot wandering," Journal of Modern Optics, vol. 63, pp. 2339–2347.

Chapter 7

Characterization of atmospheric turbulence effects and their mitigation using wavelet-based signal processing

Latsa Babu Pedireddi[1] and Balaji Srinivasan[2]

7.1 Introduction

Free space optical (FSO) communication links are attracting a lot of attention lately due to the ever-increasing bandwidth requirement coupled with the well-known last mile bottleneck [1]. Recent interest in FSO communication has also been spiked due to the demonstration of a 100 Tbps free space data link enabled by three-dimensional multiplexing of orbital angular momentum (OAM), polarization and wavelength [2]. In fact, the use of light beams with OAM has spurred worldwide interest from several groups leading to practical demonstrations including the demonstration of 400 Gbps transmission over a 120 m long link [3]. The capacity limit of such transmission systems has also been analysed recently [4].

In the above FSO links, the turbulent atmospheric channel causes fluctuations in both intensity and phase of the transmitted optical signal, which degrades link performance. In order to make FSO communications widely deployable and commercially viable, there is a clear need for characterizing the atmospheric effects on the received signal. The signal statistics gathered through such characterization with respect to the key parameters causing atmospheric turbulence may possibly contribute to mitigating the signal degradation using physical optic principles and advanced real-time signal-processing techniques. In this chapter, the key theoretical aspects of dynamic atmospheric turbulence related effects such as scintillations and beam wandering are discussed. The theoretical aspects of the major turbulent atmospheric effects scintillations, beam wandering, beam pointing stability and their simulation and experimental verification are discussed in Section 7.2. Experimental measurement of the laser beam profile at the receiver plane is discussed in Section 7.3, which will be used as the basis for the experimental characterization of the atmospheric turbulence effects later in this chapter.

[1]Ambigai Consulting Services, GmbH, Germany
[2]Department of Electrical Engineering, Indian Institute of Technology, Madras, India

The experimental work presented in this chapter is primarily focused on determination of the individual contribution of scintillations and beam wandering to the received signal variance. This chapter also deals with experimental validation of the theoretical calculations of scintillation index fluctuations and beam wandering effects on the received signal variances. Investigations on the different physical optic principles and advanced signal processing methods to improve the signal quality at the receivers for FSO links are also discussed.

The primary physical factors responsible for atmospheric turbulence are temperature, pressure, humidity and wind speed. As such, it is important to study and establish the correlation between the received signal variance and the above atmospheric parameters. Such a correlation will help to build a model through which the channel properties may be estimated based on the measurement of the atmospheric parameters and can be compensated using advanced signal processing techniques. Such analysis has been carried out over a 24 hour period to include a full day-night cycle. The analysis of signal statistics with the atmospheric parameters are discussed in Section 7.4.

There are many ways of compensating the atmospheric turbulence effects using physical optics principles and advanced signal processing techniques based on the simplified atmospheric turbulence models. Each of these techniques have their advantages as well as limitations on the receiver size and complexity. For example, it is well known that the scintillation effects may be reduced by increasing the receiver aperture size as the effects get averaged across the aperture. However, this will also increase the background radiation, which not only increases the shot noise contribution to receiver noise but also drifts with time due to their dependence on the direction of sunlight. The same applies to beam wandering effect, which requires even bigger receiver aperture than that required to compensate scintillations. The approach described in this chapter is primarily focused on the aperture averaging method to mitigate the effects due to scintillations and a novel wavelet-based signal processing to mitigate the beam wandering effects and improve the received signal quality. The wavelet-based signal processing for the turbulent distorted signals and the analysis of the signal improvements with respect to the bit error rate (BER) are discussed in Section 7.5.

7.2 Atmospheric turbulence effects

In this section, we discuss the primary atmospheric turbulence-related effects including scintillations, beam wandering and beam pointing stability. After introducing such effects, we present simulation results which are then validated by experimental results described in the following sections.

7.2.1 Scintillations

Fluctuations in received irradiance resulting from propagation through atmospheric turbulence is known as scintillation. This is just like twinkling of a star. Scintillation

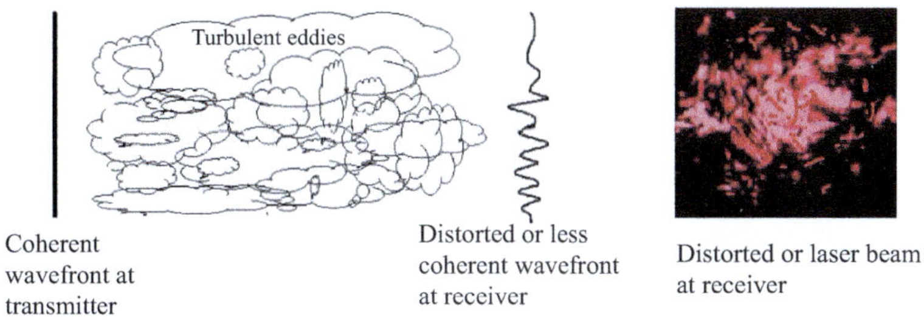

Figure 7.1 Schematic diagram illustrating the wavefront coherence degradation and laser beam distortion due to atmospheric turbulence

is basically due to random fluctuations in the wavefront at the receiver plane. Let assume that a plane wavefront is passed through a set of irregularities such as here refractive index fluctuations, then the wavefront is no longer planar and becomes less coherent as number of irregularities increases as shown in Figure 7.1 Suppose if the image of a Gaussian beam is propagated through these irregularities then the image of a Gaussian beam with distortion for weak turbulence conditions will be seen in the receiver plane as shown right side in Figure 7.1. The distortion in the beam is more for stronger turbulence conditions and it may even break up the beam in to multiple beams.

The Rytov method provides a closed form expression for the intensity fluctuations of a plane wave in weak turbulence condition for a point receiver by considering the normalized irradiance fluctuations at the receiver called scintillation index. The approximated Rytov variance is given by (7.1) [5].

$$\sigma_I^2 = 1.23 \times C_n^2 \times k^{7/6} \times L^{11/6} \tag{7.1}$$

where L is the link length (in metres). From (7.1), for any given FSO link, it is clear that the variance due to scintillations increases with increase in link length or turbulence (C_n^2). The Rytov approximation predicts that the signal variance due to turbulence can increase without limit for stronger turbulence. However, in reality the intensity fluctuations have observed to saturate [6–8]. The Rytov variance is found to be accurate only when the scintillation index is less than 0.3, or in other words when weak turbulence conditions exist [5,9,10].

7.2.1.1 Saturation of scintillations

The Rytov approximation was not able to account for the saturation effect [6,11,12]. This is because of not considering the effect of inner and outer scales of turbulence on the correlation function of the irradiance fluctuations. The outer scales effect was demonstrated by D A Dewolf [11]. In his analysis of scintillations-induced variance, he considered two parameters which describes the behaviour of the received signal, $2\alpha d$ and d/kL_0^2, where d is width of the turbulent region through which the light pass, k is the wave number, L_0 is outer scale of the turbulence, and

$$\alpha = \frac{k^2\sigma_\varepsilon^2}{16\pi}\int_0^\infty dK \cdot K\Phi_n(K,0) \approx k^2 L_0 \sigma_\varepsilon^2 \tag{7.2}$$

In (7.2), the parameter $\Phi_n(K,0)$ is the turbulence spectrum and σ_ε^2 is the variance of the dielectric permittivity. Here, the strength dependent factor $2\alpha d$ is compared with unity in order to determine whether multiple scattering must be taken into account in calculating the signal statistics at the receiver. If the flux density at the receiver is not deviating from the direction of propagation and depolarization effects are negligible, then the radiant flux density or irradiance I is proportional to the absolute square of the electric field $u(r)$. For the typical experimental conditions $l_0^2 << d/k << L_0^2$ given by [12], and by utilizing the Born series formulation of the scalar wave equation (7.3)

$$u(r) = u_0(r)[1 + B_1(r) + B_2(r) + \ldots.]B_1(r) = B_{1r}(r) + jB_{1i}(r) \tag{7.3}$$

The first Born approximation for the above experimental case is given by

$$\langle B_{1r}^2(r)\rangle = 0.033C_n^2 k^{7/6} d^{11/6} \tag{7.4}$$

The single scattering theory based on the first term of the (7.3) is not valid for $\langle B_{1r}^2(r)\rangle \geq 1$. From (7.4), if the width of the turbulent region is equal to link length then the Born approximation is the same as the Rytov solution for electric field fluctuations for weak fluctuations region. By using the high frequency perturbation method [12], the variance of the irradiance is given by [11]

$$\langle I^2\rangle = I_0^2[2 - \exp(-4\alpha' d)] \tag{7.5}$$

Here, $2\alpha' d = 2\langle B_{1r}^2\rangle$, given by [11] with a very negligible error approximation, then

$$\langle I^2\rangle = I_0^2\left[2 - \exp\left(-4\langle B_{1r}^2\rangle\right)\right] \tag{7.6}$$

by relating the Rytov approximated irradiance fluctuations to $\langle B_{1r}^2\rangle$ the log irradiance variance is given by (7.7)

$$\sigma_{lnI}^2 = ln\left[2 - exp\left(-\sigma_1^2\right)\right] \tag{7.7}$$

This equation shows that the irradiance fluctuations gets saturated at a maximum value of $ln(2)$ for strong turbulence conditions and follows the linear Rytov approximation for weak fluctuations $(\sigma_{lnI}^2 \leq 0.3)$.

7.2.1.2 Aperture averaging

The process or method of averaging the irradiance fluctuations spatially at the receiver aperture is known as aperture averaging. The analysis presented up to now is scintillations-induced variance for a point receiver. If the receiver aperture in an optical communication system is smaller than the correlation length of the irradiance fluctuations, then the aperture behaves essentially like a point receiver. However, as the aperture size increases beyond the irradiance correlation length, then the received signal is like the cumulative of the several individual point receivers, hence the receiver can find the signal at any one of those individual receivers, this means obviously a reduced variance due to scintillations. The reduction in the variance due to scintillations depends on the number of point receivers or as the aperture area is increasing. The reduction in scintillation-induced variance due to aperture averaging can be deduced from the ratio of the power fluctuations by a finite size collecting aperture to that obtained by a very small (point or aperture radius is less than the $\sqrt{\lambda L}$) aperture. The normalized variance of the power fluctuations in the receiver plane is same as the flux variance, if irradiance fluctuations defined as

$$\sigma_I^2(r) = \frac{\langle P(r)^2\rangle - \langle P(r)\rangle^2}{\langle P(r)\rangle^2} \tag{7.8}$$

Then the aperture averaging factor or coefficient A for a circular aperture of radius r is defined by the ratio $A = \sigma_I^2(r)/\sigma_I^2(0)$, where $\sigma_I^2(0)$ is the scintillation index for a point aperture receiver. Under weak irradiance fluctuations for plane wave case, a simple approximation for A is given with $<7\%$ error [12,13].

$$A = \frac{\sigma_{\ln I}^2(r, L)}{\sigma_{\ln I}^2(0, L)} = \left[1 + 1.07\left(k \times r^2/L\right)^{7/6}\right]^{-1} \tag{7.9}$$

The general behaviour of the irradiance flux variance and aperture averaging factor A are approximately same for the plane wave, spherical wave and beam wave cases. However, if the receiver aperture is high enough to receive the entire laser beam then the scintillations becomes zero. The plane wave approximation for aperture averaging factor can be used for Gaussian light beams, because the plane wave solutions gives the upper bound for scintillations. The aperture averaging factor for Gaussian beams decreases as aperture size increases. It can be seen that the Aperture averaging factor reduced to less than 4 per cent for two correlation lengths for a given link length and wavelength. Hence, the plane wave approximated solution for A can be used with less error for the case of an aperture much less than the beam diameter at the receiver.

7.2.1.3 Modification in determination of scintillation index

Atmospheric turbulence-induced image degradation and intensity fluctuations and their consequences are studied for plane wave and spherical wave approximations. Two useful parameters describing the turbulence-induced fading are the correlation length ρ_0 of intensity fluctuations, and the correlation time of intensity fluctuations t_0. The correlation length or width (ρ_0) is defined as the point of the normalized spatial covariance function. While the correlation time (t_0) is defined for temporal normalized covariance function. Generally, this is in the orders of μs. The Rytov closed form solution given by (7.1) is for weak turbulence conditions. This solution does not valid for strong turbulence conditions, because of its linear dependence on atmospheric structure parameter which describes the strength of the turbulence and

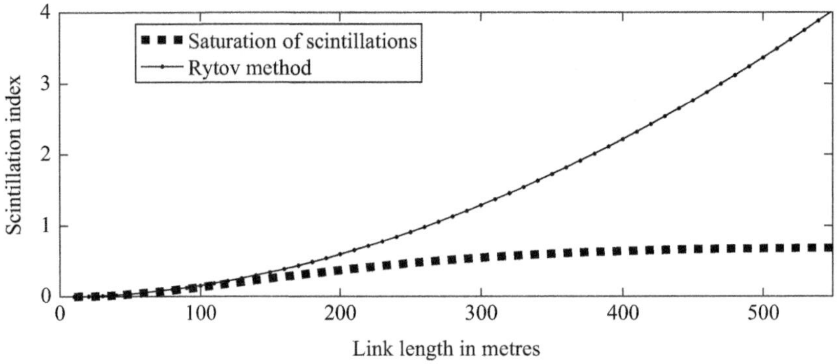

Figure 7.2 Comparison of the Rytov approximation with empirical data

link length. A long haul link provides the stronger turbulence conditions, but the reverse may not be true. The scintillations-induced variance at the receiver depends not only on scintillations saturation but also on the aperture averaging factor. Hence, the resultant scintillations-induced variance is given by the multiplication of the aperture averaging factor and the variance due to scintillations under the consideration of saturation effect as follows.

$$\sigma_{\ln I}^2 = \ln\left[2 - e^{-\sigma_I^2}\right] \times \left[1 + 1.07\left(k \times r^2/L\right)^{7/6}\right]^{-1} \tag{7.10}$$

where r is the radius of the receiver aperture. Using (7.10) and (7.1), one can determine the variance of the scintillations at the receiver as a function of the receiver aperture radius (r). Theoretical comparison of Rytov solution with the empirical relation is shown in Figure 7.2. For shorter links both the solutions have no significant change, but longer link lengths or for stronger turbulence conditions, the difference is much higher.

7.2.2 Beam wandering

In the absence of turbulence, the transmitted laser beam has an aperture diameter D would have an angular spread θ_0 approximately λ/D, where λ is the wavelength of the transmitted optical beam. This spread becomes larger at the receiver for turbulent atmospheric case, because moving turbulent eddies cause scattering of the optical fields. Assume that the observation time is $t > D/|v|$ and the intensity in the receiver plane is averaged over this interval. As such, the beam spread may be short term or long term or both. By definition when a laser beam interacts with turbulent eddies whose dimensions are larger compared to the diameter of the laser beam, the deflection of the beam is relatively low. However if such turbulent eddies are moving, then the laser wave front path is no more linear. For example point a visible laser beam on wall and throw a glass ball normal to the laser beam at different velocities, then the laser spot on the wall moves across in the direction

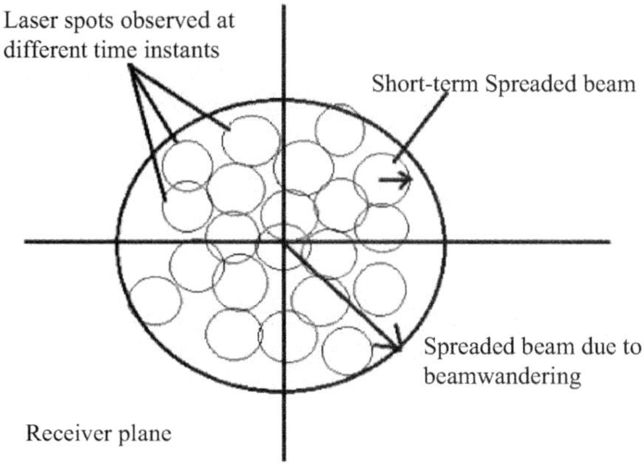

Figure 7.3 Illustration of short- and long-term beam spread due to atmospheric turbulence

glass ball is moving and the deviation seen will be more for higher velocities of the ball movement. This is because the glass ball physical position changes over time as it is moving and the time taken by the laser beam to enter and exit this glass ball. Hence the shift in the laser beam in the receiver plane will depends on the size of the turbulent eddies or the propagation time of the laser beam through these turbulent eddies and the velocity of these turbulent eddies. Such a beam spread is termed as long-term spread or beam wandering. In contrast, those eddies which are smaller compared to the beam diameter tend to broaden the beam, but do not deflect it significantly [5,9]. In general, the turbulent eddies continually flow across the laser beam with a transverse flow velocity v and cause the deflection of the laser beam in different directions with time intervals in the order of $D/|v|$. If we observe the super positioned beams over the interval greater than $D/|v|$, then resultant beam seems to be a single beam with larger diameter which is greater than the short term spread as seen from Figure 7.3.

For Gaussian beam propagation over turbulent media, with link lengths much shorter than $(k^2 \times C_n^2 \times l_0^{5/3})^{-1}$ and $\rho_0 >> D_t$, then mean square radius of the beam wander is approximated by Fante [5].

$$\langle \rho_L^2 \rangle \approx \frac{4L^2}{k^2 \times D_t^2} + \frac{D_t^2}{4}\left(1 - \frac{L}{F}\right)^2 + \frac{4L^2}{k^2 \rho_0^2} \tag{7.11}$$

where $\rho_0 = \left[0.5475 \times k^2 \times L \times C_n^2\right]^{-3/5}$. The first two terms in (7.11) represents the beam spread in vacuum and the last term represents the additional spread due to scattering of the laser beam by the turbulent eddies. The long-term beam spread is a zero mean normal random variable with variance given by (7.11). For the case of weak turbulence and shorter link lengths (less than $(k^2 \times C_n^2 \times l_0^{5/3})^{-1}$, where l_0 is the inner scale of the turbulent eddies), the variance due to beam wandering effect is strongly dependent on receiver aperture diameter.

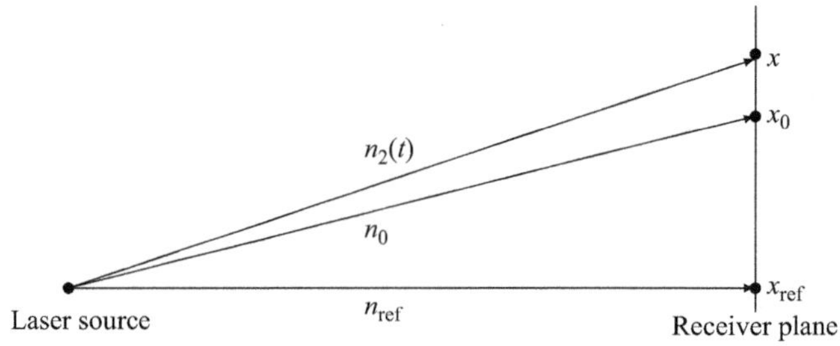

Figure 7.4 Schematic representation of tilt in the laser beam due to the changes in the atmospheric parameters

7.2.3 Beam-pointing stability

The large-scale changes in the atmospheric temperature, pressure and humidity affects the average position of the light beam measured over a period much more than $\frac{|D|}{v}$ seconds [5,10]. The atmospheric refractive index changes with time as the atmospheric parameters like temperature, pressure and humidity changes. The refractive index profile is decreasing with altitude over the day because of thermal radiations from the earth and surrounding structures like buildings. This decay in the refractive index profile causes the shift of the laser beam in the receiver plane as shown in the Figure 7.4.

Let assume that the atmospheric reference refractive index n_{ref}, when it is aligned to point at $x_{\rm ref}$ in the receiver plane. And the refractive index $n_2(t)$ as a function of the atmospheric parameters is given by (7.12) [5,9,10], can be used to calculate refractive index. The shift (x_1) in the receiver plane can be calculated theoretically by knowing the refractive index $n_2(t)$ and $n_{\rm ref}$ by using (7.13).

$$\langle n(r,t)\rangle = 1 + \frac{77.6\times 10^{-6}}{T}\times\left(1+7.52\times 10^{-3}\times\lambda^{-2}\right)\times\left(P+\frac{4810\times e}{T}\right) \tag{7.12}$$

$$x_1 = x - x_{\rm ref} = L\times\left(\sin^{-1}\left(\frac{1}{n_2(t)}\right) - \sin^{-1}\left(\frac{1}{n_{\rm ref}}\right)\right) \tag{7.13}$$

Experimentally the shift of the light beam in the receiver plane can be measured by placing a photographic plate in the receiver plane and by finding the maximum intensity points by providing the snapshots of the beam is taken for the intervals much longer than $\frac{|D|}{v}$ seconds [5,9]. The details of the experimental measurement procedure is outlined as *Positional Shift Measurement* in Section 7.4. This method can be applied for both large scale changes as well as for small scale changes (due to beam wander) depending on the average powers taken for different time intervals.

7.3 Free space optical link experimental set-up and data acquisition

Before discussing the atmospheric effects on laser beam propagation, the laser beam power profile at the receiver should be known. Theoretically, the received beam power profile is same as the transmitted. However, in reality it depends based on the divergence of the laser source used. To have a consistent experimental results and to study atmospheric turbulence effects it has to be measured experimentally and characterised under low turbulence conditions. For such characterization, it has been assumed that the received laser beam power profile is similar to that of the source. The picture of the 50 m folded outdoor FSO link experimental set-up is as shown in Figure 7.5. It is used to validate the laser beam power profile and also to capture the receiver signal data for the atmospheric turbulence effects and signal statistics analysis later in this chapter.

7.3.1 Transmitter and receiver design

The experimental set-up consists of a directly modulated optical transmitter and a low noise receiver. The optical source in the transmitter is a Vertical Cavity Surface Emitting Laser (VCSEL, HFE 4085-321 from Advanced Optical Components). The HFE 4085-321 is a high performance 850 *nm* VCSEL packaged for high speed data communications. The driver circuit is designed to protect the VCSEL by ensuring that the operating current is less than the maximum current rating of VCSEL i.e, 15 *mA* and to provide a modulation bandwidth of 10 *MHz* using a transistor BC557. The output of the VCSEL is collimated using a GS-7020-H aspheric lens from Roithner Lasertecnik. The transistor is operated in saturation and cut-off regions for logic one and logic zero transmission, respectively. The VCSEL is driven at 13.5 *mA* for logic one and zero current for logic zero transmission. The receiver consists of a Fresnel lens with a focal length of 25 *cm* which is used to focus the incoming beam on to a Si-PIN photo detector (SHF214 from

Figure 7.5 Experimental set-up of 50 m folded FSO link and data acquisition system on the terrace of Electrical Sciences Building, IIT-Madras

OSRAM). Photo current generated in the PIN photo diode is amplified in a trans-impedance configuration using a high gain FET input operational amplifier (OPA657U from Texas Instruments), which has a gain bandwidth product of 1.6 GHz.

7.3.2 Experimental set-up of 50 m folded free space optical link

The experimental set-up of a 50 m folded FSO link on the terrace of the Electrical Sciences Building at Indian Institute of Technology – Madras is shown in Figure 7.5. The transmitter is packaged in an electrically grounded aluminium box, where in the VCSEL and collimating lens are fixed on a kinematic optical mount with fine adjustable screws to control the direction of the laser beam. The transmitter box is fixed on a wooden base that is firmly seated on a rugged steel frame that ensures no displacement of the transmitted beam due to wind. The transmitter is directly modulated using 1 *kHz* square wave signal generated from a function generator (SM 5030-5 from Scientific). The modulated optical beam is directed towards a mirror firmly fixed on a wall 50 m away so that the beam is reflected towards the receiver that is placed near the transmitter. Such an arrangement eases the initial alignment process for the FSO link. The transmitter and receiver have to be placed at least at enough height (1.2 m) to avoid the effect of thermal radiation from the concrete terrace and walls of the building, which typically dies around 1 m away from the surface of the wall or floor [14].

The receiver optics and circuitry are placed inside a wooden box covered from all sides except for the laser beam entrance. Such a box ensures that the receiver picks up a minimal amount of background radiation. As in the case of the transmitter box, the receiver box is also firmly attached to a steel frame which in turn is attached to the terrace wall. The laser beam is sampled by an aperture made of black cardboard and then focused on to the photo detector using Fresnel lens. The Fresnel lens is fixed in the receiver box so as to prevent any perturbation of the lens, which may corrupt the received signal.

A weather-monitoring kit developed by the Tenet Group at IIT Madras is used for parallel monitoring of the atmospheric parameters to analyse the statistics of the atmospheric turbulence effects. Pictures of this kit for measuring the temperature, pressure, humidity and wind speed are shown in Figure 7.5. The weather monitoring station is co-located on the same terrace with the FSO link. The devices shown in the set-up are interfaced by a firmware, which provides the atmospheric data in a tabular format.

7.3.2.1 Signal capture procedure

The manual data collection from the FSO link over a day is an extremely tedious exercise. Hence, automation of the data collection has been done using a digital storage oscilloscope (DSO) and a computer as follows. The output from the receiver is connected to the DSO through a low noise co-axial cable (RG-11). The DSO samples the analogue signal at a rate controlled by the DSO time-base and the corresponding digital samples are stored in the buffer. This data is subsequently transferred to the system using a Matlab interface via a USB 2.0 cable. The experiment is carried out throughout the day, with one measurement consisting of 50 continuous frames acquired every 20 minutes. Before each measurement the link is aligned for maximum throughput so that any drift in the laser beam pointing may be compensated. In these measurements, each frame contains 2,500 samples of the signal at sampling rate of 1.25 mega samples per second. The data captured for each time frame is saved for further analysis at a later time. Assumed that the wind fluctuations are negligible at the above sampling rate. Variances due to beam wander and scintillations are measured for every 20 minutes and the atmospheric parameters are also measured at the same instant.

7.3.3 Theoretical fit to the laser beam power profile

Mathematically, the normalized laser beam profile may be approximated to be Gaussian for the transmitted beam and can be represented as

$$f(r) = \exp\left(-\frac{r^2}{2\sigma_x^2}\right) \tag{7.14}$$

where σ_x^2 is variance of the power profile. Then the normalized power received with varying aperture radius r is given by

$$\frac{P(r)}{\max(P(r))} = erf\left(\frac{r}{\sqrt{2} \times \sigma_x}\right) \tag{7.15}$$

The beam diameter is defined as twice the radius at which the normalized power ratio reaches about 0.95, that means 95 per cent of the total power is concentrated within the beam diameter. Experimentally, the laser beam profile and beam diameter can be measured by sampling the received laser beam with different aperture radii. The measured output peak to peak voltage of trans-impedence amplifier is measured with different sampling apertures and then normalized with respect to the maximum voltage. These voltage ratios are exactly same as the optical power ratios because the output voltage of the trans-impedance amplifier is proportional to the input optical power and the ratio takes out the multiplicative coefficients. Theoretical values are calculated based on the divergence of the source, which is approximately 1.1 milli-radians. The plot of normalized measured power profile and the theoretically expected profile are shown in Figure 7.6. The data shows that the beam profile at the receiver is Gaussian (same as the source) and the measured beam diameter is 11.4 *cm*, which is approximately equal to the theoretical diameter with a deviation of 3.6 per cent.

7.3.4 Controlled environment experimental set-up

In order to characterize the effect of atmospheric turbulence-induced scintillations on the received signal variance and its dependence on aperture averaging, the transmitter and receiver described earlier in this section are used. The schematic of a typical FSO link set-up is as shown in the Figure 7.7 consists of an optical

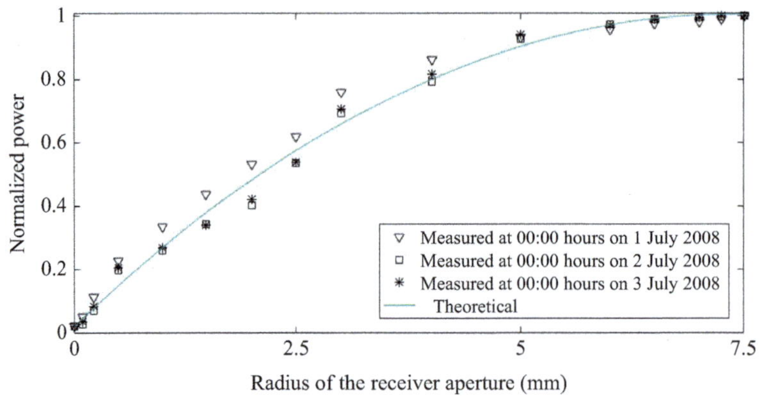

Figure 7.6 Experimental results of the beam characterization at the receiver and the corresponding Gaussian fit

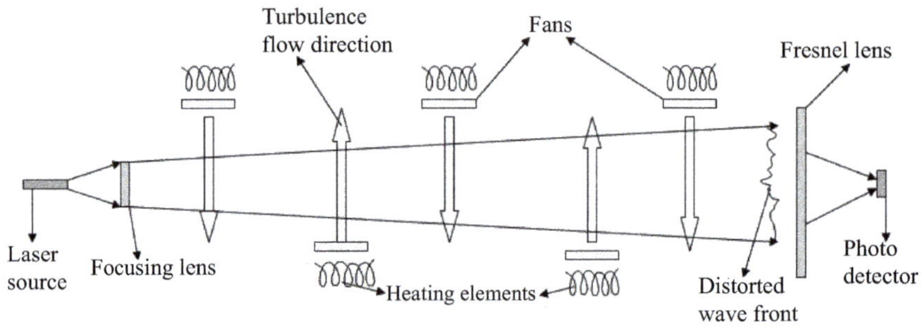

Figure 7.7 Schematic diagram of indoor experimental setup of a FSO link

wireless link with simulated turbulence channel. The atmospheric turbulence was simulated using multiple fans and heating elements. The data is captured with different aperture radii. Here, the radii used are 0.18, 0.75, 1.6, 4.25 and 6.5 mm. Here, there is no specifics on selecting the radii, it will be good for analysis if these are nearly equal to the multiples of the correlation length of irradiance fluctuations. The set-up is made on optical table which has very low vibrations, and the transmitter and receiver are fixed to the optical mounts. The experiment was repeated with out disturbing the transmitter and receiver mounts for different aperture.

The measured voltages are converted back to the Intensity by using

$$\text{Intensity} = \frac{V}{\pi r^2 \Re R_L} \tag{7.16}$$

where V is voltage measured from DSO, r the radius of the receiver aperture, $\Re$ the responsivity of the photo detector and R_L is load resistance of the trans-impedence amplifier. The variance of the log intensity is measured. The variance of the logic

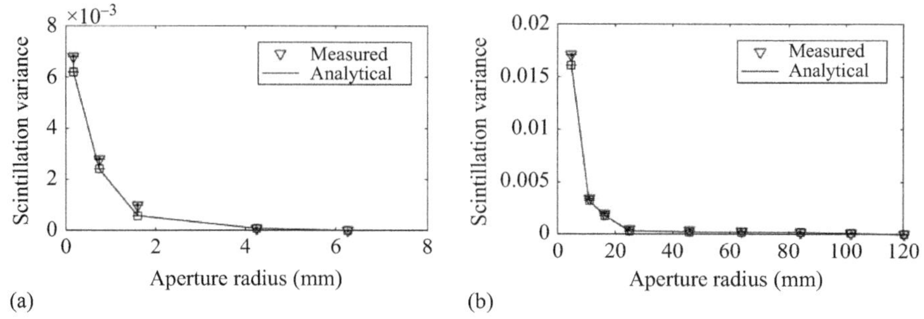

Figure 7.8 Plot of the variance due to scintillations as a function of receiver aperture for (a) indoor link, (b) outdoor link and their corresponding theoretical fit incorporating C_n^2 as the fit parameter

zero is taken out from the resultant variance of the logic one to get the variance of scintillations. Here we assumed that the photo detector output is the sum of the circuit noise plus the scintillations distorted signal. For logic zero the variance is due to noise alone and for logic one the variance is sum of the variances due to scintillations plus the circuit noise. We also assumed that the scintillations distorted signal and the receiver circuit additive white Gaussian noise are independent.

The signal variance is measured for different aperture sizes given. The wave number spectrum structure parameter C_n^2 is measured from the variance measured with 0.18 mm diameter aperture using modified scintillation index approximation given by (7.10) to be $2.1 \times 10^{-11}\ m^{-2/3}$. The experimental values of the variance due to scintillations are calculated by calculating the intensity at the receiver using the measured voltages. The theoretical values are calculated using the measured C_n^2 and the aperture radii used. The plot of experimentally measured variance in comparison to the theoretical values are given in Figure 7.8(a)-indoor link with $C_n^2 = 2.1 \times 10^{-11}\ m^{-2/3}$ and (b)-outdoor link with $C_n^2 = 3.1 \times 10^{-13}\ m^{-2/3}$).

The variance due to the beam wandering are assumed to be negligible here because the fans are operated at constant speed the outer minimal scale of the turbulence expectable is the dimension of the fan that is the diameter of the fan. Here the signal fluctuations are due to fluctuations in index of refraction produced by the heating elements and fans, which are in small scales. The experimental results are not much deviated from the theoretical estimates and are as accurate as in [12].

7.4 Experimental analysis of turbulence effects

From the frozen-in hypothesis [5,9], one can determine the spatial characteristics of the atmospheric turbulence based on its temporal characteristics. By making $r' = vt$, where v is the wind speed perpendicular to the direction of laser beam propagation. It is clear that the flow of outer scale movements are slower than the inner scales of the turbulent eddies for $t << D/|v|$ and the scintillations are due to the presence of the turbulent eddies. Following this theory, we deduce that

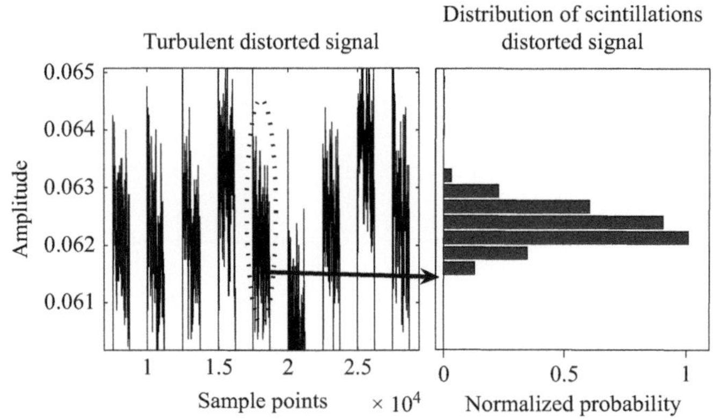

Figure 7.9 Sample received data illustrating the distortion due to scintillations within a logic one bit interval and the corresponding histogram © [15]

scintillations are not due to turbulent eddies which are larger than the laser beam diameter and the flow of eddies normal to the beam propagation. But, the beam wander solely depends on them and the beam wandering exists in the order of $D/|v|$. For most applications it is in the order of a few milliseconds. Hence, the intensity fluctuations occurring over shorter intervals (fraction of millisecond) are due to the scintillations alone. The physical means of the scintillations variance is shown in Figure 7.9. The turbulent distorted signal is zoomed for logic '1' and the distribution of the signal received for 0.5 *ms* circled samples on left is presented on right side of the Figure 7.9. Here the variance of the distribution is physically due to scintillations and due to circuit noise also. For measuring the only scintillations variance the variance of the 0.5 *ms* interval of the zero bit is subtracted.

Similarly, Figure 7.10 shows physical distortion due to the beam wander. Here also from the variance of logic one is subtracted from the logic zero for the same intervals.

7.4.1 Analysis of the beam wandering

The mean of the received signal changes for the intervals in the order of milliseconds are due to beam wander, because the ratio of the diameter of the laser beam at the receiver to the wind speed will vary around over a millisecond scale. From the theory, measured change in mean values over a millisecond intervals and then converted in to positional shifts. The positional shifts measured should have a Gaussian distribution.

To estimate the beam wander theoretically, we need to know the average refractive index structure parameter (C_n^2). This is determined by measuring the variance of irradiance over milli-second time scales followed by determining the corresponding C_n^2 through (7.10) and averaging C_n^2 values.

The theoretical simulations are made for the measured σ_1 and C_n^2 values. The experimental results are shown in Figure 7.11 for different times of a sunny day,

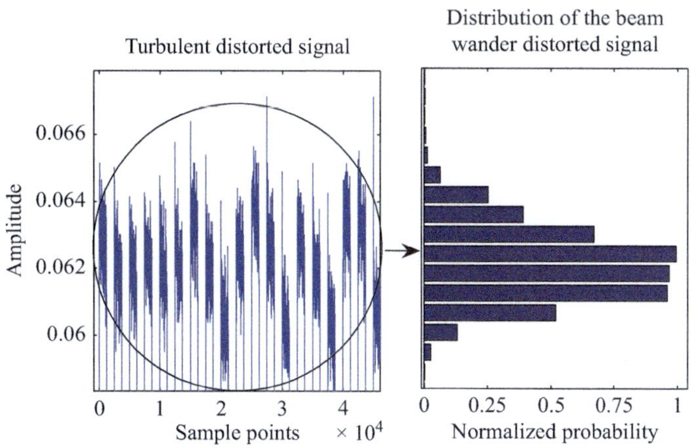

Figure 7.10 Sample received data illustrating the distortion due to beam wander over several logic one bit intervals and the corresponding histogram © [15]

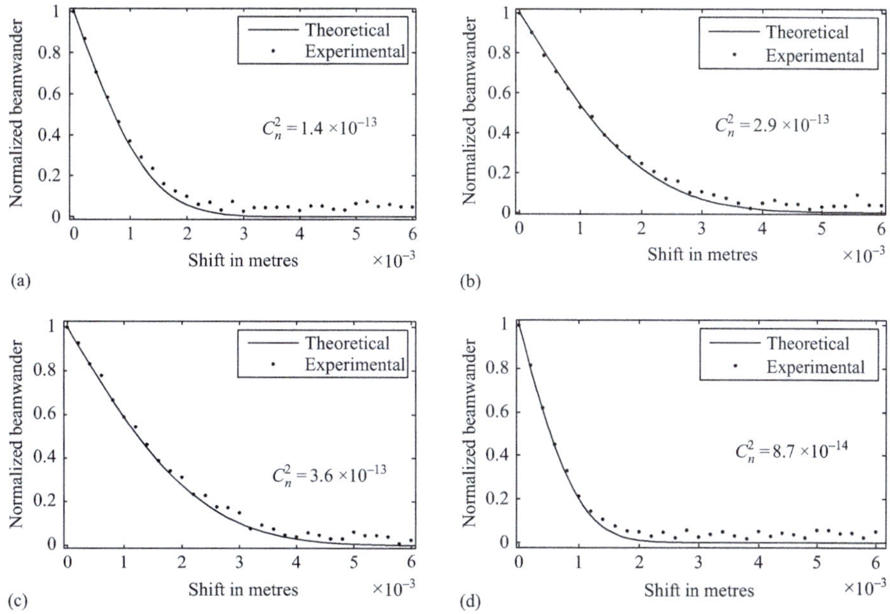

Figure 7.11 Plot of the probability of beam wander as a function of estimated beam shift measured at four different times of the day: (a) morning; (b) noon; (c) evening; and (d) night and the corresponding theoretical fit

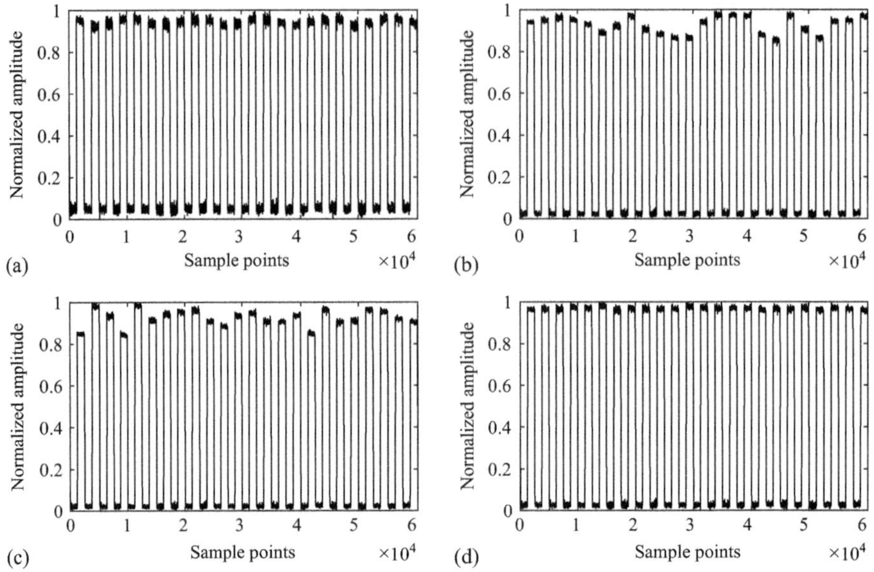

Figure 7.12 Real-time distorted signal captured at four different times of the day: (a) morning; (b) noon; (c) evening; and (d) night

with the corresponding estimated values of C_n^2. The estimated values are found to be roughly consistent with the expected turbulence in the atmosphere at different times of the day. Upon seeing the results from Figure 7.11, one can may be tempted to ask why there is a beam wander during the night also. This is because the beam wander not only depends on presence of the turbulent eddies, but also on wind velocity.

7.4.2 Signal statistics over a day and correlation with atmospheric parameters

From Figure 7.12, any offset in the time interval corresponding to zero bit is due to background radiation and must be deduced. Any signal fluctuations in the same time period may be attributed to the solar shot noise and circuit noise. These can be corrected in the data before analysing them, to determine variance due to scintillations and beam wandering.

In order to study the statistics of the link over a day, the variance of the received signal due to scintillations and beam wander should be measured continuously. The variance of the signal should be measured after compensating for any beam shift due to pointing instability, as discussed in Section 7.2.3. The compensation has to be done at each time of the experiment by adjusting the laser beam position in the receiver plane to maximize the intensity at the centre of the receiver aperture. The experimentally measured variances due to scintillations and beam wander are presented in Figure 7.13 for a sunny day. The profiles of measured variances due to scintillations and beam wander are consistent with the variance profiles shown in [13], even though the magnitudes are different (as it

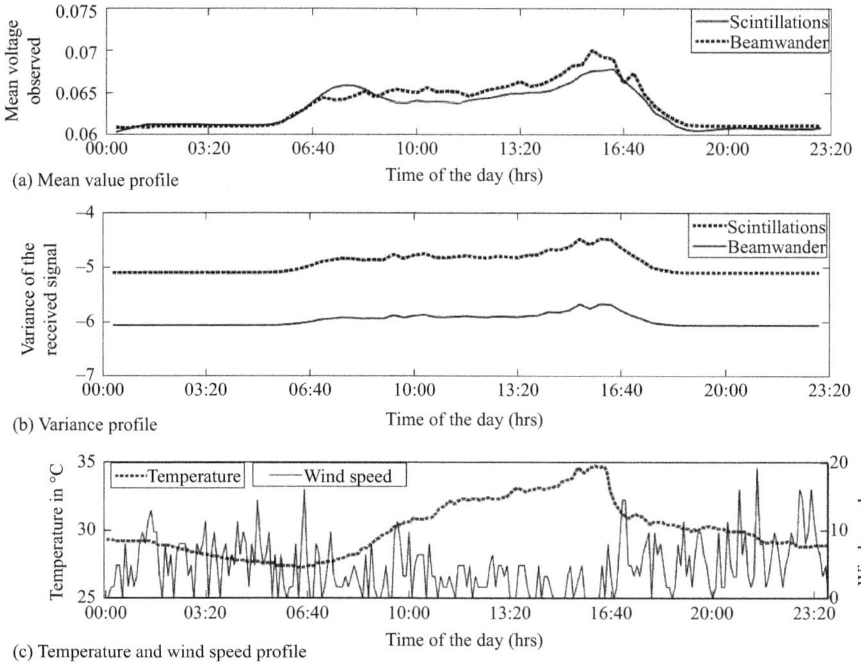

Figure 7.13 Measured receiver signal characteristics including (a) mean and (b) variance due to scintillations and beam wander, and (c) atmospheric parameters temperature and wind speed over a day

depends on the atmospheric conditions across the link). From Figures 7.12 and 7.13, the following conclusions can be made.

- The mean value of the received signal increases over a day due to background solar radiation.
- The variance also increased due to increased turbulence at warmer and windy weather conditions.
- The turbulence effects are peaking in the afternoon and minimum at night.
- The variance due to beam wander is always greater than the scintillations.

7.4.3 Correlation of turbulence-related data with atmospheric parameters

In the previous section, few statements were made on the received signal variances due to scintillations and beam wander. For better understanding of the variance fluctuations over a day, there is a need to know how the signal statistics are dependent on atmospheric parameters.

Now, let us assume that the variance due to scintillations over a day as a function of atmospheric parameters and time as follows

$$\sigma_S(t) = f(T(t), P(t), e(t), v(t)) \tag{7.17}$$

and $T(t)$, $P(t)$, $e(t)$ and $v(t)$ are the corresponding temperature, pressure, humidity and wind speed as a function of time. The functional approximation of variances with these atmospheric parametric functions gives the correlation coefficients. For example let, α_{T-S} is the correlation coefficient of scintillations variance and temperature then

$$\alpha_{T-S} = \int_0^{t_0} \sigma_s(t) \times T(t)dt \tag{7.18}$$

The correlation coefficients of variances with atmospheric parameters are shown in Table 7.1. At this point the question is that, what these correlation coefficients mean. The correlation coefficients between two functions gives the linear relation between those functions, how the variations in one of the functions affects the other function. The correlation coefficients range from -1 to 1. If correlation coefficient is zero means there is no functional approximation is possible. For $+ve$ coefficients, the functional dependence is proportionality, means the increase in one function leads to increase in the other function. For $-ve$ coefficients, increase in one function causes decrease in the other. From the correlation coefficients presented in Table 7.1, the following conclusions can be made.

- Temperature correlates well with both scintillations and beam wandering variances, because temperature is the cause for the atmospheric turbulence.
- Humidity is inversely correlated with these, because the refractive index is directly related (refer (7.12)).
- The pressure variations are less correlated because pressure variations are dissipated much faster than the effects looking for.
- The wind speed is well correlated with beam wander, as the beam wander is due to random motion of the turbulent eddies. The scintillations are also correlated with wind speed, as the wind speed results in creation of smaller turbulent eddies too.

The following are the list of measures to be considered for experimental verification of the turbulence effects.

- The external radiations effect on the receiver circuit performance.
- The building thermal radiation effects on laser beam by choosing the proper height of the transmitter and receiver.

Table 7.1 Correlation coefficients of atmospheric parameters with variance due to beam wander and scintillations

	Variance of beam wander	**Variance of scintillations**
Temperature	0.69	0.65
Humidity	−0.55	−0.51
Pressure	−0.04	−0.07
Wind speed	0.43	0.20

- The effects of wind on receiver and transmitter set-up.
- The transmitter and receiver link alignment before start of the experiment.
- Precise and accurate data acquisition.

7.4.4 Positional shift measurement

For a fixed aperture (D_0) in the receiver plane the detected power fluctuation is due to either beam wandering or scintillations. In a practical link, one should also consider tilting of the laser beam due to the relatively slow (over several minutes) variations in the refractive index of the medium. As shown in Figure 7.14, the shift of the laser beam takes the beam away from the receiver causing less power focused on to the photo detector which gives reduced voltage output of the trans-impedance amplifier. To analyse this, let the laser beam profile at the receiver plane be considered as a normalized Gaussian with zero mean (centred at the receiver aperture) and with variance σ_x^2, then the normalized power profile is mathematically represented as

$$f(x) = \exp\left(\frac{x^2}{2\sigma_x^2}\right) \tag{7.19}$$

If x_1 is the positional shift of the light beam in the receiver plane due to atmospheric parameters change, then the light beam profile becomes

$$f(x') = \exp\left(\frac{(x - x_1)^2}{2\sigma_x^2}\right) \tag{7.20}$$

Let us assume that the receiver aperture is D_0 and its position in the receiver plane is fixed, then the received power is equal to

$$P = \int_{-D_0/2}^{D_0/2} \exp\left(\frac{x^2}{2\sigma_x^2}\right) dx \tag{7.21}$$

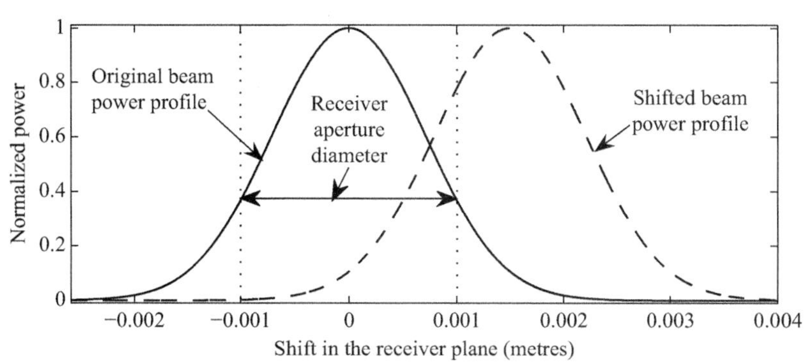

Figure 7.14 Illustration of a normalized Gaussian beam shift with respect to a fixed aperture due to atmospheric changes

Similarly, the power received with a positional shift x_1 is given by

$$P' = \int_{-D_0/2}^{D_0/2} \exp\left(\frac{(x - x_1)^2}{2\sigma_x^2}\right) dx \tag{7.22}$$

Then the ratio of the power received due to original and shifted light beams is given by

$$\frac{P'}{P} = \frac{2 \times erf\left(\frac{D}{\sqrt{2}\sigma_x}\right)}{erf\left(\frac{D+x_1}{\sqrt{2}\sigma_x}\right) + erf\left(\frac{D-x_1}{\sqrt{2}\sigma_x}\right)} \tag{7.23}$$

For small aperture diameters, by using the first two terms in the error function approximation and by substituting in (7.23) and by rewrite to determine the laser beam shift (x_1) as as a function of the measured optical power and receiver aperture diameter as follows.

$$x_1 = \sqrt{\left(\frac{1}{3}\right)\left(1 - \frac{P'}{P}\right)\left(6\sigma_x^2 - D^2\right)} \tag{7.24}$$

In general, the shift of the light beam is due to the change in the refractive index of the atmospheric channel which may be large scale or small scale, defined with respect to time scale in the order of seconds [5]. The large scale effect is due to the changes in mean index-of-refraction and the latter is due to the flow of the turbulent eddies. The mean index of refraction changes due to the atmospheric parameters, primarily due to humidity (e) and temperature (T). The measured shifts as a function of the $\frac{e}{T^2}$ is plotted in Figure 7.15 and measured shift is roughly consistent with the theoretical fit.

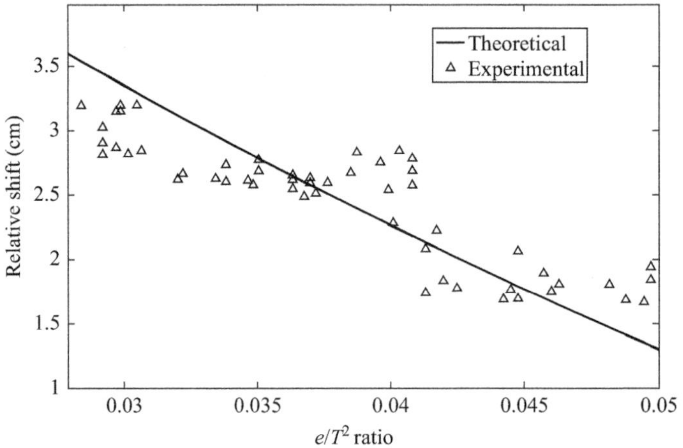

Figure 7.15 Plot of the observed beam shift as a function of e/T^2 and the corresponding theoretical fit

7.5 Turbulence effects mitigation using wavelets

The aperture averaging method reduces the turbulent-induced noises greatly by increasing receiver aperture diameter, with the assumption that the centroid of the transmitted laser beam is fixed. In reality this is not true due to the outer scale of turbulent eddies the centroid of the laser beam shifts greatly in the receiver plane, which is defined as the beam wandering effect. Aperture averaging is used to minimise the noise due to scintillations. The aperture diameter required to compensate the beam wandering effect increases with the increasing link length, hence in turn increasing the size of the receiver size, which is not a viable solution to realize commercially. This can be achieved with the simple adaptive gain control or with advanced signal-processing techniques such as wavelets. This section details, how the wavelets are used to compensate the turbulent-induced noise on FSO links, with the experimental data and how the signal quality is improved with this technique. Here, the turbulence channel is considered as a multiplicative noise. Here a typical automatic gain control (AGC) electronics is not a good solution as its gain function adds the additional white noise, which we do not have control over it.

7.5.1 Introduction to wavelet-based discrete signal processing

Let us assume that, the continuous time signal $r(t)$ is sampled to make it discrete signal $r[n]$. Then Discrete Wavelet Transform of a signal $r[n]$ is calculated by passing through a series of filters. First the samples are passed through a low-pass filter with impulse response $l[n]$, then the resulting signal is the convolution of the input signal $x[n]$ and the impulse response of the filter $l[n]$ given by [16], as shown in (7.25)

$$y[n] = x[n] * l[n] = \sum_{k=-\infty}^{\infty} x[k]h[n-k] \tag{7.25}$$

The signal is also decomposed simultaneously using a high-pass filter $h[n]$. The outputs giving the detail coefficients (from the high-pass filter) and approximation coefficients (from the low-pass filter). It is important that the two filters are related to each other and they are known as quadrature mirror filters. However, since half the frequencies of the signal have now been removed, half the samples can be discarded according to Nyquist's criteria. The filter outputs are then down sampled by 2 yielding

$$y_{\text{high}}[n] = x[n] * h[n] = \sum_{k=-\infty}^{\infty} x[k]h[2n-k] \tag{7.26}$$

$$y_{\text{low}}[n] = x[n] * l[n] = \sum_{k=-\infty}^{\infty} x[k]l[2n-k] \tag{7.27}$$

This decomposition has halved the time resolution since only half of each filter output characterizes the signal. However, each output has half the frequency band of the input, so the frequency resolution has been doubled. This decomposition is repeated to further increase the frequency resolution and the approximation coefficients decomposed with high- and low-pass filters and then down-sampled. This is represented as a binary tree with nodes representing a sub-space with a different time-frequency localization. This tree is known as a filter bank.

At each level of processing the signal is decomposed into low and high frequencies. Due to the decomposition process the input signal must be a multiple of 2^n, where n is the number of levels. For example, for a signal with 1,024 samples, frequency range 0 to f_n and 3 levels of decomposition, 4 output scales are produced. At level 1 decomposition there are 512 samples with detail components and 512 approximation coefficients with half of the frequency range each. At the second level of decomposition the input is approximation coefficients only and after processing at the second level, there are 256 samples of detailed and approximation coefficients with frequency bands $f_n/4$ to $f_n/2$ and 0 to $f_n/4$, respectively. Similarly at the third level of decomposition, the corresponding samples are 128 for each of detailed and approximated coefficients with the frequency bands of $f_n/8$ to $f_n/4$ and 0 to $f_n/8$, respectively.

7.5.2 Compensation of the atmospheric turbulence-induced distortion using wavelet-based signal processing

The optical signal propagating through the atmosphere will undergo all the effects mentioned above in addition to the attenuation due to rain and fog. All these effects will cause the fluctuation in the received optical power. These fluctuations for clear air are due to turbulence effects. Let us assume that $s(t)$ is the transmitted signal, then the received optical signal $r(t)$ can be represented as

$$r(t) = g(t) \times s(t) + n(t) \tag{7.28}$$

where $n(t)$ is a additive white Gaussian noise added due to receiver circuitry and $g(t)$ is multiplicative noise component induced due to turbulence effects. This noise component has a time dependency as similar to the turbulence. In other words $g(t)$ has the properties of scintillations and beam wandering.

A key aspect of the wavelet-based signal processing is the optimization of the level of approximation needed. The level N of approximation needed for compensating the turbulence effects depends on the sampling rate of the received signal. It was observed during the processing of the received signal for compensating the turbulent noise effects is that, if the sampling rate of the signal is ‘n’ samples/bit at the receiver, then the required approximation level N should satisfy the condition $2^{N-2} \leq n \leq 2^{N-1}$. This is obvious from that at least one sample is needed to know the atmospheric gain factor for that bit interval.

For example, if the received signal is sampled at 10 samples/bit, then the optimal compensation of the turbulent effects will be achieved with level 5 approximation. Hence, the wavelet-based signal compensation was carried out as follows: first the turbulent distorted signal is discretized by higher sampling rate (should be much higher than the data rate) and then divided in to frames.

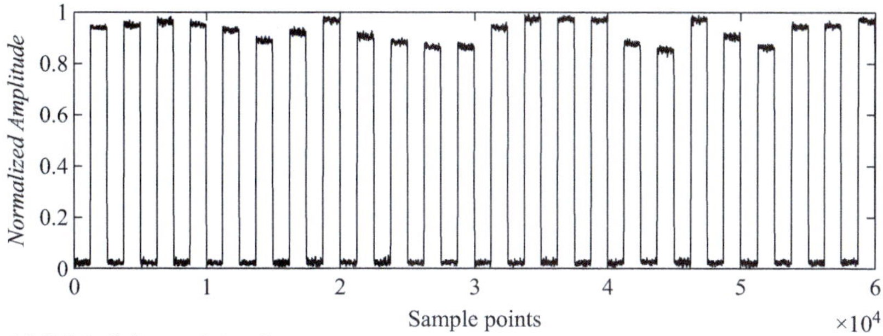

(a) Original distorted signal

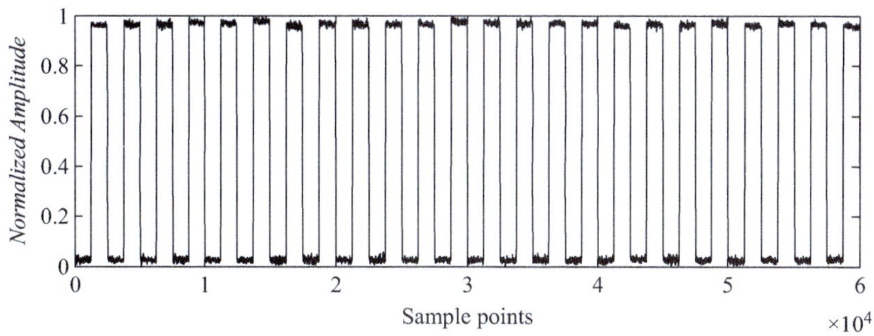

(b) Compensated signal

Figure 7.16 Turbulent distorted signal (a) and the corresponding compensated signal using fifth-order Daubechies wavelet (b)

The approximation of each frame gives the multiplicative noise of the received signal in that frame interval. Daubechies wavelet is used to calculate the approximation of each frame at level N. This approximation reproduces the fluctuations due to scintillations and beam wandering of the signal, which is proportional to the multiplicative noise $g(t)$. By dividing the original signal with this approximation component, will give the compensated signal with very low multiplicative noise. The below example shows the real time data collected over the 100 m FSO link set-up at IIT Madras with transmitted data rate at 4,420 bits per second and the received signal is sampled at 44,200 samples/s. In Figure 7.16, upper waveform shows the turbulent distorted signal $r(t)$ and lower waveform shows the enhanced signal $r(t)/a_5(t)$, where $a_5(t)$ is the 5th approximation of the turbulent distorted signal or turbulent distortion gain function. It is obvious from Figure 7.16, that the variance of the signal is reduced and also the mean value is increased. The corresponding numerical values are given in Table 7.2. The normalised signal distributions of the sampled original signal and the enhanced signal are shown in Figure 7.17. Performance comparison before and after compensation of the received signal using wavelets can be done by in comparison with different parameters like variance of the signal, average bit error rate (ABER), mean shift, BERR (bit error rate ratio) and threshold change. Sample calculated parameters of the

Table 7.2 Measured mean and standard deviation of original and compensated signals © [15]

	Bit value	Mean	Standard deviation
Original	Zero	0.36	0.10
Signal	One	1.62	0.30
Enhanced	Zero	0.37	0.10
Signal	One	1.64	0.13

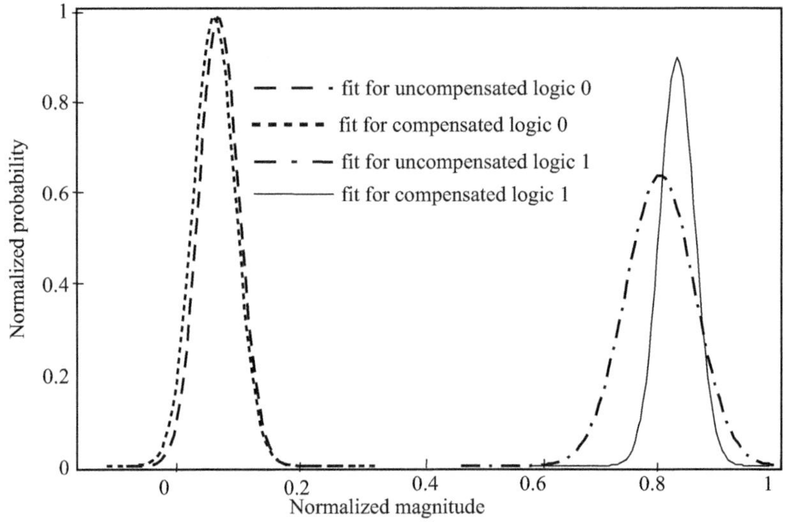

Figure 7.17 Probability density distribution of original received signal and the signal compensated using wavelet-based signal processing © [15]

original and compensated signal are given in Table 7.2. Which clearly shows the reduced standard deviation of the logic one and zero of the signal, which in turn improved correct signal detection and reduced probability of error detection.

7.5.3 Information recovery

The transmitted information can finally recovered by using an FIR differentiation filter and a three level decision device. The block diagram representation of the signal processing as shown in Figure 7.18. The sampled signal from the data capture system is framed and processed as described in Section 7.5.2. The FIR filter is basically a high-pass filter which detects the bit transitions. The expected bit transitions are 1–0, 0–1, 0–0 and 1–1. For the bit transitions 1–0 and 0–1, we will get negative and positive spikes and for the transitions 1–1 and 0–0, we should not have any transitions ideally the differentiator output should be zero, but we have the transitions due to additive white Gaussian noise with in the two threshold levels.

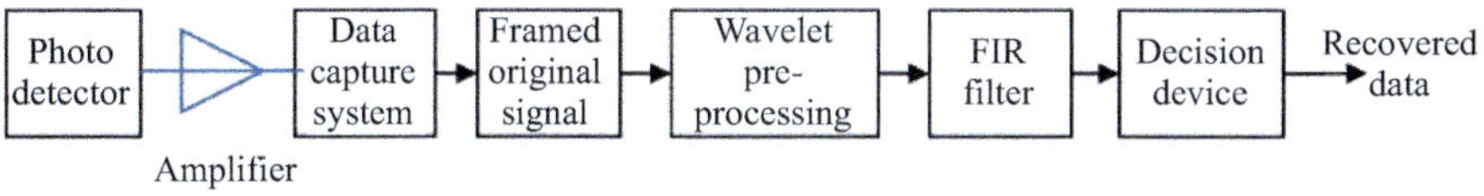

Figure 7.18 Block diagram of the steps involved in the processing of the turbulent distorted signal

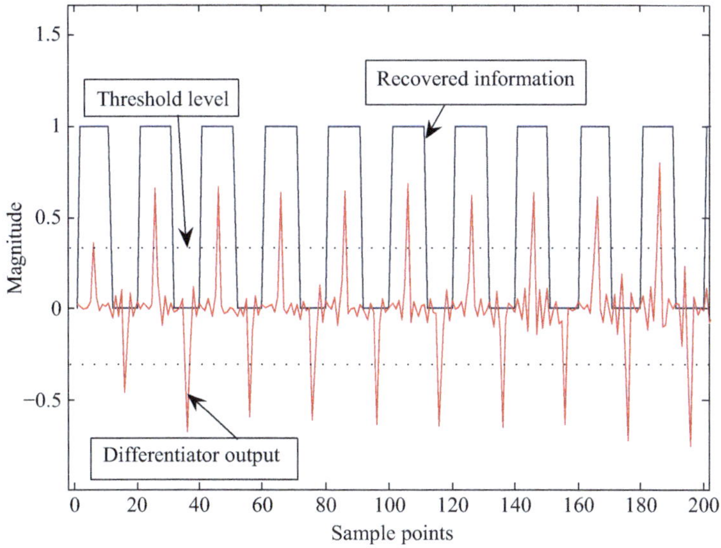

Figure 7.19 Plot of the differentiator output with threshold levels and the corresponding detected information © [15]

The decision rules are:

If $\widehat{r}(t) \geq$ threshold, then $\widehat{R} = 1$;
If $\widehat{r}(t) \leq -$ threshold, then $\widehat{R} = 0$;

else the output is same as the previous bit.

The threshold values and the differentiator output are shown in Figure 7.19 are calculated from the threshold and mean of logic one and zero. The upper threshold given by subtracting the threshold calculated from the mean of the enhanced signal. The lower threshold is quite approximately the negative of the upper threshold.

7.5.3.1 Average bit error rate ratio

Probability of error is calculated from the fit of the probability distribution function. The average probability of error is the average of making an error decision for one and zero, respectively [17,18], i.e,

$$P_e = \frac{1}{2}[P_e(1/0) + P_e(0/1)] \tag{7.29}$$

therefore,

$$P_e = \frac{1}{2}\left[Q_{1/0}\left(\frac{\text{threshold} - \mu_0}{\sigma_0}\right) + Q_{0/1}\left(\frac{\mu_1 - \text{threshold}}{\sigma_1}\right)\right] \tag{7.30}$$

For the given example, the average BER for original and compensated data are 4.204×10^{-7} and 2.33×10^{-9}, respectively. These values showing that the BER performance of the system are better than the BER values presented in [18]. Average BERR is defined as the ratio of the average BER of the original and enhanced signals. For this case $ABERR = 180.4292$. This clearly illustrates that the BER is reduced by this factor or the system performance is improved by approximately 180 times.

References

[1] Khalighi, Mohammad Ali, and Murat Uysal. "*Survey on free space optical communication: A communication theory perspective*". IEEE Communications Surveys and Tutorials, 2014, Vol. 16, No. 4, pp. 2231–2258.

[2] Huang, Hao, Guodong Xie, Yan Yan, *et al.* "*100 Tbit/s free-space data link enabled by three-dimensional multiplexing of orbital angular momentum, polarization, and wavelength*". Optics Letters, 2014, Vol. 39, No. 2, pp. 197–200.

[3] Ren, Yongxiong, Zhe Wang, Peicheng Liao, *et al.* "*Experimental characterization of a 400 Gbit/s orbital angular momentum multiplexed free-space optical link over 120 m*". Optics Letters, 2016, Vol. 41, No. 3, pp. 622–625.

[4] Ningbo Zhao, Xiaoying Li, Guifang Li, and Joseph M. Kahn. "*Capacity limits of spatially multiplexed free-space communication*". Nature Photonics, 2015, Vol. 9, No. 12, pp. 822–826.

[5] Akira Ishimaru, "*Wave propagation and scattering in random media*", IEEE Press (USA), 1978.

[6] Paul H Deitz and Neal J Wright, "*Saturation of scintillation magnitude in near-earth optical propagation*". Journal of the Optical Society of America, 1969, Vol. 59, No. 5, pp. 527–535.

[7] H Yuksel and C C Davis, "*Aperture averaging experiment for optimizing receiver design and analysing turbulence on free space optical communication links*". Proceedings of the Conference on Lasers and Electro-Optics, 2005, Vol. 1, pp. 743–745.

[8] D L Fried, "*Aperture averaging of scintillation*". Journal of the Optical Society of America, Vol. 57, No. 2, 1967.

[9] Ronald L Fante, "*Electromagnetic beam propagation in turbulent media*". Proceedings of the IEEE, 1975, Vol. 63, No. 12, pp. 1669–1692.

[10] Larry C Andrews and Ronald L Phillips, "*Laser beam propagation through random media*", Bellingham, WA: SPIE Press, Second edition 2005.

[11] D A Dewolf, "*Saturation of irradiance fluctuations due to turbulent atmosphere*". Journal of the Optical Society of America, 1968, Vol. 58, No. 4, pp. 461–466.
[12] D A Dewolf, "*Multiple Scattering in a Random Continuum*". Radio Science, 1967, Vol. 2, pp. 1379–1392.
[13] Robert S Lawrence and John W. Strohbehn, "*A survey of clear-air propagation effects relevant to optical communications*". Proceedings of the IEEE, 1970, Vol. 58, No. 10, pp. 1523–1545.
[14] B. Offerle, I. Eliasson, C. S. B. Grimmond, and B. Holmer, "*Surface heating in relation to air temperature, wind and turbulence in an urban street canyon*". Journal Boundary-Layer Meteorology, 2007, Vol. 122, No. 2, pp. 273–292.
[15] Latsa Babu P and Balaji Srinivasan, "*Characterization of atmospheric turbulence effects and their mitigation using wavelet-based signal processing*". IEEE Transactions on Communications, 2010, Vol. 58, No. 6, pp. 1795–1802.
[16] T. Stephane Mallat, "*A Wavelet our of signal processing*". Elsevier, USA, 1998.
[17] Jayasri Akella, Murat Yuksel and Shiv Kalyanaraman, "*Error analysis of multi-hop free space optical communication*". Proceedings of the IEEE International Conference on Communications, 2005, Vol. 3, pp. 1777–1781.
[18] Xiamong Zhu and Joseph M Khan, "*Performance bounds for coded free space optical communications through atmospheric turbulence channels*". IEEE Transactions on Communications, 2003, Vol. 50, No. 8, pp. 1233–1239.

Chapter 8

All-optical relay-assisted FSO systems

Norhanis Aida Mohd Nor[1,2], Zabih Ghassemlooy[1], Matej Komanec[3], Jan Bohata[3], Stanislav Zvanovec[3], Manav R. Bhatnagar[4], and Mohammad-Ali Khalighi[5]

8.1 Introduction

Free space optical (FSO) communications is an emerging high-speed, high data rate (R_d), high-bandwidth, and license-free access solution for a number of applications including the "last mile" access networks. FSO systems are highly dependent on the characteristics of the atmospheric channel. The atmospheric channel effects can be classified into absorption, scattering (caused by fog, snow, smoke, haze, aerosols, and dust particles), and turbulence-induced fading or scintillation [1] and are highly variable and unpredictable. This can cause a significant power penalty at the receiver (Rx), thus resulting in increased bit-error-rate (BER), reduced link range, temporary signal degradation and eventually leading to a link outage under the influence of deep signal fades, making it very difficult to attain the required link availability. Unlike fog, snow, and cloud, which induce beam attenuation, the atmospheric turbulence (AT), also known as scintillation, is the main source of random fluctuations of the received optical signal in terms both of intensity and phase even under clear weather conditions [2]. AT-induced signal losses will increase as the distance between the transmitter (Tx) and Rx is increased. Thus, the characteristics of AT is a critical factor in determining the FSO system performance. A number of studies have portrayed the effects of AT-induced fading on the FSO link performance analytically [3] and experimentally [4]. Moreover, various models exist to describe the stochastic behavior of AT, which can be approximately modeled by log-normal or gamma–gamma (G-G) distributions, which are mainly valid for weak-to-moderate and moderate-to-strong turbulence regimes, respectively.

In this chapter, an all-optical FSO relay-assisted system is proposed to mitigate the destructive effects of the distance-dependent AT-induced fading. Relays are inserted directly in the link in order to reduce the AT-induced link loss, thus

[1]Faculty of Engineering and Environment, Northumbria University, UK
[2]Department of Science, International Islamic University Malaysia, Kuala Lumpur, Malaysia
[3]Faculty of Electrical Engineering, Czech Technical University in Prague, Czech Republic
[4]Department of Electrical Engineering, Indian Institute of Technology-Delhi, India
[5]Aix-Marseille University, CNRS, Centrale Marseille, Institut Fresnel, Marseille, France

extending the link span and ensuring higher link availability as well as improving the overall system performance. Two all-optical relaying schemes are proposed and investigated, namely all-optical amplify-and-forward (AOAF) and all-optical regenerate-and-forward (AORF) FSO relay-assisted approaches. For the AOAF approach, the performance analysis of triple-hop AOAF FSO communications is done under the impact of nonhomogeneous atmospheric turbulence. The AORF relaying approach is then proposed to overcome the limitation imposed by AOAF system, where the signal and noise are accumulated at each relay, thus limiting the number of relay nodes that can be used.

8.1.1 Fading mitigation techniques

In order to design a reliable FSO system with maximum link availability and data throughput over the adverse atmospheric conditions, a number of channel enhancement approaches including aperture averaging [5], adaptive optics [6], error control coding [7], spatial diversity [8], and hybrid RF-FSO [9] have been reported. Aperture averaging is the simplest technique to implement, where by using a Rx aperture larger than the fading correlation length, intensity fluctuations can be averaged out over the aperture area. However, this technique has been found to be effective mostly for the weak-to-moderate turbulence regime, but not for the strong turbulence regime. The spatial diversity scheme, also known as the diversity technique in the RF technology, is a more efficient method in reducing the fading effects by employing sufficiently spaced multiple Txs/Rxs especially under strong turbulence conditions and over a long transmission span [5]. This technique has been demonstrated to provide substantial link performance improvement in spatially uncorrelated channels by employing multiple apertures at the Tx and/or the Rx that are adequately spaced as in single-input multiple-output (SIMO), multiple-input single-output (MISO) or multiple-input multiple-output (MIMO) systems [10]. The Rx collects the received optical beams from different spatial angles, so that the effective level of scintillation and probability of fade can be reduced. Ultimately, the inherent redundancy of spatial diversity has the potential to significantly enhance the system performances [11]. Apart from mitigating scintillation, this technique can also reduce the potential temporary blockage of the laser beam due to obstructions by small objects flying across the link path such as birds and drones. Furthermore, longer transmission spans can be covered under harsh weather conditions. In addition, it is much easier to provide independent aperture averaging with multiple apertures than a single aperture where the aperture size has to be far greater than the irradiance spatial coherence distance [1]. Nevertheless, employing diversity techniques also require complex electronic and processing overheads in order to accommodate the synchronization, buffering, retiming processes, and logical combining of the replicated transmission [12].

While the majority of these mitigation techniques have been proposed at the expenses of increased circuit complexity and massive signal processing requirements, more competent methods may be introduced in FSO systems in order to improve link reliability and availability while delivering high data rate. Relay-assisted FSO, also

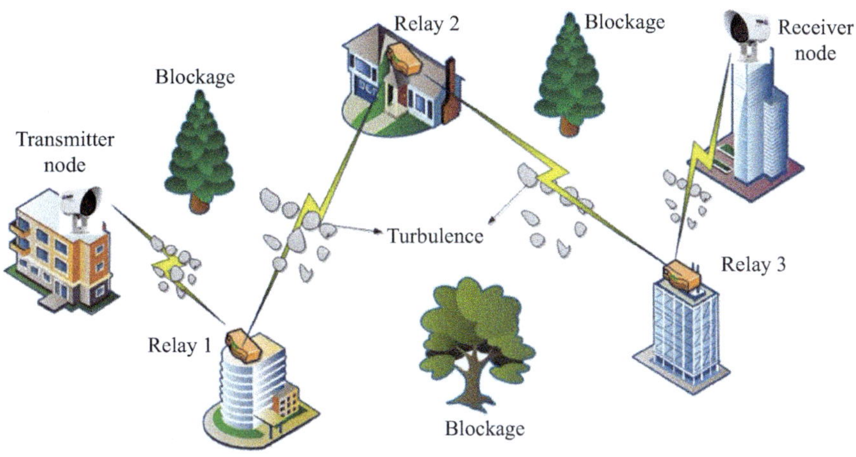

Figure 8.1 Illustration of relay-assisted FSO communications

known as multi-hop transmission (see Figure 8.1), is one of the alternative solutions, which offers spatial diversity within the network that can be adopted to combat link failure, thus substantially improving the link availability.

However, existing relay nodes are based on optical-to-electrical (OE) and electrical-to-optical (EO) conversion routing, which requires high-speed electronics and electro-optics devices, analog gain units, and digital control, which significantly increase the link latency, complexity, and last but not least the overall cost of the system [13]. Hence, with the aims of avoiding complex optoelectronic and electronic processing at each relay, and at the same time, increasing the transmission link span, this chapter investigates an alternative approach based on the link routing using all-optical FSO relay-based system. Whilst most research works on all-optical relay-assisted techniques since a few years focus on theoretical studies, this chapter addresses practical implementation, characterization, and evaluation aspects of an all-optical FSO relay system by presenting experimental evaluation of such links.

8.1.2 Relay-assisted FSO communications

Cooperative diversity is a paradigm shift in wireless technologies, which solves many shortcomings in the conventional point-to-point communications, mainly to combat fading and extend the coverage area. This technique has been introduced as an alternative way of realizing spatial diversity advantages without the need for using physical antenna arrays [14,15]. It is based on the broadcast nature of the radio frequency (RF) wireless medium and enables single-antenna users to take advantage of space diversity by sharing their physical resources through a virtual transmit and/or receive antenna array. In the FSO context, relay-assisted communication is considered in the forms of parallel or serial (also called multi-hop) relaying. It offers an efficient and low-cost solution, which mainly provides a

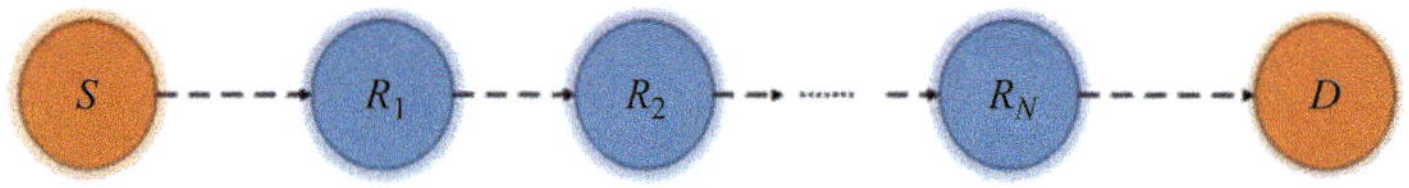

Figure 8.2 Serial relay transmission

fading reduction advantage due to the distance-dependency of turbulence-induced fading in FSO systems. The basic relay-assisted link is comprised of three main elements, a source (*S*)/Tx, a destination (*D*)/Rx, and one or more relays (*R*), as illustrated in Figure 8.1.

In RF wireless systems, relay-assisted communication takes advantage of the broadcasting nature of RF transmission, where a message transmitted from *S* to *D* can be overheard by neighboring nodes, thus enabling the retransmission of the same message to *D*. As a result, the repeated messages received by *D* will enhance the quality of signal reception since *D* essentially averages out the channel fading. The *S* node along with the neighboring nodes that are willing to share their resources, create a virtual antenna array. An advantage of this virtual multiple-aperture system is that multiple paths are spatially separated from one another, thus ensuring the independence of the corresponding fading channels. However, the broadcasting nature of RF leads to higher possibility of eavesdropping. Cooperative FSO communication was first proposed in [16] and was later adopted as an efficient solution for combating turbulence-induced fading in FSO systems [17].

Due to the stringent pointing requirement and operational implementation complexity of parallel relaying, serial relaying has been extensively studied in the literature, despite the fact that parallel relaying was shown to outperform serial relaying for higher signal-to-noise ratio (SNR) levels by providing a higher diversity gain [18,19]. Unlike wireless RF systems, the severity of the small-scale fading in the FSO channel is distance dependent. This inherent characteristic allows serial relaying FSO systems to benefit at the same time from less severe turbulence and a reduced path loss between *S* and *D* [20,21], although providing no spatial diversity benefit. In addition, multi-hop transmission offers a number of advantages including extended connectivity, no need to a direct link or a line-of-sight (LOS) path between *S* and *D*, lower required transmission power, and minimizing the need for a fixed infrastructure, and as such can be adopted in wireless local area networks and hybrid networks [22,23].

For multi-hop transmission, AF or decode-and-forward (DF) protocols in the relay nodes are among the most famous. An intensity modulated signal at *S* is transmitted to the first *R* and is processed and forwarded to the subsequent *R*s and ultimately to *D*, as shown in Figure 8.2. The asymptotic relative diversity gain of the serial relaying scheme with *N* relays can be expressed [24]:

$$D_S = (N+1)^{11/6}. \tag{8.1}$$

In [24], it was demonstrated that, the minimum outage probability for serial relaying is achieved when the relay nodes are equidistantly placed along the direct

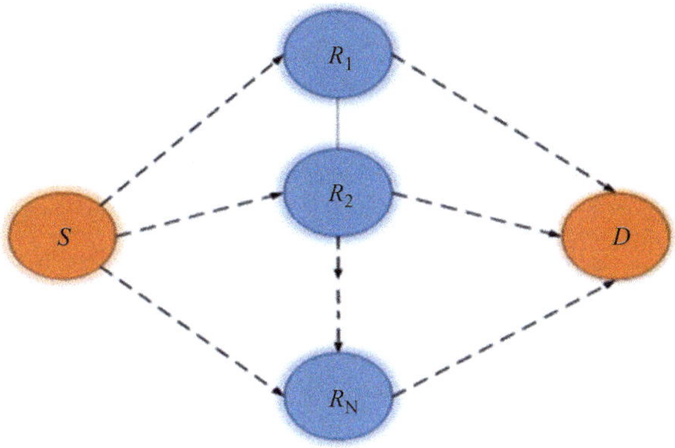

Figure 8.3 Parallel relay transmission

line from S to D. Furthermore, an outage occurs when any of the intermediate link fails. Therefore, the outage occurrences can be minimized by reducing the length of each transmission hop. However, the increase in the number of serial relaying nodes increases the system complexity and cost [25].

In parallel relay transmission, the signal propagates from S to D through parallel Rs in the same hop, as shown in Figure 8.3. In this scheme, the S is equipped with a multi-Tx laser, with each Tx pointing out in the direction of the corresponding R node. Based on the AF or DF relaying method, the R_s either amplify or decode the received signal, and retransmit it to the final destination D. At D, the signals received with the help of various combining schemes such as equal ratio combiner, fixed ratio combiner, and maximum ratio combiner. This topology is commonly used when the serial-relay transmission suffers from multipath fading effects. For this scheme, the asymptotic relative diversity gain can be expressed as [24]:

$$D_P = 2^{11/6}N. \tag{8.2}$$

Based on (8.1) and (8.2), it is clear that the number of hops (raised to the power $11/6 \approx 2$) has more impact than the number of diversity paths. Note that, in parallel relaying, if an outage occurs in one of the R_s, it does not necessarily lead to an outage in the entire relaying scheme. However, as the number of R_s increases, the parallel relaying scheme takes less advantage of the distance-dependence of fading variance than serial relaying, which makes serial relaying better than the parallel relaying in such a case.

To further boost the growth of network capacity, an all-optical (AO) network structure is currently being envisaged because of higher bandwidth requirements and new technological progress to avoid electrical termination of all connections in each network. The reason for AO systems' deployment is the fact that, it is the only available technology today that can handle the enormous amount of capacity in a cost-effective and efficient way. In traditional AF systems, relaying is accomplished by means of OE conversion, amplification in the electrical domain, and EO conversion [21,26]. On the other hand, in DF, further electrical processing is implemented following OE conversion, which involves decoding and re-encoding of the electrical signal to combat the background noise prior to re-transmission [27,28]. Therefore, there is the need for alternative schemes that can be used to improve the system performances and achieve higher data rates. In [29], a 1.6 Tbps FSO system was introduced to efficiently solve the network bottleneck issue. Therefore, an alternative solution would to be to have an AO system where all signal transmission, routing, switching, and signal processing are kept in the optical domain. This approach requires only low-speed electronic circuits to control and adjust the gain of amplifiers as well as avoids the use of complex optoelectronics devices and electronic processing at the relay nodes. Consequently, cost savings are envisaged as a result of keeping the information in the optical domain.

All-optical FSO relay-based systems are discussed in terms of two protocols - AOAF and AORF. Although the AOAF relaying technique has been widely used in fiber optic communications systems, its application in FSO systems is still in its infancy stage where the concept was tested for the first time in 2002, implementing a cascaded point-to-point located 250 m apart and using dense wavelength division multiplexing (DWDM) with R_d up to 10 Gbps [30]. However, this setup only has the cascaded relay with the Rx (i.e., there was no separation between the relay and the destination) and it neglected the effect of atmospheric channel. Following this test field, most researches have since only focused on the theoretical aspects, as the practical implementation seemed to pose an open problem. Since the AOAF scheme accumulates the noise at each relay, the performance deteriorates as the number of hops increases. Consequently, AORF has been proposed for the first time in [14], where a 2R regeneration scheme was implemented in FSO communications system to suppress the impact of background and amplified spontaneous emission (ASE) noise at each relay before retransmission to the next relay/destination. However, their work only demonstrated the superior capability of AORF compared to AOAF for relaying system (up to $N = 3$) by estimating the BER performances in mitigating the accumulation of background noise, while neglecting the impact of AT and link loss. Although it was proven that the AORF outperforms AOAF by 73% for dual-hop FSO system at a BER of 10^{-5} compared to direct transmission link [14], there has been no continuity to this work since then. Therefore, inspired by the high capability of AORF in mitigating noise, we continue the work by experimentally investigating the performance of AOAF FSO dual-hop system over various turbulence regimes.

8.2 All-optical amplify-and-forward relay-assisted systems under turbulence effects

A number of researchers have reported the reliability of relay-assisted FSO systems to broaden the coverage as well as to reduce the fading effects. The relay-assisted FSO system was first proposed in [17], where the performance of a mesh FSO network was investigated from a network capacity point of view, with the emphasis on the effectiveness of relay-assisted transmission. In [31], the BER performance of DF and AF FSO multi-hop schemes was investigated and compared to the single-hop FSO system under the atmospheric attenuation and geometric loss effects, with each hop length is fixed to 500 m. It was shown that, the mean and variance of the error rate is smaller in multi-hop systems compared to the single hop for the same link range and the launched power. In [32], the effect of AT in multi-hop FSO communications modeled by G-G distribution was outlined, where it was shown that, the outage probability of AF and DF degraded with the increasing number of hops under strong AT. The results also showed that the DF system slightly out-performed the AF-based FSO link with the performance gap increased with the growing number of hops. The capabilities of relay-based system using subcarrier phase shift keying (PSK) to combat the G-G AT fading was demonstrated in [18], by considering both the channel-state-information (CSI) and fixed-gain relays. The numerical results demonstrated that by increasing the number of hops while keeping the hop length fixed, the outage performance deteriorated accordingly. In [21], for the case of a G-G turbulence channel it was shown through the simulation that, the SNR gains of ~12 dB and ~19 dB at a target outage probability of 10^{-6} for AF and DF dual-hop serial relay-based link, respectively, were achieved compared with the direct transmission link with strong turbulence (i.e., $C_n^2 = 1 \times 10^{-14}\ \mathrm{m}^{-2/3}$ and a link range of 5 km). Using a triple-hop transmission system, the SNR improvements were ~18 dB and ~25 dB. Note that, the advantage of DF over AF is at the cost of increased system complexity i.e., the need for decoding and encoding modules at the relay nodes.

Most of reported numerical studies have focused on the conventional OE relaying scheme, where amplification is carried out in the electrical domain. However, most recently researches have focused on the provision of applying all-optical relaying system specifically for FSO communications. The concept of all-optical multi-hop FSO system was proposed in [30], where a dual-hop test-field setup was developed for a total link distance of 500 m. In the experiment, an all-optical automatic gain controller was used to adjust the optical amplifier (OA) gain to compensate for any distortion experienced by the propagating optical signal. The field test results showed that a BER of $2.1\mathrm{e}^{-12}$ was achievable at R_d of 2.5 Gbps, which reduced to a BER of $3.5\mathrm{e}^{-11}$ for R_d of 10 Gbps. However, the experiment neglected the effect of AT-induced fading. Despite that, this works broadened the horizons on all-optical multi-hop FSO communication systems, which lead to a considerable amount of research works. After about a decade, the analytical investigation on the BER performance of all-optical multi-hop FSO systems for both AF and DF was presented in [14]. The authors performed Monte Carlo simulations under weak turbulence by assuming either a fixed-gain optical amplifier or an all-optical regenerator (AOR). The study showed that, for the on–off-keying (OOK)-based system at a BER of 10^{-5} under weak turbulence (i.e., $C_n^2 = 1 \times 10^{-14}\ \mathrm{m}^{-2/3}$), the SNR gain of 6.6 dB was achieved for the all-optical dual-hop AF relay, and the total links span was increased by 0.9 km and 1.9 km for AF and DF two-hops systems, respectively, compared to a 3 km long of the direct transmission. However, the authors underestimated the effect of ASE in their system.

Following [14], numerous studies have been reported to explain the significance of adopting all-optical FSO relay-based systems specifically using an erbium-doped fiber amplifier (EDFA) based on the AF relay [22]. In particular, in [22], the outage probability analysis for a dual-hop and multi-hop system using EDFA over the lognormal fading was investigated through numerical simulations. They showed that, an all-optical relay-based FSO link outperformed the electrical relay-based link by reducing the outage probability from 10^{-2} to 10^{-5} at -130 dBJ of background radiation energy based on a 1,550 nm laser source, a 2 Gbps OOK data format, and a 5 km link span with $C_n^2 = 1.7 \times 10^{-14}$ m$^{-2/3}$ and the atmospheric attenuation of 4.2 dB/km. The authors also considered all relevant noise sources including thermal noise, background radiation, ASE, beat noise, and signal dependent noise. Furthermore, the link performance for all-optical relaying, improved with the increasing number of relays m until $m = 8$. Increasing m to up 10 resulted in the link's performance degradation. Further outage performance analyses of dual-hop all-optical AF relaying taking into account the effects of ASE and degree-of-freedom (DoF) was reported in [33]. The authors stated DoF (i.e., $M = 1$, 10, 100, and 1,000) is an important parameter in quantify the ratio of optical filter bandwidth to the electrical bandwidth and should be in the order of 1,000 unless a narrowband optical filter is employed (for an ideal case, $M = 1$). The numerical results showed that SNR reduced from 14.7 dB ($M = 1$) to 14.4 dB, 13.2 dB and 10.3 dB for $M = 10{,}100$, 1,000, respectively, at a target outage probability of 10^{-6}, compared to a direct transmission at a wavelength λ of 1,550 nm is transmitted over a link span of 10 km with $C_n^2 = 10^{-14}$ m$^{-2/3}$ and visibility of 10 km.

In [34], an all-optical AF dual-hop FSO system employing a 4-pulse position modulation (4-PPM) operating at λ of 1,550 nm over the G-G turbulence channel was reported showing that the required transmit power was reduced in proportion to the optical amplifier gain. The authors theoretically showed the benefit of combining relaying and multiuser diversity. For example, for an amplifier gain of 10 dB, the transmit power per bit of 10 dBm was conserved at a target BER of 10^{-6} over a link span of 3 km. In [35], a novel cumulative density function (CDF) analysis for the end-to-end SNR for relay-assisted subcarrier intensity modulation (SIM)-based all-optical FSO networks over the G-G fading channel was reported where both the variable AF gain and the pointing error effect were taken into consideration. The authors theoretically showed the benefit of combining relaying and multiuser diversity.

Recently, in [36], a novel relaying technique based on combining EDFA with an optical hard-limiter (OHL) to improve the BER performance were reported. In these approaches OHL was placed before EDFA to mitigate the accumulated background noise. The analyses carried out were based on Monte Carlo simulations for the multi-hop system with moderate turbulence using the G-G distribution. The results showed that, the proposed setup improved the BER performance and extended the link distance by removing the accumulative background noise over multiple relays. For instance, for a 2.5 Gbps OOK-based link of 5 km long and with $C_n^2 = 5 \times 10^{-14}$ m$^{-2/3}$ a transmit power gain of 5 dB was at a target BER of 10^{-6} compared to the system employing a conventional EDFA for the triple-hop relaying system with turbulence. Nevertheless, the drawback of the proposed technique was the BER performance was dependent on the threshold level of OHL, and the optimal threshold levels must be set based on the strength of turbulence and the background noise level, which are not straightforward.

Recently, a hybrid RF-FSO scheme was considered to further enhance the system performance when adopted between source-to-relay and relay-to-destination [37]. The authors derived analytical expressions for the probability density function (PDF) and CDF, and investigated the outage probability and the BER performance of a dual-hop relaying system. However, almost all works reported on this topic are theoretical in nature with very little or no experimental verifications. Therefore, this chapter will further investigate both theoretically and experimentally the performance of all-optical relaying system employing OA as the relaying technique.

8.2.1 All-optical amplify-and-forward

The AOAF protocol is the simplest relaying protocol, which relies on amplifying the received signal and forwarding it to the next destination. The basic configuration of the AOAF relaying technique is illustrated in Figure 8.4. Each relay is composed of all-optical elements, which are optical lens, optical fiber, and OA. Two optical lenses (normally convex lenses) are placed before and after the OA, acting as the Rx and Tx, receiving and forwarding the beams to the next relay or toward the destination, respectively.

The basic relay communication networks contain of a three-terminal communication channel, which are the Tx/source (*S*), the relay (*R*), and the destination (*D*). In a AOAF multi-hop relaying scheme, at least a single OA is used prior to retransmission of the signal to the next relay/destination.

8.2.1.1 Performance analysis of triple-hop AOAF FSO

This section outlines the end-to-end mathematical frameworks for the SNR and the BER for triple-hop AOAF-based intensity modulation-direct detection (IM-DD) FSO link over the G-G turbulence channel. The simulation results are shown in the next section and validated by experimental results. Note, a number of assumptions made including (i) no direct link between *S* and *D;* (ii) at each hop, the relay amplifies the received signal and forward it to the next *R* or to *D*; and (iii) channels are stationary, memoryless, and ergodic with independent but not necessarily identically distributed fading statistics. The schematic system block diagram is illustrated in Figure 8.5.

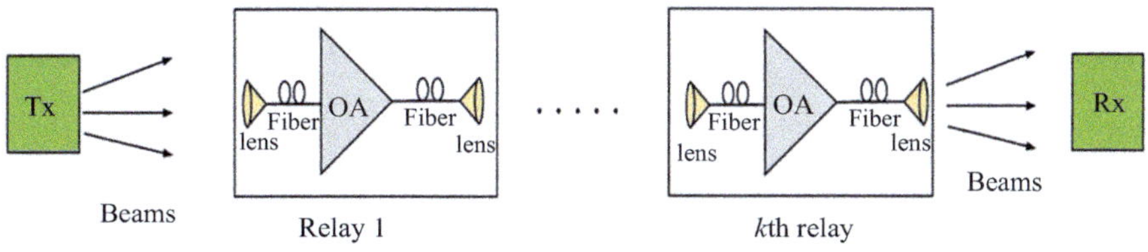

Figure 8.4 Basic configuration of AOAF relay-based system; OA: optical amplifier

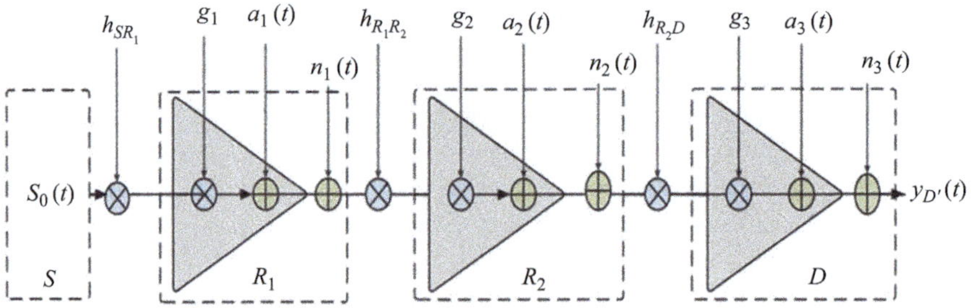

Figure 8.5 Schematic block diagram of a triple-hop all-optical OAF relay-assisted FSO

The signal $s_0(t)$ is transmitted from S to R_1, and the received signal at R_1 is given by:

$$y_{R_1}(t) = h_{SR_1} s_0(t) + n_1(t) \tag{8.3}$$

where h_{SR_1}, is the channel gain from S to R_1, $n_1(t)$ is the noise at R_1, which can be considered as a combination of the ambient light, thermal, dark, and shot noise, which is modeled as an additive white Gaussian noise (AWGN) with zero mean and a power spectral density of N_0 [1]. $y_{R_1}(t)$ is afterwards amplified using an OA and then retransmitted to R_2 via the second free space turbulence channel. The received signal at R_2 is given as:

$$y_{R_2}(t) = h_{R_1R_2}\left(g_1 y_{R_1}(t) + a_1(t)\right) + n_2(t) \tag{8.4}$$

where a_1 is the ASE noise of OA at R_1 and $n_2(t)$ is the noise at R_2. In most previous works reported on relay-assisted FSO networks, the OA gain g was assumed to be fixed [14]. However, in this work we define g for jth OA relay ($j = 1, 2, 3$) as [35]:

$$g_j = \sqrt{\frac{P_{R_j}}{h_{j-1,j}^2 P_S^2 + N_0}}, \tag{8.5}$$

where P_s and P_R are the transmit and received powers at S and R/D (i.e., either R_1, R_2, or D), respectively. So, for the whole transmission system, the input–output relation is represented by:

$$y_D^{'}(t) = h_{\text{eq}} s_0(t) + n_{\text{eq}}(t), \tag{8.6}$$

where $h_{\text{eq}} = g_1 g_2 g_3 h_{SR_1} h_{R_1R_2} h_{R_2D}$ and n_{eq} is the AWGN with zero mean and variance σ^2 given by:

$$\sigma^2 = \frac{{g_1}^2{g_2}^2{g_3}^2\sigma_{n_1}^2 + {g_2}^2{g_3}^2\sigma_{a_1}^2 + {g_2}^2{g_3}^2\sigma_{n_2}^2 + {g_3}^2\sigma_{a_2}^2 + {g_3}^2\sigma_{n_3}^2 + \sigma_{a_3}^2}{{g_1}^2{g_2}^2{g_3}^2 + 2{g_2}^2{g_3}^2 + 2{g_3}^2 + 1} \tag{8.7}$$

where $\sigma_{a_1}^2$, $\sigma_{a_2}^2$, and $\sigma_{a_3}^2$ are the variance of ASE noise of OA at R_1, R_2, and R_3, respectively, while $\sigma_{n_1}^2$, $\sigma_{n_2}^2$, and $\sigma_{n_3}^2$ are the variance of the background noise at R_1, R_2, and R_3, accordingly.

The BER of the dual-hop system can be written as:

$$P_{e2} = \left(\frac{(\alpha\beta)^{\frac{\alpha+\beta}{2}}}{\Gamma(\alpha)\Gamma(\beta)}\right)^2 \frac{(8\sigma_2^2)^{\frac{\alpha+\beta+2}{4}}}{4\sqrt{2\pi\sigma_2^2}(\mathcal{g}_1\mathcal{g}_2)^{\frac{\alpha+\beta}{2}}2^2(2\pi)^2} \times G_{3,10}^{8,3}\left(\frac{8\sigma_N^2(\alpha\beta)^4}{2^8\mathcal{g}_1^1\mathcal{g}_2^2}\middle| \begin{matrix}\frac{2-\alpha-\beta}{4}, & \frac{2-\alpha-\beta}{4}, & \frac{4-\alpha-\beta}{4} \\ \mathcal{g}^2, \frac{-\alpha-\beta}{4}, & \frac{2-\alpha-\beta}{4} & \end{matrix}\right), \quad (8.8)$$

Similarly, the BER for a triple-hop system is given as:

$$P_{e3} = \left(\frac{(\alpha\beta)^{\frac{\alpha+\beta}{2}}}{\Gamma(\alpha)\Gamma(\beta)}\right)^3 \frac{(8\sigma^2)^{\frac{\alpha+\beta+2}{4}}}{4\sqrt{2\pi\sigma^2}(\mathcal{g}_1\mathcal{g}_2\mathcal{g}_3)^{\frac{\alpha+\beta}{2}}8(2\pi)^3} \times G_{3,14}^{12,3}\left(\frac{8\sigma^2(\alpha\beta)^6}{2^{12}(\mathcal{g}_1\mathcal{g}_2\mathcal{g}_3)^2}\middle| \begin{matrix}\frac{2-\alpha-\beta}{4}, \frac{2-\alpha-\beta}{4}, \frac{4-\alpha-\beta}{4} \frac{\alpha-\beta}{4}, \frac{\alpha-\beta+2}{4}, \\ \frac{\beta-\alpha}{4}, \frac{\beta-\alpha+2}{4}, \frac{\alpha-\beta}{4}, \frac{\alpha-\beta+2}{4}, \frac{\beta-\alpha}{4}, \frac{\beta-\alpha+2}{4}, \\ \frac{\alpha-\beta}{4}, \frac{\alpha-\beta+2}{4}, \frac{\beta-\alpha}{4}, \frac{\beta-\alpha+2}{4}, \frac{-\alpha-\beta}{4}, \frac{2-\alpha-\beta}{4}\end{matrix}\right). \quad (8.9)$$

By carefully examining (8.8) and (8.9), we can further generalize the BER expression for N-hop link as:

$$P_e = \left(\frac{(\alpha\beta)^{\frac{\alpha+\beta}{2}}}{\Gamma(\alpha)\Gamma(\beta)}\right)^N \frac{(8\sigma^2)^{\frac{\alpha+\beta+2}{4}}}{4\sqrt{2\pi\sigma_N^2}(\mathcal{g}_N)^{\frac{\alpha+\beta}{2}}2^N(2\pi)^N} \times G_{3,4N+2}^{4N,3}\left(\frac{8\sigma_N^2(\alpha\beta)^{2N}}{2^{4N}\mathcal{F}_N{}^2}\middle| \begin{matrix}\frac{2-\alpha-\beta}{4}, \frac{2-\alpha-\beta}{4}, \frac{4-\alpha-\beta}{4} \\ \mathcal{G}^N, \frac{-\alpha-\beta}{4}, \frac{2-\alpha-\beta}{4}\end{matrix}\right) \quad (8.10)$$

where $\mathcal{F}_N = \mathcal{g}_1\mathcal{g}_2\cdots\mathcal{g}_N$, $\mathcal{G}^N$ implies that $\mathcal{G}$ is repeated N-times, and σ_N^2 denotes the equivalent variance for N-hop. Note that, if we consider the value of all gains to be equal as $\mathcal{g}$ then $\mathcal{F} = \mathcal{g}^N$.

8.2.2 AOAF numerical analysis

This section presents numerical simulation using MATLAB and Optisystem software of the FSO relay-assisted system with no turbulence and the G-G turbulence.

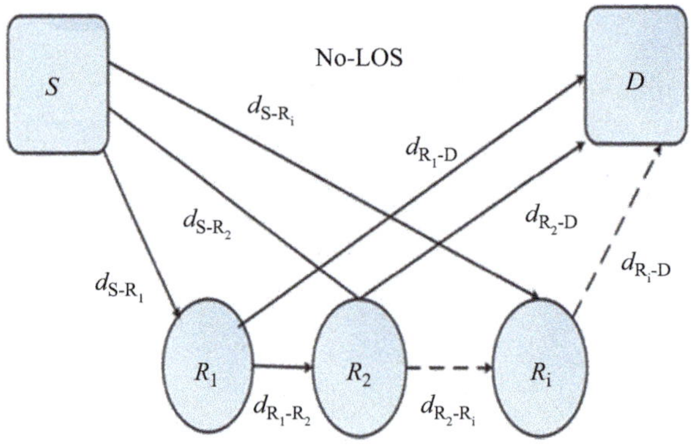

Figure 8.6 FSO relay-assisted transmission systems [38]

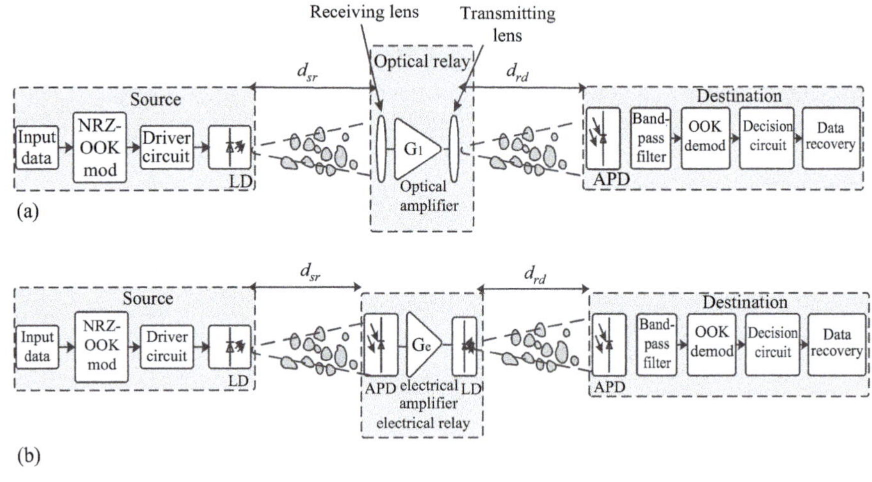

Figure 8.7 Block diagram of AF dual-hop relay-assisted system: (a) all-optical relaying and (b) electrical relaying [38]

Throughout the analysis, we consider IM-DD-based FSO system, where the information at the source S is transmitted to the destinations D via a number of all-optical relay R_i, where $i \in \{1, 2, \ldots n\}$, as illustrated in Figure 8.6. As shown in Figure 8.6, there are no LoS paths between S and D. We assume that, the channels between S, the ith relay R_i and D are stationary, memoryless, and ergodic with independent but not necessarily identically distributed fading statistics.

The numerical analysis starts by comparing the performance of a dual-hop relaying system in terms of the BER and SNR for signal amplification in optical and electrical domains under different turbulence regimes as shown in Figure 8.7(a) and (b), respectively. Focusing particularly on an all-optical system using OA, we investigate the BER performance of the EDFA and the semiconductor optical amplifier (SOA), which are the most common OA used in relaying systems. Following that, we also consider the effect of aperture averaging as the additional way to combat turbulence.

At the Tx side, a pseudorandom bit sequence of $2^{16} - 1$ bit length in the non-return zero (NRZ)-OOK signal format is generated for IM of a 1,550 nm wavelength laser source at R_d of 1 Gbps. Following propagation via the free space channel, the optical signal is amplified in the optical domain at the relay node for further transmission over the channel; see Figure 8.6(a). At the Rx, the optical signal is then detected using an Avalanche photodiode (APD). Following filtering and demodulation, a decision circuit is used to compare the detected signal with a threshold level to regenerate the original data stream. In the electrical-based relay, we have used OE and EO conversion modules to regenerate the signal for further transmission over the free space channel, see Figure 8.6(b). In order to make a fair comparison, we have fixed the output power of the EDFA set 2 dBm and the noise figure of EDFA was 4.0 dB as in [39]. For the case of electrical amplifications, we set the amplifier gain G to 10 dB in order to make sure that the output power of the amplifier is approximately 2 dBm, to compensate the fading variance through the channel. We also modeled the background radiation beat noise at the electrical relay as AWGN. The overall key system parameters are depicted in Table 8.1. The relay node is located at equidistance between S and D. In order to vary the SNR, the optical transmit power P_{tx} was changed with an optical attenuator (placed after the laser diode) to sweep the input power P_{in} to the amplifier. We considered three turbulence levels of weak, moderate, and strong corresponding to σ_R^2 of 0.2, 1.6, and 3.5, respectively [1].

The BER performance of a direct transmission link and the dual-hop FSO system with a fixed relay gain g_R for both optical (i.e., EDFA) and electrical amplifiers for two turbulence regimes σ_R^2 of 0.2 and 1.6 is shown in Figure 8.8. As expected, for all link configurations, the BER performance degrades with the AT strength. The results also show the potential of the relaying system in combating AT-induced fading compared to the direct link with no relay. Furthermore, the results reveal that all-optical relaying outperforms the electrical relaying scheme, which also imposes additional complexity and bandwidth restrictions to the system [40]. For example, at

Table 8.1 Dual-hop FSO system parameters

Parameter	**Value**	**Unit**
	FSO Systems	
Bit rate	1×10^9	bit/s
Total link range	1,000	m
Wavelength	1,550	nm
Transmit power	5	dBm
Tx aperture	20	mm
Rx aperture	100	mm
Beam divergence	2	mrad
APD responsivity	1.0	A/W

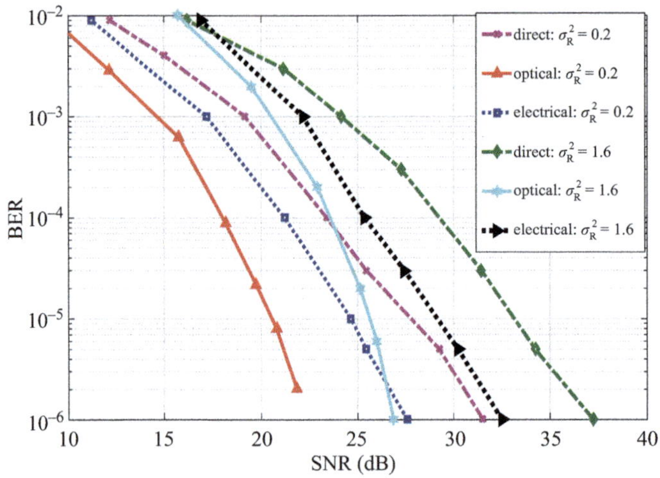

Figure 8.8 BER vs. SNR of direct transmission and dual-hop AF system for optical (EDFA-based) and electrical relaying systems for turbulence $\sigma_R^2 = 0.2$ and $\sigma_R^2 = 1.6$ [38]

a target BER of 10^{-5} there are SNR gains of ~3 dB and ~7.5 dB for the links with optical and electrical relaying, respectively, compared to the direct transmission link at $\sigma_R^2 = 0.2$. While at $\sigma_R^2 = 1.6$, the SNR gains of ~5.5 dB and ~9 dB are observed for the respective optical and electrical relaying systems when compared to the direct link. However, at higher BERs, we notice that the difference between the plots is much reduced for the two systems. This is due to the effect of the beat noise due to O/E and E/O conversions. Note that, the main noise sources are the shot noise and the electrical thermal noise, which are more dominant than the EDFA's ASE noise for the optical relaying case [41].

Figure 8.9 shows the simulated BER as a function of SNR for NRZ-OOK and RZ-OOK for the triple-hop FSO link with no turbulence and with turbulence (i.e., $\sigma_R^2 = 3.5$). Insets represent eye-diagrams and Q-factors for several values of SNR as an alternative indicator of the link performance. These results clearly illustrate the adverse effect of turbulence on the system under test, which significantly degrades the BER performance and the opening of eye-diagrams compare to the case with no turbulence. The plots also show that RZ-OOK offers improved BER performance and higher Q-factor compared to the NRZ-OOK. For example, at a target BER of 10^{-6} there is a SNR gain of ~1.3 dB for RZ-OOK compared to considering NRZ-OOK with no turbulence. However, with turbulence (of $\sigma_R^2 = 3.5$) there are ~3.5 dB and ~5.0 dB of SNR penalties for RZ-OOK and NRZ-OOK, respectively. In addition, it is observed from the figure that, for a BER of 10^{-7} Q-factors are 4.86 and 5.06 for NRZ-OOK and RZ-OOK, respectively, over the turbulence regime. This performance improvement is achieved at the cost of doubling the transmission bandwidth [42].

Figure 8.10 presents the numerically simulated BER as a function of SNR for single, dual-hop, and triple-hop FSO links for NRZ-OOK and RZ-OOK over a strong

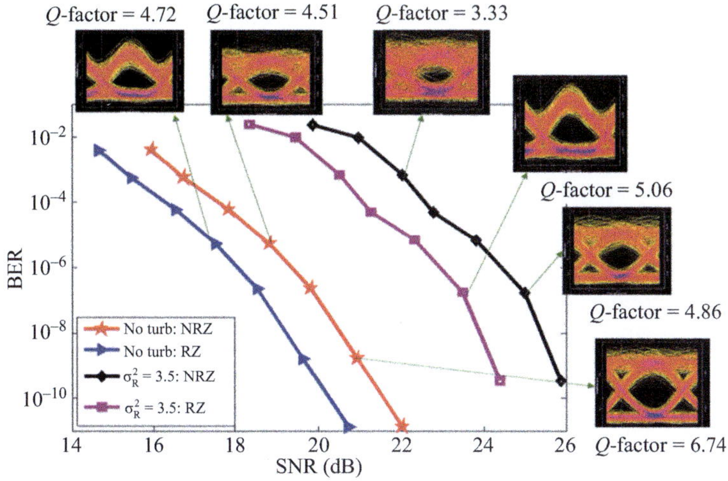

Figure 8.9 Numerical simulation of BER vs. SNR for NRZ and RZ triple-hop FSO links over no turbulence and turbulence $\sigma_R^2 = 3.5$ [43]

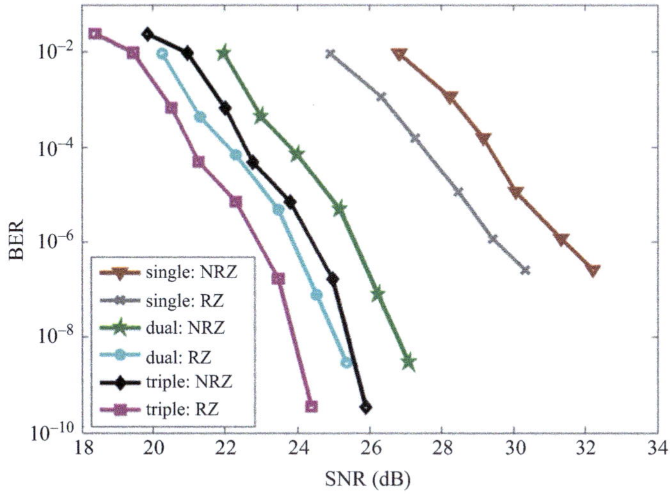

Figure 8.10 Numerical simulation of BER vs. SNR for single, dual, and triple-hop FSO links of NRZ-OOK and RZ-OOK with turbulence $\sigma_R^2 = 3.5$ [43]

turbulence regime $\sigma_R^2 = 3.5$. As predicted, the BER performance improves as the number of the hops increases from single-to-triple hop links regardless of the modulation scheme. Likewise, the plots confirm once again the advantage of RZ over NRZ. As it can be seen from the figure, RZ-OOK outperforms NRZ-OOK by ~1.9, ~1.7, and ~1.5 dB in terms of the SNR gain for single, dual, and triple-hop links, respectively. Furthermore, at a target BER of 10^{-6} and by considering RZ-OOK, dual-hop and triple-hop links provide SNR gains of ~5.0 dB and ~6.2 dB, respectively, compared to direct transmission. These substantial improvements show the potential of multi-hop FSO transmission in combating the turbulence effect.

8.2.3 Experimental analysis for single, dual-hop, and triple-hop AF systems

Outdoor FSO measurement is challenging and time consuming especially to carry out full system performance evaluation under all weather conditions. However, using purpose indoor-built laboratory testbed we were able to carry out comprehensive measurement campaign under a controlled environment over a short time scale. In order to increase the reliability of measuring results, each measurement was performed three times.

The schematic diagrams of proposed indoor experimental single, dual-hop, and triple-hop AF FSO relay-assisted systems setup are illustrated in Figure 8.11(a), (b), and (c), respectively. The systems are composed of a combination of FSO links of a total span of 6.6 m, EDFAs, single-mode fibers (SMFs), and optical components. As shown in Figure 8.11(a), at the Tx, a signal source/BER tester (BERT-VeEX VEPAL TX300) is used to generate a pseudo-random binary sequence (PRBS) d_i in the NRZ-OOK format at R_{d} of 10 Gbps for IM of a 1,550 nm distributed feedback (DFB) laser diode. An optical digital attenuator and a 50/50 coupler are used to control the level of transmit power P_s and split the signal into two, for both transmission and monitoring using an optical spectrum analyzer (OSA). The laser output is then launched into the free space channel via a gradient index (GRIN) fiber collimator (Thorlabs 50-1550A-APC) with a clear aperture diameter lens of 0.18 cm. In order to focus the optical beam and minimize the divergence of the optical beam, two plano convex lenses Lens 1 and Lens 2 with a focal length f of 10 cm were placed after the Tx and before the Rx, respectively. At the Rx, a GRIN fiber collimator is used to couple the received optical beam into the SMF the output of which is amplified using an EDFA. The signal is then applied via a 50/50 coupler to an OSA and a BERT for the link assessment in terms of the BER and SNR. Both dual-hop and a triple-hop AF FSO relay system were implemented based on Figure 8.11(a).

In a controlled indoor environment using a dedicated atmospheric chamber, the strength of turbulence is controlled by blowing hot air from two 2 kW external heating fans in the direction perpendicular to the propagating optical beam along the transmission path. The heating sources were positioned near and/or away from the FSO Tx or Rx. In order to maintain a constant temperature gradient within the chamber along the propagation path the temperature was kept at the room temperature of ~25 °C and within the range of 25 °C to 80 °C, respectively, with the wind speed of <10 m/s. The strength of AT was varied by means of controlling the output of the fan heaters and thus varying the temperature profile. Moreover, two plastic barriers were placed at the ends of optical table in order to maintain the temperature profile along the chamber. To continuously monitor the temperature profile along the channel and ultimately be able to determine C_n^2, the instantaneous

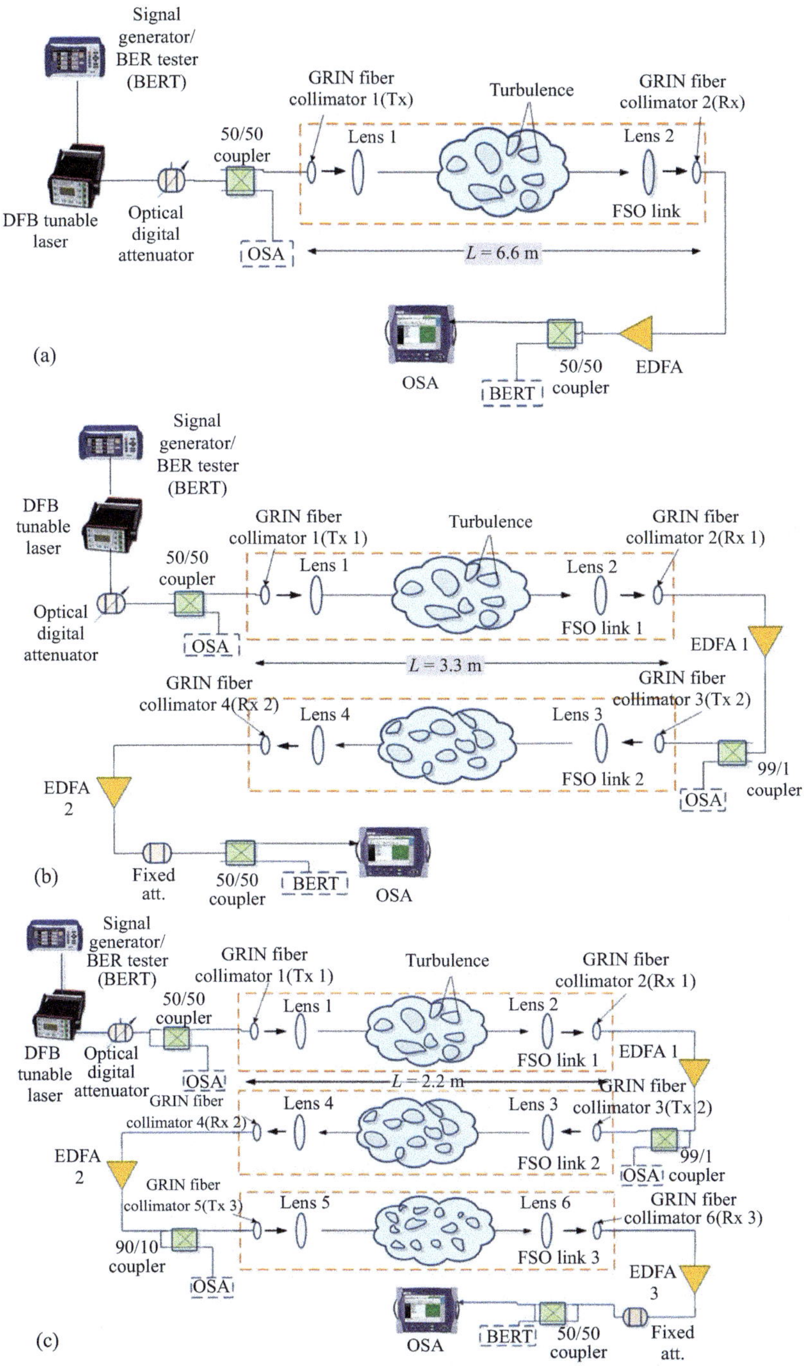

Figure 8.11 Block diagram of the experimental setup of: (a) direct link; (b) dual-hop; and (c) triple-hop AF FSO relay-assisted systems [44]

Table 8.2 Main parameters used in the experiment

	Parameter	Value
Transmitter	Data rate and signaling scheme	10 Gbps NRZ-OOK
	Laser type	DFB
	Wavelength	1,550 nm
	Output power	1 dBm
	Grin-lens aperture	0.18 cm
	Beam divergence	0.25 mrad
Lens	Clear aperture	2.54 cm
	Focal distance	10 cm
EDFA	Max. output power	21 dBm
	Noise figure	< 7.5 dB
Receiver	Clear aperture	0.18 cm
	Rx type	PIN
	Rx sensitivity	-16 to -1 dBm
	Responsivity	0.6 A/W
SMF-28	Mode-field diameter @1,550 nm	10.4 ± 0.8 μm
	Attenuation @ 1,550 nm	$\leq$ 0.2 dB/km

temperatures were recorded every 4 seconds throughout the experiment using 20 temperature sensors positioned at 0.28 cm apart along the channel (i.e., between the Tx and the Rx). While carrying out the experimental measurement with no turbulence, we ensured that the temperature along the channel was kept at the room temperature. With turbulence, we continuously recorded the temperature at 20 different positions T_1-T_{20} along the channel. All the main parameters adopted in the experimental setup are summarized in Table 8.2.

The scintillation effect due to AT is simulated by exploiting the dependence of the channel index of refraction on the temperature variations. As aforementioned, the turbulence simulation process involves blowing hot air using two heaters fans at two different locations into the chamber. Figure 8.12 shows the histograms of C_T^2, where the number of occurrence is normalized to the unity, which represents the occurrence PDF. By fitting the C_T^2 experimental data to the G-G distribution curve, we can see a good match between theoretical and measurement data. From the fit, we can confirm that the C_n^2 will be as accurate as possible.

Figure 8.13 depicts the experimental and theoretical results for the BER performance against the SNR for single, dual-hop, and triple-hop FSO links for $C_n^2 = 5.4 \times 10^{-12}$ m$^{-2/3}$. As a benchmark, the measured BER of a triple-hop link with no turbulence is also shown in the figure. The results confirm the potential of the relaying system in mitigating the AT-induced fading compared to the direct single link. As observed, there is a trade-off between the number of relays and the average P_s to achieve a particular BER. For example, for measured plots at a BER of 10^{-5} the SNR gains are ~2.8 dB and ~4.8 dB for dual-hop and triple-hop, respectively, compared to the single FSO link. As mentioned before, these SNR gains are because of reduced effects of AT as the link span is shortened in relayed-based systems. The marginal differences between the theoretical and measured plots in

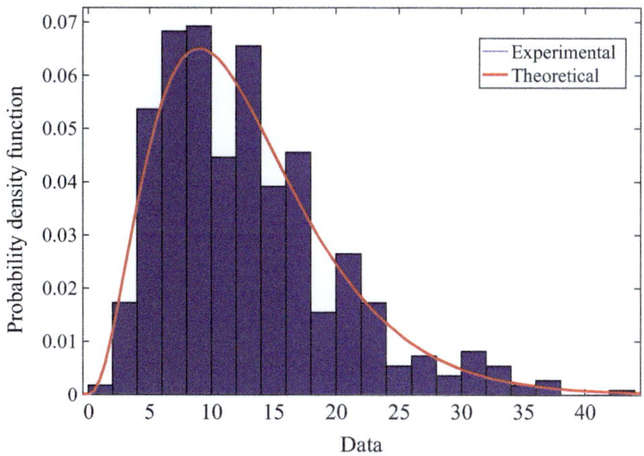

Figure 8.12 Example of data sets for C_T^2 distribution

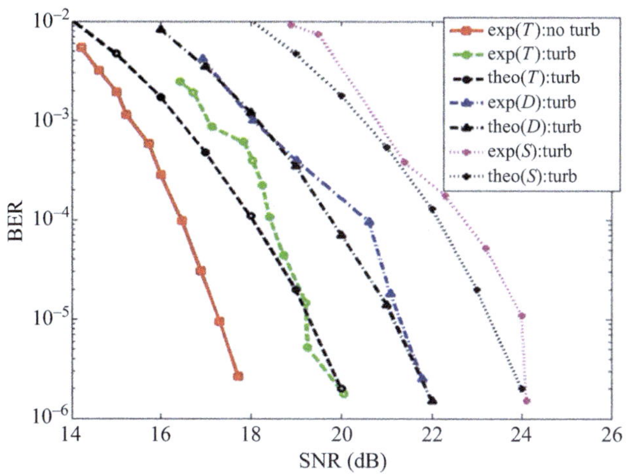

Figure 8.13 Experimental (exp) and theoretical (theo) results for BER vs. SNR for single- (S), double- (D), and triple- (T) hop links for $C_n^2 = 5.4 \times 10^{-12}$ m$^{-2/3}$, and an exp. T link with no turbulence [44]

terms of SNR less than 0.5 dB at higher values of BER stem from the fact that no ideal theoretical turbulence flow can be experienced in real practical environment. Furthermore, at a BER of 10^{-5} and $C_n^2 = 5.4 \times 10^{-12}$ m$^{-2/3}$, the SNR penalty for the triple-hop link is ~2.0 dB compare to link with no turbulence. The SNR penalty is approximately 2.0 dB less compared to the link with $C_n^2 = 3.8 \times 10^{-10}$ m$^{-2/3}$.

Moreover, Figure 8.14 illustrates the measured and theoretical BER performance against the SNR for the case of triple-hop FSO link for different turbulence levels.

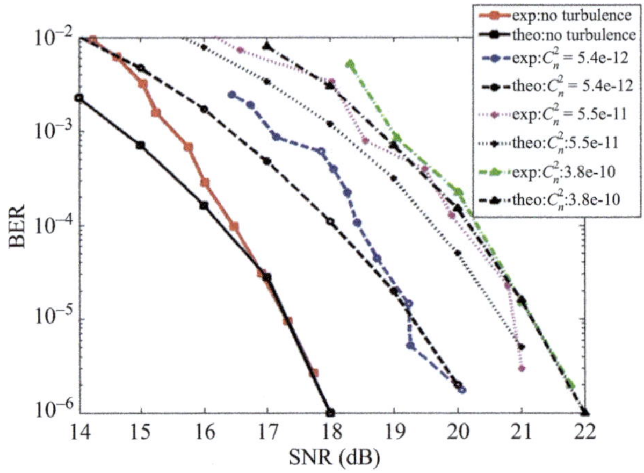

Figure 8.14 Experimental (exp) and theoretical (theo) results for BER vs. SNR for T-hop link with and without turbulence [44]

We can observe reasonable match between the analytically simulated and measured plots. However, for lower values of SNR, the analytical results slightly differ from the measured data owing to the OA noise. The BER performance degrades with the AT strength. For example, at a BER of 10^{-5} the SNR penalties are ~ 2.0 dB, 3.7 dB, and ~4.0 dB for C_n^2 of 5.4×10^{-12} m$^{-2/3}$, 5.5×10^{-11} m$^{-2/3}$, and 3.8×10^{-10} m$^{-2/3}$, respectively, compared to the link with no turbulence. Note, at higher levels of turbulence the beam front of propagating optical wave experiences higher level of intensity and phase fluctuations, which ultimately leads to much wider, scattered and random optical beam patterns at the Rx as discussed in [45]. This characteristic is best described by Kolmogrov's theory of turbulence where the refractive index changes in the order of several parts per million in every 1 K variation in the atmospheric temperature [46].

Finally, in order to provide more insight into the practical application of the system, we further consider the extension of the link span to 500 m, which is compatible with the last mile network access typical requirements in urban areas [1]. Figure 8.15 shows the simulated BER performance for single, dual-hop, and triple-hop all-optical relay-assisted FSO links for C_n^2 of 10^{-12} and 10^{-16} m$^{-2/3}$, representing typical strong and weak turbulence regimes in an outdoor environment [47]. The plots confirm that the proposed scheme can successfully be adopted in a real practical environment, to ensure link availability and quality performance under turbulence conditions. For example, at a target BER of 10^{-3} the triple-hop link offers SNR gains of ~4 dB and ~5 dB for $C_n^2 = 10^{-16}$ and 10^{-12}m$^{-2/3}$, respectively, compared to the single link transmission case.

Eye-diagrams for the single-hop, dual-hop, and triple-hop FSO links over $C_n^2 = 5.4 \times 10^{-12}$ m$^{-2/3}$ for BER values of 10^{-6}, 10^{-4}, and 10^{-2}, respectively, as shown in Figure 8.16.

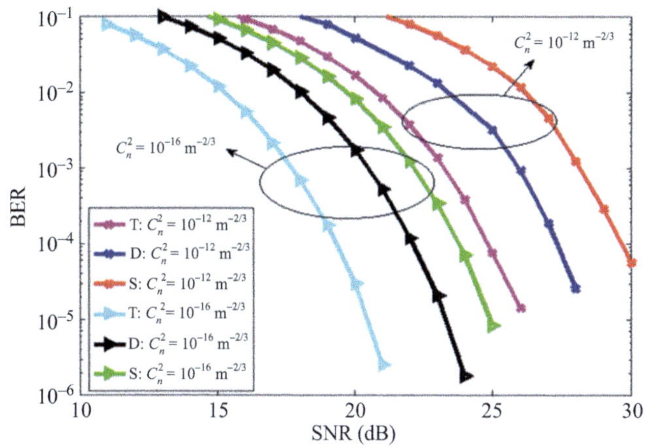

Figure 8.15 Simulated BER performance for all-optical single- (S), double- (D), and triple- (T) hop FSO links for C_n^2 of 10^{-12} $m^{-2/3}$ and 10^{-16} $m^{-2/3}$ and a total link span of 500 m [44]

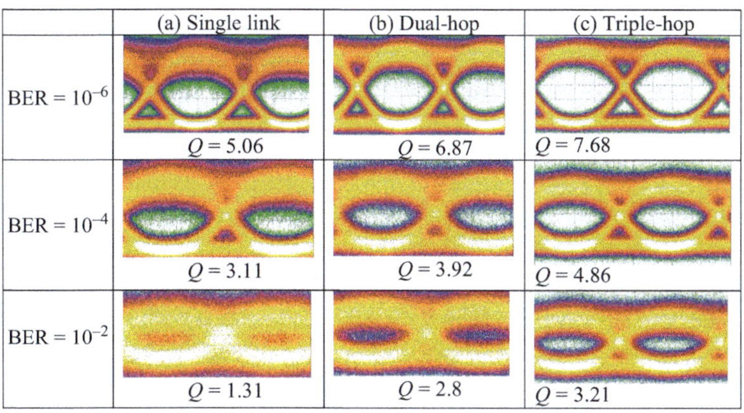

Figure 8.16 Measured eye-diagrams of received FSO system for: (a) single link; (b) dual-hop; and (c) triple-hop for BER values of 10^{-6}, 10^{-4}, and 10^{-2}, respectively, over $C_n^2 = 5.4 \times 10^{-12}$ $m^{-2/3}$

8.3 All-optical regenerate-and-forward relaying technique

When an optical signal propagates through a free-space or an optical fiber channel, several signal impairments take place that degrade the optical signal-to-noise ratio (OSNR). These impairments originate from the accumulating noise; from the channel fading effect; as well as from the presence of ASE noise and last but not least from fiber nonlinearities. Even though AOAF can be easily implemented, it is limited by the abovementioned impairments and especially by the cumulative background and ASE

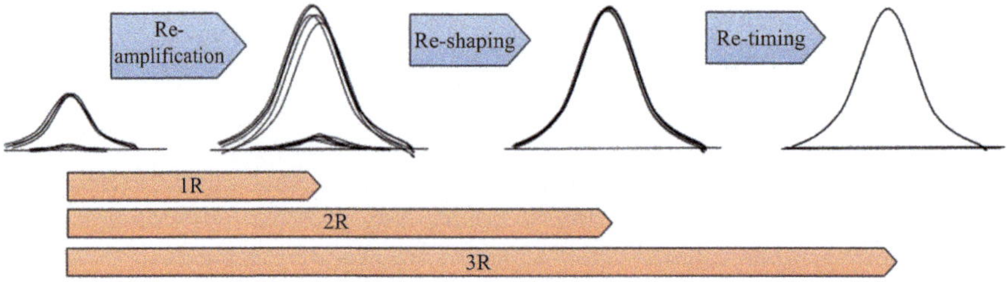

Figure 8.17 A signal regeneration scheme illustrating the re-amplification, re-shaping, and re-timing operations

noise generated by optical amplifiers. Thus, there is a limited number of relays that can be employed, since the performance of the system will start to degrade at some point. AORF is introduced in this chapter as means to overcome this particular challenge by eliminating the background and ASE noise at each relay and enabling longer transmission link distance for the transmitted signal. The ability to regenerate signals ensures maximum flexibility in the network design and architecture [48].

All-optical regeneration (AOR) is at the forefront of current research interests in optical communications systems, due to the high demand for migrating traditional electronic processes into the optical domain. Optical amplification has already been utilized for a number of applications in current optical networks, as the use of EDFAs, SOAs, and Raman amplifiers is now commonplace. With the development of long-haul optical telecommunication systems working at high data rates (40 Gbps and beyond) [49], performing AOR has become of great importance in combating cumulative noise, further improving the signal quality, and increasing the transmission link distance. The three basic signal processing functions of (i) Re-amplification; (ii) Re-shaping; and (iii) Re-timing are required to form either a 1R, 2R (re-amplification and reshaping), or 3R (re-amplification, reshaping, and retiming) regenerator (see Figure 8.17). In other words, 1R can also be referred to as the AOAF technique. Reshaping is used to suppress the noise accumulated and to improve the extinction ratio in an intensity modulation signal. Re-timing requires extracting the signal clock from the input data signal and transferring it to a newly generated pulse train [50]. However, AOR technology is still considered to be in its infancy, and as such, more research is required before the technology is ready for wide scale commercial deployment.

8.3.1 Self-phase modulation-based 2R regenerator

Although all-optical 3R regenerators have demonstrated great performance and versatility, 2R regenerators can be used beneficially in medium distance transmission systems [51]. Several AOR architectures that rely on optical nonlinear effects have been proposed [52–54] to replace the widespread optoelectronic regenerators.

Among these phenomena, self-phase modulation (SPM) is an often-utilized effect due to its simplicity, since neither a pump source, nor a probe source is required. In general, 2R regenerators lead to an extinction ratio improvement and

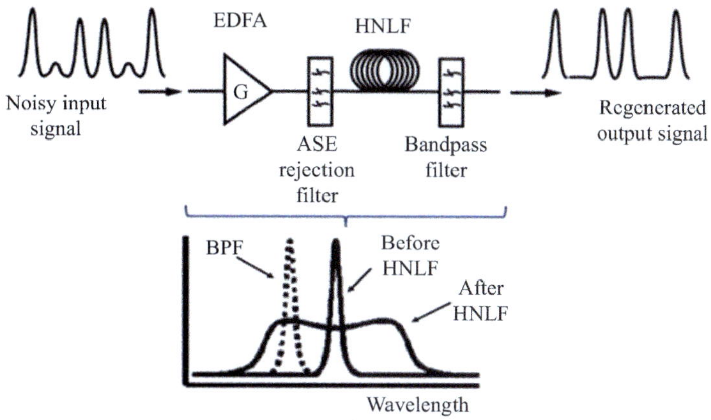

Figure 8.18 Schematic diagram of an SPM-based 2R regenerator

an amplitude noise reduction [55]. SPM-based 2R regenerator proposed in [56] is an attractive regeneration choice due to its experimental setup simplicity and robustness, where its function is based on spectral broadening followed by an offset spectral filtering. Furthermore, the quasi-instantaneous response time of the Kerr nonlinearities in optical fibers [57] provides the capability of regenerating high data rate signals. General theoretical guidelines have been proposed in [57] to design a high-performance 2R AOR based on SPM. A set of general scaling rules of the 2R SPM-based AOR has been reported in [58]. The guidelines were further analyzed in [59] to optimize the system performance of Mamyshev-type AOR. Additional literature [60–63] studied various aspects of the 2R SPM-based design.

Figure 8.18 illustrates the schematic diagram of a 2R SPM-based AOR. The AOR normally consists of three main elements, which are EDFA, highly nonlinear fiber (HNLF), and optical bandpass filter (OBPF). At the beginning of regeneration, the degraded (noisy) optical signal is fed into an EDFA to boost the peak power to a power level that results in sufficient spectral broadening via the SPM effect. The amplified signal is then passed through an ASE rejection filter to eliminate the out-band ASE noise added by the booster EDFA. Following that, the filtered signal is injected into the HNLF, where it experiences SPM-induced spectral broadening. An OBPF is subsequently used at the output of the HNLF as a reshaping element, whose center λ is slightly shifted from the input signal wavelength. The bandwidth of the output OBPF sets the output pulse width and needs to be optimized carefully.

OBPF acts an optical decision gate to discriminate ones and zeros based on the difference in SPM-induced spectral broadening [64]. Only ones (pulses with sufficient power) produce sufficient spectral broadening, which can pass through the OBPF. This results in a nonlinear transfer function for the regenerator, and therefore, suppression of the noise present in the former signal. Clearly, this nonlinear transfer function depends strongly on the launched power, filter offset, and HNLF (length, and dispersion) [65]. This 2R AOR improves the extinction ratio and the optical SNR of a degraded signal, and have been demonstrated in highly nonlinear silica-based fibers to achieve a million kilometers of error free transmission without electrical conversion [66].

Note that, all of the aforementioned investigations were based on optical fiber communications. To the best of the author's knowledge, only the study reported in [14] implemented a 2R Mamyshev-type AOR for a FSO communication link, acting as the relaying system. The authors in that study theoretically analyzed the improvement that the regenerator offers in mitigating the accumulation of background noise. The results showed that for the dual-hop FSO system, the maximum accessible transmission distance increased from 0.8 km to 1.9 km for the all-optical AF system and the 2R regenerator relaying-based system, respectively, compared to the direct transmission system. In addition, it was shown that when the relay was increased from single to triple-hop, the regenerator was able to eliminate the background noise at each relay by offering a power gain improvement of 1.4 dB, 2.9 dB, and 5.3 dB, respectively, compared to the all-optical AF system at a BER of 10^{-5}. However, only the background noise was considered as the limiting factor neglecting the effect of the AT on the system performances. Moreover, details of the analysis were not sufficient and the investigation was purely analytical with no experimental work.

8.3.2 Experimental analysis of AORF FSO

In this section, the performance of the novel dual-hop AORF SPM-based 2R regenerator as part of the FSO relay-assisted system with and without turbulence is discussed. Experimental demonstrations were conducted in controlled indoor laboratory environment. The measurements were carried out several times in order to mimic the real outdoor environment and ensure that the analysis is valid for real-case scenarios. AT generation in a dedicated chamber was used similarly for the analysis of the FSO AOAF system performance. Figure 8.18 illustrates the experimental block diagram of the proposed AORF dual-hop FSO relay-assisted communication system, based on a 2R AOR, under the influence of atmospheric channel turbulence effects. The system contained four main stages, namely signal transmission by a Tx, signal degradation by turbulence effects, signal regeneration by means of an SPM-based 2R AOR, and an Rx. The total link span was 4.4 m. At the Tx, a pulsed mode-locked laser was operated at the operating wavelength λ of 1,547 nm with a repetition rate of 10 GHz. Output pulses had a Gaussian profile, with a full-width at half-maximum (FWHM) pulse width of 2.2 ps, and emulated an RZ-OOK signal. The pulsed signal was then fed into a 99/1 fiber coupler, with 1% of the signal sent to an OSA for monitoring. The remaining 99% of the signal passed through a pre-amp EDFA, and a variable attenuator to enable adjustment of the signal SNR. The output signal was then launched into the free space channel via a GRIN fiber collimator with a clear aperture diameter lens of 0.18 cm. Two plano-convex lenses (CL 1 and CL 2) with a focal length f of 10 cm were placed after the Tx and before the Rx, respectively, in order to focus the optical beam and minimize its divergence. The degraded signal was injected into the AOR for signal reshaping.

The AOR consisted of EDFA and HNLF, followed by OBPF. Since the received power after the FSO turbulence channel was at a level of −15 dBm, a pre-amp EDFA

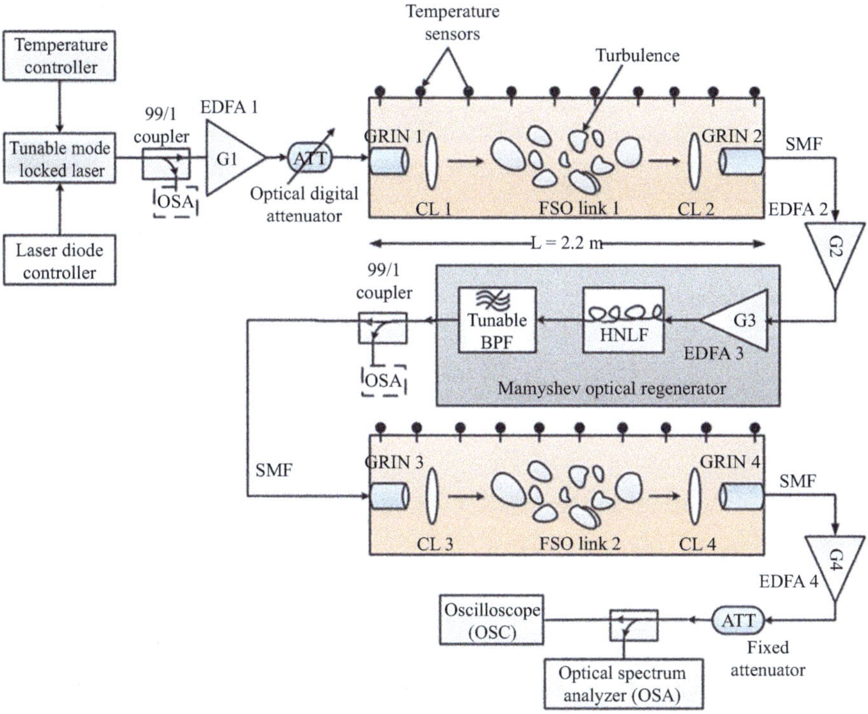

Figure 8.19 Block diagram of the experimental setup of the all-optical 2R regenerative FSO relay-based system. CL: Convex lens [67]

was placed in front of the booster EDFA. The launch output power of the booster EDFA was then adjusted to a power level sufficient to induce SPM spectral broadening in the HNLF. The amplified signal was then fed into a 500-m long HNLF. The HNLF had a chromatic dispersion of −2.5 ps/nm/km, a dispersion slope of 0.019 ps/nm^2/km, an attenuation loss of 0.88 dB/km, and the nonlinearity coefficient of 11.5 W^{-1}km^{-1} at 1,550 nm. At the output of the HNLF, a wavelength and bandwidth tunable OBPF (bandwidth adjustable range of 0.2–6.0 nm) was detuned with respect to the input signal λ in order to transmit the SPM-broadened signal (logical ones) and reject the original spectrum, thus providing an excellent optical control and a flat-top passband shape. Following regeneration, the signal is passed through a 99/1 fiber coupler prior to another free space channel transmission. A similar setup to that used in Link 1 above was used for the Link 2 transmission (see Figure 8.19).

At the Rx, another GRIN fiber collimator was used to couple the received optical beam into an SMF, and feed it into a pre-amp EDFA for further amplification. The signal was then passed into a 90/10 coupler, in which 90% of the signal went to an OSC, and the remaining 10% was fed to an OSA for link assessment.

The turbulence channel was implemented in the controlled indoor environment using the identical setup as discussed in Chapter 2.3 using with 20 temperature

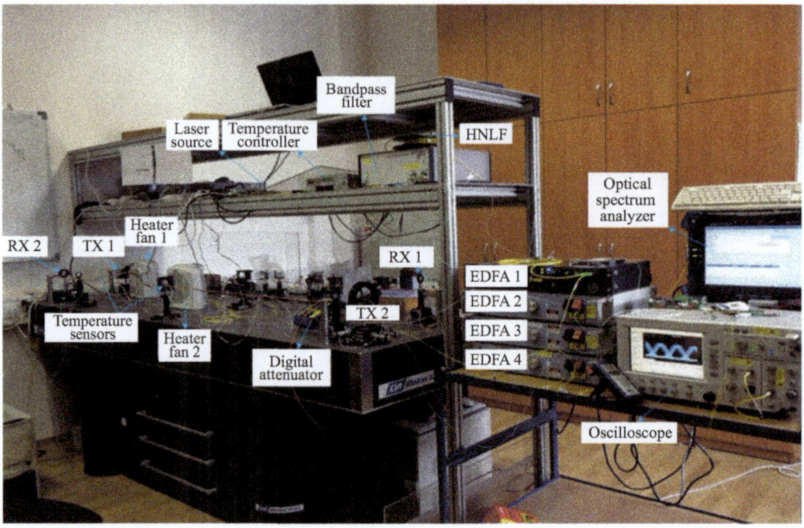

Figure 8.20 Snapshot of the indoor laboratory in which the all-optical 2R regenerative FSO relay-based system was implemented [67]

sensors (see Figure 8.19). A snapshot of the indoor laboratory in which experimental setup of the all-optical dual-hop 2R regenerative FSO relay-based system was implemented is depicted in Figure 8.20. Results confirming short pulse regeneration have been published in [67]. It has been validated, the scheme recovers even the picosecond pulses (i.e., 2 ps) corresponding to a maximal potential bit rate of 350 Gbps under AT influences. More than 3 dB gain has been reached in comparison to the AOAF method [67].

8.4 Conclusions

Relay-assisted FSO communication is one of the alternative mitigation solutions that offers diversity within the networks, which can be adopted to combat link failure thus substantially improve the link availability and distance. To mitigate the degrading effect of the turbulence-induced fading, all-optical relay-assisted system was chosen as the potential solution. This technique elegantly exploit the distance-dependent of the turbulence effect, while maintaining a high data rate.

Recently, relay-assisted or multi-hop transmission systems have been introduced as a powerful fading mitigation tool as an alternate option in realizing the spatial diversity advantages. AOAF FSO relay-assisted for the FSO network was done by using an optical amplifier at the relay, to boost the received signal and forward it to the next relay/destination. The performance analysis of single, dual-hop, and triple-hop AOAF FSO were comprehensively discussed theoretically and experimentally under the effects of turbulence channels. A mathematical framework for the end-to-end SNR and BER performance for the general multi-hop relay-assisted FSO link was also outlined, considering the G-G distribution model. The novel experimental demonstrations were discussed and results presented a good agreement between the analytical calculations and measurements, particularly at relatively high SNRs.

Even though AOAF is simple and straightforward to be implemented, optical amplifiers generate excess noise which is subsequently amplified and limits the link transmission span. Therefore, AORF was proposed to overcome this particular issue by eliminating the amplifier cumulative noise and regenerate the signal at each relay, before re-transmitting to the next relay or destination. This technique significantly increases the total transmission distance and ensures the maximum flexibility in network design by providing greater tolerance in adding or dropping channels.

References

[1] Z. Ghassemlooy, W. Popoola, and S. Rajbhandari, Optical wireless communications: System and channel modelling with MATLAB. United Kingdom: CRC Press Taylor and Francis Group, 2013.

[2] L. C. Andrews and R. L. Phillips, Laser beam propagation through random media, 2nd Ed. Washington, USA: SPIE Press, 2005.

[3] A. García-Zambrana, C. Castillo-Vázquez, and B. Castillo-Vázquez, "On the capacity of FSO links over gamma-gamma atmospheric turbulence channels using OOK signaling," Eurasip J. Wirel. Commun. Netw., 2010.

[4] I. E. Lee, Z. Ghassemlooy, W. P. Ng, and M. Uysal, "Performance analysis of free space optical links over turbulence and misalignment induced fading channels," 2012 8th Int. Symp. Commun. Syst. Networks Digit. Signal Process., pp. 1–6, Jul. 2012.

[5] M.-A. Khalighi, N. Schwartz, N. Aitamer, and S. Bourennane, "Fading reduction by aperture averaging and spatial diversity in optical wireless systems," J. Opt. Commun. Netw., vol. 1, no. 6, pp. 580–593, 2009.

[6] F. E. Zocchi, "A simple analytical model of adaptive optics for direct detection free-space optical communication," Opt. Commun., vol. 248, no. 4–6, pp. 359–374, 2005.

[7] M. Uysal, J. Li, and M. Yu, "Error rate performance analysis of coded free-space optical links over gamma-gamma atmospheric turbulence channels," IEEE Trans. Wirel. Commun., vol. 5, no. 6, pp. 1229–1233, 2006.

[8] G. Yang, M.-A. Khalighi, Z. Ghassemlooy, and S. Bourennane, "Performance evaluation of receive-diversity free-space optical communications over correlated Gamma-Gamma fading channels," Appl. Opt., vol. 52, no. 24, pp. 5903–5911, 2013.

[9] M. Ahmad and O. Awwad, "Synergies of radio frequency and free space optics communication: New hybrid solutions for next generation wireless mesh networks," Int. J. Comput. Networks, vol. 4, no. 4, pp. 135–155, 2012.

[10] S. Navidpour, M. Uysal, and M. Kavehrad, "BER performance of free-space optical transmission with spatial diversity," IEEE Trans. Wirel. Commun., vol. 6, no. 8, pp. 2813–2819, 2007.

[11] T. A. Tsiftsis, H. G. Sandalidis, G. K. Karagiannidis, and S. Member, "Optical wireless links with spatial diversity over strong atmospheric turbulence channels," IEEE Trans. Wirel. Commun., vol. 8, no. 2, pp. 951–957, 2009.

[12] D. K. Borah, A. C. Boucouvalas, C. C. Davis, S. Hranilovic, and K. Yiannopoulos, "A review of communication-oriented optical wireless systems," EURASIP J. Wirel. Commun. Netw., vol. 1, no. 1, pp. 1–28, 2012.

[13] S. Kazemlou, S. Hranilovic, and S. Kumar, "All-optical multihop free-space optical communication systems," J. Light. Technol., vol. 29, no. 18, pp. 2663–2669, 2011.

[14] A. Sendonaris, E. Erkip, and B. Aazhang, "User cooperation diversity-Part I: System description," IEEE Trans. Commun., vol. 51, no. 11, pp. 1927–1938, 2003.

[15] A. Sendonaris, E. Erkip, and B. Aazhang, "User cooperation diversity-Part II: Implementation aspects and performance analysis," IEEE Trans. Commun., vol. 51, no. 11, pp. 1939–1948, 2003.

[16] A. Acompora and S. Krishnamurthy, "A broadband wireless access network based on mesh-connected free space optical links," IEEE Pers. Commun., vol. 6, no. 10, pp. 62–65, 1999.

[17] C. Abou-rjeily and S. Haddad, "Inter-relay cooperation: A new paradigm for enhanced relay-assisted FSO communications," IEEE Trans. Commun., vol. 62, no. 6, pp. 1970–1982, 2014.

[18] S. M. Aghajanzadeh and M. Uysal, "Outage performance and DMT analysis of DF parallel relaying in FSO IM/DD communications," in IEEE Vehicular Technology Conference (VTC Fall), 2012, vol. 2, pp. 1–5.

[19] M. Safari and M. Uysal, "Relay-assisted free-space optical communication," IEEE Trans. Wirel. Commun., vol. 7, no. 12, pp. 5441–5449, 2008.

[20] E. Bayaki, D. S. Michalopoulos, and R. Schober, "EDFA-based all-optical relaying in free-space optical systems," IEEE Trans. Commun., vol. 60, no. 12, pp. 3797–3807, 2012.

[21] M. A. Khalighi and M. Uysal, "Survey on free space optical communication: A communication theory perspective," IEEE Commun. Surv. Tutorials, vol. 16, no. 4, pp. 2231–2258, 2014.

[22] X. Tang, Z. Wang, Z. Xu, and Z. Ghassemlooy, "Multihop free-space optical communications over turbulence channels with pointing errors using heterodyne detection," IEEE/OSA J. Light. Technol., vol. 32, no. 15, pp. 2597–2604, 2014.

[23] C. K. Datsikas, K. P. Peppas, N. C. Sagias, and G. S. Tombras, "Serial free-space optical relaying communications over gamma-gamma atmospheric turbulence channels," J. Opt. Commun. Netw., vol. 2, no. 8, p. 576, 2010.

[24] M. A. Kashani, M. Safari, and M. Uysal, "Optimal relay placement and diversity analysis of relay-assisted free-space optical communication systems," Opt. Commun. Netw., vol. 5, no. 1, pp. 37–47, 2013.

[25] M. Dohler and Y. Li, Cooperative communications: Hardware, channel and PHY. United Kingdom: Wiley, 2010.

[26] A. Nasri, R. Schober, and I. F. Blake, "Performance and optimization of amplify-and-forward cooperative diversity systems in generic noise and interference," IEEE Trans. Wirel. Commun., vol. 10, no. 4, pp. 1132–1143, 2011.

[27] M. R. Bhatnagar, "Performance analysis of decode-and-forward relaying in gamma-gamma fading channels," IEEE Photonics Technol. Lett., vol. 24, no. 7, pp. 545–547, 2012.

[28] M. R. Bhatnagar, "Average BER analysis of relay selection based decode-and-forward cooperative communication over gamma-gamma fading FSO links," 2013 IEEE Int. Conf. Commun., pp. 3142–3147, Jun. 2013.

[29] G. Parca, "Optical wireless transmission at 1.6-Tbit/s (16×100 Gbit/s) for next-generation convergent urban infrastructures," Opt. Eng., vol. 52, no. 11, p. 116102, 2013.

[30] U. Adar, T. Kuzniz, I. D. Haber, and N. Budin, "An all-optical multi-hop (cascaded) high bit rate wireless communication field trial," in Optical Fiber Communication Conference and Exhibit (OFC), 2002, no. 2001, pp. 446–448.

[31] J. Akella, M. Yuksel, and S. Kalyanaraman, "Error analysis of multi-hop free-space optical communication," in IEEE International Conference on Communications ICC, 2005, pp. 1777–1781.

[32] T. A. Tsiftsis, H. G. Sandalidis, G. K. Karagiannidis, and N. C. Sagias, "Multihop free-space optical communications over strong turbulence channels," in IEEE International Conference on Communications, 2006, pp. 2755–2759.

[33] M. A. Kashani, M. M. Rad, M. Safari, and M. Uysal, "All-optical amplify-and-forward relaying system for atmospheric channels," IEEE Commun. Lett., vol. 16, no. 10, pp. 1684–1687, 2012.

[34] P. V. Trinh, A. T. Pham, H. T. T. Pham, and N. T. Dang, "BER analysis of all-optical AF dual-hop FSO systems over gamma-gamma channels," IEEE 4th Int. Conf. Photonics, no. 1, pp. 175–177, Oct. 2013.

[35] L. Yang, X. Gao, and M. Alouini, "Performance analysis of relay-assisted all-optical FSO networks over strong atmospheric turbulence," J. Light. Technol., vol. 32, no. 23, pp. 4011–4018, 2014.

[36] P. V Trinh, S. Member, N. T. Dang, A. T. Pham, and S. Member, "All-optical relaying FSO systems using EDFA combined with optical hard-limiter over atmospheric turbulence channels," J. Light. Technol., vol. 33, no. 19, pp. 4132–4144, 2015.

[37] G. T. Djordjevic, M. I. Petkovic, A. M. Cvetkovic, and G. K. Karagiannidis, "Mixed RF/FSO relaying with outdated channel state information," J. Sel. Areas Commun., vol. 33, no. 9, pp. 1935–1948, 2015.

[38] N. A. M. Nor, Z. Ghassemlooy, S. Zvanovec, M. Khalighi, and M. R. Bhatnagar, "Comparison of optical and electrical based amplify-and-forward relay-assisted FSO links over gamma-gamma channels," 10th International Symposium on Communication Systems, Networks and Digital Signal Processing – Proceeding. CSNDSP 2016, IEEE. 2016, s. 1–5.

[39] C. Lin, Optical components for communications: Principles and applications. New York: Springer, 2004.
[40] K. Prabu, D. S. Kumar, and T. Srinivas, "Performance analysis of FSO links under strong atmospheric turbulence conditions using various modulation schemes," Optik (Stuttg)., vol. 125, no. 19, 2014.
[41] Le Nguyen Binh, Optical fiber communications systems: Theory and practice with MATLAB and simulink models. CRC Press Taylor and Francis Group, 2010.
[42] M. Ijaz, Z. Ghassemlooy, S. Ansari, *et al.*, "Experimental investigation of the performance of OOK-NRZ and RZ modulation techniques under controlled turbulence channel in FSO systems," in 5th International Symposium on Telecommunications (IST), 2010, pp. 59–64.
[43] N. A. M. Nor, Z. Ghassemlooy, S. Zvanovec, and M. Khalighi, "Performance analysis of all-optical amplify-and-forward FSO relaying over atmospheric turbulence," in IEEE Student Conference on Research and Development, 2015, no. 1, pp. 1–5.
[44] N. A. M. Nor, Z. Ghassemlooy, J. Bohata, *et al.*, "Experimental investigation of all-optical relay-assisted 10 Gbps FSO link over the atmospheric turbulence channel," IEEE Journal of Lightwave Technology, vol. 35, no. 1, pp. 45–53, 2017.
[45] M. Abaza, R. Mesleh, A. Mansour, and E. M. Aggoune, "Spatial diversity for FSO communication systems over correlated atmospheric turbulence channels," in IEEE Wireless Communications and Networking Conference (WCNC), 2014, pp. 382–387.
[46] I. E. Lee, Z. Ghassemlooy, W. P. Ng, M.-A. Khalighi, and S.-K. Liaw, "Effects of aperture averaging and beam width on a partially coherent Gaussian beam over free-space optical links with turbulence and pointing errors," Appl. Opt., vol. 55, no. 1, p. 1, 2016.
[47] M. A. Kallistratova and D. F. Timanovskiy, "The distribution of the structure constant of refractive index fluctuations in the atmospheric surface layer," Repr. from Izv. Atmos. Ocean Phys., vol. 7, no. 1, pp. 73–75, 1990.
[48] S. Wabnitz and B. J. Eggleton, All-optical signal processing: Data communication and storage. Springer Series in Optical Sciences, 2015.
[49] C. Finot, T. N. Nguyen, J. Fatome, *et al.*, "Numerical study of an optical regenerator exploiting self-phase modulation and spectral offset filtering at 40 Gbit/s," Opt. Commun., vol. 281, no. 8, pp. 2252–2264, 2008.
[50] H. Padda, "Noise modelling for 2R optical regenerator systems," University of Nottingham, 2009.
[51] S. S. Hnaung, "Design and implementation of 10 Gbps all-optical 2R regenerator," vol. 4, no. 6, pp. 3–7, 2014.
[52] P. P. Baveja, D. N. Maywar, A. M. Kaplan, and G. P. Agrawal, "Self-phase modulation in semiconductor optical amplifiers: Impact of amplified spontaneous emission," IEEE J. Quantum Electron., vol. 46, no. 9, pp. 1396–1403, 2010.

[53] I. Tomkos, C. Kouloumentas, and S. Tsolakidis, "Performance studies of multi-wavelength all-optical 2R regeneration subsystems based on highly non-linear fibers," Proc. 2007 9th Int. Conf. Transparent Opt. Networks, Ict. 2007, vol. 1, pp. 132–135, 2007.
[54] G. Xu, Z. Li, T. Yang, and C. Huang, "Analysis of a 2R all-optical regenerator utilizing cascaded four-wave mixing in a highly nonlinear fiber," Optik (Stuttg)., vol. 124, no. 21, pp. 4758–4761, 2013.
[55] J.-C. Simon, M. Gay, L. Bramerie, *et al.*, "Long distance transmission using optical regeneration," in Optical Fiber Communication Conference, 2008, pp. 8–10.
[56] P. V Mamyshev, "All-optical data regeneration based on self-phase modulation effect," in European Conference on Optical Communications, 1998, vol. 1, no. September, pp. 475–476.
[57] L. A. Provost, C. Finot, P. Petropoulos, K. Mukasa, and D. J. Richardson, "Design scaling rules for 2R-optical self-phase modulation-based regenerators," Opt. Express, vol. 15, no. 8, pp. 5100–5113, 2007.
[58] L. A. Provost, C. Finot, K. Mukasa, P. Petropoulos, and D. J. Richardson, "Generalisation and experimental validation of design rules for self-phase modulation-based 2R-regenerators," in Conference on Optical Fiber Communication and the National Fiber Optic Engineers, 2007, pp. 6–8.
[59] P. Johannisson and M. Karlsson, "Characterization of a self-phase-modulation-based all-optical regeneration system," IEEE Photonics Technol. Lett., vol. 17, no. 12, pp. 2667–2669, 2005.
[60] M. Aoudeh and J. C. Cartledge, "Impact of residual dispersion and ASE noise on the performance optimization of all-optical regenerators utilizing self-phase modulation in a highly nonlinear fiber," IEEE J. Sel. Top. Quantum Electron., vol. 12, no. 4, pp. 717–725, 2006.
[61] Y. Yang and C. Lou, "Experimental investigation of the influence of filters on regeneration performance in self-phase modulation based regenerator," Microw. Opt. Technol. Lett., vol. 49, no. 1, pp. 192–195, 2007.
[62] T. Her, T. Leng, G. Raybon, *et al.*, "Enhanced 40-Gbit/s receiver sensitivity with all-fiber optical 2R regenerator," in Technical Digest Conference Lasers and Electro-Optics Society, 2002, pp. 534–535.
[63] M. Rochette, J. N. Kutz, J. L. Blows, D. Moss, J. T. Mok, and B. J. Eggleton, "Bit-error-ratio improvement with 2R optical regenerators," IEEE Photonics Technol. Lett., vol. 17, no. 4, pp. 908–910, 2005.
[64] M. Karimi and M. Nasiri-kenari, "Free space optical communications via optical amplify-and-forward relaying," J. Light. Technol., vol. 29, no. 2, pp. 242–248, 2011.
[65] S. Arnon, J. R. Barry, G. K. Karagiannidis, R. Schober, and M. Uysal, Advanced optical wireless communication systems. New York, USA: Cambridge University Press, 2012.
[66] T.-H. Her, G. Raybon, and C. Headley, "Optimization of pulse regeneration at 40 Gb/s based on spectral filtering of self-phase modulation in fiber," IEEE Photonics Technol. Lett., vol. 16, no. 1, pp. 200–202, 2004.

[67] G. Raybon, Y. Su, J. Leuthold, *et al.*, "40 Gbit/s Pseudo-linear transmission over one million kilometers," in Optical Fiber Communication Conference, 2002.

[68] N. A. M. Nor, M. Komanec, J. Bohata, Z. Ghassemlooy, M. R. Bhatnagar, and S. Zvanovec, "An experimental all-optical relay-assisted FSO link with reconstruct and forward scheme transmitting ultra-short pulses," IEEE Journal of Lightwave Technology, submitted.

Chapter 9
Optical spatial diversity for FSO communications

Sujan Rajbhandari[1] and Zahir Ahmad[1]

9.1 Introduction

The performance of the terrestrial free-space optical (FSO) communication is adversely affected by atmospheric conditions such as fog, rain, dust particles, and smoke. Even in the clear atmospheric conditions, the FSO link experiences severe fading due to the turbulence effect. Turbulence is induced by random variations in refractive index due to temperature and pressure fluctuation along the FSO beam propagation. Turbulence can introduce a deep fading (20 to 30 dB) as a result the outage probability increases significantly. In order to mitigate the channel fading, the receiver or transmitter diversities or a combined transmitter and receiver diversity (known as multiple-input multiple-output (MIMO) techniques) are often considered along with other techniques such as modulation, temporal and wavelength diversities. In this chapter, the FSO link performance in the presence of the turbulence fading using the transmitter and receiver diversity is discussed.

9.2 Outdoor channel

Understanding the channel characteristics is a very important aspect to design any communication systems. In this section, some of the factors which degrade the performance of an outdoor optical wireless communication channel will be discussed. By nature, the outdoor optical wireless channel is a complex environment compared to the indoor channels because of the dynamic nature of the atmosphere. Two most important factors which cause substantial degradation of optical beam propagation in outdoor channels are visibility effect caused by fog and fluctuation of phase and amplitude caused by atmospheric turbulence.

[1]School of Computing, Electronics and Mathematics, Coventry University, UK

9.3 Visibility and fog models

Fog is the dominant atmospheric particles which heavily attenuates the optical wave propagation in free space. The definition of fog is when visibility decreases to less than 1 km, due to the cloud at the ground level [1]. Water droplets are the main elements of fog and measured by the set of parameters such as particle size, temperature, liquid water content (LWC), humidity, etc. Although, there are other atmospheric particles such as dust, snow, rain, etc. exist in the atmosphere, the effect of fog is severe compared to the rest. The main reason for this is the size of the fog particles which is close to the wavelength of light used for FSO wireless communication system [2]. Depending on the physical process of saturation of the air, fog can be classified into radiation fog and advection fog. Radiation fog is seen in the winter when land cools down in the night time due to the thermal radiation of the air close to the surface. Because of this, air could not hold the moisture resulting in condensation and fog. Most of the time, this fog disappears after the sunrise. The particle size distribution for radiation fog is 1–3 μm and LWC ranges from 0.01 to 0.1 g/m^3 [3]. The visibility for this kind of fog is up to 500 m which mainly affects the communication when the shorter wavelength is used. Advection fog is formed when moist air passed over a cool surface. It is defined when LWC is measured higher than 0.20 g/m^3 with a particle diameter close to 20 μm. The visibility range for this kind of fog is considered to be 200 m. There are other types of fog named as valley and upslope fog, which are localised based and can be seen near to valley and hills.

The effect of fog in FSO communication can be studied using different models reported in the literature. Some of these models are summarised in Sections 9.3.1 and 9.3.2.

9.3.1 Kruse model

Predicting fog-induced attenuation is a challenging task since fog particle distribution and concentration varies depending on the location. Therefore, empirical models are the best way to predict attenuation characteristic. Visibility (V) data collected from experiments can be used to measure the fog density. Koschmieder law is one of the well-known methods which can be used to predict the density of fog. The meteorological visibility V in km can be defined as a function of atmospheric attenuation coefficient β_λ and transmittance threshold T_{th} level of the original visual contrast (100%) along the propagation path. Therefore, V can be defined as

$$V = \frac{10\log(T_{th})}{\beta_\lambda} \tag{9.1}$$

Here, the value of T_{th} changes from 0.0077 to 0.06. For a specific atmospheric environment, larger meteorological visual range (MVR) can be achieved by a smaller value of T_{th}.

Kruse developed a new model by considering the scattering coefficient of fog which depends on the wavelength of the propagating optical beam [4]. The model is defined as

$$V(\text{km}) = \frac{10\log_{10}T_{\text{th}}}{\beta_\lambda}\left(\frac{\lambda}{\lambda_o}\right)^{-\delta} \tag{9.2}$$

Here, the maximum spectrum of the solar band is defined as λ_o and the particle size-related coefficient is defined by δ. The value of δ depends on the distribution of particle size which was determined from the experimental data and defined as [4]:

$$\delta = \begin{cases} 1.6 & \text{for } V > 50 \text{ km} \\ 1.3 & \text{for } 6 \text{ km} < V < 50 \text{ km} \\ 0.585V^{1/3} & \text{for } 0 \text{ km} < V < 6 \text{ km} \end{cases} \tag{9.3}$$

Kruse model was developed for the haze particles which consist of small aerosols and particle size considered for this model was smaller than the wavelength in visible light and infrared [5]. Fog attenuation caused by the large particle size was not considered by the Kruse model, therefore, this model is not suitable when visibility is less than 1 km.

9.3.2 Kim model

Kim has expanded the value of δ from the Kruse model for $V < 1$ km and suggested that β_λ is wavelength independent for $V < 0.5$ km [1]. The calculation in this model was based on the Mie theory and considered the effect of dense haze and fog. The modified value of δ achieved in the Kim model is as follows [6]:

$$\delta = \begin{cases} 1.6 & \text{for } V > 50 \text{ km} \\ 1.3 & \text{for } 6 \text{ km} < V < 50 \text{ km} \\ 0.16V + 0.34 & \text{for } 1 \text{ km} < V < 6 \text{ km} \\ V - 0.5 & \text{for } 0.5 \text{ km} < V < 1 \text{ km} \\ 0 & \text{for } V < 0.5 \text{ km} \end{cases} \tag{9.4}$$

Since the attenuation coefficient is not dependent on the wavelength of the propagation wave, therefore, according to the Kim model there is no advantage of using higher wavelength compared to the lower one for the free space optical wireless communication in the dense fog condition.

9.3.3 Naboulsi model

Naboulsi developed models to measure the fog attenuation in the spectral band between 0.69 and 1.55 μm in the visible ranges of 0.05–1 km. Two separate models have been proposed for convection and advection fog environment. The proposed convection fog attenuation model is given by

$$\alpha_{con}(\lambda) = 10\log_e\left(\frac{0.11478\lambda + 3.8367}{V}\right) \tag{9.5}$$

and the proposed advection fog attenuation model is given by

$$\alpha_{adv}(\lambda) = 10\log_e\left(\frac{0.18126\lambda^2 + 0.13709\lambda + 3.7205}{V}\right) \tag{9.6}$$

Here, V is given in km and λ is in μm. As shown, the attenuation depends on the wavelength where attenuation increases as wavelength increases. This is opposite to the Kim and Kruse model which suggest that attenuation decrease when wavelength increases.

9.4 Wavelength diversity to mitigate fog

The atmospheric attenuation due to fog and other particles depends on the operating wavelength. For example, Figure 9.1 shows the attenuation (dB/km) against visibility for operating wavelength of 830 nm and 1,550 nm based on Kruse model (9.2). The figure clearly demonstrates that fog attenuation is significantly lower for longer wavelength (1,550 nm). For example, attenuation at the visibility of 1 km is ~13.4 dB/km and ~9.3 dB/km for wavelengths of 830 and 1,550 nm, respectively. The attenuation difference is significantly higher when the fog is dense (at a low visibility level). At the visibility of 0.1 km, attenuation difference is ~24 dB (~128 dB/km @ 1,550 nm and ~152 dB/km @ 830 nm). This clearly indicates that selection of appropriate operating wavelength is important for 99.999% link availability, especially in the place where the occurrence of fog is very frequent.

The wavelength dependence of the fog attenuation is different to characterised due to rare occurrence of the event. In order characterised fog attenuation in a controlled laboratory environment, indoor FSO chamber can be used. The

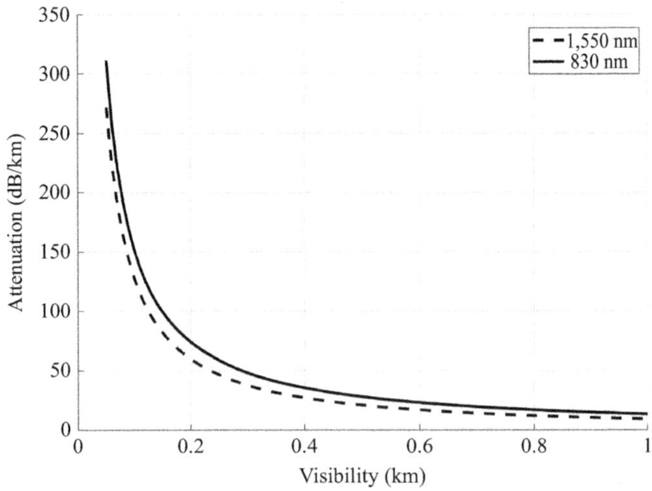

Figure 9.1 Attenuation versus the visibility using Kruse model

schematic of one of the such FSO chamber at Northumbria University to characterise atmospheric FSO system is given in Figure 9.2. Three lasers operating at 550 nm, 830 nm and 1,550 nm are operated simultaneously. Received optical power was continuously measured at a link distance of 6 m. The 550 nm laser was used for estimating the visibility. The 830 nm and 1,550 nm were used to estimate the wavelength dependent attenuation in the presence of fog. Fog was induced to the chamber at various rate to vary the visibility.

Figure 9.3(a) shows the measured attenuation (dB/km) for 830 nm and 1,550 nm. As predicted by Kruse model, there is a significant difference in the fog attenuation between the two operating wavelengths. Figure 9.3(b) shows the average difference in attenuation as a function of visibility. As predicted from Kruse model (see Figure 9.1), the average difference in the attenuation increase with the decreasing visibility. However, the measurement shows a higher difference than the one predicted by Kruse model. Nonetheless, these results (theoretical and simulation) clearly demonstrates that longer wavelength (in this case 1,550 nm) is more favourable in fog conditions. The study carried out by Harris *et al.* [7] also demonstrated the advantage of wavelength diversity to overcome the fog effect for FSO links.

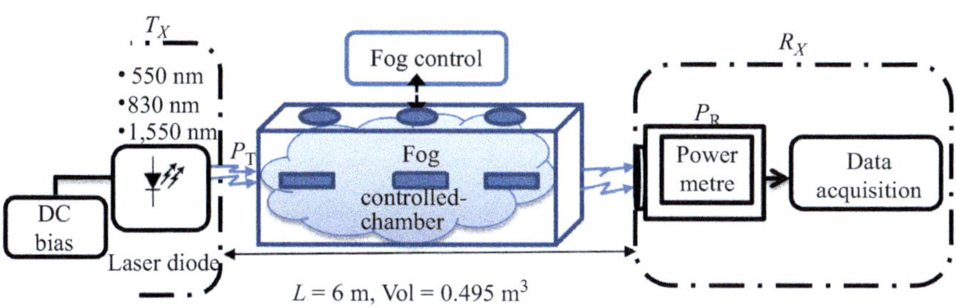

Figure 9.2 Experimental setup to measure fog attenuation and visibility

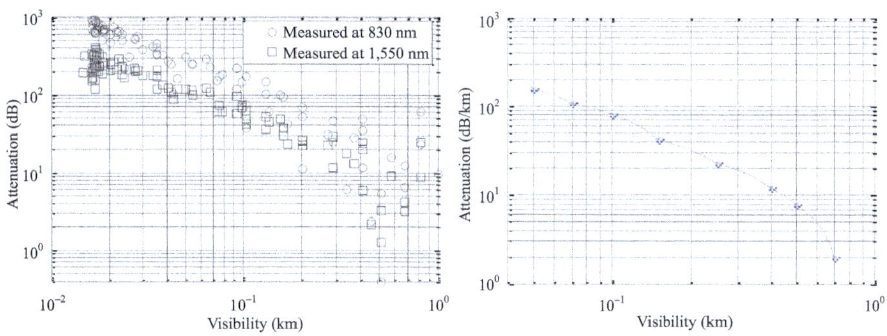

Figure 9.3 (a) Comparison of the measured loss (dB/km) for 830 and 1,550 nm and (b) average of their loss difference as a function of measured visibility (km)

9.5 Atmospheric turbulence model and mitigation

Atmospheric turbulence (scintillation) severely affects the optical wave propagation in FSO communications. The air close to the earth surface is generally warner compared to the higher altitude because of the absorption of solar radiation by Earth's surface. This fluctuation of air's temperature and density causes a random phenomenon named as atmospheric turbulence. As temperature changes, the refractive index of the air particles also changes and the medium acts as a combination of different cells known as eddies. Because of this, the propagating optical beam travelling through a turbulent medium suffers random variation in amplitude and phase which eventually results fading and degrades the performance of the communication.

The relation between the propagation path of optical wave and refractive index caused by the air temperature and density is given by [7]:

$$n(r,t) = n_0 + n_t(r,t); \tag{9.7}$$

where n_0 is the average refractive index and $n_t(r,t)$ is the component related to turbulence.

The refractive index structure parameter C_n^2 is widely used to measure the atmospheric turbulence strength. Mainly three different factors such as atmospheric pressure, temperature structure constant and small scale temperature changes are considered to find C_n^2 as given below:

$$C_n^2 = \left(86 \times 10^{-6} \frac{P}{T^2}\right)^2 C_T^2 \quad \text{for} \quad \lambda = 850\,\text{nm} \tag{9.8}$$

Here, P is atmospheric pressure in millibar, T is absolute temperature, C_T^2 is the temperature structure constant. The value of the refractive index structure C_n^2 can vary from 10^{-17} m$^{-2/3}$ to 10^{-12} m$^{-2/3}$ depending on the strength of turbulence.

For a long distance FSO communication, the temporal log-amplitude variance σ_x^2 which is known as the Rytov parameter can be predicted using the variation of C_n^2. The relation between these two is given as [3,8]:

$$\sigma_x^2 = 0.56k^{7/6} \int_0^{L_p} C_n^2(x)\left(L_p - x\right)^{5/6} dx \tag{9.9}$$

Here, k is known as spatial wave number and the value of k is $(2\pi/\lambda)$, λ is the wavelength of the propagating wave and L_p is the propagation length.

The refractive index structure parameter C_n^2 is constant for a turbulent medium when fields propagates horizontally and for this kind of scenario the normalised variance of a plane wave is given by:

$$\sigma_I^2 = 1.23 C_n^2 k^{7/6} {L_p}^{11/6} \tag{9.10}$$

9.5.1 *Lognormal turbulence model*

Lognormal turbulence model is a well-known method to measure the weak atmospheric turbulence and has been used to analyse different FSO wireless channels.

Optical wave collides with different types of elements when travel through a turbulent medium. Each of this element contributes random scattering and phase delay to the optical beam. The distribution of the amplitude fluctuation is found to be lognormal. Therefore, the probability density function (PDF) of the received irradiance I due to turbulence can be found as:

$$p(I) = \frac{1}{\sqrt{2\pi\sigma_I^2}}\frac{1}{I}\exp\left\{-\frac{\left(ln\left(\frac{I}{I_0}\right)+\frac{\sigma_I^2}{2}\right)^2}{2\sigma_I^2}\right\}, \tag{9.11}$$

Here, I_0 is defined as the irradiance when no turbulence is present.

9.5.2 The gamma–gamma turbulence model

For strong turbulence prediction, Andrew *et al.* [3] proposed the gamma–gamma turbulence model which is based on the modulation process. The normalised received irradiance for this method is defined as the product of two independent random processes that arise from the large- and small-scale turbulent eddies. If I_x represents the large-scale turbulent eddies and I_y represent the small-scale turbulent eddies, then received irradiance I can be expressed as:

$$I = I_x I_y \tag{9.12}$$

where PDFs of I_x and I_y are defined as:

$$\begin{aligned} p(I_x) &= \frac{\alpha(\alpha I_x)^{\alpha-1}}{\Gamma(\alpha)}\exp(-\alpha I_x);\ I_x > 0; \quad \alpha > 0 \\ p(I_y) &= \frac{\beta(\beta I_y)^{\beta-1}}{\Gamma(\beta)}\exp(-\beta I_y)\ I_y > 0; \quad \beta > 0 \end{aligned} \tag{9.13}$$

After re-arranging the above equations, $p(I/I_x)$ can be written as:

$$p(I/I_x) = \frac{\beta(\beta I/I_x)^{\beta-1}}{I_x\Gamma(\beta)}\exp(-\beta I/I_x)I > 0 \tag{9.14}$$

The gamma–gamma irradiance distribution function can be obtained by taking the average value of conditional probability $p(I/I_x)$ over the statistical distribution of I_x, which is given as:

$$\begin{aligned} p(I) &= \int_0^\infty p(I/I_x)p(I_x)dI_x \\ p(I) &= \frac{2(\alpha\beta)^{(\alpha+\beta)/2}}{\Gamma(\alpha)\Gamma(\beta)}I^{\left(\frac{\alpha+\beta}{2}\right)-1}K_{\alpha-\beta}\left(2\sqrt{\alpha\beta I}\right);\quad I > 0 \end{aligned} \tag{9.15}$$

Here, α is known as the large-scale eddies of the scattering process and β is the small-scale eddies of the scattering process. $K_n(\cdot)$ represent the modified Bassel function of the second kind of order n, and the gamma function is represented by $\Gamma(\cdot)$.

For a plane wave optical radiation, α and β can be expressed as a function of atmospheric conditions by the given equation:

$$\alpha = \left[\exp\left(\frac{0.49\sigma_l^2}{\left(1 + 1.11\sigma_l^{12/5}\right)^{7/6}} \right) - 1 \right]^{-1}$$
$$\beta = \left[\exp\left(\frac{0.51\sigma_l^2}{\left(1 + 0.69\sigma_l^{12/5}\right)^{5/6}} \right) - 1 \right]^{-1} \tag{9.16}$$

and the atmospheric turbulence index is defined as:

$$\sigma_N^2 = \exp\left[\frac{0.49\sigma_l^2}{\left(1 + 1.11\sigma_l^{12/5}\right)^{7/6}} + \frac{0.51\sigma_l^2}{\left(1 + 0.69\sigma_l^{12/5}\right)^{5/6}} \right] - 1 \tag{9.17}$$

9.6 Turbulence-induced fading mitigation methods

The performance of the FSO channel is severally limited by the atmospheric turbulence. If the size of the receiver aperture is significantly smaller than the beam diameter, the received signal experience intensity fluctuations due to turbulence as shown in Figure 9.4. Due to this turbulence-induced fading, the performance of FSO communication suffers severely.

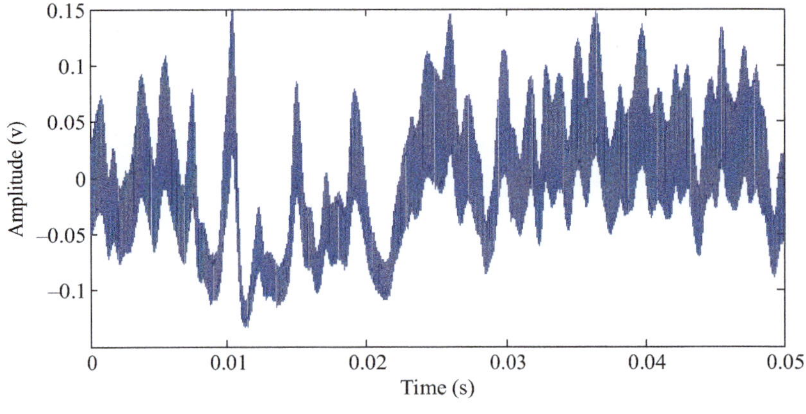

Figure 9.4 Received OOK signal at 1 Mbps in the presence of turbulence ($\sigma_l^2 = 0.54$)

Figure 9.5 shows the measured bit error rate (BER) against the scintillation index (SI) for ON–OFF keying (OOK) and binary phase shift keying (BPSK). The BER rate of OOK modulation schemes suffer severely in the turbulence channel. Depending upon the strength of the turbulence (which is generally measured using Rytov variance), the power penalty due to turbulence fading can be as high as ~26 dB for weak turbulence regime for OOK with a fixed threshold value. The power penalty further increases with the strength of the turbulence, and it is not feasible to obtain a reasonable BER using OOK with a fixed threshold [9].

Alternatively, decision making based on the adaptive threshold (threshold level at the decision-making stage is varied based on the turbulence strength) is considered. The adaptive threshold offers an improved performance and often required to avoid the BER floor. However, the adaptive scheme requires the channel-state information (CSI) to efficiently adapt the threshold. Though the turbulence varies relatively slow, the adaptive threshold increases the system complexity due to necessity to estimate CSI. However, adaptive thresholding provides a limited gain in the strong turbulence regime. Alternatively, modulation schemes that carry information in the intensity level such as pulse position modulation (PPM) with soft decoding, subcarrier intensity modulation (SIM) [10] and polarisation-based modulation [11] can improve the performance. As shown in Figure 9.5, BPSK-SIM scheme offer significant resilience to turbulence in comparison to OOK as the information is carried in the phase in BPSK-SIM. Recently, a hybrid modulation scheme combining PPM and SIM was suggested to take unique advantage of these schemes to overcome turbulence-induced fading. The experimental results presented in [12] shows that the hybrid modulation scheme offers improved performance in comparison to OOK. However, selection of an appropriate modulation scheme will not always guarantee the link availability especially in the strong turbulence regime and hence wavelength diversity [13], route diversity [14] and aperture averaging [15,16] have also been studied in combination with modulation

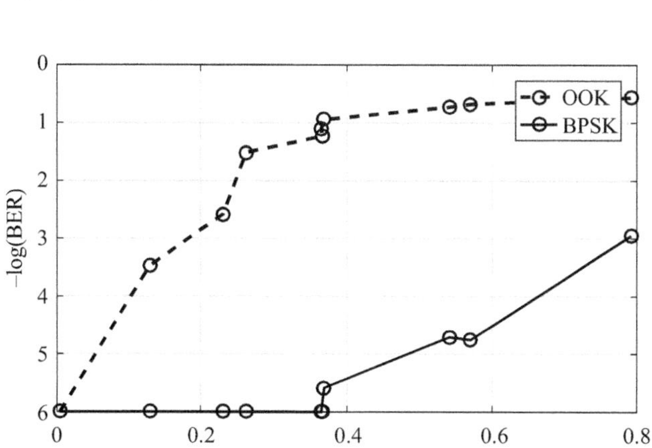

Figure 9.5 The measured bit error rate against SI for OOK and BPSK modulation schemes

scheme. Though the wavelength diversity enhances the performance, the complexity and cost increase significantly due to the requirement of multiple lasers and receiver operating at different wavelengths [13]. Route diversity scheme as suggested in [14] relies on the availability of alternative FSO links as in mesh network communication. The performance gain of the route diversity depends on channel correlation and the best performance is obtained if the channels are isolated i.e. geographically well separated. However, there has been very limited work reported so far on the performance of FSO with the route diversity. The most reported turbulence mitigation techniques in the literature by far is the spatial diversity. In the following section, turbulence mitigation techniques based on aperture averaging and spatial diversity is outlined.

9.6.1 Aperture averaging

The aperture averaging is the simplest and very cost-effective turbulence-induced fading mitigation method. If the receiver collection area is greater than the transverse coherence distance, the turbulence-induced fading can be effectively reduced, improving the BER performance. The aperture averaging factor, which is widely used to quantify the fading reduction, is given by [3]:

$$A = \frac{\sigma_I^2(D)}{\sigma_I^2(0)} \tag{9.18}$$

where $\sigma_I^2(D)$ and $\sigma_I^2(0)$ denote the SI for a receiver aperture of diameter D and a 'point receiver' ($D \approx 0$), respectively.

The SI with aperture $\sigma_I^2(D)$ depends on the Rytov variance and type of wavefront. For example, for a plane wave, the SI for a receiver with aperture D is given by:

$$\sigma_I^2(D) = \exp\left[\frac{0.49\sigma_R^2}{\left(1 + 0.65d^2 + 1.11\sigma_R^{12/5}\right)^{7/6}} + \frac{0.51\sigma_R^2\left(1 + 0.69\sigma_R^{12/5}\right)^{-5/6}}{1 + 0.90d^2 + 0.62d^2\sigma_R^{12/5}}\right] - 1 \tag{9.19}$$

where σ_R is the Rytov variance for the plane wave and parameter d is defined as:

$$d = \sqrt{\frac{kD^2}{4L}} \tag{9.20}$$

where L is the propagation distance and k is the spatial frequency.

The aperture averaging factor under a weak irradiance with the plane wave can be approximated as [3,17]:

$$A \approx \left[1 + 1.062\left(\frac{D^2k}{4L}\right)\right]^{-7/6} \tag{9.21}$$

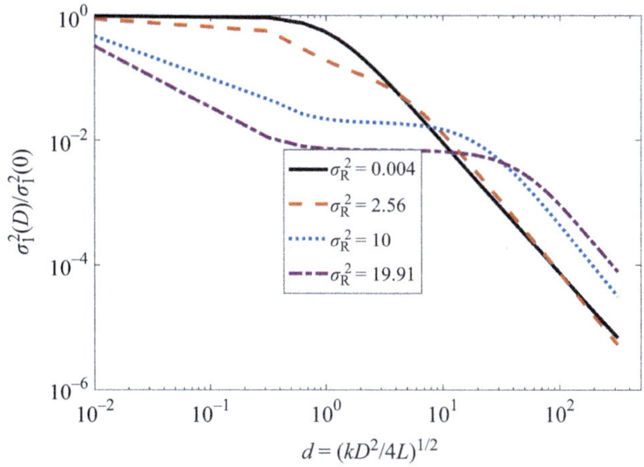

Figure 9.6 Aperture averaging factor for a plane wave propagation

Figure 9.6 shows the aperture averaging factor for a various ratio of $kD^2/4L$. For a weak turbulence level ($\sigma_R^2 = 0.04$,) the averaging gain is almost 1 for small aperture size $kD^2/4L < 1$. However, the gain decreases exponentially for $kD^2/4L > 1$. As the turbulence strength increases, the effectiveness of the aperture averaging is reduced and hence the gain is 1 only for $kD^2/4L \ll 1$. This clearly indicates that the variance σ_R^2 decrease with increasing aperture size. However, the aperture size must be significantly large to mitigate the strong turbulence. Furthermore, a larger aperture size provides higher optical gain, improving the signal-to-noise ratio (SNR) and hence a better BER performance can be obtained. However, it may not always be possible to increase the receiver aperture diameter beyond a certain limit. A larger aperture will not only incur a higher cost but also reduces the field-of-view (FOV), increasing the possibility of pointing error. Hence, spatial diversity is often the preferred choice even though a single receiver gives better performance than an array of receivers whose total area is the same as that of the single receiver [18]. In the following section, the spatial diversity is discussed in detailed.

9.6.2 Spatial diversity

As discussed in the previous section, the aperture averaging to combat turbulence may not always be feasible as the spatial coherence distance is of the order of centimetres [19]. Hence, a large single aperture is replaced by a spatially separated array of small apertures (either at the transmitter (transmitter diversity) or at the receiver (receiver diversity)). In order to maximise the spatial diversity gain, the separation between the multiple elements must be greater than the coherence length so that multiple copies of uncorrelated signals are received. The spatial diversity can be applied at the transmitter (multiple-input single-output – MISO) or at the

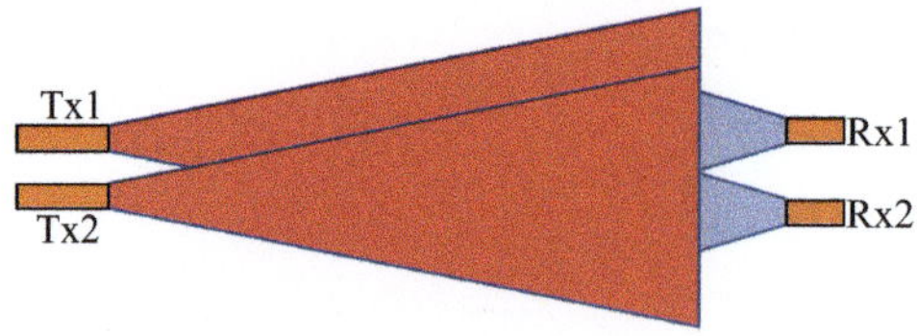

Figure 9.7 A simplified schematic of a spatial multiplexing scheme

receiver (single-input multiple-output – SIMO) or at both sides (multiple-input multiple-output – MIMO).

A simplified diagram of spatial diversity is shown in Figure 9.7. The figure shows that two transmitters and two receivers. For the diversity schemes, there must be at least two (transmitter or/and receiver) elements. In the MISO system, the multiple transmitters will transmit multiple copies of the same data (also known as repetition coding (RC)). Whereas, in SIMO, the multiple receiver receives the copies of the same data from the same transmitter. It is shown in [20] that a significant improvement in BER performance can be obtained by using MISO or SIMO. In the MIMO, the individual transmitter sends an independent information stream in parallel through the spatial channels and increasing spectral efficiency as well as provides resilience to the turbulence.

In the MISO system, also sometimes known as repetition coding [21], multiple transmitters are modulated using the same data and optical signal are spatially superposed. The transmitted optical power per source with a given aperture in the FSO system is strictly limited by the eye-safety regulation. The multiple transmitters (transmitting the same signal) are a in order to provide a better used in the FSO to provide a better link margin, to increase link distance, and/or to increase resilience to attenuation and turbulence-induced fading [20,22].

The SIMO scheme utilises a single transmitter and multiple independent receivers. If the separation between the receiver elements is more than few centimetres, it is expected that output of these receivers will not experience the same turbulence as the spatial coherence length of the atmospheric channel is only a few centimetres. Hence, the performance of the FSO system under atmospheric channel can be improved by combining these received signals [20]. The received signals are linearly combined (as shown in Figure 9.8) before signal demodulation. The combined received signal is given by:

$$i_T(t) = \sum_{i=1}^{\mathcal{N}} a_i i_i(t) \tag{9.22}$$

where N is the total number of the receiver elements and a_i is the individual weight. The signal can be combined using equal gain combining (EGC), maximum ratio combining (MRC) or selection combining (SelC).

In the EGC technique, the photocurrents is summed them coherently with equal weights i.e. $\{a_i\}_{i=1}^{\mathcal{N}} = 1$. In the MRC, the gain factor a_i is proportional to the received signal strength i.e. higher weight is given to the stronger signal. In the SelC algorithm,

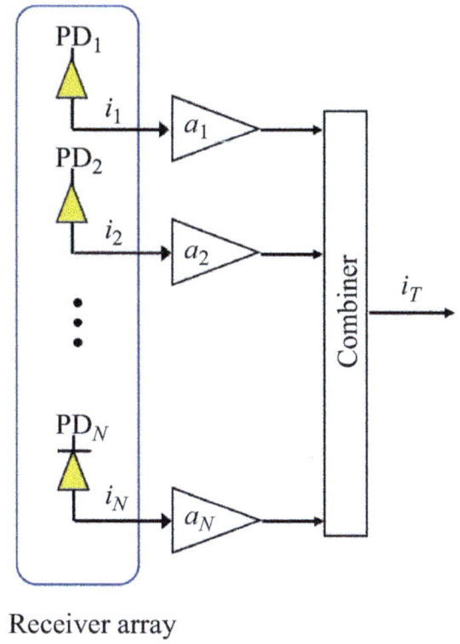

Figure 9.8 The schematic of the linear diversity combination at the receiver

the strong signal (i.e. the signal with highest SNR) is selected disregarding other receivers. The MRC is often the optimal spatial combing algorithm [23]. However, MRC is also the most complexed as the algorithm requires to have the knowledge of the individual channel irradiance or estimate the received optical signal strength.

Figure 9.9 shows the diversity gain using EGC, MRC and SelC for various turbulence level for BPSK-SIM. (The detailed mathematical derivation is provided in [9].) This figures clearly implies that the diversity gain increases (irrespective of the linear combination algorithms) as the turbulence level increases. For example, diversity gain with two photodiodes are ~2 dB to ~8 dB for $\sigma_l^2 = 0.2^2$ and 1, respectively, for EGC. This clearly demonstrates the importance of the diversity schemes. Among the three linear combiners, the MRC perform the best followed by EGC. Difference in diversity gain for MRC and EGC is relatively small for wear turbulence regime, which increase with the increase in the turbulence strength. The study in [24] also concluded that MRC offer improved performance in comparison to EGC in a Gamma–Gamma channel. However, the diversity gain for MRC is relatively small [25], but at the cost of significant computational complexity. Hence, EGC is more attractive for practical implementation due to its simplicity [24]. The figures also demonstrate that the diversity gain increase with the number of the receiver. However, the optimum number of receiver is two to four, depending upon the strength of the turbulence [9]. Beyond four receivers, the diversity gain flattens out indicating lower diversity gain per added receiver.

9.6.3 MIMO system

One of the key problems with the previously described spatial diversity schemes is their lack of improvement in spectral efficiency even though the hardware cost and complexity increase with the number of transmitter/receiver elements. The MIMO

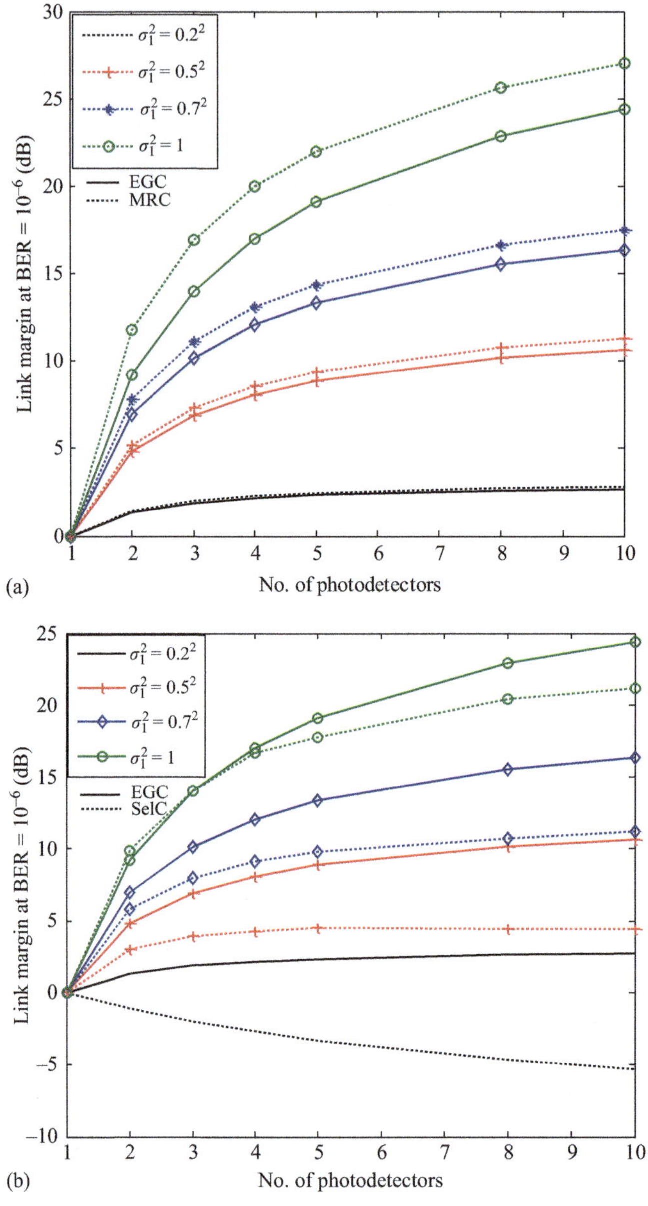

Figure 9.9 BPSK-SIM diversity link margin against the number of photodetectors for various turbulence levels and a BER of 10^{-6} for (a) EGC and MRC and (b) EGC and SelC (adopted from [9])

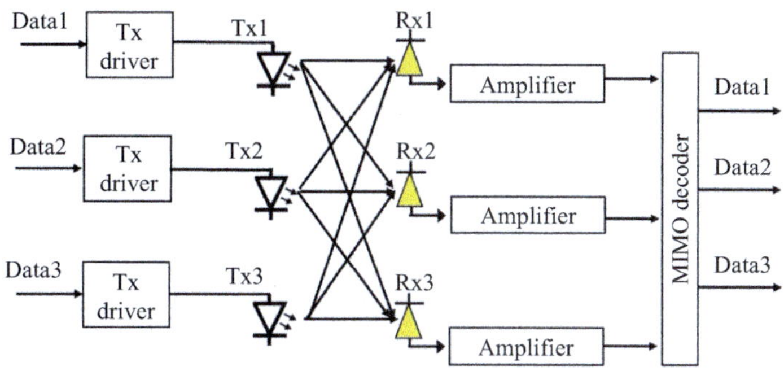

Figure 9.10 The schematic of the optical MIMO system

system overcomes this shortcoming by offering almost linear increment in the spectral efficiency with the number of transmitter/receiver elements by parallel transmission of data.

A simplified schematic of the optical MIMO scheme is shown in Figure 9.10 with M transmitters and N ($>M$) receivers. In the MIMO system, the individual transmitter is modulated by independent data and transmitted through the same physical channel. At the receiver, decoding algorithms are applied to separate these channels and decode individual data stream.

The free-space optical MIMO can be imaging and non-imaging [26]. In the imaging MIMO, a 2-D planner photodiode array is used with an imaging optical lens (as in digital cameras). The imaging MIMO systems can use a singlet convex or aspheric lens or a combination of lenses. In the non-imaging MIMO, receiver consists of the individual detector with individual optical concentrators. The imaging MIMO tends to be more compact due to the single optical lens and spatial data density tends to be very high. However, to the best knowledge of the authors' knowledge, there has not been any report of the performance analysis of the imaging MIMO system for FSO system in turbulence environment. In outdoor FSO, the goal of the MIMO system is not only to increase the data rate but also to mitigate the turbulence. As discussed in the aperture averaging section, the use of large aperture to mitigate the turbulence may not be the optimum. This may be the prime reason for adopting the non-imaging MIMO for FSO [27,28]. Furthermore, the individual receiver in the non-imaging MIMO has individual lens, which offers the flexibility of adjusting/optimising the spatial separation between the receivers for the optimum performance.

Though there can be significant inter-channel interference in an imaging/non-imaging MIMO system, the receiver can still recover the individual channel provided that the channel-matrix **H** is well conditioned. The channel matrix **H**-matrix is defined as:

$$\mathbf{H} = \begin{bmatrix} h_{11} & h_{12} & \cdots & h_{1M} \\ h_{21} & h_{22} & \ldots & h_{2M} \\ \vdots & \vdots & \ddots & \vdots \\ h_{N1} & h_{N2} & \ldots & h_{NM} \end{bmatrix} \tag{9.23}$$

where h_{ij} is the channel gain from the jth transmitter to the ith receiver elements.

Hence, the received signal at any discrete-time instance k for a MIMO system is given by:

$$\mathbf{y}(k) = \mathbf{H}\mathbf{x}(k) + \mathbf{n}(k) \tag{9.24}$$

where $\mathbf{x}(k)$ is a $(M \times 1)$ transmitted signal vector, $\mathbf{y}(k)$ is a $(N \times 1)$ received signal vector and $\mathbf{n}(k)$ is a $(N \times 1)$ noise vector and.

The transmitted signal can be estimated by applying 'pseudo-inverse' of the channel $\mathbf{H}$ matrix $\mathbf{W}_{ZF}$ to the received signal $\mathbf{y}$. The 'pseudo-inverse' matrix $\mathbf{W}_{ZF}$ is given by:

$$\mathbf{W}_{ZF} = \left(\mathbf{H}^H\mathbf{H}\right)^{-1}\mathbf{H}^H \tag{9.25}$$

where $\mathbf{H}^H$ is the conjugated transpose of $\mathbf{H}$.

This approach is called a zero-forcing (ZF) algorithm. The ZF algorithm relies on the existence of the well-conditioned matrix so that the inverse can be computed. Unlike indoor visible light communication (VLC), the outdoor FSO system is predominately line-of-sight (LOS) link with a very limited multipath. Hence, the channel matrix is expected to be well-condition though there is very limited work to characterise the channel H matrix in diverse environmental condition.

Though the ZF algorithm is simple to implement, this algorithm neglects the presence of noise. Hence, its performance is poor in the presence of significant noise i.e. when the SNR is low. A more robust equaliser is based on the minimum mean square error (MMSE) algorithm which jointly minimises the noise and interference. The MMSE equaliser is obtained by minimising the mean square error (MSE) defined as $E\left[\|x - W_{MMSE}^{H} y\|^2\right]$. For $N > M$, $\mathbf{W}_{MMSE}$ is given by [29]:

$$\mathbf{W}_{MMSE} = \left(\mathbf{H}^H\mathbf{H} + \frac{1}{SNR}\mathbf{I}\right)^{-1}\mathbf{H}^H \tag{9.26}$$

where $\mathbf{I}$ is a $M \times M$ identity matrix.

The MMSE and ZF algorithms tend to coverage at high. Though linear equalisers (MMSE and ZE) are relatively simple to implement, the performance of them is severely affected if the channel is ill-conditioned. The ultimate MIMO decoding algorithm is maximum-likelihood-decoding (MLD) algorithm. The MLD algorithm compares the received signal with all possible transmitted signal and estimates the transmitted sequence based on the following minimum distance criterion.

$$\arg\min\|y(k) - \mathbf{H}x(k)\|^2 \tag{9.27}$$

The MLD achieves the best performance irrespective of the channel matrix. However, the computational complexity of the MLD increases exponential with MIMO order and constellation size. Thus, MLD is impractical for a higher MIMO order and large constellation. Alternatively, successive interference cancellation schemes such as the vertical Bell Labs layered space-time algorithm (V-BLAST) can be adopted [30]. The V-BLAST is a recursive algorithm. The strongest signal is estimated first, and then the inference from this symbol is removed from remaining symbols. Then, the next strongest symbol is estimated followed by interference removal and so on. The computation complexity of V-BLAST algorithm is comparable to linear algorithms with the error performance almost comparable to MLD technique.

Though the optical MIMO systems has potential to increase the spectral efficiency and has been successfully implemented in indoor VLC system (see [31,32]), the MIMO in the FSO are generally used to mitigate the turbulence-induced fading only [33]. Hence, the transmitter employing RC and linear combiner algorithm at the receiver are commonly deployed [24,25], [34–36].

A MIMO system with RC to mitigate the turbulence-induced fading was studied in [34] with aperture averaging. It is shown that with a small MIMO order (2×2) system, the BER performance can be very close the ideal additive white Gaussian noise (AWGN) channel performance in a weak turbulence regime. However, in a strong turbulence regime, 7×7 MIMO system was required to achieve a BER performance close to the AWGN system. However, in all the cases, MIMO provided significantly improved performance in comparison to the SISO system.

9.7 Conclusion

The performance of FSO system is significantly affected by the atmospheric fog and turbulence. The dense fog induced significant attenuation (>100 dB/km) limiting the link range to less than 100 m. Turbulence induces temporal fading which also limits the availability of FSO channel. Hence, the adverse atmospheric effect in FSO system needs to be mitigated to ensure the availability of the FSO channel. It is demonstrated practically that the attenuation due to fog is dependent upon the wavelength and longer wavelength tends to have lower attenuation. Hence, wavelength diversity is considered to mitigate the fog effect along with spatial diversity. On the other hand, spatial diversity (transmitter or receiver or both) can be used to overcome the turbulence-induced fading. A simple repetition coding with equal gain combiner can provide almost optimal performance. It is also observed that the MIMO system in FSO is mostly used for mitigating the turbulence-induced fading. The spectral efficiency increment of optical MIMO has not been fully exploited, though the MIMO system has the potential to mitigate the turbulence and at the same time increase the spectral efficiency.

References

[1] M. Al Naboulsi and H. Sizun, "Fog attenuation prediction for optical and infrared waves," *Opt. Eng.*, vol. 23, no. 2, pp. 319–329, 2004.

[2] A. K. Majumdar and J. C. Ricklin, "Effects of the atmospheric channel on free-space laser communications," in *Free-Space Laser Communications V*, vol. 5892, D. G. Voelz and J. C. Ricklin, Eds. 2005, p. 8920K–1–58920K–16.

[3] L. C. Andrews and R. L. Phillips, *Laser beam propagation through random media*, Second. Washington: SPIE Press, 2005.

[4] P. W. Kruse, L. D. McGlauchlin, and R. B. McQuistan, *Elements of infrared technology: Generation, Transmission, and Detection*. New York: John Wiley and Sons, 1962.

[5] M. A. Esmail, H. Fathallah, and M.-S. Alouini, "Analysis of fog effects on terrestrial Free Space optical communication links," in *2016 IEEE International Conference on Communications Workshops (ICC)*, 2016, pp. 151–156.

[6] I. I. Kim, B. McArthur, and E. Korevaar, "Comparison of laser beam propagation at 785 nm and 1550 nm in fog and haze for optical wireless communications," *SPIE Proceeding Opt. Wirel. Commun. III*, vol. 4214, pp. 26–37, 2001.

[7] L. C. Andrews, R. L. Phillips, and C. Y. Hopen, *Laser beam scintillation with applications*. Bellingham: SPIE, 2001.

[8] H. Kaushal, V. K. Jain, and S. Ka, "Effect of atmospheric turbulence on acquisition time of ground to deep space optical communication system," *Int. J. Electr. Comput. Eng.*, vol. 4, no. 12, pp. 730–734, 2009.

[9] Z. Ghassemlooy, W. O. Popoola, and S. Rajbhandari, *Optical wireless communications – System and channel modelling with Matlab*, 1st ed. Boca Raton, FL: CRC Press, 2012.

[10] W. Popoola O., Z. Ghassemlooy, and E. Leitgeb, "BER and outage probability of DPSK subcarrier intensity modulated free space optics in fully developed speckle," *J. Commun. North Am.*, vol. 4, no. 8, pp. 546–554, 2009.

[11] X. Tang, Z. Ghassemlooy, S. Rajbhandari, W. O. Popoola, and C. G. Lee, "Coherent heterodyne multilevel polarization shift keying with spatial diversity in a free-space optical turbulence channel," *J. Light. Technol.*, vol. 30, no. 16, pp. 2689–2695, 2012.

[12] S. Rajbhandari, Z. Ghassemlooy, P. A. A. Haigh, T. Kanesan, and T. Xuan, "Experimental error performance of modulation schemes under a controlled laboratory turbulence FSO channel," *Light. Technol. J.*, vol. 33, no. 1, pp. 244–250, 2015.

[13] V. Xarcha, A. N. Stassinakis, H. E. Nistazakis, *et al.*, "Wavelength diversity/or free space optical systems: Performance evaluation over log normal turbulence channels," in 2012 19th International Conference on Microwaves, Radar and Wireless Communications, MIKON 2012, 2012, vol. 2, pp. 678–683.

[14] S. Zvanovec, J. Perez, Z. Ghassemlooy, S. Rajbhandari, and J. Libich, "Route diversity analyses for free-space optical wireless links within turbulent scenarios," *Opt. Express*, vol. 21, no. 6, pp. 7641–50, 2013.

[15] M. A. Khalighi, N. Schwartz, N. Aitamer, and S. Bourennane, "Fading reduction by aperture averaging and spatial diversity in optical wireless systems," *IEEE/OSA J. Opt. Commun. Netw.*, vol. 1, no. 6, pp. 580–593, 2009.

[16] F. S. Vetelino, C. Young, L. Andrews, and J. Recolons, "Aperture averaging effects on the probability density of irradiance fluctuations in moderate-to-strong turbulence," *Appl. Opt.*, vol. 46, no. 11, pp. 2099–2108, 2007.

[17] M. A. Khalighi, N. Aitamer, N. Schwartz, and S. Bourennane, "Turbulence mitigation by aperture averaging in wireless optical systems," in *2009 10th International Conference on Telecommunications*, Zagreb, Croatia, 2009, pp. 59–66.

[18] P. Gopal, V. K. Jain, and S. Kar, "Performance improvement of FSO satellite downlink using aperture averaging and receiver spatial diversity," *IET Optoelectron.*, vol. 10, no. 4, pp. 119–127, 2016.

[19] X. Zhu and J. M. Kahn, "Free-space optical communication through atmospheric turbulence channels," *IEEE Trans. Commun.*, vol. 50, no. 8, pp. 1293–1300, 2002.

[20] T. A. Tsiftsis, H. G. Sandalidis, G. K. Karagiannidis, and M. Uysal, "Optical wireless links with spatial diversity over strong atmospheric turbulence channels," *IEEE Trans. Wirel. Commun.*, vol. 8, no. 2, pp. 951–957, 2009.

[21] T. Fath and H. Haas, "Performance comparison of MIMO techniques for optical wireless communications in indoor environments," *IEEE Trans. Commun.*, vol. 61, no. 2, pp. 733–742, 2013.

[22] A. García-Zambrana, C. Castillo-Vázquez, B. Castillo-Vázquez, and A. Hiniesta-Gómez, "Selection transmit diversity for FSO links over strong atmospheric turbulence channels," *IEEE Photonics Technol. Lett.*, vol. 21, no. 14, pp. 1017–1019, 2009.

[23] A. Tsiatmas, F. M. J. Willems, and S. Baggen, "Optimum diversity combining techniques for visible light communication systems," in *Globecom 2014 Workshop – Optical Wireless Communications*, 2014, vol. 1, pp. 541–546.

[24] E. Bayaki, R. Schober, and R. K. Mallik, "Performance analysis of MIMO free-space optical systems in gamma-gamma fading," *IEEE Trans. Commun.*, vol. 57, no. 11, pp. 3415–3424, 2009.

[25] E. J. Lee and V. W. S. Chan, "Part 1: Optical communication over the clear turbulent atmospheric channel using diversity," *IEEE J. Sel. Areas Commun.*, vol. 22, no. 9, pp. 1896–1906, 2004.

[26] S. Rajbhandari, J. J. D. McKendry, J. Herrnsdorf, *et al.*, "A review of gallium nitride LEDs for multi-gigabit-per-second visible light data communications," *Semicond. Sci. Technol.*, vol. 32, no. 2, p. 023001, 2017.

[27] B. Braua and D. Barua, "Channel capacity of MIMO FSO under strong turbulent conditions," *Int. J. Electr. Comput. Sci.*, vol. 11, no. 2, pp. 1–5, 2011.

[28] S. M. Navidpour, M. Uysal, and L. Jing, "BER performance of MIMO free-space optical links," *60th IEEE Veh. Technol. Conf.*, vol. 5, pp. 3378–3382, 2004.

[29] Y. Jiang, M. K. Varanasi, and J. Li, "Performance analysis of ZF and MMSE equalizers for MIMO systems: An in-depth study of the high SNR regime," *IEEE Trans. Inf. Theory*, vol. 57, no. 4, pp. 2008–2026, 2011.

[30] G. J. Foschini and M. J. Gans, "On limits of wireless communications in a fading environment when using multiple antennas," *Wirel. Pers. Commun.*, vol. 6, no. 3, pp. 311–335, 1998.

[31] S. Rajbhandari, H. Chun, G. Faulkner, *et al.*, "Multi-gigabit integrated MIMO visible light communication system: Progress and updates," in *Summer Topicals Meeting Series (SUM), 2015*, 2015, pp. 230–231.
[32] X. Zhang, K. Cui, H. Zhang, and Z. Xu, "Capacity of MIMO visible light communication channels," *Photonics Soc. Summer Top. Meet. Ser. 2012 IEEE*, vol. 4, pp. 159–160, 2012.
[33] M. A. Khalighi and M. Uysal, "Survey on free space optical communication: A communication theory perspective," *Commun. Surv. Tutorials, IEEE*, vol. 16, no. 4, pp. 2231–2258, 2014.
[34] Z. Hajjarian and J. Fadlullah, "MIMO free space optical communications in turbid and turbulent atmosphere," vol. 4, no. 8, pp. 524–532, 2009.
[35] N. Cvijetic, S. G. Wilson, and M. Brandt-Pearce, "Performance bounds for free-space optical MIMO systems with APD receivers in atmospheric turbulence," *IEEE J. Sel. Areas Commun.*, vol. 26, no. 3, pp. 3–11, 2008.
[36] S. M. Navidpour, M. Uysal, and M. Kavehrad, "BER performance of free-space optical transmission with spatial diversity," *IEEE Trans. Commun.*, vol. 6, no. 8, pp. 2813–2819, 2007.

Chapter 10

Analysis of the effects of aperture averaging and beam width on a partially coherent Gaussian beam over free-space optical communication links

It Ee Lee[1] and Zabih Ghassemlooy[2]

10.1 Introduction

The evolution of the wireless communications standards into the fourth generation (4G) and beyond has witnessed recent rapid progress in information and communication technologies, which in turn revealed a multitude of mobile broadband facilities, such as ultra-broadband internet access, Internet protocol (IP) telephony, gaming services, streamed multimedia applications and high-definition television (HDTV) broadcasting services [1,2]. The exponential surge in the commercial demand to pursue unlimited high-speed and ubiquitous broadband wireless access, to accommodate the ever-increasing utilization of internet and multimedia services among individual mobile users and residential and enterprise clusters, has spurred prodigious growth in Internet traffic demand in the recent decade, scaling much faster than the prediction of Moore's law [3,4]. As a result, data transmission rates have been rising at approximately a factor of ten every five years, which corresponds to a tremendous growth in the overall traffic volume by several hundred-fold within the next decade [5], in order to be able to transport an exponentially increasing amount of (available) data to end users within an acceptable amount of time, inadvertently resulting in severe congestion of the radio frequency (RF) spectrum and wireless traffic bottleneck [6]. Complementing the existing wireless RF solutions, free-space optical (FSO) communications is poised to become a promising broadband wireless access (BWA) candidate to resolve the existing 'last mile' access network problems [7–9]. This is mainly due to the vastly attractive features of such technology option, which include [10]:

(i) No licensing requirements or tariffs for its utilization;
(ii) Virtually unlimited bandwidth for providing near-optimal capacity and supporting high-speed applications;

[1]Faculty of Engineering, Multimedia University, Malaysia
[2]Faculty of Engineering and Environment, Northumbria University, UK

(iii) Extensive link range in excess of 5 km;
(iv) High energy efficiency due to low power consumption, reduced interference and fading immunity;
(v) High scalability and reconfigurability;
(vi) Secure and reliable data transmission medium;
(vii) Minimal cost of deployment; and
(viii) Reduced time-to-market.

Terrestrial FSO communication is a line-of-sight (LOS) technology based upon the transmission of optical laser beam waves operating at typical wavelengths of 850 nm and 1,550 nm through the atmosphere, in order to establish wireless communication for providing larger bandwidths and high-data-rate transfer of information [7,11]. The deployment of FSO communication systems is particularly advantageous in metropolitan areas, geographically challenging terrains and underserved rural areas lacking broadband network connectivity [12,13]; whereby the installation of optical-fibre links is least feasible or impossible, as evident from the numerous case studies brought forward by LightPointe™ [14]. With the emergence of powerful and efficient optoelectronic components and advanced communication techniques, current state-of-the-art FSO prototypes have demonstrated data transmission through the atmospheric channel at ultra-high speed up to terabits per second (Tbps) [15–20]. In [21], Arimoto presented the design and demonstration of a novel FSO communication terminal, featuring fast and accurate fine tracking system and transparent connectivity to single-mode fibre lines; which represents a significant breakthrough in the stability, reliability and capacity of terrestrial FSO communications. This state-of-the-art compact FSO terminal has demonstrated an aggregated 320 (8×40) gigabits per second (Gbps) transmission at a link distance of 212 m in [17], and then successfully upgraded to an astounding 1.28 Tbps (32×40 Gbps) in [18], which is by far the highest achievable capacity reported for terrestrial FSO communication systems. Furthermore, the transmission of multiple RF signals carrying various broadband wireless services including the integrated services digital broadcasting-terrestrial (ISDB-T) signals over a 1 km turbulent FSO link based on the advanced dense wavelength-division multiplexing (DWDM) technique have been demonstrated in [22]. The proposed work unfolds new insight on the system design, operation and performance characteristics relevant in implementing economical and reliable alternative broadband wireless technology for complementing optical-fibre networks, especially in environments where the deployment of optical-fibre links are not feasible.

FSO communications remains as one of the least commercially deployed broadband technologies, albeit its tremendous potential in resolving the ‘last mile’ issues. The widespread deployment of such BWA solution is hampered by the combined effects from numerous factors, which include: the atmospheric channel which is highly variable, unpredictable and vulnerable to different weather conditions, such as scattering, absorption and turbulence [6,11,23]; and the presence of pointing errors (PEs) [13,24,25]. Aerosol scattering and absorption due to rain, snow and fog result in significant optical power attenuation, beam spreading and link distance reduction, severely impairing the system performance with an increase in the link error probability; thus causing the FSO system to fall short of the desired carrier-grade availability of 99.999% under heavy, visibility-limiting weather conditions [26]. In particular, Kim and Korevaar reported in [27] that the atmospheric attenuation of laser beam is a random function of the weather, which can vary from 0.2 dB/km in exceptionally clear weather to 350 dB/km in very dense fog.

Laser beams propagating through the atmospheric turbulent channel are highly susceptible to the adverse effects of scintillation and beam wander, which are natural phenomena commonly observed in terrestrial FSO communication systems, due to the refractive index variations along the transmission paths caused by inhomogeneities in both temperature and pressure of the atmosphere [11]. Correspondingly, this produces random fluctuations in both temporal and spatial domains of the received irradiance, known as channel fading, whereby the FSO links may suffer temporary signal degradation or complete system annihilation under the influence of deep signal fades [28,29]. Modelling of the fading effects in an atmospheric turbulent channel using the extended Huygens-Fresnel principle [11] shows that large transient dips in the optical signals typically last approximately 1–100 milliseconds, which may result in the loss of potentially up to 10^9 consecutive bits at a transmission rate of 10 Gbps [28].

Misalignment-induced fading is another non-negligible effect in FSO systems, as optical terminals typically installed on high-rise buildings are susceptible to building sway, while continuous precise pointing is required to establish link connectivity for successful data transmission with minimum error probability, particularly when narrow beam divergence angle and receiver field-of-view (FOV) are employed [13]. Under the influence of wind loads, thermal expansions and weak earthquakes, building sway causes vibrations in both the transmitter and receiver, in which the stochastic process deviates the optical wave propagation path from the common LOS, thus resulting in decrease of the average received signal [13,30,31]. In addition, PEs can arise due to mechanical misalignment, errors in tracking systems, or presence of mechanical vibrations within the system [25].

Through an in-depth survey of the relevant literatures, it is evident that a substantial number of analytical and simulation studies [32–35] have isolated the contributing effects of different atmospheric channel conditions resulting from atmospheric loss, turbulence-induced fading and PE loss for the sake of analytical simplicity. These channel impairments must be collectively taken into account using a combined channel fading model, in order to describe the optical channel characteristics with better accuracy. In addition, the limiting cases of unbounded plane- or spherical-wave approximations are insufficient to characterize the propagation properties of the optical beam wave through random turbulent medium, particularly for the case of spatially partially coherent Gaussian laser beam, whereby the spatial coherence and beam divergence of the optical laser source are concerned. Moreover, the conventional Rytov-based scintillation model associated with the above-mentioned approximations may not accurately reflect the irradiance fluctuations, mainly because the beam wander effect contributing to a widening of the long-term beam profile and PEs is not considered in the limiting Rytov theory. In this research work, the above concerns are appropriately addressed and taken into account through the joint adoption of a combined optical slow-fading channel model and Gaussian-Schell beam model; which in turn substantiate the development of a comprehensive design benchmark encompassing important link design criteria, thereby creating a holistic perspective for optimal planning and design of terrestrial FSO communication links. Through reliable theoretical analysis and simulation studies, feasible mitigation techniques and/or optimization methods can be proposed and examined with better accuracy, in order to promote effective mitigation of the adverse atmospheric channel effects.

The remainder of this chapter is organized as follows: Section 10.2 reviews the relevant literatures pertaining to the investigation and performance analysis of spatially partially coherent Gaussian beams in FSO systems, and highlights the main contributions of our proposed work. To substantiate the design, analysis and optimization of FSO systems for long-distance terrestrial applications, the key aspects of the FSO technology are discussed in Section 10.3, which include: the built-up of single-input single-output (SISO) horizontal-path FSO communication link (Section 10.3.1); the Gaussian-beam wave model (Section 10.3.2) in characterizing the propagation of optical laser beam in free-space; the characteristics and modelling of atmospheric propagation channel effects (Section 10.3.3); and the analytical model in describing the aperture-averaging phenomenon (Section 10.3.4). Next, Section 10.4 describes the important performance metrics considered in our numerical analysis encompassing the bit-error rate (BER), the probability of outage, and the average capacity for both cases of known and unknown channel-state information (CSI) at the receiver. In Section 10.5, we examine the significance and impact of the aperture-averaging phenomenon and the relevant link design considerations on a partially coherent beam (PCB) for long-distance horizontal-path FSO transmissions. In particular, the notable observations and findings from the outage analysis demonstrate the necessity of optimization methods to minimize FSO link outages under different atmospheric channel conditions. Then, joint investigation of the effects of PCB and aperture averaging are presented in Section 10.6, showing that an improvement in the average channel capacity can be achieved with the introduction of an enlarged receiver aperture, albeit without knowledge of the channel state conditions. In Section 10.7, several appealing observations pertaining to the alteration of the PCB characteristics are made, which in turn reveal that beam width optimization is a feasible approach in promoting capacity enhancement for long-distance horizontal FSO communication links. Finally, our concluding remarks are highlighted in Section 10.7.

10.2 Background and motivation

The propagation properties and performance of spatially partially coherent Gaussian laser beams through random turbulent media have been substantially explored thus far [36–38], which revealed the benefits of mitigating the turbulence-induced scintillations and PEs. A partially coherent laser beam can be generated by passing an initially spatially coherent laser transmitter through a phase diffuser, which partially destroys the spatial coherence of the information-bearing laser beam, thereby increasing the receiver beam size while still retaining its beamlike (i.e., highly directional) properties [36]. This approach inadvertently results in the reduction of the average received power, thus implying the need for careful investigation and optimization of the PCB through free-space, such that the inevitable trade-off between the reduction in the scintillations and PEs, and the reduction in the mean received irradiance can be achieved conveniently.

The adverse effects of atmospheric loss, turbulence-induced channel fading and PEs on the performance and design of FSO links have been extensively investigated from the information theory perspective in a large body of literature. In these relevant studies, a variety of channel fading models have been considered in the evaluation of the error performance and channel capacities of the system under study. In particular, Farid and Hranilovic [24] presented a statistical model for the optical intensity fluctuation at the receiver due to the combined influences of turbulence and PEs, which include the effects of beam width, detector size and jitter variance. It was shown that beam width optimization provides large gains in the FSO channel capacity subject to outage. In [39], Sandalidis proposed several optimization models to limit the PE loss, taking into account various metrics such as the beam width, mean electrical SNR, normalized jitter, outage probability and BER. Similar approach to [24] was also noted in [40–42], whereby the authors evaluated the error performance by taking into account both misalignment- and turbulence-induced fluctuations. However, the link design criteria were not emphasized in the analytical studies, thus lacking a holistic perspective for optimal planning and design of FSO links. The above theoretical treatments are based upon the classical unbounded plane- or spherical-wave approximations (the latter often taken as a point source), which are the limiting forms of the lowest-order Gaussian-beam wave model [11]. These simplified field models are insufficient to characterize the propagation properties of the optical wave through random turbulent medium, particularly for the case of spatially partially coherent Gaussian beam, whereby focusing and diverging characteristics are important [36]. In addition, the conventional Rytov-based scintillation model associated with these approximations may not accurately reflect the irradiance fluctuations, particularly in applications involving long-distance horizontal transmissions and strong atmospheric turbulence [43]. This is mainly because the influence of beam wander attributed to the presence of large-scale turbulent eddies, which contributes to a widening of the long-term beam profile and PEs, is not considered in the limiting Rytov theory [11,43].

Furthermore, the effects of aperture averaging on PCB in mitigating the scintillations and PEs for achieving higher mean SNR have not been thoroughly examined [25,44]. In [25], Borah and Voelz demonstrated the effect of PEs on the capacities of FSO links affected by the atmospheric turbulence, taking into account a Gaussian-beam wave model and wave-optics-based simulation approach; but the effects of larger apertures and optimization method were not considered here. In [43], Ren *et al.* investigated the impact of beam wander on the average and outage capacities of the FSO link, through the direct inclusion of this effect in the traditional scintillation theory; and showed that an optimum transmitter beam radius can be selected for best achievable channel capacities. Nevertheless, the effects of PEs and aperture averaging were not considered in their work. Therefore, it is vital to investigate the combined effects of PCB and aperture averaging in horizontal (terrestrial) FSO links, in the presence of turbulence and PEs; which then substantiates the study of beam width optimization.

In our proposed work, the performance of partially coherent FSO communication links is investigated from the information theory perspective, taking into account the impairments resulting from atmospheric loss, turbulence and PEs. In particular, a spatially partially coherent Gaussian-beam wave and important link design criteria are jointly considered, in which the latter consists of the receiver aperture dimension and its resulting aperture averaging effect, transmitter beam width, link range, knowledge of CSI, and weather conditions. The optical channel characteristics is analytically described using a combined optical slow-fading channel model, for evaluating the important performance metrics of the FSO link – the BER, average channel capacity and outage probability. These performance indicators resemble the error performance, practically achievable data rate averaged over all fading states, and probability of the system being in an outage for achieving a specified rate of transmission, respectively. The log-normal and gamma-gamma distributions are utilized in modelling the stochastic behaviour of optical irradiance under weak-to-moderate and moderate-to-strong atmospheric turbulence conditions, respectively. The lowest-order Gaussian-beam wave model characterizes the propagation properties of the optical signal through random turbulent medium; taking into account the beam spreading (i.e., diverging) and focusing of the PCB, and the scintillation and beam wander effects arising from the turbulent eddies. Correspondingly, this study presents a holistic perspective for optimal planning and design of horizontal FSO links employing spatially partially coherent laser beams.

From the outage analysis, it is evident that a lower outage probability can be achieved with the introduction of a larger receiver aperture due to the aperture-averaging effect, which effectively mitigates the PE loss and scintillation-induced optical intensity fluctuations. In addition, outage capacity optimization is best performed through proper selection of the transmitter beam width and the receiver aperture size for a known laser wavelength and/or increasing the transmit power. Under the light fog condition, complete system outage occurs at link distances exceeding 1.5 km, and is non-recoverable albeit optimizing the system parameters. Furthermore, it is demonstrated that the average channel capacity can be significantly improved due to the aperture-averaging effect, albeit without knowledge of the channel state conditions; and the PCB experiences substantial alteration in its characteristics when propagating through free-space. Several interesting observations are made here, showing that the optimum beam width can be made smaller to maximize the average capacity, albeit in the presence of turbulence-induced scintillations and PEs, given the condition that either one or both of these adversities are least dominant. In the strong turbulence regime, the beam width must be increased accordingly with respect to the PE loss, where such incremental trend changes from a linear to an exponential behaviour. Hence, beam width optimization presents a feasible approach in promoting capacity enhancement for long-distance horizontal FSO links; since the optimum beam width of the laser transmitter is susceptible to the combined effects of turbulence and PEs, and must be adjusted accordingly to obtain optimality.

10.3 An overview of free-space optical communications

10.3.1 System description

Figure 10.1 illustrates the block diagram of a SISO horizontal FSO communication link, employing intensity modulation with direct detection (IM/DD) and the non-return-to-zero on–off keying (NRZ-OOK) technique. In principle, the information-bearing electrical signals are modulated onto the instantaneous intensity of a collimated Gaussian-beam wave operating in the lowest-order transverse electromagnetic (TEM_{00}) mode [45,46], and then goes through a phase diffuser [36,38]. The latter alters the divergence of the optical beam while retaining its beamlike (i.e., highly directional) properties [11,37]. Next, the PCB propagates along a horizontal path through a turbulent channel with additive white Gaussian noise (AWGN), in the presence of beam extinction and PEs, in which the turbulence is assumed to be uniform across the link [6]. At the receiving-end, the optical signals are collected by a finite Gaussian lens [11,24] before being focused onto a photodetector, which in turn converts the received optical intensities into a resulting photocurrent.

The received signal y_κ can be described by the conventional channel model [29,34]:

$$y_\kappa = h\gamma x_\kappa + n_o \tag{10.1}$$

where h is the channel state, γ is the detector responsivity (in A/W), $x_\kappa \in \{0, 2P_{\text{FSO}}\}$ is the optical power of the transmitted signal which corresponds to the message signal m_κ taking values 0 or 1, P_{FSO} is the average transmitted optical

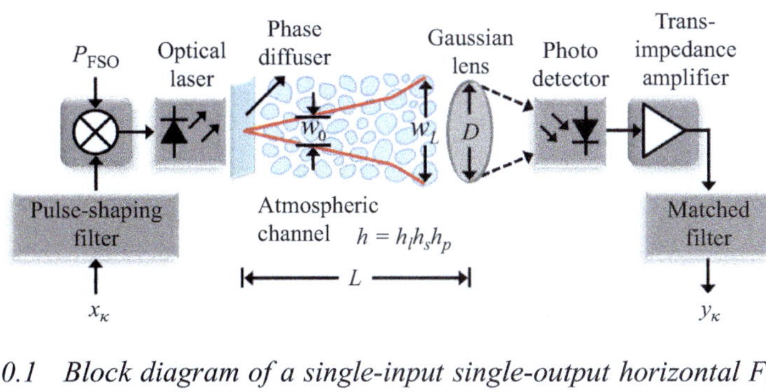

Figure 10.1 Block diagram of a single-input single-output horizontal FSO communication link

power, and n_o is signal-independent AWGN with zero mean and variance σ_n^2. The noise n_o arises from various sources, such as shot noise caused by the signal itself and/or ambient light, dark current noise, and electrical thermal noise [6]. For a slow-fading channel with OOK signalling, the received electrical signal-to-noise ratio (SNR) is defined as [24,39]:

$$SNR(h) = \frac{2P_{\text{FSO}}^2 \gamma^2 h^2}{\sigma_n^2} \tag{10.2}$$

and is a fluctuating term (i.e., an instantaneous value) due to the influence of h, which is chosen from the random ensemble according to the distribution $f_h(h)$.

The channel state h models the optical intensity fluctuations resulting from atmospheric loss, and turbulence- and misalignment-induced channel fading, which can be described as [24,41]:

$$h = h_l h_s h_p \tag{10.3}$$

where h_l, h_s and h_p denote the attenuation due to beam extinction arising from both scattering and absorption, scintillation effects, and geometric spread and PEs, respectively. The attenuation h_l is a deterministic component which exhibits no randomness in its behaviour, thus acting as a fixed scaling factor for a long period of time (i.e., on the order of hours) that stands in contrast to the bit intervals in ranges of nanoseconds or less [29]. On the other hand, both h_s and h_p are time-variant factors, exhibiting variations in the fading channel on the order of milliseconds, in which their stochastic behaviour are described by their respective distributions [25,40,42].

10.3.2 Gaussian-beam wave

The propagation of optical beams in free-space can be described by the plane wave, spherical wave and Gaussian-beam wave models, in which the latter is most often preferred and introduced in optical wave propagation analysis. This is mainly because the simplified field models based upon the unbounded plane and spherical wave approximations are insufficient to characterize the propagation properties of the optical beam waves, particularly when focusing and diverging characteristics are important. The free-space propagation of a lowest-order Gaussian-beam wave (TEM_{00} wave) is considered, with properties defined as follows:

1. The exit aperture of the transmitter is located in the plane $z = 0$;
2. The intensity distribution of the optical laser beam in the z-plane is Gaussian with an effective beam radius (spot size) w_0 (in m), which represents the beam width at which the field amplitude falls to 1/e of its peak on the beam axis, as depicted in Figure 10.2; and
3. The phase front radius of curvature F_0 (in m) is taken to be parabolic, in which the three possible scenarios of $F_0 = \infty, F_0 > 0$ and $F_0 < 0$ correspond to collimated, convergent and divergent beam forms, respectively (see Figure 10.3) [11].

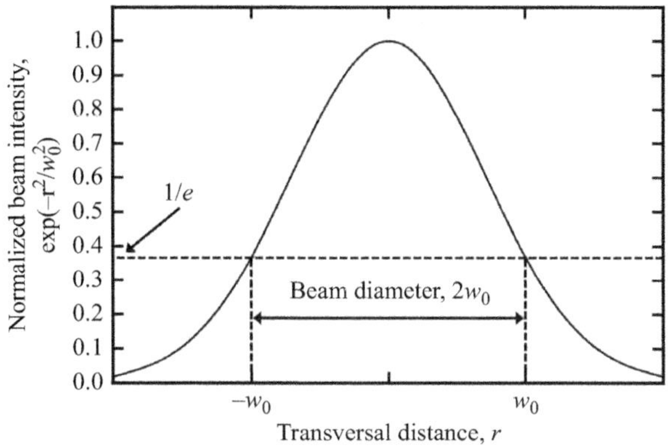

Figure 10.2 *Optical intensity profile of a lowest-order Gaussian-beam wave showing the beam width (i.e., spot size) w_0 at which the field amplitude falls to 1/e of its peak on the beam axis [11,48]*

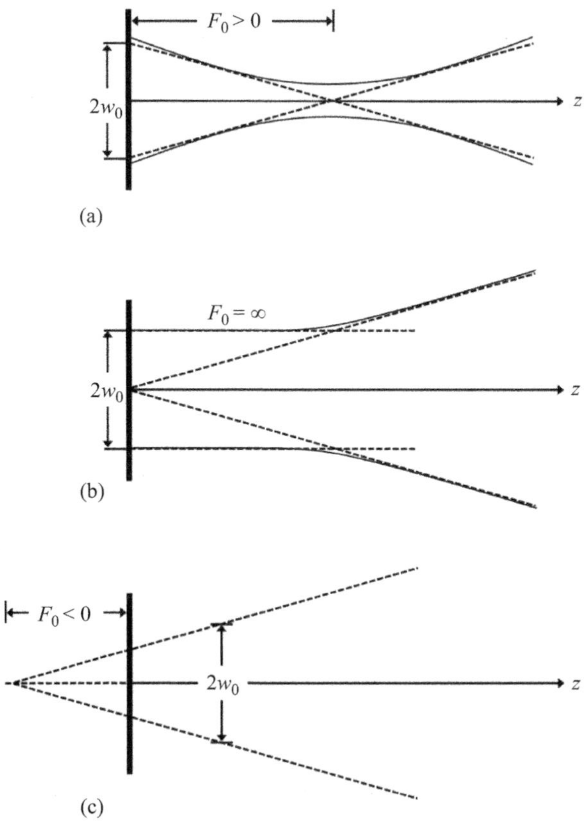

Figure 10.3 *Lowest-order Gaussian-beam wave with an effective beam radius w_0 and phase front radius of curvature F_0 under three possible scenarios of (a) collimated ($F_0 = \infty$); (b) convergent ($F_0 > 0$); and (c) divergent ($F_0 < 0$) beams [11]*

At $z = 0$, the free-space electric field of the optical beam wave having unit amplitude and propagating predominantly along the z-axis is given by [46,47]:

$$z = 0:\ U(r,0) = \exp\left[-\left(\frac{1}{w_0^2} + \frac{jk}{2F_0}\right)r^2\right] = \exp\left(-\frac{1}{2}\alpha_0 k r^2\right) \tag{10.4}$$

where $r = |\mathbf{r}|$ is the magnitude of the transversal distance from the beam centre line, $k = 2\pi/\lambda$ is the optical wave number with λ being the laser wavelength, $j = \sqrt{-1}$, and α_0 (in m^{-1}) is a complex parameter related to the transmitter beam width and phase front radius of curvature according to the relation:

$$\alpha_0 = \frac{2}{kw_0^2} + j\frac{1}{F_0}. \tag{10.5}$$

After propagating through a distance L from the transmitter, the free-space optical field in the receiver plane at $z = L$ becomes [49]:

$$z = L:\ U(\rho, L) = \frac{\exp(jkL)}{\Theta_n + j\Lambda_n}\exp\left[-\left(\frac{1}{w_L^2} + \frac{jk}{2F_L}\right)\rho^2\right], \tag{10.6}$$

where the non-dimensional input (transmitter) beam parameters are given by [49,50]:

$$\Theta_n = 1 - \frac{L}{F_0} \quad \text{and} \quad \Lambda_n = \frac{2L}{kw_0^2}. \tag{10.7}$$

The curvature parameter Θ_n characterizes the refractive (focusing) properties in terms of deviation of the wave-front curvature from the optimal focusing condition $F_0 = L$, whereas the Fresnel ratio Λ_n depicts the diffractive changes in the on-axis amplitude of the Gaussian beam. The beam width w_L and phase front radius of curvature F_L at the output (receiver) plane are expressed in terms of the normalized components (in (10.7)) as:

$$w_L = w_0\sqrt{\Theta_n^2 + \Lambda_n^2}, \text{ and} \tag{10.8}$$

$$F_L = \frac{L(\Theta_n^2 + \Lambda_n^2)}{\Theta_n(1 - \Theta_n) - \Lambda_n^2}, \text{ respectively.} \tag{10.9}$$

In the case of a collimated beam, the parameters w_L and F_L can be described as a function of L through the well-known expressions below [51]:

$$w_L = w_0\sqrt{1 + \frac{\lambda L}{\pi w_0^2}}, \text{ and} \tag{10.10}$$

$$F_L = L\sqrt{1 + \frac{\pi w_0^2}{\lambda L}}. \tag{10.11}$$

The irradiance or intensity of the optical wave is a function of the axial z and radial r distances, and can be found by taking the squared magnitude of the field. Correspondingly, the average intensity at the receiver is given by [11,51]:

$$I(r,L) = | U(r,L) |^2 = I(0,L)\exp\left[\frac{-2r^2}{w_L^2}\right] \tag{10.12}$$

where $I(0, L)$ resembles the on-axis irradiance with relation as follow:

$$I(0,L) = \frac{w_0^2}{w_L^2} = \frac{1}{\Theta_n^2 + \Lambda_n^2}. \tag{10.13}$$

10.3.3 Free-space optical communication channel

10.3.3.1 Atmospheric loss

Attenuation (or extinction) of the optical beam wave limits the intensity of a propagating optical laser beam, mainly due to the presence of a vast variety of particulate composition within the earth's atmosphere, which include nitrogen (N_2), oxygen (O_2), argon (Ar), ozone (O_3), water (H_2O), carbon dioxide (CO_2) and other small particles generated by combustion, dust, debris and soil [48,52,53]. These atmospheric particles result in the adverse effects of light absorption by gaseous molecules, and Rayleigh and Mie scattering contributed by gaseous molecules or aerosol particles suspended in the air.

In particular, the absorption process involves the capturing of electromagnetic radiation by a particle, and conversion of the photon energy into the kinetic energy of the particle's molecules, thus implying that absorption is a mechanism occurring from the heating of the atmosphere. Atmospheric absorption is a strong function of wavelength, in which the absorption by O_2 and O_3 essentially eliminates propagation of radiation at wavelengths below 0.2 mm, whereas very little absorption is observed at the visible wavelength window (i.e., 400 to 700 nm).

The Rayleigh and Mie scattering phenomena significantly deteriorate electromagnetic waves in the visible and infrared wavelength bands, when the optical radiation propagates through air molecules, aerosols and particulate matters, such as fog, haze, smoke and dust. The occurrence of Rayleigh scattering (also known as molecular scattering) is mainly contributed by air molecules and haze particles, which are smaller in comparison to the wavelength λ of electromagnetic radiation (see Figure 10.4(a)); and hence, this scattering effect is negligible at wavelengths greater than ~3 μm, since the scattering coefficient varies in accordance to λ^{-4} (a relation known as the Rayleigh law). On the other hand, Mie scattering (or aerosol scattering) is caused by particles comparable in size to or greater than the radiation wavelength; in which this scattering effect is concentrated in the forward direction (as depicted in Figure 10.4(b)), and the losses decrease rapidly with increasing wavelength, eventually approaching the Rayleigh scattering case. Since the operating wavelengths of terrestrial FSO links (i.e., 785, 850 and 1,550 nm) are chosen to fall within the transmission windows of the atmospheric absorption spectra, the light absorption process does not have much contribution to the atmospheric loss [26,54]. The scattering effects typically dominate the total attenuation coefficient, in which

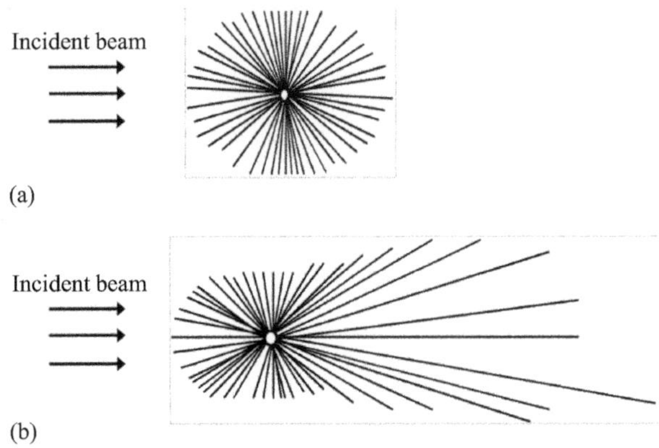

Figure 10.4 Distribution of the electromagnetic radiation resulting from (a) Rayleigh or molecular scattering; and (b) Mie or aerosol scattering [11,48]

the type of scattering (i.e., Rayleigh/molecular or Mie/aerosol) is determined by the size of the atmospheric particulate with respect to the laser wavelength of operation, as described by the dimensionless size parameter α_m of the Mie theory [55]:

$$\alpha_m = \frac{2\pi r_p}{\lambda}, \tag{10.14}$$

where r_p denotes the radius of the particle.

The atmospheric loss (or transmittance) is considered as a fixed scaling factor over multiple fading states of the FSO channel; and results from the combined deterministic effects of absorption and scattering, which can be modelled by the exponential Beers-Lambert law [56]:

$$h_l = e^{-\sigma L}, \tag{10.15}$$

where σ denotes a wavelength- and weather-dependent attenuation coefficient, with typical values as follows: clear air = 0.1 (0.43 dB/km); haze = 1 (4.3 dB/km); and fog = 10 (43 dB/km). For clear and foggy weather conditions, the attenuation coefficient can be determined from the visibility data through Kim's model [26]:

$$\sigma = \frac{3.91}{V}\left(\frac{\lambda}{550}\right)^{-q_s}, \tag{10.16}$$

where V is the visibility (in km), λ is the laser wavelength (in nm) and q_s is a parameter related to the particle size distribution and visibility defined by:

$$q_s = \begin{cases} 1.6, V > 50 \text{ km (high visibility)} \\ 1.3, 6 \text{ km} < V < 50 \text{ km (average visibility)} \\ 0.16V + 0.34,\ 1 \text{ km} < V < 6 \text{ km (haze visibility)} \\ V - 0.5, 0.5 \text{ km} < V < 1 \text{ km (mist visibility)} \\ 0.585V^{1/3}, V < 0.5 \text{ km (fog visibility).} \end{cases} \tag{10.17}$$

10.3.3.2 Optical turbulence in the atmosphere

Through multiple decades of extensive theoretical investigations [57–62], it is apparent that optical turbulence (i.e., random variations in the index of refraction) resulting from inhomogeneities in the temperature and pressure of the atmosphere contributes to the occurrence of various adverse channel effects, which include: (1) spreading of the optical laser beam beyond that due to pure diffraction (which reduces the spatial power density incident on the receiver); (2) random motion of the instantaneous beam centroid about the receiver (known as beam wander); (3) random fluctuations in the irradiance of the signal-carrying laser beam (termed as scintillations); (4) phase fluctuations; and (5) loss of spatial coherence. In principle, beam wander is primarily caused by turbulent eddies with scales sizes larger than the diameter of the optical beam, typically occurring near the transmitter; whereas the scintillation effect arises when the turbulence scale sizes are on the order of the first Fresnel zone. These turbulence-induced effects work in tandem to create random signal losses at the optical receiver; thereby degrading the error performance due to signal fading, and resulting in complete system annihilation under more severe channel conditions. The theoretical background, characteristics and analytical approach pertaining to the above-mentioned effects are introduced and discussed in detail here.

Beam wander

In the presence of atmospheric turbulence, a finite optical laser beam is susceptible to random deflections when propagating through free-space, which aggravates the spreading of the optical beam due to large random inhomogeneities of the atmosphere with scale sizes bounded above by the outer scale L_0, thus causing the resultant beam to become highly skewed from Gaussian (see Figure 10.5(a)). This phenomenon is known as beam wander, which can be explained by the random displacement of the instantaneous beam centroid (i.e., point of maximum irradiance termed as hot spot) off the boresight, observed over short time periods in the receiver plane [63–67]. From a statistical point-of-view, the beam wander effect can be characterized by the variance of the hot spot displacement along an axis, or equivalently by the variance of the magnitude of the hot spot displacement.

Earlier studies [68] have shown that the principle of reciprocity can be applied to the performance of a telescope as measured by its effective coherence size in atmospheric turbulence, whether it is functioning as part of a transmitter or as part of a receiver. By invoking the reciprocity principle, beam wander at the receiver plane can be modelled as if it arises from a random tilt angle at the transmitter plane, in which the random movement of the short-term beam (i.e., instantaneous hot spot) depicted by the shaded circular regions in Figure 10.5(b) leads to the large outer circle over a long time period. The resultant beam width (observed at the receiver plane) is termed as the long-term spot size w_{LT}, which forms the basis for the variance of beam wander fluctuations, the square of which yields:

$$w_{\mathrm{LT}}^2 = w_L^2(1 + T_{\mathrm{Tot}}) = w_L^2\left(1 + 1.33\sigma_R^2\Lambda_L^{5/6}\right) \tag{10.18}$$

where Λ_L denotes the Fresnel ratio at the output plane, as defined by (10.45).

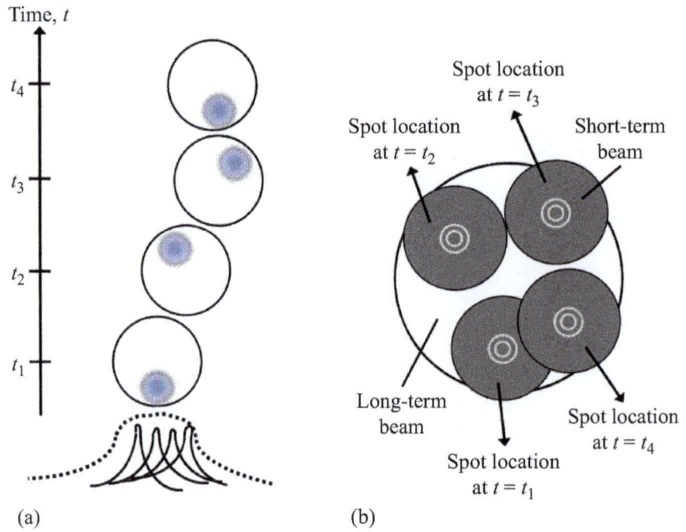

Figure 10.5 (a) Random deflections of the optical laser beam due to the beam wander effect are observed by the movement of hot spot (i.e., instantaneous centre of maximum irradiance) within the beam, causing the resultant beam to become highly skewed from Gaussian; and (b) illustration of the long-term beam spot size resulting from the random motion of the short-term beam in the receiver plane with respect to time, as indicated by the shaded circles [11]

Based upon the concept of short- and long-term spot size, w_{LT}^2 can be expressed in the form:

$$w_{\mathrm{LT}}^2 = w_L^2 + w_L^2 T_{\mathrm{SS}} + w_L^2 T_{\mathrm{LS}} \tag{10.19}$$

where the term $T_{\mathrm{Tot}} = T_{\mathrm{SS}} + T_{\mathrm{LS}}$ can be partitioned into a sum of small-scale (T_{SS}) and large-scale (T_{LS}) contributions. It is apparent that the long-term spot size w_{LT} arises from the effects of pure diffraction spreading (as described by the first term w_L^2), and turbulent cells (or known as eddies) of all scale sizes due to random inhomogeneities of the atmosphere. Turbulent eddies with scale sizes smaller than the beam diameter produce small-scale beam diffraction, as interpreted by the second term $w_L^2 T_{\mathrm{SS}}$; in which the combined diffractive effects of $w_L^2(1 + T_{\mathrm{SS}})$ (i.e., sum of the first and second term) is defined as 'beam breathing'. On the other hand, turbulent scale sizes larger than the beam diameter lead to refractive effects known as beam wander, which causes random displacement of the instantaneous centre of the incident optical beam in the receiver plane, as denoted by the final term $w_L^2 T_{\mathrm{LS}}$.

In accordance with Fante's [61] relation between the long-term beam radius w_{LT}, short-term beam radius w_{ST}, and beam wander variance $\langle r_c^2 \rangle$ of a Gaussian-beam wave, the above analytic expression is given by:

$$w_{\mathrm{LT}}^2 = w_{\mathrm{ST}}^2 + \langle r_c^2 \rangle. \tag{10.20}$$

Relating (10.20) to (10.19), the short-term beam radius w_{ST} can be identified by the relation:

$$w_{\mathrm{ST}} = w_L\sqrt{1+T_{\mathrm{SS}}}. \tag{10.21}$$

In the case of an infinite outer scale, the expressions for w_{ST} and $\langle r_c^2\rangle$ for a collimated and focused beam is given by:

$$w_{\mathrm{ST}} = \begin{cases} w_L\sqrt{1+1.33\sigma_R^2\Lambda_L^{5/6}\left[1-0.66\left(\dfrac{\Lambda_n^2}{1+\Lambda_n^2}\right)^{1/6}\right]} \text{ (collimated beam)} \\ w_L\sqrt{1+0.35\sigma_R^2\Lambda_L^{5/6}} \text{ (focused beam).} \end{cases} \tag{10.22}$$

$$\langle r_c^2\rangle = \begin{cases} 2.42C_n^2L^3w_0^{-1/3} \text{ (collimated beam)} \\ 2.72C_n^2L^3w_0^{-1/3} \text{ (focused beam).} \end{cases} \tag{10.23}$$

To substantiate the interpretation of jitter-induced PEs, the root mean square (RMS) beam wander displacement of a collimated beam is presented in the form:

$$\sqrt{\langle r_c^2\rangle} = 0.69L\left(\frac{\lambda}{2w_0}\right)\left(\frac{2w_0}{r_0}\right)^{5/6} \tag{10.24}$$

where $(\lambda/2w_0)$ is termed as the diffraction angle of the transmitted optical beam, and $r_0 = \left(0.16C_n^2k^2L\right)^{-3/5}$ denotes the Fried's parameters, or the atmospheric coherence width of a reciprocal propagating point source from the receiver at distance L. It should be noted that only the constant is different for a focused beam; and the beam wander effect arises from random tilt phase fluctuations averaged over the transmitter aperture on the order of $(2w_0/r_0)^{5/3}$.

Beam jitter is defined as the whole movement of the short-term optical beam around its unperturbed position in the receiver plane, due to random inhomogeneities of the atmosphere bounded above by r_0 ($r_0 \ll L_0$); thereby acting like an effective wavefront tilt at the transmitter, which can be significantly smaller than the wavefront tilt associated with the random fluctuations of hot spot displacement. The net result of total beam wander is a widening of the long-term beam profile near the boresight, which in turn leads to a slightly 'flattened' beam as shown in Figure 10.6(a), in comparison to the conventional Gaussian-beam profile (dashed curve) as suggested by Rytov theory. The implication of the resultant flattened beam profile is an 'effective PE' σ_{pe} corresponding to the 'beam centre' (i.e., area within the circumference of the lightly shaded circle) as depicted by Figure 10.6(b), which creates an increase in the longitudinal scintillation index that has not been considered in the first-order Rytov theory. In the case of a Gaussian-beam wave obeying the condition $2w_0 < r_0$, the scintillation effect resulting from the jitter-induced PEs can be explained by the presence of turbulent cells near the transmitter, with scale sizes on the order of the beam diameter ($2w_0$) and larger up to the atmospheric coherence width r_0. When $2w_0 \geq r_0$, the coherence width r_0 diminishes.

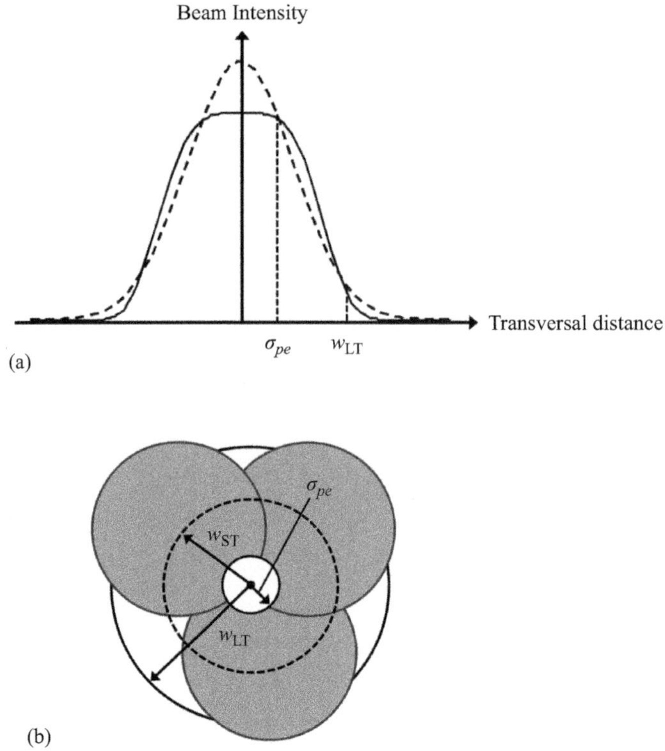

Figure 10.6 (a) The resultant flattened beam profile (solid line) as a function of the radial distance due to the widening of the long-term beam near the boresight, in comparison to the conventional Gaussian-beam profile (dashed line) as suggested by Rytov theory and (b) the net result is an effective rms PE σ_{pe} of the optical laser beam, as depicted by the small circle in the centre. The long-term beam (i.e., the larger hidden circle) is perceived as the superimposition of short-term beam over a long time period (as indicated by the shaded circles) [11]

In the case of a collimated beam ($\Theta_0 = 1$), the jitter-induced PE variance σ_{pe}^2 is given by:

$$\sigma_{pe}^2 = 0.48\left(\frac{\lambda L}{2w_0}\right)^2\left(\frac{2w_0}{r_0}\right)^{5/3}\left[1-\left(\frac{C_r^2 w_0^2/r_0^2}{1+C_r^2 w_0^2/r_0^2}\right)^{1/6}\right] \tag{10.25}$$

and for a focused beam ($\Theta_0 = 0$), the comparable expression is:

$$\sigma_{pe}^2 \cong 0.54\left(\frac{\lambda L}{2w_0}\right)^2\left(\frac{2w_0}{r_0}\right)^{5/3}\left[1-\frac{8}{9}\left(\frac{C_r^2 w_0^2/r_0^2}{1+0.5C_r^2 w_0^2/r_0^2}\right)^{1/6}\right], \tag{10.26}$$

where the parameter C_r is a scaling constant typically on the order $C_r \sim 2\pi$.

Depending on the value of the ratio $(2w_0/r_0)$, the asymptotic behaviour can be deduced to give:

$$\sigma_{pe}^2 \sim \begin{cases} \left(\dfrac{\lambda L}{2w_0}\right)^2 \left(\dfrac{2w_0}{r_0}\right)^{5/3}, & 2w_0/r_0 \ll 1 \\ \left(\dfrac{\lambda L}{2w_0}\right)^2 \left(\dfrac{r_0}{2w_0}\right)^{1/3}, & 2w_0/r_0 \gg 1. \end{cases} \tag{10.27}$$

which is applicable to both collimated and focused beams. As a consequence, even though there can be significant beam wander in some cases (i.e., with smaller optical beams), the related PE variance in both asymptotic cases will tend to zero.

Scintillation

The inter-mixing of turbulent winds within the atmosphere with the always-present vertical moisture and temperature gradients due to the Sun's heating of the Earth's surface creates optical turbulences in the atmospheric channel. The resultant irregularities in the space-time redistribution of the refractive index of the atmosphere (appearing in the form of eddies or cells, termed as optical turbules) are responsible for random fluctuations in the intensity of the optical laser beam, a phenomenon commonly known as scintillations. Theoretical treatments pertaining to the temporal irradiance fluctuations are largely based upon the classical Kolmogorov theory of turbulence [11], which relies on the dimensional analysis of isotropic velocity fields and inertial-sub-range predictions of velocity spectra and velocity structure functions. Through this statistical approach, it was discovered that a subclass of optical turbules within the inertial sub-range, defined as the range of turbulent cell sizes bounded above by the outer scale L_0 and below by the inner scale l_0, share a degree of statistical consistency; thus allowing further mathematical simplifications and approximations.

Since optical turbulence is fundamentally a nonlinear process, the energy cascade theory of turbulence (depicted in Figure 10.7) can be conveniently adopted

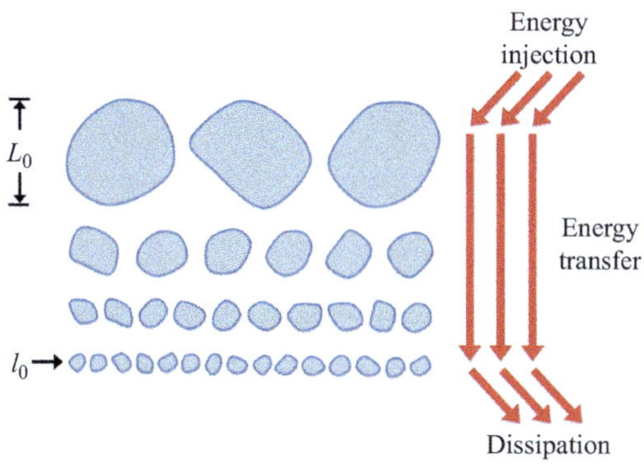

Figure 10.7 Conceptual illustration of the Kolmogorov energy cascade theory of turbulence. Within the inertial sub-range, the family of turbulent eddies bounded above by the outer scale L_0 and below by the inner scale l_0 exhibits statistical homogeneity and isotropy [11]

to describe the structure of atmospheric turbulence. In the presence of turbulent winds with increasing velocities, local unstable air masses (conceptualized as 'eddies') are created in the atmosphere, with characteristic dimensions slightly smaller than, and independent of, the parent flow. Under the influence of inertial forces originating from wind shear or convection, the turbulent eddies of larger scale size break up into smaller turbulent eddies, in order to form a continuum of eddy size for the transfer of energy from a macroscale L_0 (i.e., outer scale of turbulence) to a microscale l_0 (i.e., inner scale of turbulence). The family of turbulent eddies bounded above by the outer scale L_0 and below by the inner scale l_0 constitutes the inertial sub-range, as derived from the Kolmogorov theory; and exhibits the important characteristics of statistical homogeneity and isotropy. This implies that the mean value of the stochastic field is constant; and the point-to-point correlations between random fluctuations in the field are independent of the chosen observation points, while showing dependency only upon the magnitude of their vector separation.

The longitudinal structure function of wind velocity (parallel to the vector $\mathbf{R_P}$ connecting two observation points) is given by:

$$D_{RR}(R_p) = \langle (V_1 - V_2)^2 \rangle = \begin{cases} C_V^2 l_0^{-4/3} R_p^2, & 0 \ll R_p \ll l_0 \\ C_V^2 R_p^{2/3}, & l_0 \ll R_p \ll L_0 \end{cases} \tag{10.28}$$

where V_1 and V_2 represent the velocity components at two points separated by distance R_p, and C_V^2 denotes the velocity structure constant (in units of $m^{4/3}s^{-2}$). Within the boundaries of the inertial sub-range ($l_0 \ll R_p \ll L_0$), the longitudinal structure function is characterized by the universal 2/3 power law; whereas a quadratic behaviour is noted at small-scale sizes of turbulent cells ($R_p \ll l_0$). The velocity structure constant C_V^2 is a measure of the total amount of energy in the turbulence, and is related to the average energy dissipation rate ε (in units of m^2/s^3) through the relation:

$$C_V^2 = 2\varepsilon^{2/3}. \tag{10.29}$$

l_0 (in m) is typically on the order of the Kolmogorov microscale η_K:

$$l_0 \sim \eta_K = \left(\frac{\nu_K^3}{\varepsilon} \right)^{1/4}, \tag{10.30}$$

where ν_K resembles the kinematic viscosity. The inverse relationship between l_0 and ε reveals that the strong turbulence has smaller inner scales, whereas the weak turbulence has larger inner scales. On the other hand, L_0 is a function of $\varepsilon^{1/2}$; hence, varies in direct proportion to the strength of turbulence. In the case of statistically homogeneous and isotropic temperature fluctuations, the Kolmogorov theory is extended these conservative passive scalars (which do not involve the exchange of energy with the velocity turbulence) to define the associated structure functions as follows:

$$D_T(R_p) = \langle (T_1 - T_2)^2 \rangle = \begin{cases} C_T^2 l_0^{-4/3} R_p^2, & 0 \ll R_p \ll l_0 \\ C_T^2 R_p^{2/3}, & l_0 \ll R_p \ll L_0 \end{cases} \tag{10.31}$$

where T_1 and T_2 are the temperature at two points separated by distance R_p, and C_T^2 is the temperature structure constant (in units of $\deg^2 m^{-2/3}$). l_0 of the small-scale temperature fluctuations, which forms the lower boundary of inertial-convective range ($l_0 \ll R_p \ll L_0$) has the same order of magnitude as the inner scale for velocity fluctuations, with the expression given by [69]:

$$l_0 = 5.8\left(\frac{D_H{}^3}{\varepsilon}\right)^{1/4}, \tag{10.32}$$

where D_H is the diffusivity of heat in air (in units of $m^2 s^{-1}$).

The combined effects of small-scale temperature fluctuations and turbulent eddies induce a random behaviour in the atmospheric index of refraction n, which can be described by the expression:

$$n(\mathbf{R_p}, t) = n_0 + n_1(\mathbf{R_p}, t). \tag{10.33}$$

At a point $\boldsymbol{R}_\mathbf{p}$ in space and time t, $n_0 = \langle n(\boldsymbol{R}_\mathbf{p}, t) \rangle \cong 1$ indicates the mean value of the refractive index; and $n_1(\boldsymbol{R}_\mathbf{p}, t)$ represents the random deviation of $n(\boldsymbol{R}_\mathbf{p}, t)$ from the corresponding mean, thus giving $\langle n_1(\boldsymbol{R}_\mathbf{p}, t) \rangle = 0$. Since time variations in the atmospheric index of refraction can be suppressed in the treatment of optical wave propagation (i.e., the optical beam wave maintains a single frequency as it propagates), (10.33) can be written in the form:

$$n(\boldsymbol{R}_\mathbf{p}) = n_0 + n_1(\boldsymbol{R}_\mathbf{p}), \tag{10.34}$$

in which $n(\boldsymbol{R}_\mathbf{p})$ has been normalized by its mean value n_0.

For visible and infrared wavelengths, the atmospheric index of refraction are related to the temperature and pressure fluctuations according to [70]:

$$\begin{aligned} n(\boldsymbol{R}_\mathbf{p}) &= 1 + 77.6 \times 10^{-6}\left(1 + 7.52 \times 10^{-3}\lambda^{-2}\right)\frac{P(\boldsymbol{R}_\mathbf{p})}{T(\boldsymbol{R}_\mathbf{p})} \\ &\cong 1 + 79 \times 10^{-6}\frac{P(\boldsymbol{R}_\mathbf{p})}{T(\boldsymbol{R}_\mathbf{p})} \end{aligned} \tag{10.35}$$

where P is the pressure (in millibars), and T is the temperature (in Kelvin). The above approximation is based upon the assumption that $\lambda \sim 0.5$ μm, since wavelength dependence is considerably small for optical frequencies; and the variations in the optical signal resulting from absorption or scattering by molecules or aerosols are not considered here. Furthermore, pressure fluctuations are typically assumed negligible, thus implying that index-of-refraction disturbances associated with the visible and near-infrared region of the spectrum are predominantly due to random temperature fluctuations.

For statistically homogeneous and isotropic turbulence, the related structure function exhibits the asymptotic behaviour as follows:

$$D_n(R_p) = \begin{cases} C_n^2 l_0^{-4/3} R_p^2, & 0 \ll R_p \ll l_0 \\ C_n^2 R_p^{2/3}, & l_0 \ll R_p \ll L_0. \end{cases} \tag{10.36}$$

The refractive index structure parameter C_n^2 (in unit of $\mathrm{m}^{-2/3}$) is physically interpreted as a measure of the strength of the fluctuations in the index of refraction, and is expressed through the relation (as inferred directly from (10.35)):

$$C_n^2 = \left(79 \times 10^{-6} \frac{P}{T^2}\right)^2 C_T^2 \tag{10.37}$$

where C_T^2 denotes the temperature structure constant, which can be determined from (10.31). The C_n^2 values typically range from $10^{-17}\ \mathrm{m}^{-2/3}$ for weak turbulence scenarios and up to $10^{-13}\ \mathrm{m}^{-2/3}$ in the case of strong turbulences; and are essentially taken to be constant over short time intervals at a fixed propagation distance and height above ground level. For vertical or slant propagation paths, the refractive index structure parameter varies as a function of height above ground, which can be determined from the Hufnagel-Valley model [71]. This case does not fall within the present scope of discussion, since this research work is based upon long-distance horizontal (terrestrial) FSO communication links. The scintillation index [6,64] presents a viable statistical quantitative measure of the magnitude of atmospheric turbulence-induced irradiance fluctuations (i.e., the level of scintillations) given by:

$$\sigma_I^2 = \frac{\langle I^2 \rangle - \langle I \rangle^2}{\langle I \rangle^2} = \frac{\langle I^2 \rangle}{\langle I \rangle^2} - 1 \tag{10.38}$$

where I is the irradiance of the optical laser beam, and $\langle \cdot \rangle$ denotes ensemble averaging, equivalent to long-time averaging with the assumption of an ergodic process. The classification of atmospheric turbulence-induced scintillations (i.e., weak, moderate and strong turbulences) is commonly distinguished through the values of the Rytov variance:

$$\sigma_R^2 = 1.23 C_n^2 k^{7/6} L^{11/6} \tag{10.39}$$

where C_n^2 is the strength of turbulence, as defined by (10.37). Weak irradiance fluctuation conditions are typically associated with $\sigma_R^2 < 1$, in which the scintillation index is proportional to the Rytov variance, physically representing the random disturbances in the optical intensity of an unbounded plane wave. On the other hand, moderate fluctuation scenarios are described by $\sigma_R^2 \sim 1$, whereas strong fluctuations are associated with $\sigma_R^2 \gg 1$; and the so-called saturation regime corresponds to the asymptotic limit of $\sigma_R^2 \to \infty$.

Atmospheric turbulence effects on partially coherent Gaussian beam

Recall the characteristics of a lowest-order Gaussian-beam wave travelling through free-space, as discussed in Section 10.3.2. To examine the propagation effects of a partially coherent Gaussian laser beam in the presence of atmospheric turbulence, a Gaussian Schell beam model [11,37,38] is considered, having a Gaussian amplitude distribution with effective beam radius (spot size) w_0 at the exit aperture of the optical transmitter. After travelling through the atmospheric turbulent channel over a link distance L, the receiving beam size is given by:

$$w_L = w_0\sqrt{\Theta_n^2 + \zeta\Lambda_n^2}, \tag{10.40}$$

in which the normalized components (i.e., Θ_n and Λ_n) are defined by (10.7). The global coherence parameter is a measure of the global degree of coherence of light across each transversal plane along the propagation path, and is defined by:

$$\zeta = \zeta_s + \frac{2w_0^2}{\rho_0^2}. \tag{10.41}$$

It is related to the source coherence parameter:

$$\zeta_s = 1 + \frac{2w_0^2}{l_c^2}, \tag{10.42}$$

and the coherence length of a spherical wave:

$$\rho_0 = \left[0.55C_n^2k^2L\right]^{-3/5} \tag{10.43}$$

where l_c is the spatial coherence length. Parameter ζ_s describes the degree of partial (spatial) coherence of the source laser beam at the transmitter, in which $\zeta_s = 1$ for a coherent laser beam and $\zeta_s > 1$ for a partially coherent beam [6].

The point-receiver scintillation index can be conveniently expressed as a sum of radial and longitudinal components to give [11,37]:

$$\sigma_I^2(0) \cong 4.42\sigma_R^2\Lambda_L^{5/6}\frac{\sigma_{pe}^2}{w_L^2} + 3.86\sigma_R^2\left\{0.40\left[(1+2\Theta_L)^2 + 4\Lambda_L^2\right]^{5/12} \times \cos\left[\frac{5}{6}\tan^{-1}\left(\frac{1+2\Theta_L}{2\Lambda_L}\right)\right] - \frac{11}{16}\Lambda_L^{5/6}\right\} \tag{10.44}$$

where σ_{pe}^2 is the jitter-induced PE variance (see (10.25)–(10.27)), and σ_R^2 is the Rytov variance for a plane wave as defined by (10.39). The receiver beam parameters are:

$$\Theta_L = 1 + \frac{L}{F_L} \quad \text{and} \quad \Lambda_L = \frac{2L}{kw_L^2} \tag{10.45}$$

where the phase front radius of curvature for a PCB at the receiver is defined by [36]:

$$F_L = \frac{L(\Theta_n^2 + \zeta\Lambda_n^2)}{\phi\Lambda_n - \zeta\Lambda_n^2 - \Theta_n^2}, \quad \text{with } \phi \equiv \frac{\Theta_n}{\Lambda_n} - \frac{\Lambda_n w_0^2}{\rho_0^2}. \tag{10.46}$$

In the limiting cases of a plane wave ($\Theta_L = 1$, $\Lambda_L = 0$) and a spherical wave ($\Theta_L = \Lambda_L = 0$), the radial term vanishes and the longitudinal term reduces, respectively, to the well-known results [62]:

$$\sigma_{I,\text{pl}}^2(0) = \sigma_R^2 = 1.23C_n^2k^{7/6}L^{11/6}(\text{plane wave}), \tag{10.47}$$

$$\sigma_{I,\text{sp}}^2(0) = 0.4\sigma_R^2 = 0.5C_n^2k^{7/6}L^{11/6}(\text{spherical wave}). \tag{10.48}$$

The aperture-averaging effect (to be discussed in further detail in Section 10.3.4) of a non-point receiver aperture is taken into account through the aperture-averaged scintillation index given by:

$$\sigma_I^2(D) = A_g\sigma_I^2(0) \tag{10.49}$$

where A_g denotes the aperture-averaging factor as defined by (10.90), and D is the receiver aperture diameter. The aperture-averaged scintillation index for the limiting plane-wave and spherical-wave models take on the forms:

Plane wave:

$$\sigma_{I,\text{pl}}^2(D) = \exp\left[\frac{0.49\sigma_R^2}{\left(1 + 0.65d^2 + 1.11\sigma_R^{12/5}\right)^{7/6}} + \frac{0.51\sigma_R^2\left(1 + 0.69\sigma_R^{12/5}\right)^{-5/6}}{1 + 0.90d^2 + 0.62d^2\sigma_R^{12/5}}\right] - 1, \tag{10.50}$$

Spherical wave:

$$\sigma_{I,\text{sp}}^2(D) =$$
$$\exp\left[\frac{0.49\sigma_{I,\text{sp}}^2(0)}{\left(1 + 0.18d^2 + 0.56\sigma_{I,\text{sp}}^{12/5}(0)\right)^{7/6}} + \frac{0.51\sigma_{I,\text{sp}}^2(0)\left(1 + 0.69\sigma_{I,\text{sp}}^{12/5}(0)\right)^{-5/6}}{1 + 0.90d^2 + 0.62d^2\sigma_{I,\text{sp}}^{12/5}(0)}\right] - 1, \tag{10.51}$$

where $d = \sqrt{kD^2/4L}$ denotes the scaled aperture size.

Distribution models for the irradiance

Turbulence-induced scintillation effect resulting from beam propagation through the atmosphere causes significant power losses at the optical receiver and fading of the detected signal below a detectable threshold, thereby imposing severe diminishing effects on the performance of FSO communication systems. Myriads of theoretical and experimental studies have been carried out within the research community, in order to develop tractable and reliable mathematical models for the irradiance probability density function (PDF), capable of predicting the stochastic nature of the optical intensity fluctuations and deducing the reliability of the system under most possible atmospheric conditions. While numerous irradiance models have been proposed with varying degrees of success in predicting the observed phenomena, the log-normal distribution is the most widely adopted model for the PDF of the randomly fading irradiance signal under weak-to-moderate turbulence conditions due to its simplicity, albeit underestimating the peak irradiance of the PDF and the behaviour in the distribution tails, particularly with increasing turbulence strength [11]. The PDF of the irradiance intensity in the turbulent medium is given by:

$$f_{h_s}(h_s) = \frac{1}{h_s \sigma_I(D)\sqrt{2\pi}} \exp\left\{ -\frac{\left[ln(h_s) + \frac{1}{2}\sigma_I^2(D)\right]^2}{2\sigma_I^2(D)} \right\} \tag{10.52}$$

where $\sigma_I^2(D)$ is obtained accordingly by using (10.49) for a spatially partially coherent Gaussian laser beam, taking into account the aperture-averaging effect. Alternatively, (10.50) and (10.51) can be employed for the limiting plane-wave and spherical-wave models, respectively.

Figure 10.8 presents the variation in the log-normal distribution for different values of Rytov variance σ_R^2 and aperture-averaged scintillation index $\sigma_I^2(D)$ in the weak turbulence regime, in which the corresponding parameters are common measures for the strength of the atmospheric turbulence-induced irradiance fluctuations. The C_n^2 values ranging from 10^{-15} to 10^{-13} m$^{-2/3}$ are considered here, with the assumption of $D = 200$ mm. It is evident that the distribution becomes more skewed with longer tails towards the infinity direction for greater values of σ_R^2 and $\sigma_I^2(D)$, which denotes the extent of fluctuation of the irradiance as the channel inhomogeneity increases [72].

To address the strong fluctuations or multiple scattering regimes, the statistical PDF models (of the optical field amplitude or irradiance) based upon heuristic

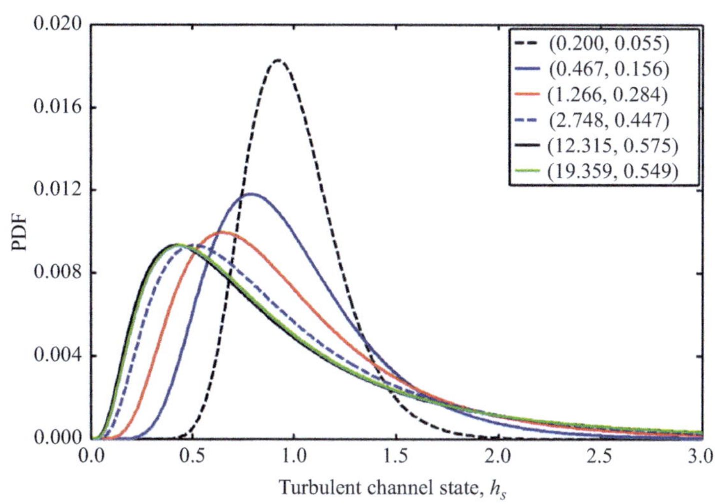

Figure 10.8 Variations of the log-normal PDF for different values of (Rytov variance σ_R^2, scintillation index $\sigma_I^2(D)$) (see figure legend) based on the spherical-wave model, with $D = 200$ mm and C_n^2 values ranging from 10^{-15} to 10^{-13} m$^{-2/3}$ [72]

arguments and observed experimental data, which relate the discrete scattering regions in the turbulent medium to the individual phase front inhomogeneities in the optical beam wave, are found to be more reliable by showing good agreement with actual field measurements. In the far saturation regime with sufficiently large number of discrete scattering regions, the radiation field of the wave is approximately zero-mean Gaussian; and thus, the irradiance statistics are governed by the negative exponential distribution. The gamma-gamma distribution is a more recent fading model which has evolved from an assumed modulation process, whereby the (normalized) irradiance $I = XY$ can be partitioned into factors arising from the large-scale (X) and small-scale (Y) scintillations under moderate-to-strong turbulence scenarios. Consider the large-scale and small-scale irradiance fluctuations are governed by gamma distributions (due to their excellent approximation in propagation problems involving intensity or amplitude) [73]:

$$p_X(X) = \frac{\alpha(\alpha X)^{\alpha-1}}{\Gamma(\alpha)}\exp(-\alpha X), \quad \alpha > 0, X > 0; \text{ and} \tag{10.53}$$

$$p_Y(Y) = \frac{\beta(\beta Y)^{\beta-1}}{\Gamma(\beta)}\exp(-\beta Y), \quad \beta > 0, Y > 0 \tag{10.54}$$

where $\Gamma(\cdot)$ denotes the gamma function [74]. The conditional PDF $p_Y(I \mid X)$ can be obtained by fixing X and representing $Y = I/X$, to give:

$$p_Y(I|X) = \frac{\beta(\beta I/X)^{\beta-1}}{X\Gamma(\beta)}\exp(-\beta I/X), \quad I > 0 \tag{10.55}$$

in which X is the (conditional) mean value of I. Correspondingly, by forming the average of (10.55) over the distribution in (10.53), the unconditional irradiance distribution $p(I)$ can be defined by:

$$\begin{aligned} p(I) &= \int_0^\infty p_Y(I|X)p_X(X)dX \\ &= \frac{2(\alpha\beta)^{(\alpha+\beta)/2}}{\Gamma(\alpha)\Gamma(\beta)} I^{(\alpha+\beta)/2-1} K_{\alpha-\beta}\left(2\sqrt{\alpha\beta I}\right), \quad I > 0 \end{aligned} \tag{10.56}$$

where $K_{\alpha-\beta}$ resembles the modified Bessel function of the second kind of order $(\alpha - \beta)$ [74]. Relating the normalized intensity I of the optical laser beam to the instantaneous turbulent channel state h_s, the gamma-gamma distribution is given by:

$$f_{h_s}(h_s) = \frac{2(\alpha\beta)^{(\alpha+\beta)/2}}{\Gamma(\alpha)\Gamma(\beta)} h_s^{(\alpha+\beta)/2-1} K_{\alpha-\beta}\left(2\sqrt{\alpha\beta h_s}\right). \tag{10.57}$$

Parameters α and β correspond to the effective number of large- and small-scale turbulent eddies, respectively, relating to the scattering effects of the atmospheric turbulent environment, and are valid with the relationship [11]:

$$\sigma_I^2(D) = \frac{1}{\alpha} + \frac{1}{\beta} + \frac{1}{\alpha\beta} \tag{10.58}$$

in which

$$\alpha = \frac{1}{\sigma_X^2} = \frac{1}{\exp(\sigma_{\ln X}^2) - 1}, \quad \text{and} \tag{10.59}$$

$$\beta = \frac{1}{\sigma_Y^2} = \frac{1}{\exp(\sigma_{\ln Y}^2) - 1}. \tag{10.60}$$

For the Gaussian-beam wave model with aperture-averaging, in the presence of beam wander effects, the limiting case of the large-scale and small-scale log variances with $l_0 = 0$ and $L_0 = 0$ take the forms [11]:

$$\sigma_{\ln X}^2(D) = 4.42\sigma_R^2 \Lambda_L^{5/6} \frac{\sigma_{pe}^2}{w_L^2} + \frac{0.49\left(\dfrac{\Omega_L - \Lambda_L}{\Omega_L + \Lambda_L}\right)^2 \sigma_B^2}{\left[1 + \dfrac{0.4(2-\overline{\Theta}_L)(\sigma_B/\sigma_R)^{12/7}}{(\Omega_L + \Lambda_L)\left(\dfrac{1}{3} - \dfrac{1}{2}\overline{\Theta}_L + \dfrac{1}{5}\overline{\Theta}_L^2\right)^{6/7}} + 0.56(1+\Theta_L)\sigma_B^{12/5}\right]^{7/6}}, \quad \text{and} \tag{10.61}$$

$$\sigma_{\ln Y}^2(D) = \frac{(0.51\sigma_B^2)/\left(1 + 0.69\sigma_B^{12/5}\right)^{5/6}}{1 + \left[1.20(\sigma_R/\sigma_B)^{12/5} + 0.83\sigma_R^{12/5}\right]/(\Omega_L + \Lambda_L)} \tag{10.62}$$

where $\overline{\Theta}_L = 1 - \Theta_L$, σ_R^2 is defined previously in (10.39), and the non-dimensional parameter characterizing the spot radius of the collecting lens is:

$$\Omega_L = \frac{16L}{kD^2}. \tag{10.63}$$

The approximated Rytov variance for collimated and divergent Gaussian beams is:

$$\sigma_B^2 \cong 3.86\sigma_R^2 \left\{ 0.40\left[(1+2\Theta_L)^2 + 4\Lambda_L^2\right]^{5/12} \cos\left[\frac{5}{6}\tan^{-1}\left(\frac{1+2\Theta_L}{2\Lambda_L}\right)\right] - \frac{11}{16}\Lambda_L^{5/6} \right\}. \tag{10.64}$$

10.3.3.3 Pointing errors

A statistical misalignment-induced fading model developed in [24], based on an earlier work in [30], provides a tractable pdf for describing the stochastic behaviour of the PEs. The model assumes a circular detection aperture of diameter D, and a Gaussian spatial intensity profile of beam waist radius w_L on the receiver plane. In addition, both the elevation and horizontal displacement (sway) are considered independent and identically Gaussian distributed with variance σ_s^2. Correspondingly, the pdf of h_p is given by:

$$f_{h_p}(h_p) = \frac{\xi^2}{A_0^{\xi^2}} h_p^{\xi^2-1}, \quad 0 \le h_p \le A_0 \tag{10.65}$$

where

$$\xi = \frac{w_{z_{\mathrm{eq}}}}{2\sigma_{\mathrm{pe}}}, \tag{10.66}$$

is the ratio between the equivalent beam radius at the receiver and the PE displacement (jitter) standard deviation at the receiver; and the remaining parameters are defined by [24]:

$$A_0 = \left[\operatorname{erf}(v_p)\right]^2, \tag{10.67}$$

$$w_{z_{eq}} = w_L \sqrt{\frac{\sqrt{\pi}\operatorname{erf}(v_p)}{2v\exp\left(-v_p^2\right)}}, \quad \text{and} \tag{10.68}$$

$$v_p = \frac{\sqrt{\pi}D}{2\sqrt{2}w_L}. \tag{10.69}$$

10.3.3.4 Combined channel fading model

The combined channel fading model of $h = h_l h_s h_p$ can be expressed as [24,42]:

$$f_h(h) = \int f_{h|h_s}(h|h_s) f_{h_s}(h_s) dh_s, \tag{10.70}$$

where $f_{h|h_s}(h \mid h_s)$ is the conditional probability given a turbulence state h_s. Since h_l is deterministic and acts as a scaling factor, the resulting conditional distribution can be expressed as:

$$\begin{aligned} f_{h|h_s}(h|h_s) &= \frac{1}{h_s h_l} f_{h_p}\left(\frac{h}{h_s h_l}\right) \\ &= \frac{\xi^2}{A_0^{\xi^2} h_s h_l} \left(\frac{h}{h_s h_l}\right)^{\xi^2-1}, \quad 0 \le h \le A_0 h_s h_l. \end{aligned} \tag{10.71}$$

Substituting (10.71) into (10.70) yields:

$$f_h(h) = \frac{\xi^2}{(A_0 h_l)^{\xi^2}} h^{\xi^2-1} \int_{h/A_0 h_l}^{\infty} h_s^{-\xi^2} f_{h_s}(h_s) dh_s. \tag{10.72}$$

By substituting the proper atmospheric turbulence models $f_{h_s}(h_s)$ (as defined by (10.52) and (10.57)) into (10.72) and re-arranging the resultant expressions, the channel state distributions of $h = h_l h_s h_p$ for both the weak and strong turbulence regimes are given by (10.73) and (10.74), respectively. For the weak atmospheric turbulence conditions:

$$f_h(h) = \frac{\xi^2}{(A_0 h_l)^{\xi^2}} h^{\xi^2-1} \times \int_{h/A_0 h_l}^{\infty} \frac{1}{h_s^{\xi^2+1} \sigma_I(D)\sqrt{2\pi}} \exp\left[- \frac{\left[ln(h_s) + \frac{1}{2}\sigma_I^2(D) \right]^2}{2\sigma_I^2(D)} \right] dh_s. \tag{10.73}$$

In the strong turbulence regime:

$$f_h(h) = \frac{2\xi^2 (\alpha\beta)^{(\alpha+\beta)/2}}{(A_0 h_l)^{\xi^2} \Gamma(\alpha)\Gamma(\beta)} h^{\xi^2-1} \times \int_{h/A_0 h_l}^{\infty} h_s^{(\alpha+\beta)/2-1-\xi^2} K_{\alpha-\beta}\left(2\sqrt{\alpha\beta h_s}\right) dh_s. \tag{10.74}$$

10.3.4 Aperture averaging phenomenon

The receiver aperture in an optical communication system behaves essentially like a 'point aperture', when its aperture dimension is smaller than the correlation width of the irradiance fluctuations. When the aperture size increases beyond the irradiance correlation width, the receiver detects several correlation patches and the scintillation level measured in the image plane begins to decrease, thereby resulting in higher mean SNR. This phenomenon is known as the aperture-averaging effect, whereby enlarged receiver apertures are intentionally utilized in direct detection systems, in order to enhance the system performance through the reduction of scintillations. Measurements from early astronomical studies by Mikesell *et al.* [75,76] revealed that aperture averaging potentially causes a shift in the relative frequency content of the irradiance power spectrum towards lower frequencies; in essence, averaging out the fastest fluctuations and thus mitigating the adverse effects of scintillation.

In principle, the reduction of scintillation level resulting from the aperture-averaging effect can be deduced from the ratio of received optical power fluctuations by a finite-size collecting aperture to that obtained by a very small (i.e., point) aperture [11]. Correspondingly, the normalized variance of the power fluctuations in the receiver plane is the same as the flux variance of irradiance fluctuations defined by:

$$\begin{aligned} \sigma_I^2(D) &= \frac{\langle P_R{}^2 \rangle - \langle P_R \rangle^2}{\langle P_R \rangle^2} \\ &= \frac{16}{\pi D^2} \int_0^D \rho B_I(xD, L) \left[\cos^{-1}\left(\frac{\rho}{D}\right) - \frac{\rho}{D}\sqrt{1 - \frac{\rho^2}{D^2}} \right] d\rho \end{aligned} \tag{10.75}$$

where P_R is the received optical power, D is the circular aperture diameter, $B_I(\rho, L)$ denotes the irradiance covariance function in the pupil plane, and the terms in brackets arise from the modulation transfer function of the circular aperture. By defining the variable $\delta = \rho/D$, (10.75) can be simplified to give the expression as follows:

$$\sigma_I^2(D) = \frac{16}{\pi}\int_0^1 \delta B_I(\delta D, L)\left[\cos^{-1}\delta - \delta\sqrt{1-\delta^2}\right]d\delta. \quad (10.76)$$

For a circular aperture of diameter D, the aperture-averaging factor (or coefficient) is defined by:

$$A_g = \frac{\sigma_I^2(D)}{\sigma_I^2(0)} = \frac{16}{\pi}\int_0^1 \delta b_I(\delta D, L)\left[\cos^{-1}\delta - \delta\sqrt{1-\delta^2}\right]d\delta, \quad (10.77)$$

which typically ranges between 0 and 1, with A_g values approaching these extreme values implying complete aperture-averaging and no aperture-averaging, respectively; $\sigma_I^2(0) = B_I(0, L)$ is the scintillation index for a point aperture ($D = 0$), and the normalized covariance function is obtained through the relation:

$$b_I(\rho, L) = \frac{B_I(\rho, L)}{B_I(0, L)}. \quad (10.78)$$

10.3.4.1 Extended Huygens–Fresnel principle

After propagating a distance L through a random medium, the complex field of the optical wave can be represented, using the Huygens-Kirchhoff principle, as:

$$U(\rho, L) = \iint d^2\mathbf{r}G(\mathbf{r}, \rho, L)U(\mathbf{r}, 0) \quad (10.79)$$

where $U(\mathbf{r}, 0)$ is the field at the transmitter plane $z = 0$, defined by (10.4). The paraxial (parabolic) wave equation derived from the Helmholtz equation has the Green's function solution:

$$G(\mathbf{r}, \rho, L) = \frac{-jk}{2\pi L}\exp\left[jkL + \frac{jk}{2L}\mid \rho - \mathbf{r} \mid^2 + \Psi(\mathbf{r}, \rho)\right], \quad (10.80)$$

where $\Psi(r, \rho)$ resembles the random part of the complex phase of a spherical wave propagating in the turbulent medium from the point $(\mathbf{r}, 0)$ to the point (ρ, L). Using this Green's-function solution, (10.79) can be expressed in terms of the Huygens-Fresnel integral [58]:

$$U(\rho, L) = \frac{-jk}{2\pi L}\exp(jkL)\iint d^2\mathbf{r}U(\mathbf{r}, 0)\exp\left[\frac{jk}{2L}\mid \rho - \mathbf{r} \mid^2 + \Psi(\mathbf{r}, \rho)\right]. \quad (10.81)$$

The Huygens–Fresnel principle [51] states that each point on a wave front generates a spherical wave; in which the envelope of these secondary waves constitutes a new wave front, and their superposition produces the wave in another plane. The expression in (10.81) is commonly known as the extended Huygens-Fresnel principle, and is applicable to both the weak and strong turbulence fluctuation regimes.

10.3.4.2 Spatial covariance of irradiance fluctuations

Classical studies have characterized wave propagation through atmospheric turbulence in terms of the complex mutual coherence function:

$$\Gamma_I(\rho_1,\rho_2;\tau) = \langle E(\rho_1;\tau)E^*(\rho_2;t+\tau)\rangle \tag{10.82}$$

where ρ_1 and ρ_2 are the observation points in the receiver plane with a time delay τ, $\langle\cdot\rangle$ denotes ensemble averaging, E is the optical frequency electric field, and E^* resembles the complex conjugate field. On the contrary, recent theoretical treatments [36,37,77,78] pertaining to the propagation of partially coherent Gaussian laser beam in the presence of atmospheric turbulence, have placed much emphasis on the analysis of cross-spectral density function:

$$W(\rho_1,\rho_2;\nu) = \langle U(\rho_1;\nu)U^*(\rho_2;\nu)\rangle, \tag{10.83}$$

which is the temporal Fourier transform of $\Gamma_I(\rho_1,\rho_2;\tau)$. The cross-spectral density function, which obeys the Helmholtz equation, is a measure of the correlation between the fluctuations of two field components at the same frequency ν of a partially coherent laser beam. If the field is strictly monochromatic (i.e., having sufficiently narrow band), such that the condition below is satisfied:

$$\frac{|\rho_2-\rho_1|}{c} \ll \frac{1}{\Delta\nu}, \tag{10.84}$$

then both characterizations give identical results; in which the typical values of $|\rho_2-\rho_1|$ are on the order of 10 cm, requiring that $\Delta\nu < 1$ GHz.

At the input plane, a phase diffuser is placed over the exit aperture of a quasi-monochromatic Gaussian laser transmitter, such that the resultant field of the optical wave emerging from the diffuser can be modelled as:

$$\widetilde{U}(\mathbf{r},0) = U(\mathbf{r},0)\exp[j\varphi_d(\mathbf{r})] \tag{10.85}$$

where $U(r,0)$ denotes the deterministic field entering the diffuser defined by (10.4), and $\exp[j\varphi_d(r)]$ resembles the small random perturbation introduced by the phase diffuser. Based on the assumption that the ensemble average of the spatially dependent random phases introduced by the diffuser is Gaussian and depends only on the separation distance and not on the actual location on the diffuser, the cross-spectral density at the transmitter can be expressed in the form of a Gaussian Schell-model beam:

$$\begin{aligned} W(\mathbf{r}_1,\mathbf{r}_2,0) &= \langle\widetilde{U}(\mathbf{r}_1,0)\widetilde{U}^*(\mathbf{r}_2,0)\rangle \\ &= U(\mathbf{r}_1,0)U^*(\mathbf{r}_2,0)\langle\exp[j\varphi_{d1}(\mathbf{r}_1)]\exp[-j\varphi_{d2}(\mathbf{r}_2)]\rangle \\ &= U(\mathbf{r}_1,0)U^*(\mathbf{r}_2,0)\exp\left[\frac{-(\mathbf{r}_1-\mathbf{r}_2)^2}{2\sigma_g^2}\right] \end{aligned} \tag{10.86}$$

where σ_g^2 is the variance of the Gaussian describing the ensemble average of the random phases (i.e., the partial coherence properties of the effective transmitted source), as determined by the characteristics of the diffuser. This variance can be interpreted to represent the number of 'speckle cells' (i.e., statistically independent patches) at the diffuser, whereby each speckle cell essentially behaves like an independent source term. Weak diffusers ($\sigma_g^2 \gg w_0^2$) typically have only one speckle cell (implying a coherent wave), whereas many speckle cells will normally exist for the case of a strong diffuser.

Consider the behaviour of $W(\rho_1, \rho_2; \nu = \nu_0)$, where ν_0 is the centre frequency of the quasi-monochromatic Gaussian laser beam. The cross-spectral density at the receiver can be represented as:

$$\begin{aligned} W(\rho_1,\rho_2,L) &= \langle U(\rho_1,L)U^*(\rho_2,L)\rangle \\ &= \frac{1}{(\lambda L)^2}\iiiint \mathrm{d}^2\mathbf{r}_1\mathrm{d}^2\mathbf{r}_2 W(\mathbf{r}_1,\mathbf{r}_2,0) \\ &\quad \times\langle \exp[\Psi(\mathbf{r}_1,\rho_1)+\Psi^*(\mathbf{r}_2,\rho_2)]\rangle \\ &\quad \times\exp\left\{\frac{jk}{2L}\left[(\rho_1-\mathbf{r}_1)^2-(\rho_2-\mathbf{r}_2)^2\right]\right\}. \end{aligned} \tag{10.87}$$

The spatial covariance of irradiance fluctuations can be estimated through the relation:

$$B_I(\rho_1,\rho_2,L) = |\,\Gamma(\rho_1,\rho_2,L)\,|^2 \tag{10.88}$$

in which the mutual coherence function $\Gamma_I(\rho_1,\rho_2,L)$ can be approximated by the cross-spectral density function, i.e., $\Gamma_I(\rho_1,\rho_2,L) \cong W(\rho_1,\rho_2,L)$. By normalizing the spatial covariance of irradiance fluctuations $B_I(\rho_1,\rho_2,L)$ yields:

$$\begin{aligned} b_I(\rho,L) &= \frac{B_I(\rho,L)}{B_I(0,L)} \\ &= \exp\left\{-\frac{\rho^2}{\rho_0^2}\left[2+\frac{\rho_0^2}{w_0^2\Lambda_n^2}-\frac{\rho_0^2\phi^2}{w_L^2}\right]\right\} = \mu(2\rho,L) \end{aligned} \tag{10.89}$$

where $\mu(\rho,L)$ denotes the complex degree of spatial coherence, and $\rho = |\,\rho_1-\rho_2\,|$. It is noted that the normalized spatial covariance of irradiance fluctuations has an analytic solution in terms of the complex degree of coherence $\mu(\rho,L)$ evaluated at twice the radial separation distance.

Substituting (10.89) into (10.77) results in an easily evaluated expression for the aperture-averaging factor [37]:

$$\begin{aligned} A_g &= \frac{16}{\pi}\int_0^1 \delta d\delta \exp\left\{\frac{-D^2\delta^2}{\rho_0^2}\left[2+\frac{\rho_0^2}{w_0^2\Lambda_n^2}-\frac{\rho_0^2\phi^2}{w_L^2}\right]\right\} \\ &\quad \times\left[\cos^{-1}(\delta)-\delta\sqrt{1-\delta^2}\right]. \end{aligned} \tag{10.90}$$

10.4 Performance analysis

The BER, probability of outage and average channel capacity are the three key application-specific performance indicators identified and employed throughout the proposed studies, in order to present a comprehensive performance evaluation of the partially coherent FSO links, which are commercially deployed to complement and/or serve numerous wireless communication purposes [79]. In a binary-encoded IM/DD system, the BER is typically used to quantify the probability of making a bit decision error, where the optical signals (transmitted as a pulse of light to indicate a '1' bit, and no light output for a '0' bit) are susceptible to various channel impairments and noise contributions [6,42,80]. For non-real-time data services, the average channel capacity determines the maximum achievable information rate averaged over all possible fading states; since the FSO channel is time-varying due to the turbulence- and PE-induced fluctuations, hence the information carrying capacity is also transient [33,35,81]. In the case of block fading (or quasi-static fading) channels, where the channel or the received irradiance intensity remains unchanged for a significantly long period of time, system outages will occur when the irradiance intensity falls below a threshold [24,82]. Correspondingly, the outage capacity is a useful capacity measure, which presents the maximum rate of transmission, such that the probability of the system being in an outage meets a specified value [32].

10.4.1 Bit-error rate

For an IM/DD FSO communication link using OOK modulation, the BER can be expressed as [6,25,42,80,83]:

$$P_e = P_X(0)P_X(e|0) + P_X(1)P_X(e|1) \tag{10.91}$$

where $P_X(0)$ and $P_X(1)$ denote the probabilities of transmitting '0' and '1' bits, respectively, and $P_X(e \mid 0)$ and $P_X(e \mid 1)$ resemble the conditional bit error probabilities when the transmitted bit is '0' or '1'. Conditioned on the channel state h, the following relation is then obtained:

$$P_X(e|0,h) = P_X(e|1,h) = Q\left(\sqrt{SNR(h)}\right) \tag{10.92}$$

where $Q(\cdot)$ is the Gaussian-Q function given by:

$$Q(y) = \frac{1}{\sqrt{2\pi}} \int_y^{\infty} \exp\left(\frac{-t^2}{2}\right) dt. \tag{10.93}$$

With reference to the expression for $SNR(h)$ defined by (10.2), it should be noted that the total noise contributions (i.e., shot noise, dark current noise, thermal noise and background noise) are taken into account by the statistical noise variance σ_n^2, and is assumed constant in the corresponding SNR calculations [6,24,34]. Considering the symmetry of the problem with $P_X(0) = P_X(1) = 0.5$ and $P_X(e|0) = P_X(e|1)$, the average BER can be determined by averaging over the PDF of h [41]:

$$P_e = \int_0^{\infty} f_h(h) Q\left(\sqrt{SNR(h)}\right) dh. \tag{10.94}$$

10.4.2 Probability of outage

The instantaneous capacity corresponding to a channel state h is given by [24,25,43]:

$$C(SNR(h)) = \sum_{x=0}^{1} P_X(x) \int_{-\infty}^{+\infty} f(y|x,h) \times \log_2 \left[\frac{f(y|x,h)}{\sum_{m=0,1} f(y|x=m,h) P_X(m)} \right] dy \tag{10.95}$$

where $P_X(x)$ is the probability of the bit being one ($x = 1$) or zero ($x = 0$), with $P_X(x=0) = P_X(x=1) = 0.5$, and $f(y|x,h)$ can be expressed as:

$$f(y|x,h) = \begin{cases} \dfrac{1}{\sqrt{2\pi\sigma_n^2}} \exp\left[-\dfrac{y^2}{2\sigma_n^2} \right], & x = 0 \\ \dfrac{1}{\sqrt{2\pi\sigma_n^2}} \exp\left[-\dfrac{(y - 2P_{\text{FSO}}\gamma h)^2}{2\sigma_n^2} \right], & x = 1. \end{cases} \tag{10.96}$$

Since the optical slow-fading channel is random and remains unchanged over a long block of bits, the time-varying channel capacity will not be sufficient to support a maximum data rate, when the instantaneous SNR falls below a threshold [25,82]. Such occurrence, known as system outage, can result in a loss of potentially up to 10^9 consecutive bits at a data rate of 10 Gbps, under deep fades scenarios that may last for ~1–100 ms [28]. In this case, the outage probability is an appropriate performance measure of the capacity, which represents the probability that the instantaneous channel capacity C falls below a transmission rate R_0, given by the relation:

$$P_{\text{out}} = \text{Prob}(C(SNR(h)) < R_0). \tag{10.97}$$

Since $C(\cdot)$ is monotonically increasing with SNR, P_{out} is the cumulative density function of h evaluated at $h_0 = \sqrt{C^{-1}(R_0)\sigma_n^2 / 2P_{\text{FSO}}^2\gamma^2}$, thus can be determined equivalently from the expression [24]:

$$P_{\text{out}} = \int_0^{h_0} f_h(h) dh. \tag{10.98}$$

10.4.3 Average channel capacity

The average channel capacity represents the practically achievable information carrying rate (in bits per channel use) through the time-varying fading channel with an arbitrarily small probability of detection error. This can be approached by using channel codes of large block lengths, in order to capture the effects of the channel variations through the codewords [25]. The average capacity $\langle C\rangle$ for a binary-input continuous-output channel is defined as the maximum mutual information between the input to the channel x, and output from the channel y, where the maximum is taken over all input distributions [82,84]. For a known channel at the receiver:

$$\langle C\rangle = \sum_{x=0}^{1} P_X(x) \int_{-\infty}^{\infty}\int_{0}^{\infty} f(y|x,h) f_h(h) \times \log_2\left[\frac{f(y|x,h)}{\sum_{m=0,1} f(y|x=m,h)P_X(m)}\right] dhdy \tag{10.99}$$

where $f(y|x,h)$ can be expressed as:

$$f(y|x,h) = \begin{cases} \dfrac{1}{\sqrt{2\pi\sigma_n^2}} \exp\left[-\dfrac{y^2}{2\sigma_n^2}\right], & x = 0 \\ \dfrac{1}{\sqrt{2\pi\sigma_n^2}} \exp\left[-\dfrac{(y - 2P_{\mathrm{FSO}}\gamma h)^2}{2\sigma_n^2}\right], & x = 1. \end{cases} \tag{10.100}$$

In the case of unknown channel at the receiver:

$$\langle C\rangle = \sum_{x=0}^{1} P_X(x) \int_{-\infty}^{\infty} f(y|x) \log_2\left[\frac{f(y|x)}{\sum_{m=0,1} f(y|x=m)P_X(m)}\right] dy \tag{10.101}$$

where and $f(y|x)$ can be expressed as:

$$f(y|x) = \begin{cases} \dfrac{1}{\sqrt{2\pi\sigma_n^2}} \exp\left[-\dfrac{y^2}{2\sigma_n^2}\right], & x = 0 \\ \dfrac{1}{\sqrt{2\pi\sigma_n^2}} \displaystyle\int_0^{\infty} \exp\left[-\dfrac{(y - 2P_{\mathrm{FSO}}\gamma h)^2}{2\sigma_n^2}\right] f_h(h)dh, & x = 1. \end{cases} \tag{10.102}$$

10.5 Outage analysis

Recall the proposed system model described in Section 10.3.1, with the block diagram of the SISO horizontal FSO communication link as presented in Figure 10.1. The FSO link design analysis is carried out under the influence of different operating conditions, such as the weather effects as signified by the visibility V and turbulence strength C_n^2, the PE loss with jitter variance σ_{pe}^2, and the propagation distance L. In addition, system design considerations, which include the transmitter beam width w_0, receiver aperture diameter D, laser wavelength λ, data rate R_0, transmitted optical power P_{FSO}, and knowledge of CSI are investigated and compared with respect to the performance metrics. The relevant parameters considered in the performance analysis are provided in Table 10.1; and assumed for the numerical results presented here, unless otherwise specified. These nominal

Table 10.1 Parameters of the partially coherent FSO communication system

Parameter	**Symbol**	**Typical value**
Laser wavelength	λ	1,550 nm
Average transmitted optical power	P_{FSO}	10 mW
Photodetector responsivity	γ	0.5 A/W
Noise variance	σ_n^2	10^{-14} A^2
Receiver diameter	D	200 mm
Spatial coherence length	l_c	1.38 mm
Nominal beam width	w_0	50 mm
PE-induced jitter standard deviation	σ_{pe}	30 cm

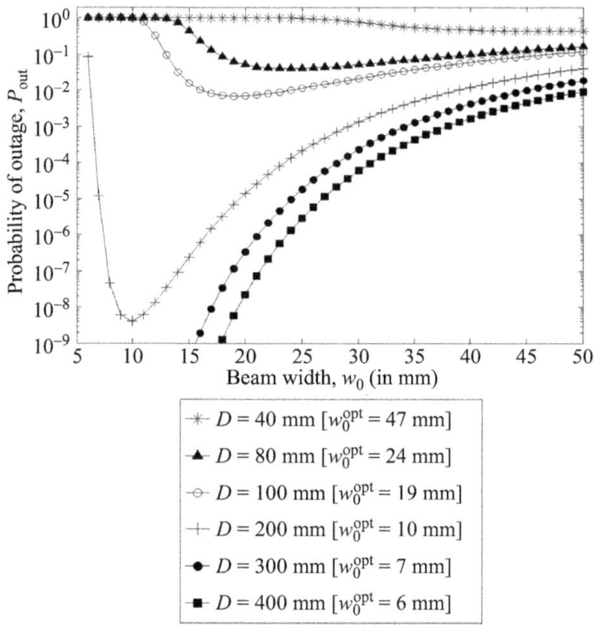

Figure 10.9 Probability of outage in terms of the transmitter beam width under the light fog condition, for $L = 1.0$ km *and* $R_0 = 0.5$ bits/channel use. *For the different receiver aperture sizes D under examined, the optimum beam width* w_0^{opt} *have been identified and shown here*

settings are selected by carefully reviewing the relevant parameter values, which have been widely considered and scrutinized [24,36–39,44,85,86].

10.5.1 Outage probability under light fog condition

Figure 10.9 illustrates the probability of outage P_{out} in terms of the transmitter beam width w_0 for an achievable rate $R_0 = 0.5$ bits/channel use, under the light

Table 10.2 Weather-dependent parameters considered in the outage analysis

Weather conditions	Visibility, V (km)	Atmospheric turbulence strength, C_n^2 ($m^{-2/3}$)	Rytov variance, σ_R^2 (at 1.0 km)
Light fog	0.642	5.0×10^{-15}	0.1
Clear weather	10.0	5.0×10^{-14}	1.0

fog condition at $L = 1.0$ km; and for a variety of receiver aperture diameter D. The simulation settings and weather-dependent parameters considered in the outage analysis are defined in Tables 10.1 and 10.2, respectively. Recalling the combined channel fading model in Section 10.3.3.4, P_{out} is obtained by substituting (10.73) into (10.98) for a range of w_0 values varying between 4 mm and 100 mm at a step size of 1 mm, whereby the weak turbulence case (i.e., $\sigma_R^2 = 0.1$) is considered here. From Figure 10.9, it is noted that lower P_{out} can be achieved with larger values of D as a result of the aperture-averaging effect, which is evident from the reduction in the scintillation index. For instance, the system performance improves by ~8 orders of magnitude for $D = 200$ mm at $w_0 = 10$ mm, as compared to the remaining cases of D with smaller collecting areas, which experience complete system annihilation. In addition, the numerical results show that the best performance can be obtained through the selection of an optimum beam width w_0^{opt} for a considered D value; and suggest that the best combination of D and w_0^{opt} enables optimal link design, as w_0^{opt} decreases with larger D. At $L > 1.5$ km, the FSO link experiences complete system outage, which is non-recoverable albeit optimizing the system design parameters. This is mainly because the absorption and scattering effects due to the presence of atmospheric particles attenuates and destroys the optical beams propagating at longer link distances.

Next, Figure 10.10 depicts the trade-off between the probability of outage P_{out} and the maximum achievable transmission rate R_0, under the light fog condition at $L = 1.0$ km and for $\lambda = \{850, 1,064, 1,550\}$nm. Based upon the assumed values of $D = 200$ mm and $w_0 = 10$ mm, P_{out} is determined through the substitution of (10.73) into (10.98) for all the considered R_0 values ranging from 0 to 1. It is observed that FSO links operating at $\lambda = 1{,}550$ nm can achieve a significantly larger R_0 with reduced aperture-averaged scintillation index $\sigma_I^2(D)$, as compared to smaller laser wavelengths. For instance, at $P_{out} = 10^{-6}$, an achievable rate in excess of 0.7 bits/channel use is attainable at $\lambda = 1{,}550$ nm, while much smaller R_0 of 0.15 bits/channel use and 0.45 bits/channel use are achieved at $\lambda = 850$ nm and $\lambda = 1,064$ nm, respectively. This implies that the adopted wavelength is an important link design criterion for maximizing the transmission rate with low outage probability.

10.5.2 Outage probability under clear weather condition

Figure 10.11 presents the probability of outage P_{out} against the transmitter beam width w_0 for $R_0 = 0.5$ bits/channel use, under the clear weather condition at $L = 3.5$ km (see Table 10.2), where different values of D are considered. In the case of a stronger turbulence with $\sigma_R^2 = 1.0$, P_{out} is obtained by substituting (10.74) into (10.98) for different w_0 values ranging from 2 mm to 200 mm at a fixed

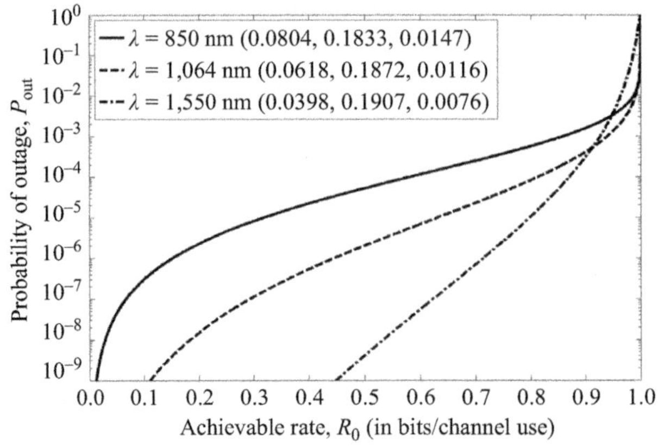

Figure 10.10 Trade-off between the outage probability and maximum achievable rate under the light fog condition, for $L = 1.0$ km, $D = 200$ mm, $w_0 = 10$ mm, *and* $\lambda = \{850, 1,064, 1,550\}$mm. *Numerical values in the figure legend refer to* $\left(\sigma_I^2(0), A_g, \sigma_I^2(D)\right)$

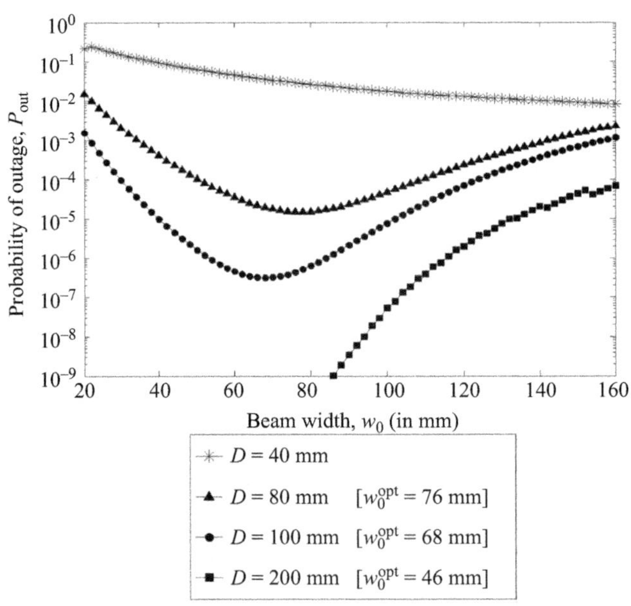

Figure 10.11 Probability of outage as a function of the transmitter beam width under the clear weather scenario, for $L = 3.5$ km and $R_0 = 0.5$ bits/channel use. *The optimum beam width is identified and shown for various receiver aperture sizes, except for* $D = 40$ mm *which does not have an optimal value*

interval of 2 mm. Under high-visibility atmospheric conditions, FSO links can achieve greater propagation distances. This is mainly because the optical beams are less susceptible to attenuation caused by absorption and scattering, as indicated by a higher h_l (= 0.7006), albeit requiring a larger w_0 to mitigate the impact of scintillation due to the strong turbulence and misalignment-induced fading. It is evident from the numerical results that a larger w_0^{opt} of 46 mm is required for $D = 200$ mm, as compared to the light fog condition (see Figure 10.9), in order to produce a more coherent laser beam with greater resilience to the scintillation effect. This in turn results in a larger beam spot size at the receiver, which mitigates the jitter-induced PE loss while enabling effective collection of the optical signal by the enlarged receiver aperture. On the other hand, the optimal value is generally in excess of 60 mm for smaller aperture sizes under investigation, except for $D = 40$ mm, in which significant reduction in P_{out} is unattainable albeit manifold increase in w_0.

The trade-off between the probability of outage P_{out} and the maximum achievable transmission rate R_0 for $\lambda = \{850, 1,064, 1,550\}$nm, is also examined for the high-visibility strong-turbulence scenario as shown in Figure 10.12. Taking into account the assumed parameters of $L = 3.5$ km, $D = 100$ mm, and $w_0 = 50$ mm, P_{out} is found through the substitution of (10.74) into (10.98) for the different R_0 values varying between 0 and 1. The advantage of adopting a particular laser wavelength on the FSO link performance is less distinctive in this case. This observation can be explained by the fact that the optical channel fading characteristics have altered the propagation properties of the laser beam, thereby concluding that the optimization of capacity metrics is best achieved through proper selection of D and w_0 for a known λ and/or increasing the transmit power P_{FSO}.

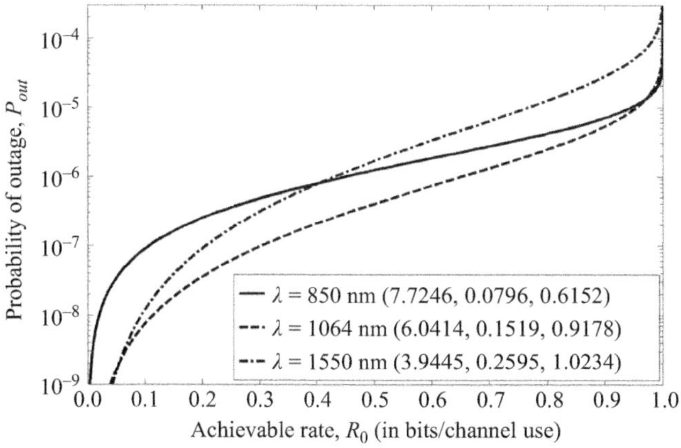

Figure 10.12 Trade-off between the outage probability and maximum achievable rate under the clear weather scenario, for $L = 3.5$ km, $D = 100$ mm, $w_0 = 50$ mm, *and* $\lambda = \{850, 1,064, 1,550\}$mm. *Numerical values in the figure legend refers to* $\left(\sigma_I^2(0), A_g, \sigma_I^2(D)\right)$

10.6 Analysis of the aperture averaging effect

The effect of aperture-averaging on horizontal FSO communication links with spatially partially coherent laser beams is studied here, in order to examine the two distinctive advantages of using enlarged receiver apertures, while taking into account numerous link design considerations. In principle, aperture averaging reduces the turbulence-induced scintillation level through the shifting of the relative frequency content of the irradiance power spectrum towards lower frequencies, which averages out the fastest fluctuations; and potentially mitigates the PE loss, as can be observed from the reduction in the normalized jitter $2\sigma_{pe}/D$.

10.6.1 Error performance due to atmospheric effects

Figure 10.13 illustrates the average BER $\langle BER \rangle$ in terms of the average transmitted optical power P_{FSO}, under different light fog scenarios at $L = 1.0$ km, and for a range of D. The simulation settings and weather-dependent parameters considered in the aperture averaging studies are defined in Tables 10.1 and 10.3, respectively. In the occurrence of low-visibility conditions with typically weak turbulences (i.e., $\sigma_R^2 \ll 1.0$), $\langle BER \rangle$ is calculated for range of P_{FSO} values by employing (10.73) and (10.93) in (10.94). It is noted that a lower $\langle BER \rangle$ can be achieved with much

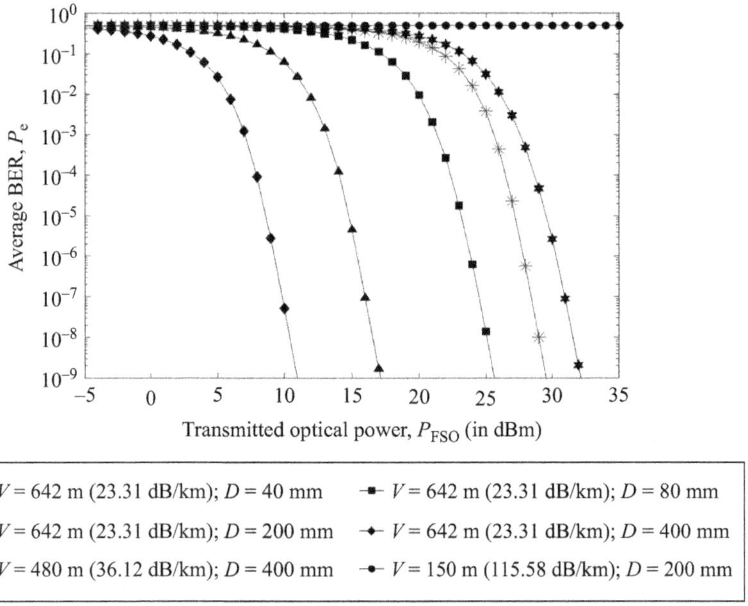

Figure 10.13 Average BER in terms of the transmitted optical power under different light fog conditions at $L = 1.0$ km. *In the figure legend, the numerical values (in parenthesis) refer to the attenuation coefficient* σ

Table 10.3 Weather-dependent parameters considered in the aperture-averaging studies

Weather conditions	**Visibility, V (km)**	**Atmospheric turbulence strength, C_n^2 ($m^{-2/3}$)**	**Link distance, L (km)**	**Rytov variance, σ_R^2**
Light fog	0.642	5.0×10^{-15}	1.0	0.10
Moderate fog	0.480	2.0×10^{-15}	1.0	0.04
Dense fog	0.150	1.0×10^{-15}	1.0	0.02
Clear weather	10.0	5.0×10^{-14}	1.0	1.00
			4.5	15.69
			7.5	40.02
		3.1230×10^{-16}		0.25
		1.1244×10^{-14}	7.5	9.00
		1.8739×10^{-14}		15.00

smaller transmit power requirement by utilizing receiver apertures of larger D; in which a substantial P_{FSO} reduction of > 20 dB can be made possible with $D = 400$ mm, for $\langle BER \rangle = 10^{-9}$ and $V = 0.642$ km, as compared to the case of $D = 40$ mm. In addition, the system under study would require higher P_{FSO} under the moderate fog condition ($V = 0.480$ km), to compensate for the increase in atmospheric attenuation ($\sigma = 36.12$ dB/km); in which a greater P_{FSO} requirement of ~13 order-of-magnitude is observed for $\langle BER \rangle = 10^{-9}$, albeit the introduction of an enlarged receiver aperture with $D = 200$ mm. Under the most extreme low-visibility case ($V = 0.150$ km, $\sigma = 115.58$ dB/km), the FSO link experiences complete system annihilation, which is non-recoverable albeit deliberately increasing the values of D and P_{FSO}.

Next, Figure 10.14 depicts the variation of the average BER $\langle BER \rangle$ against the average transmitted optical power P_{FSO}, for the clear weather conditions with $V = 10.0$ km. In particular, the fading reduction resulting from aperture averaging is evaluated for a range of considered D values, with $C_n^2 = 5.0 \times 10^{-14} m^{-2/3}$ $(\sigma_R^2 = 1.00)$ at $L = 1.0$ km; and the turbulence-induced scintillation effect is also observed for two other cases of L at 4.5 km $(\sigma_R^2 = 15.69)$ and 7.5 km $(\sigma_R^2 = 40.02)$. Refer Table 10.3 for complete details of the relevant weather-dependent parameters. In the event of stronger turbulences with $\sigma_R^2 \geq 1.0$ under clear weather conditions, $\langle BER \rangle$ is determined for range of P_{FSO} values by taking into account (10.74) and (10.93) in (10.94). In general, the FSO link has a lower P_{FSO} requirement as compared to the low-visibility cases (Figure 10.13) for the same system configuration, which is mainly due to the significantly lower atmospheric attenuation of $\sigma = 0.4508$ dB/km. The introduction of a receiver aperture with larger D enhances the link performance, in which P_{FSO} reduction in excess of 25 dB is observed for $\langle BER \rangle = 10^{-9}$ with $D = 400$ mm, as compared to the case of $D = 40$ mm. At greater link distances, the increase in scintillation level (indicated by larger σ_R^2) results in the skewing of the mean BER curve, exhibiting much smaller step size in $\langle BER \rangle$ reduction with respect to the same increment in the P_{FSO}. These observations are particularly prominent in the cases of smaller D, which can

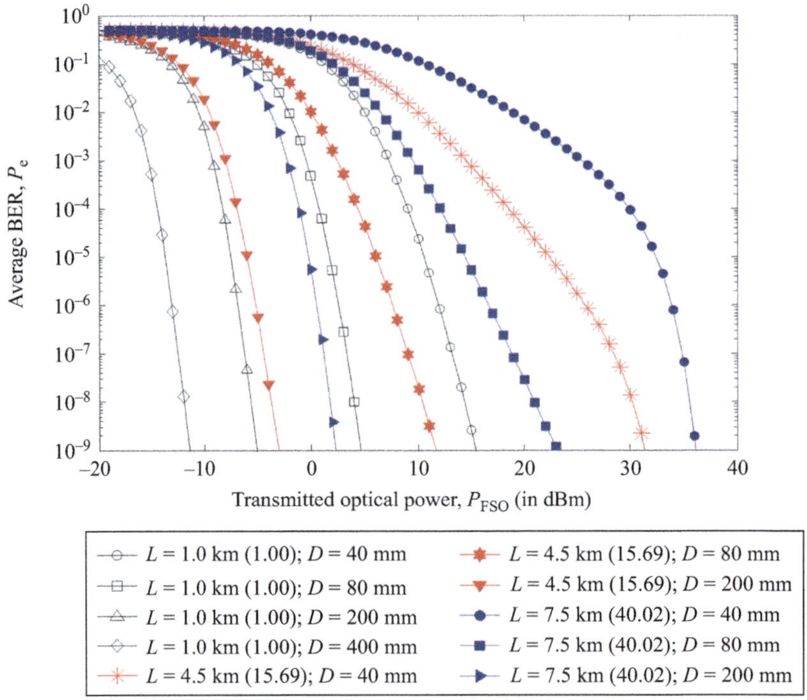

Figure 10.14 Average BER in terms of the transmitted optical power in the strong turbulence regime at $L = \{1.0, 4.5, 7.5\}$ km. *In the figure legend, the numerical values (in parenthesis) refer to the Rytov variance* σ_R^2

be explained by the alteration in the irradiance profile from a normally distributed pattern towards a distribution with longer tails in the infinite direction. The skewness of the density distribution denotes the extent of the optical intensity fluctuations as the channel inhomogeneity increases.

Figure 10.15(a) presents the average BER $\langle BER \rangle$ as a function of the average electrical SNR $\langle SNR \rangle$, under the light fog condition with $V = 0.642$ km and $C_n^2 = 5.0 \times 10^{-15} \text{m}^{-2/3}$ ($\sigma_R^2 = 0.1$) at $L = 1.0$ km (see Table 10.3); taking into account the effect of D and comparing with respect to the AWGN (i.e., no turbulence) case. The calculation of $\langle BER \rangle$ presented in Figure 10.15(a) and 10.15(b) involves the same approach used for Figures 10.13 and 10.14, respectively, whereas the $\langle SNR \rangle$ values are found by integrating (10.2) over all the possible channel fading states h. The numerical results show that a twofold reduction in the mean BER is observed for $\langle SNR \rangle = 15$ dB with $D = 400$ mm, as compared to the case of $D = 40$ mm; thus bringing the resulting $\langle BER \rangle$ curve closer to the ideal non-turbulent case, albeit deviating from the reference with a 2-dB gap in the mean SNR for achieving a $\langle BER \rangle$ of 10^{-9}. Performance enhancement due to aperture averaging is further prevalent in the strong turbulence regime ($V = 10.0$ km, $\sigma_R^2 = 1.0$ at $L = 1.0$ km) as depicted in Figure 10.15(b), whereby the resulting

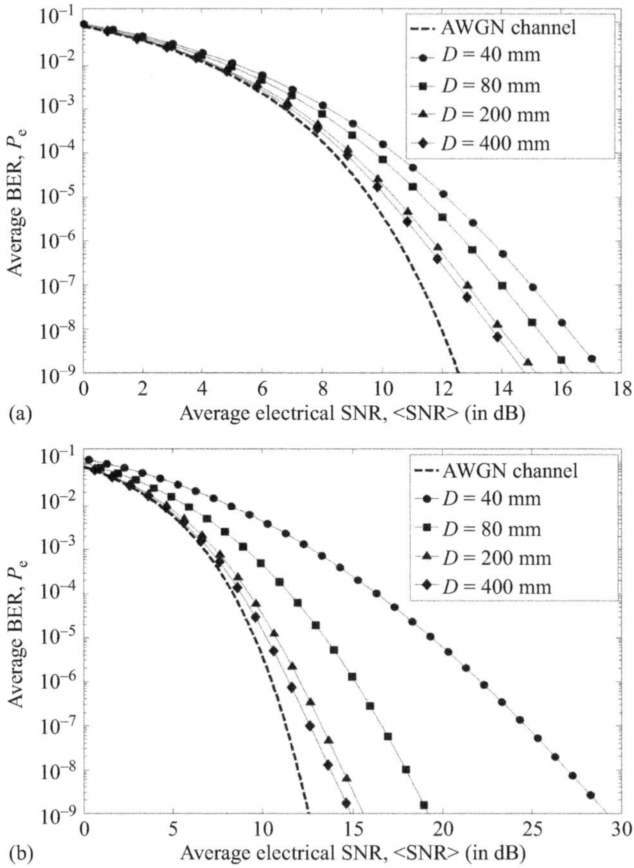

Figure 10.15 Average BER as a function of the average electrical SNR under the (a) light fog and (b) clear weather conditions at $L = 1.0$ km, *taking into account the effect of D and comparing with the AWGN (non-turbulent) case*

scintillation reduction potentially decreases the average BER by more than four order-of-magnitude for $\langle SNR \rangle > 15$ dB with $D = 400$ mm, as compared to the case of $D = 40$ mm. In effect, these observations can be justified by the shifting of the relative frequency content of the irradiance power spectrum toward lower frequencies due to the aperture averaging effect; in essence, averaging out the fastest fluctuations, thereby resulting in scintillation reduction and lower BER attainment [11,87].

10.6.2 Average channel capacities due to channel state information

Figure 10.16(a) presents the average channel capacity $\langle C \rangle$ as a function of the average electrical SNR $\langle SNR \rangle$ for $D = \{40, 80, 200, 400\}$mm, assuming the CSI is known to the receiver. The light fog condition is considered here, with $V = 0.642$ km,

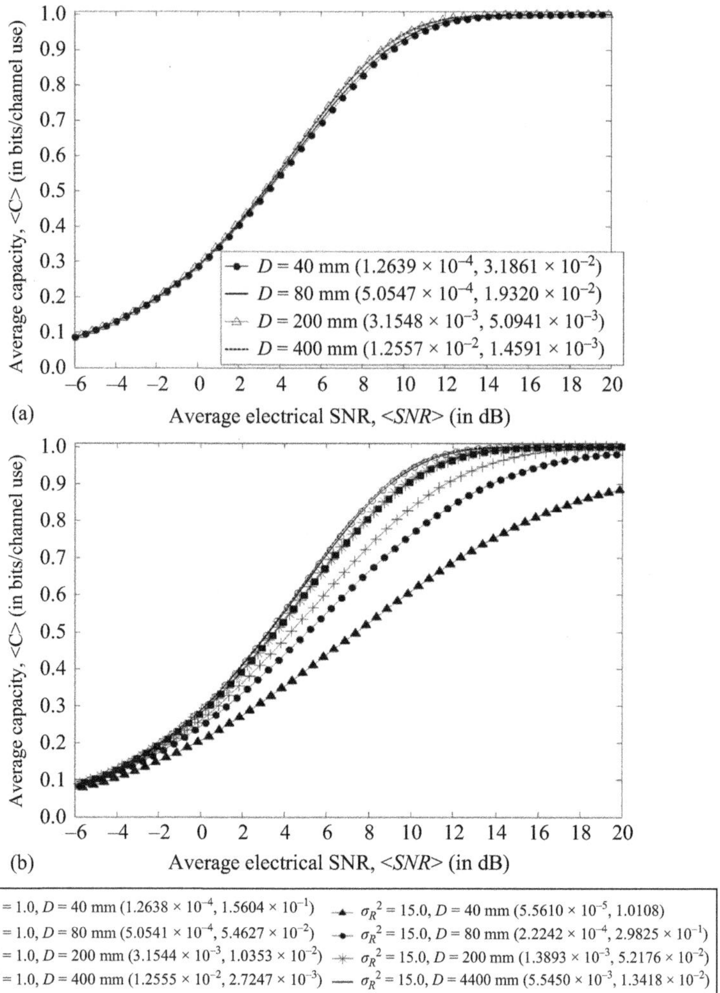

Figure 10.16 Average channel capacity as a function of the average electrical SNR for different receiver aperture dimensions, under the (a) light fog ($V = 0.642$ km, $\sigma_R^2 = 0.1$ at $L = 1.0$ km); and (b) clear weather $V = 10.0$ km, $\sigma_R^2 = 1.0$ at $L = 1.0$ km, $\sigma_R^2 = 15.0$ at $L = 7.5$ km) conditions. The CSI is assumed known to the receiver

$C_n^2 = 5.0 \times 10^{-15}\text{m}^{-2/3}$ ($\sigma_R^2 = 0.1$) at $L = 1.0$ km (see Table 10.3). In the case of a known channel at the receiver under weak turbulence condition, $\langle C \rangle$ is determined by substituting (10.73) and (10.100) into (10.99) for different values of $\langle SNR \rangle$. Under the visibility-impaired weak turbulence condition, a faster increase in the average capacity with respect to the $\langle SNR \rangle$ is observed, in which the maximum channel capacity is approached at higher mean SNR values in excess of 14 dB. In addition, near-identical behaviour is noted for all the cases of D under investigation, albeit observing an increase in the fraction of collected power A_0 and a reduction in the aperture-averaged scintillation level $\sigma_I^2(D)$ (see figure legend). These observations reveal that $\langle C \rangle$ is almost independent of D, since the effects of scintillations and PEs are less dominant, as compared to the beam extinction resulting from the scattering and absorption of atmospheric particles, which can be compensated by increasing the transmit optical power of the laser beam.

The corresponding results for the clear weather scenarios are illustrated in Figure 10.16(b), with $V = 10.0$ km, $C_n^2 = 5.0 \times 10^{-14}\text{m}^{-2/3}$ at $L = 1.0$ km ($\sigma_R^2 = 1.0$), and $C_n^2 = 1.8739 \times 10^{-14}\text{m}^{-2/3}$ at $L = 7.5$ km ($\sigma_R^2 = 15.0$) (see Table 10.3). In the event of a known channel at the receiver for various moderate-to-strong turbulence cases, $\langle C \rangle$ is determined by substituting (10.74) and (10.100) into (10.99) for different values of $\langle SNR \rangle$. In comparison to the light fog condition (Figure 10.16(a)), it is evident that the average channel capacity is more susceptible to the adverse effects of atmospheric turbulence, which imposes a higher scintillation level and PE loss, as signified by a larger $\sigma_I^2(D)$ and ξ values, respectively. At $\sigma_R^2 = 1.0$, the average capacity increases at a relatively slower rate for $D = 40$ mm, as compared to the remaining cases; and then approaches the maximum channel capacity for $\langle SNR \rangle > 16$ dB. It is evident that maximum capacity enhancement can be attained with larger receiver apertures of $D = \{200, 400\}$mm, in which near-optimal channel capacities of more than 0.99 bits/channel use are achieved for $\langle SNR \rangle \geq 14$ dB. Furthermore, a larger penalty on the FSO channel capacity is observed at a longer link distance of $L = 7.5$ km with $\sigma_R^2 = 15.0$, which is particularly pronounced for larger $\langle SNR \rangle$ values; where it is shown that there is significant variation in $\langle C \rangle$ for all the considered D values, with a maximum deviation of 0.31 bits/channel use at $\langle SNR \rangle = 14$ dB. It is noted that the average capacity increases at a much slower rate for the cases of $D = \{40, 80\}$mm, and approaches their respective values well below the maximum capacity. However, improvement in the channel capacity of more than 25% can be made possible for $\langle SNR \rangle$ in excess of 10 dB with $D = 400$ mm.

Figure 10.17 depicts the average capacity $\langle C \rangle$ against the average electrical SNR $\langle SNR \rangle$ for the weak and moderate-to-strong turbulence regimes, with $C_n^2 = 3.1230 \times 10^{-16}\text{m}^{-2/3}$ ($\sigma_R^2 = 0.25$) and $C_n^2 = 1.1244 \times 10^{-14}\text{m}^{-2/3}$ ($\sigma_R^2 = 9.00$), respectively (see Table 10.3). The effect of PEs, as signified by the normalized jitter $2\sigma_{pe}/D$, is examined for a variety of D values, with $L = 7.5$ km and $V = 10$ km; and for the cases of known and unknown channels at the receiver. In the weak turbulence regime (Figure 10.17(a)), a faster increase in the average capacity with respect to the SNR is observed, showing less variation between the known and unknown channel cases for all the considered D values; in which the maximum channel capacity is approached at higher SNR values in excess of 16 dB. The variation between known and unknown channels is larger for the moderate-to-strong turbulence scenario (Figure 10.17(b)), and is prevalent under the influence of PEs with $2\sigma_{pe}/D = 15.0$; where it is noted that an average capacity gap of ~0.15 bits/channel use at $\langle SNR \rangle = 14$ dB. This implies that greater penalty will be imposed upon the channel capacity under the combined effects of turbulence and PEs, and without knowledge of the channel state conditions. Nevertheless, the introduction of

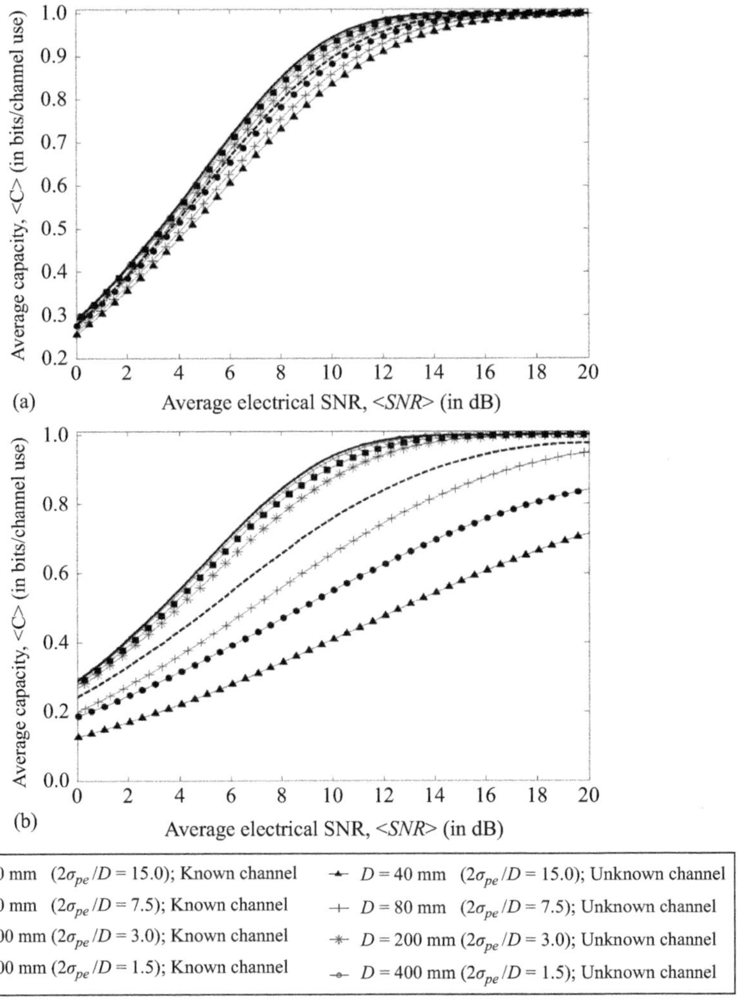

Figure 10.17 Average channel capacity in terms of the average electrical SNR for the (a) weak ($\sigma_R^2 = 0.25$) and (b) moderate-to-strong ($\sigma_R^2 = 9.00$) turbulence regimes, where $V = 10.0$ km and $L = 7.5$ km. The effect of PEs, as signified by the normalized jitter, is examined for the cases of known and unknown channels at the receiver

larger receiver aperture promotes capacity enhancement and potentially minimizes the observed gap; in which a notable improvement in $\langle C \rangle$ of ~0.46 bits/channel use can be achieved at $\langle SNR \rangle = 14$ dB with $D = 400$ mm, albeit not knowing the channel state at the receiver. In addition, $\langle C \rangle$ exhibits near-identical behaviour with negligibly small variation for both channel cases with $D = 400$ mm; thus approaching an asymptotic limit of 1.0 bits/channel use for $\langle SNR \rangle \geq 14$ dB.

10.7 Beam width optimization

10.7.1 Dependence on link design criteria

In Figure 10.18, the relationship between the transmitter beam width w_0 on the average channel capacity $\langle C \rangle$ (see (10.101) and (10.102)) is examined, in the presence of atmospheric turbulence and PEs, given that the channel state is unknown to the receiver. The weak ($\sigma_R^2 = 0.25$) and moderate-to-strong ($\sigma_R^2 = 9.00$) turbulence conditions are considered here, for $L = 7.5$ km, $V = 10.0$ km and $\langle SNR \rangle = 14$ dB. The simulation settings and weather-dependent parameters considered in the present beam width optimization studies are defined in Tables 10.1 and 10.4, respectively.

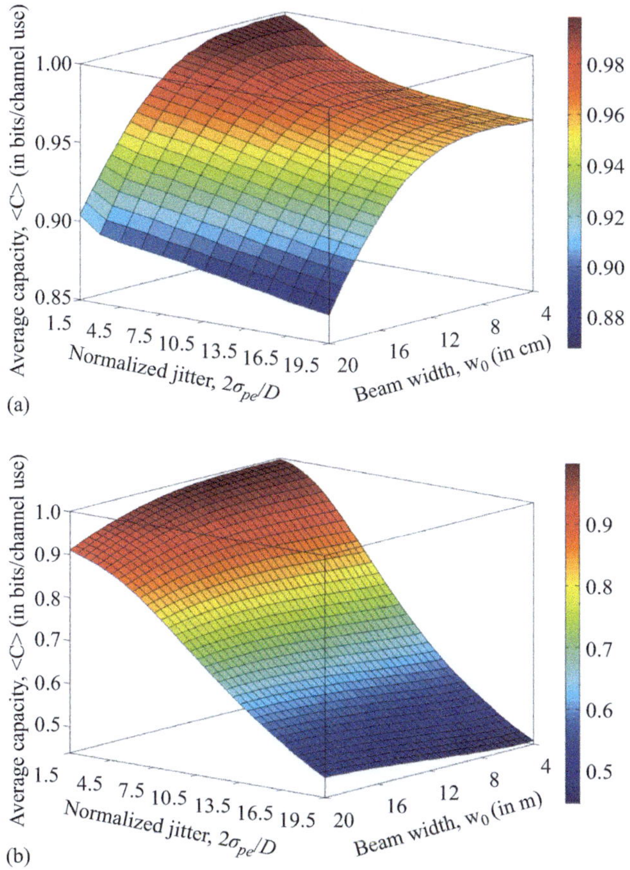

Figure 10.18 Relationship between the average channel capacity, transmitter beam width and PE loss, for $V = 10.0$ km*,* $L = 7.5$ km*,* $\langle SNR \rangle = 14$ dB*, and an unknown channel at the receiver. The (a) weak (*$\sigma_R^2 = 0.25$*) and (b) moderate-to-strong (*$\sigma_R^2 = 9.00$*) turbulence cases are considered*

Table 10.4 Weather-dependent parameters considered in the beam width optimization studies

Visibility, V (km)	**10.0**		
Atmospheric loss, σ	**0.4665**		
	Atmospheric turbulence strength, C_n^2 ($m^{-2/3}$)		
Link distance, L (km) / **Rytov variance, σ_R^2**	**4.5**	**7.5**	**18.0**
0.10	3.1869×10^{-16}	1.2493×10^{-16}	2.5097×10^{-17}
0.15	4.7804×10^{-16}	1.8739×10^{-16}	3.7645×10^{-17}
0.20	6.3739×10^{-16}	2.4986×10^{-16}	5.0191×10^{-17}
0.25	7.9659×10^{-16}	3.1232×10^{-16}	6.2739×10^{-17}
0.30	9.5609×10^{-16}	3.7479×10^{-16}	7.5287×10^{-17}
0.40	1.2748×10^{-15}	4.9971×10^{-16}	1.0038×10^{-16}
0.50	1.5935×10^{-15}	6.2464×10^{-16}	1.2548×10^{-16}
0.60	1.9122×10^{-15}	7.4957×10^{-16}	1.5058×10^{-16}
0.70	2.2309×10^{-15}	8.7450×10^{-16}	1.7567×10^{-16}
0.80	2.5496×10^{-15}	9.9942×10^{-16}	2.0077×10^{-16}
0.90	2.8683×10^{-15}	1.1244×10^{-15}	2.2586×10^{-16}
1.00	3.1869×10^{-15}	1.2493×10^{-15}	2.5097×10^{-16}
1.50	4.7804×10^{-15}	1.8739×10^{-15}	3.7645×10^{-16}
2.00	6.3739×10^{-15}	2.4986×10^{-15}	5.0191×10^{-16}
3.00	9.5609×10^{-15}	3.7478×10^{-15}	7.5287×10^{-16}
4.00	1.2748×10^{-14}	4.9971×10^{-15}	1.0038×10^{-15}
5.00	1.5935×10^{-14}	6.2464×10^{-15}	1.2548×10^{-15}
6.00	1.9122×10^{-14}	7.4957×10^{-15}	1.5058×10^{-15}
7.00	2.2309×10^{-14}	8.7450×10^{-15}	1.7567×10^{-15}
8.00	2.5496×10^{-14}	9.9942×10^{-15}	2.0077×10^{-15}
9.00	2.8683×10^{-14}	1.1244×10^{-14}	2.2586×10^{-15}
10.00	3.1870×10^{-14}	1.2493×10^{-14}	2.5097×10^{-15}
12.00	3.8244×10^{-14}	1.4991×10^{-14}	3.0115×10^{-15}
14.00	4.4618×10^{-14}	1.7490×10^{-14}	3.5134×10^{-15}
16.00	5.0992×10^{-14}	1.9989×10^{-14}	4.0153×10^{-15}
18.00	5.7366×10^{-14}	2.2487×10^{-14}	4.5172×10^{-15}
20.00	6.3740×10^{-14}	2.4986×10^{-14}	5.0191×10^{-15}
22.00	7.0114×10^{-14}	2.7484×10^{-14}	5.5211×10^{-15}
24.00	7.6488×10^{-14}	2.9983×10^{-14}	6.0230×10^{-15}
26.00	8.2862×10^{-14}	3.2481×10^{-14}	6.5249×10^{-15}
28.00	8.9236×10^{-14}	3.4980×10^{-14}	7.0268×10^{-15}
30.00	9.5610×10^{-14}	3.7478×10^{-14}	7.5287×10^{-15}
32.00	1.0198×10^{-13}	3.9977×10^{-14}	8.0307×10^{-15}
34.00	1.0836×10^{-13}	4.2476×10^{-14}	8.5326×10^{-15}
36.00	1.1473×10^{-13}	4.4974×10^{-14}	9.0345×10^{-15}

It is evident that the FSO channel capacity is susceptible to the adverse effects of turbulence-induced scintillations and PEs, in which this phenomenon is particularly prevalent in the moderate-to-strong turbulence case (Figure 10.18(b)); whereby the FSO link suffers performance degradation with a significant reduction of 0.55 bits/channel use, under the severe PE loss of $2\sigma_{pe}/D = 19.5$ for $w_0 = 4$ cm, as compared to the weak turbulence condition (Figure 10.18(a)) with a negligibly small variation of < 0.05 bits/channel use. Another interesting observation is made here, showing that the FSO channel capacity is in fact dependent on the beam width; and changing the w_0 value affects the achievable capacity under different atmospheric channel conditions. Correspondingly, this reveals the importance of finding an optimal w_0 value and adjusting the parameter to maximize $\langle C \rangle$; and hence, w_0 is a vital parameter in the FSO link design consideration.

In Figure 10.19, the present analysis is extended to investigate the effects of beam width on the average channel capacity under the influence of a variety of turbulence strength (as signified by σ_R^2); taking into account the clear weather condition at $L = \{4.5, 7.5, 18.0\}$ km, with $V = 10$ km, $2\sigma_{pe}/D = 12.0$ and $\langle SNR \rangle = 14$ dB. At $L = 4.5$ km (Figure 10.19(a)), near-optimal $\langle C \rangle$ in excess of 0.85 bits/channel use can be approached in the weak turbulence regime ($\sigma_R^2 \leq 1.0$), with optical beams of smaller size ($w_0 = \{3, 8\}$cm); whereas performance degradation is observed with increasing beam width. For instance, the average capacity suffers a substantial reduction of 0.48 bits/channel use when w_0 is intentionally increased from 3 cm to 24 cm at $\sigma_R^2 = 0.1$; and exhibits a rather persistent behaviour at $\langle C \rangle \approx 0.5$ bits/channel use, for σ_R^2 values ranging from 0.1 to 10.0, and increases thereafter albeit giving the worst performance for all the considered w_0 values. At a longer link distance of $L = 7.5$ km (Figure 10.19(b)), it is observed that smaller beam widths ($w_0 = \{5, 10\}$cm) are desirable at $\sigma_R^2 \leq 1.0$ with $\langle C \rangle > 0.83$ bits/channel use, but gives the lowest average capacity attainment in the stronger turbulence regime of $\sigma_R^2 \geq 10.0$. In particular, it is noted that the optimal $\langle C \rangle$ can be achieved with $w_0 = 20$ cm in the intermediate turbulence regime of $\sigma_R^2 = 2.0$ to 7.0. The numerical results presented for the case of $L = 18.0$ km (Figure 10.19(c)) also exhibits similar behaviour, showing that $\langle C \rangle > 0.90$ bits/channel use can be achieved with $w_0 = 20$ to 40 cm for $\sigma_R^2 \leq 5.0$, whereas performance degradation is observed for the remaining cases of $w_0 = \{60, 90\}$ cm, and vice versa. These observations can be explained by the fact that the PCB suffers substantial alteration in its beam profile/characteristics when propagating through free-space, which is mainly contributed by the combined effects of the turbulence-induced scintillation and beam wander; thereby resulting in the variation of the channel capacity for different σ_R^2. In addition, link distance remains as another important link design criterion, since greater L corresponds to stronger turbulence; which in turn requires a larger optimal w_0 value for the cases of $L = 7.5$ km (Figure 10.19(b)) and $L = 18.0$ km (Figure 10.19(c)), as compared to $L = 4.5$ km (Figure 10.19(a)), under a significant PE loss of $2\sigma_{pe}/D = 12.0$, as evident from the observations made for $\sigma_R^2 \geq 10.0$. Therefore, beam width optimization is a feasible approach in promoting capacity enhancement for long-distance terrestrial FSO links, since larger transmitter beam radius improves the average channel capacity only in the stronger turbulence regime, while imposing severe performance degradation under weak-to-moderate turbulence conditions, and vice versa.

10.7.2 Optimum beam width

The characteristics of the optimum beam width w_0^{opt} for finding the best achievable average capacity $\langle C \rangle$ is examined, under the combined effects of different

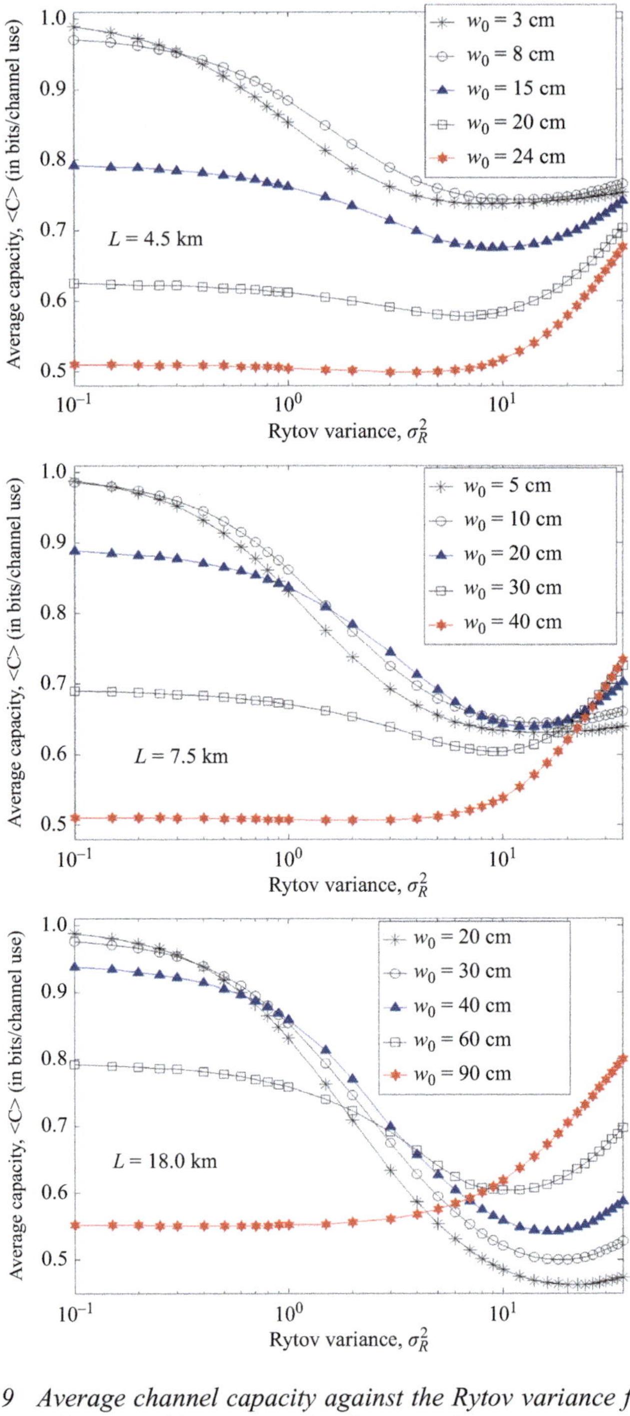

Figure 10.19 Average channel capacity against the Rytov variance for a variety of beam width settings. The clear weather scenario is considered at (a) $L = 4.5$ km, *(b)* $L = 7.5$ km, *and (c)* $L = 18.0$ km, *where* $V = 10.0$ km, $2\sigma_{pe}/D = 12.0$, *and* $\langle SNR \rangle = 14$ dB

atmospheric channel conditions that are likely to occur in practice. These parameters are determined through an exhaustive search over their discrete sets, with the results as readily presented in Figure 10.20. Figure 10.20(a) shows the variation of w_0^{opt} in terms of the turbulence strength (as signified by σ_R^2), for $L = 7.5$ km, $V = 10$ km, $\langle SNR \rangle = 14$ dB and different normalized jitter values of $2\sigma_{pe}/D = \{6.0, 9.0, 12.0, 15.0\}$; and the corresponding optimized $\langle C \rangle$ is depicted in Figure 10.20(b). In general, the w_0^{opt} values are relatively smaller when the effects of turbulence and PEs are less severe; whereby it is observed that w_0^{opt} is typically less than 15 cm for $\sigma_R^2 \leq 1.5$ and/or $2\sigma_{pe}/D \leq 6.0$, which corresponds to an optimal capacity in excess of 0.88 bits/channel use. In the weak turbulence

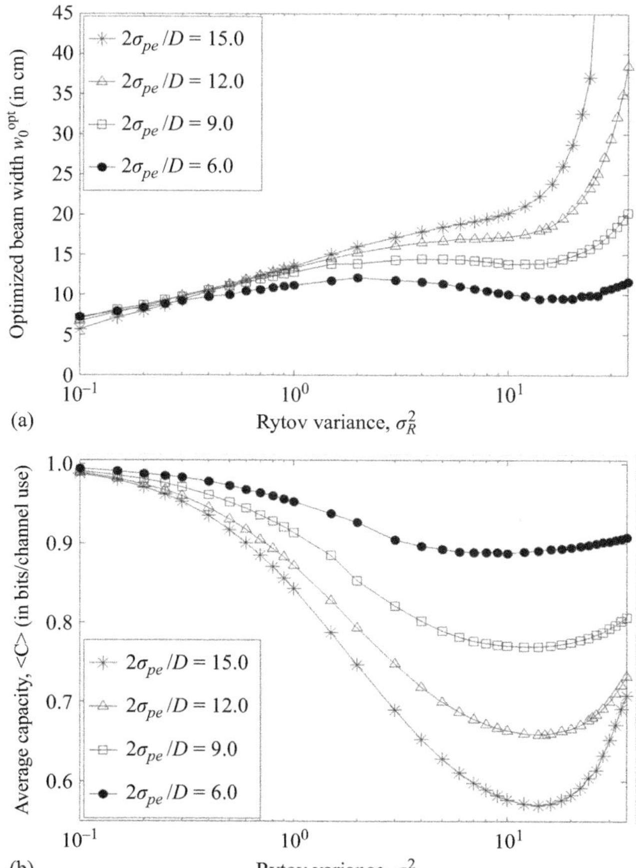

Figure 10.20 The optimal (a) beam width and (b) average channel capacity with respect to the Rytov variance under the clear weather condition, for $2\sigma_{pe}/D = \{15.0, 12.0, 9.0, 6.0\}$ *at* $\langle SNR \rangle = 14$ dB*; where* $V = 10.0$ km, $L = 7.5$ km, *and the channel state is unknown to the receiver*

regime, w_0^{opt} exhibits a linear incremental trend for all cases of $2\sigma_{pe}/D$ when $\sigma_R^2 \leq 2.0$; and shows vastly contrasting characteristics thereafter. An appealing observation is made here, showing a reduction in w_0^{opt} from 7.25 cm to 5.65 cm, as $2\sigma_{pe}/D$ varies from 6.0 to 15.0 at $\sigma_R^2 = 0.1$; thereby revealing the fact that the transmitter beam width can be made smaller albeit with increasing PE loss. Moreover, in the intermediate turbulence regime with σ_R^2 ranging from 2.0 to 10.0, w_0^{opt} exhibits small steps of decrease in its value for $2\sigma_{pe}/D = \{6.0, 9.0\}$; but shows a linear increase for the remaining cases, where the PE loss is more severe. This phenomena can be explained by the fact that the combined effects of turbulence-induced scintillations and PEs can result in an increase in the effective beam radius, given the condition that either one or both of these adversities are least dominant; which in turn reduces the sensitivity to optical intensity fluctuations, and thus a smaller w_0^{opt} value for maximizing $\langle C \rangle$. Nevertheless, w_0^{opt} must be increased accordingly for the strong turbulence case with $\sigma_R^2 > 10.0$, in which such incremental trend changes from a more linear to an exponential behaviour with larger $2\sigma_{pe}/D$; since increasing the receiver beam size of a PCB enables the reduction in the scintillations and PEs. As evident from Figure 10.20(b), the FSO channel capacity is highly susceptible to the adverse effects of atmospheric turbulence, and suffers greater performance degradation for larger PE loss albeit performing beam width optimization; whereby a notable reduction in $\langle C \rangle$ is observed for all the considered $2\sigma_{pe}/D$ values, with a maximum variation of 0.11, 0.22, 0.33 and 0.42 bits/channel use, respectively. Furthermore, a slight improvement in the FSO channel capacity is noticed for $\sigma_R^2 > 10.0$, since increasing the beam width and transmit optical power (to give $\langle SNR \rangle = 14$ dB) alters the PCB into a relatively coherent laser beam, which is more resilient to the strong turbulence conditions.

10.8 Conclusions

This chapter has examined the performance of partially coherent FSO communication links from the information theory perspective, taking into account the adverse effects of atmospheric loss, turbulence-induced scintillations and PEs. In particular, a spatially partially coherent Gaussian-beam wave and important link design criteria have been jointly considered, in which the latter consists of the receiver aperture dimension and its resulting aperture averaging effect, transmitter beam width, link range, knowledge of CSI and weather conditions. By using the combined optical slow-fading channel model to describe the optical channel characteristics, a comprehensive analysis of the error performance, average channel capacity and outage probability of the FSO system have been presented. Moreover, the lowest-order Gaussian-beam wave model has been introduced in the proposed study, to characterize the propagation properties of the optical signal through random turbulent medium in an accurate manner; taking into account the diverging and focusing of the PCB, and the scintillation and beam wander effects arising from the atmospheric turbulent eddies. Correspondingly, the proposed study have presented a holistic perspective for optimal planning and design of horizontal FSO links employing spatially partially coherent laser beams.

It has been demonstrated that a lower outage probability and improved BER can be achieved with the introduction of a larger receiver aperture due to the aperture-averaging effect, which effectively mitigates the scintillations and PE loss. From the outage analysis, it is evident that there exist an optimum beam width for a considered receiver aperture size and known laser wavelength, which must be selected according to changing weather condition for optimizing the outage capacity. While higher transmit power potentially enhances the error performance of the FSO system for moderate-to-high visibility scenarios, complete system outage occurs at link distances greater than 1.5 km under the visibility-limiting light fog condition, and is non-recoverable albeit optimizing the system parameters. In the aperture-averaging studies, the numerical results have shown that greater penalty will be imposed upon the average channel capacity under the combined effects of turbulence and PEs, particularly when the channel state is unknown at the receiver. With the introduction of an enlarged receiver aperture, a notable average capacity improvement of up to 0.46 bits/channel use can be achieved for a mean SNR of 14 dB in the moderate-to-strong turbulence regime, albeit without knowledge of the channel state conditions. Furthermore, it has been noted that the PCB properties are substantially altered when propagating through free-space, and revealed the importance of finding an optimal beam width to maximize the average capacity. Several appealing observations have been made in this study, showing that the beam width can be reduced to improve the FSO channel capacity, albeit in the presence of turbulence-induced scintillations and PEs, given the condition that either one or both of these adversities are least dominant. Nevertheless, the beam width must be increased accordingly in the strong turbulence regime, whereby such incremental trend changes from a more linear to an exponential behaviour; since increasing the beam width and transmit optical power (to give a mean SNR of ~14 dB) alters the PCB into a relatively coherent laser beam, thus becoming more resilient to the strong turbulence conditions. Therefore, beam width optimization is a feasible approach in promoting capacity enhancement for long-distance terrestrial FSO links; since the beam parameter is subject to the combined effects of turbulence and PEs, and its optimal value must be adjusted according to varying channel conditions.

References

[1] R. Baldemair, E. Dahlman, G. Fodor, *et al.*, "Evolving wireless communications: addressing the challenges and expectations of the future," *IEEE Veh. Technol. Mag.*, vol. 8, pp. 24–30, 2013.

[2] L. Hanzo, H. Haas, S. Imre, D. O'Brien, M. Rupp, and L. Gyongyosi, "Wireless myths, realities, and futures: From 3G/4G to optical and quantum wireless," *Proc. IEEE*, vol. 100, pp. 1853–1888, 2012.

[3] Ericsson. (2018). Ericsson mobility report: On the pulse of the networked society. [Online]. Available: https://www.ericsson.com/en/mobility-report (Accessed June 18, 2018).

[4] G. Fettweis and E. Zimmermann, "ICT energy consumption – trends and challenges," in *Proc. 11th International Symposium on Wireless Personal Multimedia Communications (WPMC 2008)*, 2008, pp. 1–4.
[5] A. Fehske, G. Fettweis, J. Malmodin, and G. Biczok, "The global footprint of mobile communications: The ecological and economic perspective," *IEEE Commun. Mag.*, vol. 49, pp. 55–62, 2011.
[6] A. K. Majumder, "Free-space laser communication performance in the atmospheric channel," in *Free-Space Laser Communications: Principles and Advances* A. K. Majumder and J. C. Ricklin, Eds., ed New York, NY: Springer-Verlag, 2008, pp. 57–108.
[7] H. Willebrand and B. Ghuman, *Free-space optics: Enabling optical connectivity in today's networks*. Indianapolis, IN: Sams Publishing, 2002.
[8] T. H. Carbonneau and D. R. Wisely, "Opportunities and challenges for optical wireless: The competitive advantage of free space telecommunications links in today's crowded marketplace," in *Proc. SPIE, Wireless Technologies and Systems: Millimeter-Wave and Optical*, 1997, pp. 119–128.
[9] K. Tsukamoto, A. Hashimoto, Y. Aburakawa, and M. Matsumoto, "The case for free space," *IEEE Microw. Mag.*, vol. 10, pp. 84–92, 2009.
[10] A. Mahdy and J. S. Deogun, "Wireless optical communications: A survey," in *Proc. IEEE Wireless Communications and Networking Conference (WCNC)*, 2004, pp. 2399–2404.
[11] L. C. Andrews and R. L. Philips, *Laser Beam Propagation through Random Media*. Bellingham, WA: SPIE Optical Engineering Press, 2005.
[12] K. Kazaura, K. Wakamori, M. Matsumoto, T. Higashino, K. Tsukamoto, and S. Komaki, "RoFSO: A universal platform for convergence of fiber and free-space optical communication networks," *IEEE Commun. Mag.*, vol. 48, pp. 130–137, 2010.
[13] D. Kedar and S. Arnon, "Urban optical wireless communication networks: The main challenges and possible solutions," *IEEE Commun. Mag.*, vol. 42, pp. S2–S7, 2004.
[14] Lightpointe (2018). https://www.lightpointe.com/solutions.html [Online] (Accessed June 18, 2018).
[15] D.-Y. Song, Y.-S. Hurh, J.-w. Cho, *et al.*, "4 × 10 Gb/s terrestrial optical free space transmission over 1.2 km using an EDFA preamplifier with 100 GHz channel spacing," *Opt. Express*, vol. 7, pp. 280–284, 2000.
[16] M.-C. Jeong, J.-S. Lee, S.-Y. Kim, *et al.*, "8 × 10-Gb/s terrestrial optical free-space transmission over 3.4 km using an optical repeater," *IEEE Photon. Technol. Lett.*, vol. 15, pp. 171–173, 2003.
[17] Y. Arimoto, M. Presi, V. Guarino, *et al.*, "320 Gbit/s (8 × 40 Gbit/s) double-pass terrestrial free-space optical link transparently connected to optical fibre lines," in *Proc. 34th European Conference on Optical Communication (ECOC 2008)*, 2008, pp. 1–2.
[18] E. Ciaramella, Y. Arimoto, G. Contestabile, *et al.*, "1.28 terabit/s (32×40 Gbit/s) WDM transmission system for free space optical communications," *IEEE J. Sel. Areas Commun.*, vol. 27, pp. 1639–1645, 2009.

[19] P.-L. Chen, S.-T. Chang, S.-T. Ji, *et al.*, "Demonstration of 16 channels 10 Gb/s WDM free space transmission over 2.16 km," in *Proc. 2008 Digest of the IEEE/LEOS Summer Topical Meetings*, 2008, pp. 235–236.

[20] A. Bekkali, C. Ben Naila, K. Kazaura, K. Wakamori, and M. Matsumoto, "Transmission analysis of OFDM-based wireless services over turbulent radio-on-FSO links modeled by gamma-gamma distribution," *IEEE Photonics J.*, vol. 2, pp. 510–520, 2010.

[21] Y. Arimoto, "Compact free-space optical terminal for multi-gigabit signal transmissions with a single-mode fiber," in *Proc. SPIE, Free-Space Laser Communication Technologies XXI*, 2009, pp. 719908(1)–(9).

[22] C. B. Naila, K. Wakamori, M. Matsumoto, A. Bekkali, and K. Tsukamoto, "Transmission analysis of digital TV signals over a Radio-on-FSO channel," *IEEE Commun. Mag.*, vol. 50, pp. 137–144, 2012.

[23] Z. Ghassemlooy, W. Popoola, and S. Rajbhandari, *Optical Wireless Communications: System and Channel Modelling with MATLAB*. Boca Raton, FL: Taylor & Francis Group, 2012.

[24] A. A. Farid and S. Hranilovic, "Outage capacity optimization for free-space optical links with pointing errors," *J. Lightwave Technol.*, vol. 25, pp. 1702–1710, 2007.

[25] D. K. Borah and D. G. Voelz, "Pointing error effects on free-space optical communication links in the presence of atmospheric turbulence," *J. Lightw. Technol.*, vol. 27, pp. 3965–3973, 2009.

[26] I. I. Kim, B. McArthur, and E. Korevaar, "Comparison of laser beam propagation at 785 nm and 1550 nm in fog and haze for optical wireless communications," in *Proc. SPIE, Optical Wireless Communications III*, 2001, pp. 26–37.

[27] I. I. Kim and E. Korevaar, "Availability of free space optics (FSO) and hybrid FSO/RF systems," in *Proc. SPIE, Optical Wireless Communications IV*, 2001, pp. 84–95.

[28] E. J. Lee and V. W. S. Chan, "Part 1: optical communication over the clear turbulent atmospheric channel using diversity," *IEEE J. Sel. Areas Commun.*, vol. 22, pp. 1896–1906, 2004.

[29] X. M. Zhu and J. M. Kahn, "Free-space optical communication through atmospheric turbulence channels," *IEEE Trans. Commun.*, vol. 50, pp. 1293–1300, 2002.

[30] S. Arnon, "Effects of atmospheric turbulence and building sway on optical wireless-communication systems," *Opt. Lett.*, vol. 28, pp. 129–131, 2003.

[31] S. Arnon, "Optimization of urban optical wireless communication systems," *IEEE Trans. Wireless Commun.*, vol. 2, pp. 626–629, 2003.

[32] H. E. Nistazakis, T. A. Tsiftsis, and G. S. Tombras, "Performance analysis of free-space optical communication systems over atmospheric turbulence channels," *IET Commun.*, vol. 3, pp. 1402–1409, 2009.

[33] H. E. Nistazakis, E. A. Karagianni, A. D. Tsigopoulos, M. E. Fafalios, and G. S. Tombras, "Average capacity of optical wireless communication

systems over atmospheric turbulence channels," *J. Lightw. Technol.*, vol. 27, pp. 974–979, 2009.

[34] A. Garcia-Zambrana, C. Castillo-Vazquez, and B. Castillo-Vazquez, "On the capacity of FSO links over gamma-gamma atmospheric turbulence channels using OOK signaling," *EURASIP Journal on Wireless Communications and Networking*, vol. 2010, pp. 127657(1)–(9), 2010.

[35] J. Anguita, I. Djordjevic, M. A. Neifeld, and B. Vasic, "Shannon capacities and error-correction codes for the optical atmospheric channel," *OSA J. Opt. Networking*, vol. 4, pp. 586–601, 2005.

[36] J. C. Ricklin and F. M. Davidson, "Atmospheric turbulence effects on a partially coherent Gaussian beam: implications for free-space laser communication," *J. Opt. Soc. Am. A*, vol. 19, pp. 1794–1802, 2002.

[37] J. C. Ricklin and F. M. Davidson, "Atmospheric optical communication with a Gaussian Schell beam," *J. Opt. Soc. Am. A*, vol. 20, pp. 856–866, 2003.

[38] O. Korotkova, L. C. Andrews, and R. L. Phillips, "Model for a partially coherent Gaussian beam in atmospheric turbulence with application in Lasercom," *Opt. Eng.*, vol. 43, pp. 330–341, 2004.

[39] H. G. Sandalidis, "Optimization models for misalignment fading mitigation in optical wireless links," *IEEE Commun. Lett.*, vol. 12, pp. 395–397, 2008.

[40] H. G. Sandalidis, "Coded free-space optical links over strong turbulence and misalignment fading channels," *IEEE Trans. Commun.*, vol. 59, pp. 669–674, 2011.

[41] H. G. Sandalidis, T. A. Tsiftsis, and G. K. Karagiannidis, "Optical wireless communications with heterodyne detection over turbulence channels with pointing errors," *J. Lightw. Technol.*, vol. 27, pp. 4440–4445, 2009.

[42] H. G. Sandalidis, T. A. Tsiftsis, G. K. Karagiannidis, and M. Uysal, "BER performance of FSO links over strong atmospheric turbulence channels with pointing errors," *IEEE Commun. Lett.*, vol. 12, pp. 44–46, 2008.

[43] Y. Ren, A. Dang, B. Luo, and H. Guo, "Capacities for long-distance free-space optical links under beam wander effects," *IEEE Photon. Technol. Lett.*, vol. 22, pp. 1069–1071, 2010.

[44] D. K. Borah and D. G. Voelz, "Spatially partially coherent beam parameter optimization for free space optical communications," *Opt. Express*, vol. 18, pp. 20746–20758, 2010.

[45] A. Ishimaru, *Wave Propagation and Scattering in Random Media*. New York, NY: Academic, 1978.

[46] A. Ishimaru, "The beam wave case in remote sensing," in *Laser Beam Propagation in the Atmosphere*, J. W. Strohbehn, Ed., ed New York, NY: Springer, 1978.

[47] A. Ishimaru, "Wave propagation and scattering in random media and rough surfaces," *Proc. IEEE*, vol. 79, pp. 1359–1366, 1997.

[48] J. C. Ricklin, S. M. Hammel, F. D. Eaton, and S. L. Lachinova, "Atmospheric channel effects on free-space laser communication," in *Free-Space Laser Communications: Principles and Advances* A. K. Majumder and J. C. Ricklin, Eds., ed. New York, NY: Springer-Verlag, 2008, pp. 9–56.

[49] W. B. Miller, J. C. Ricklin, and L. C. Andrews, "Log-amplitude variance and wave structure function: a new perspective for Gaussian beams," *J. Opt. Soc. Am. A*, vol. 10, pp. 661–672, 1993.
[50] L. C. Andrews, W. B. Miller, and J. C. Ricklin, "Geometrical representation of Gaussian beams propagating through complex paraxial optical systems," *Appl. Opt.*, vol. 32, pp. 5918–5929, 1993.
[51] B. E. A. Saleh and M. C. Teich, *Fundamentals of Photonics*, 2nd ed. Hoboken, New Jersey: John Wiley & Sons, 2007.
[52] R. Ramirez-Iniguez, S. M. Idrus, and Z. Sun, *Optical Wireless Communications: IR for Wireless Connectivity*. Boca Raton, FL: Taylor & Francis, 2008.
[53] R. M. Gagliardi and S. Karp, *Optical Communications*, 2nd ed. Danvers, MA: John Wiley & Sons, Inc., 1995.
[54] S. Bloom, E. Korevaar, J. Schuster, and H. Willebrand, "Understanding the performance of free-space optics [Invited]," *J. Opt. Netw.*, vol. 2, pp. 178–200, 2003.
[55] J. M. Wallace and P. V. Hobbs, *Atmospheric Science: An Introductory Survey*, 2nd ed. Burlington, MA: Academic Press, 2006.
[56] H. Weichel, *Laser Beam Propagation in the Atmosphere*. Bellingham, WA: SPIE Optical Engineering Press, 1990.
[57] R. S. Lawrence and J. W. Strohbehn, "A survey of clear-air propagation effects relevant to optical communications," *Proc. IEEE*, vol. 58, pp. 1523–1545, 1970.
[58] R. F. Lutomirski and H. T. Yura, "Propagation of a finite optical beam in an inhomogeneous medium," *Appl. Opt.*, vol. 10, pp. 1652–1658, 1971.
[59] A. M. Prokhorov, F. V. Bunkin, K. S. Gochelashvily, and V. I. Shishov, "Laser irradiance propagation in turbulent media," *Proc. IEEE*, vol. 63, pp. 790–811, 1975.
[60] R. L. Fante, "Electromagnetic beam propagation in turbulent media," *Proc. IEEE*, vol. 63, pp. 1669–1692, 1975.
[61] R. L. Fante, "Electromagnetic beam propagation in turbulent media: an update," *Proc. IEEE*, vol. 68, pp. 1424–1443, 1980.
[62] A. Ishimaru, *Wave Propagation and Scattering in Random Media*. Piscataway, NJ: IEEE Press, 1997.
[63] J. H. Churnside and R. J. Lataitis, "Wander of an optical beam in the turbulent atmosphere," *Appl. Opt.*, vol. 29, pp. 926–930, 1990.
[64] R. Esposito, "Power scintillations due to the wandering of the laser beam," *Proc. IEEE*, vol. 55, pp. 1533–1534, 1967.
[65] T. Chiba, "Spot dancing of the laser beam propagated through the turbulent atmosphere," *Appl. Opt.*, vol. 10, pp. 2456–2461, 1971.
[66] D. L. Fried, "Statistics of laser beam fade induced by pointing jitter," *Appl. Opt.*, vol. 12, pp. 422–423, 1973.
[67] P. J. Titterton, "Power reduction and fluctuations caused by narrow laser beam motion in the far field," *Appl. Opt.*, vol. 12, pp. 423–425, 1973.
[68] D. L. Fried and H. T. Yura, "Telescope-performance reciprocity for propagation in a turbulent medium," *J. Opt. Soc. Am.*, vol. 62, pp. 600–602, 1972.

[69] R. J. Hill and S. F. Clifford, "Modified spectrum of atmospheric temperature fluctuations and its application to optical propagation," *J. Opt. Soc. Am.*, vol. 68, pp. 892–899, 1978.

[70] J. C. Owens, "Optical refractive index of air: Dependence on pressure, temperature and composition," *Appl. Opt.*, vol. 6, pp. 51–59, 1967.

[71] G. C. Valley, "Isoplanatic degradation of tilt correction and short-term imaging systems," *Appl. Opt.*, vol. 19, pp. 574–577, 1980.

[72] I. E. Lee, Z. Ghassemlooy, W. P. Ng, and S. Rajbhandari, "Fundamental analysis of hybrid free space optical and radio frequency communication systems," in *Proc. 12th Annual Post Graduate Symposium on the Convergence of Telecommunications, Networking and Broadcasting (PGNet 2011)*, 2011.

[73] M. A. Al-Habash, R. L. Phillips, and L. C. Andrews, "Mathematical model for the irradiance probability density function of a laser beam propagating through turbulent media," *Opt. Eng.*, vol. 40, pp. 1554–1562, 2001.

[74] M. Abramowitz and I. A. Stegun, *Handbook of Mathematical Functions with Formulas, Graphs, and Mathematical Tables*. New York, NY: Dover Publications, 1965.

[75] A. H. Mikesell, A. A. Hoag, and J. S. Hall, "The scintillation of starlight," *J. Opt. Soc. Am.*, vol. 41, pp. 689–691, 1951.

[76] S. H. Reiger, "Starlight scintillation and atmospheric turbulence," *Astron. J.*, vol. 68, pp. 395–406, 1963.

[77] A. T. Friberg and R. J. Sudol, "Propagation parameters of gaussian Schell-model beams," *Opt. Commun.*, vol. 41, pp. 383–387, 1982.

[78] L. Mandel and E. Wolf, *Optical Coherence and Quantum Optics*. Cambridge, UK: Cambridge University Press, 1995.

[79] V. W. S. Chan, "Free-space optical communications," *J. Lightw. Technol.*, vol. 24, pp. 4750–4762, 2006.

[80] S. M. Navidpour, M. Uysal, and M. Kavehrad, "BER performance of free-space optical transmission with spatial diversity," *IEEE Trans. Wireless Commun.*, vol. 6, pp. 2813–2819, 2007.

[81] L. Jing and M. Uysal, "Achievable information rate for outdoor free space optical communication with intensity modulation and direct detection," in *Proc. IEEE Global Telecommunications Conference (GLOBECOM 2003)*, 2003, pp. 2654–2658.

[82] E. Biglieri, J. Proakis, and S. Shamai, "Fading channels: Information-theoretic and communications aspects," *IEEE Trans. Inf. Theory*, vol. 44, pp. 2619–2692, 1998.

[83] J.-B. Wang, M. Sheng, X. Song, Y. Jiao, and M. Chen, "Comments on BER performance of FSO links over strong atmospheric turbulence channels with pointing errors," *IEEE Commun. Lett.*, vol. 16, pp. 22–23, 2012.

[84] T. M. Cover and J. A. Thomas, *Elements of Information Theory*, Second ed.: Wiley-Interscience, 2006.

[85] C. Liu, Y. Yao, Y. X. Sun, J. J. Xiao, and X. H. Zhao, "Average capacity optimization in free-space optical communication system over atmospheric

turbulence channels with pointing errors," *Opt. Lett.*, vol. 35, pp. 3171–3173, 2010.

[86] J. Cang and X. Liu, "Average capacity of free-space optical systems for a partially coherent beam propagating through non-Kolmogorov turbulence," *Opt. Lett.*, vol. 36, pp. 3335–3337, 2011.

[87] M. A. Khalighi, N. Schwartz, N. Aitamer, and S. Bourennane, "Fading reduction by aperture averaging and spatial diversity in optical wireless systems," *IEEE/OSA J. Opt. Commun. Netw.*, vol. 1, pp. 580–593, 2009.

Chapter 11

Relaying techniques for free space optical communications

Mohammadreza Aminikashani[1] *and Murat Uysal*[1]

11.1 Introduction

Free-space optical (FSO) communication is a line-of-sight technology that uses lasers for wireless connectivity. Terrestrial FSO links are considered a powerful complementary and/or alternative technology to radio frequency (RF) and fiber-optic counterparts for a wide range of applications including cellular backhaul, local area network (LAN)-to-LAN connectivity and redundant link. A major degrading factor in FSO links is the atmospheric turbulence-induced fading particularly for long link ranges. Relay-assisted FSO systems have been introduced in the literature as an effective method to extend coverage and mitigate the effects of fading.

This chapter provides a comprehensive overview of the relaying techniques for FSO communications considering various deployment configurations (i.e., serial, parallel, or mesh), and cooperation protocols (i.e., amplify-and-forward, decode-and-forward, all-optical relaying). Serial relaying (multi-hop transmission) is typically used in wireless RF communication to broaden the signal coverage for limited-power transmitters and does not offer increase in spatial diversity order. However, unlike the RF channel, the fading variance of FSO channel is distance-dependent. This feature allows serial FSO relaying bring performance improvements through inherent diversity gains besides extension in the link range. Parallel relaying is an alternative configuration where the source is equipped with a multi-laser transmitter with each of the transmitter pointing out in the direction of a corresponding relay node. This induces an artificial broadcasting which is normally not possible with a single transmitter in FSO communication due to the line-of-sight nature. As a more general configuration, an FSO mesh network can be further considered based on a combination of serial and parallel relaying.

The above deployment configurations can be used in conjunction with amplify-and-forward (AF) and decode-and-forward (DF) relaying [1–4]. In DF relaying, the relay decodes the signal after direct detection, modulates it, and

[1]Department of Electrical and Electronics Engineering, Ozyegin University, Turkey

retransmits it to the next relay or the destination in a dual-hop scenario. On the other hand, in AF relaying, the relay does not perform any decoding on the received signal and, after multiplication with a proper energy scaling term, simply forwards it to the next relay or the destination. The majority of the existing works on AF relaying systems build on the assumption that relays employ optical-to-electrical (OE) and electrical-to-optical (EO) convertors. The actual advantage of AF relaying over the DF counterpart emerges if its implementation avoids the requirement for high-speed (at the order of GHz) electronics and electro-optics. This becomes possible with all-optical AF relaying where the signals are processed in optical domain and the relay requires only low-speed electronic circuits to control and adjust the gain of amplifiers [3]. Therefore, EO/OE domain conversions are eliminated, allowing an efficient implementation. In this chapter, we will present a comparative performance evaluation of FSO systems with AF and DF as well as discuss all-optical AF relaying.

11.2 System and channel model

In the first part of this chapter, we consider a relay-assisted FSO communication system in which the transmitted signal from a source node propagates through N relays before detection at the destination node. The relaying nodes are equipped with single transmit/receive apertures and either AF or DF mode. We consider three main deployment configurations, namely, serial, parallel and mesh relaying, illustrated, respectively, in Figures 11.1–11.3.

In serial relaying (see Figure 11.1), the source node transmits a modulated signal to a relay node. Assuming DF relaying, the relay decodes the signal after

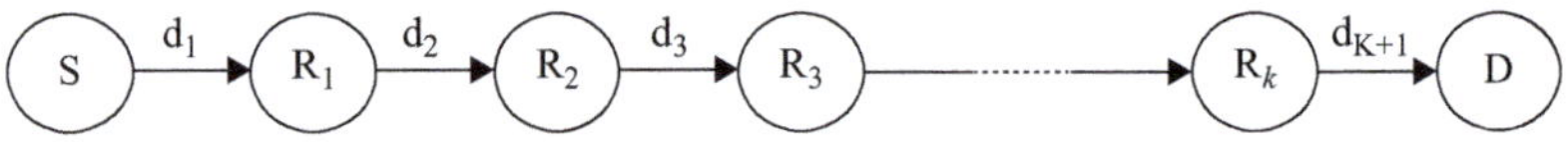

Figure 11.1 FSO serial relaying configuration

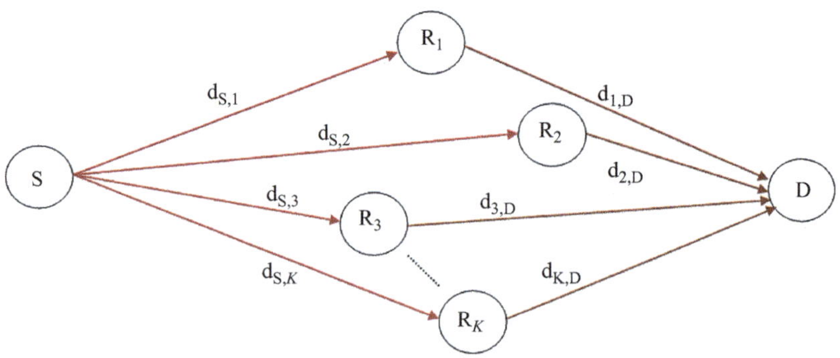

Figure 11.2 FSO parallel relaying configuration

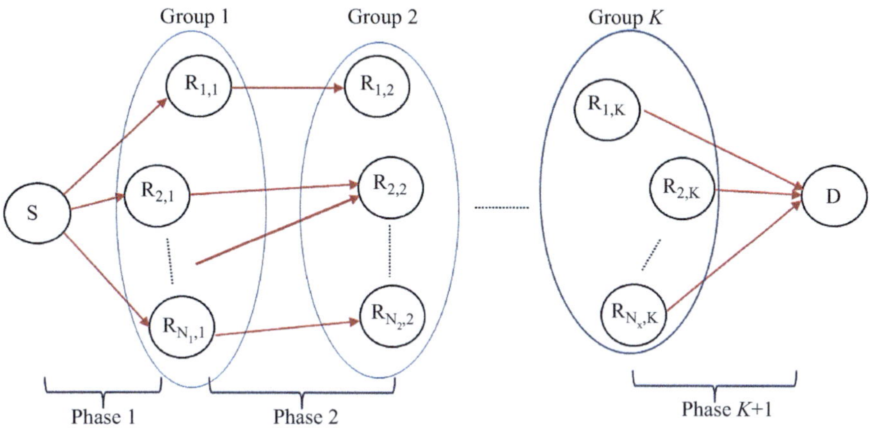

Figure 11.3 FSO multi-hop parallel relaying configuration

direct detection, modulates it, and retransmits it to the next relay (or the destination in a dual-hop scenario). On the other hand, if AF relaying is used, the relay does not perform any decoding on the received signal and, after multiplication with a proper energy scaling term, simply forwards it to the next relay. This continues until the source's data arrives at the destination node.

It is obvious that broadcasting is not possible due to the nature of FSO communication. Therefore, in parallel relaying (see Figure 11.2), an artificial broadcasting is created by equipping the source with a multi-laser transmitter where each of the lasers pointing out in the direction of a corresponding relay node. The source transmits the same signal to N relays. At each relay, based on the AF or DF relaying method, the signal is either decoded and retransmitted or scaled and forwarded to the destination. It should be noted that unlike wireless RF communication, distributed space-time block coding across relays is not required as the received diffraction patterns at the destination will be orthogonal if the transmit apertures are sufficiently separated.

Since FSO is a line-of-sight technology, a practical FSO mesh network would be likely to deploy a combination of serial and parallel relaying, i.e., multi-hop parallel relaying, as illustrated in Figure 11.3. Similar to the previous methods, the source data are transmitted to the destination node via N relays. However, in this method, the nearby relays are grouped together resulting in K groups. The source node is equipped with a multi-laser transmitter with each of them points out in the direction of a corresponding relay node within the first group. Note that since the number of the relays at one group might be less than the number of the relays at its previous group, all the relays (except those in the first group) and the destination node are supposed to have a sufficiently large aperture allowing the simultaneous detection of several diffraction patterns transmitted from different relay nodes.

Due to ease and low cost of implementation, intensity-modulation direct-detection (IM/DD) FSO systems are considered in this chapter. IM/DD systems uses intensity modulation techniques such as on–off keying (OOK) or pulse position modulation (PPM), which information is contained in the power variation of the transmitted field [5]. Without loss of generality, we assume the deployment of binary pulse position modulation (BPPM). In BPPM, information is conveyed with the presence or absence of light where optical transmitter is "on" during one half of the BPPM bit interval (i.e., "signal slot") and is "off" during the other half (i.e., "non-signal slot").

When a laser beam propagates through the atmosphere, the optical field is attenuated as a result of scattering and absorption. The path loss of the optical link with length is defined as [6]

$$L(d) = e^{-\rho d} \frac{A_{TX} A_{RX}}{(\lambda d)^2} \tag{11.1}$$

where A_{TX}, A_{RX}, λ are, respectively, the transmitter aperture area, the receiver aperture area, the wavelength, and ρ denotes the attenuation coefficient which is made up of scattering and absorption components.

In addition to the path loss, the optical links experience fading due to the turbulent atmosphere. Under the assumption of weak turbulence conditions, the turbulence-induced fading amplitude is modeled as a log-normal random variable. Therefore, the fading amplitude can be represented as $|\alpha| = \exp(\chi)$, where χ is normally distributed with mean μ_χ and variance σ_χ^2. The fading amplitude is normalized such that $E\left[|\alpha|^2\right] = 1$ implying $\mu_x = -\sigma_x^2$. This ensures that the fading does not attenuate or amplify the average power [7]. Therefore, the probability distribution function (pdf) of $|\alpha|$ is given by

$$p(|\alpha|) = \frac{1}{|\alpha|\sqrt{2\pi\sigma_\chi^2}} \exp\left(-\frac{\left(\ln(|\alpha|) + \sigma_\chi^2\right)^2}{2\sigma_\chi^2} \right) \tag{11.2}$$

Based on the Rytov theory and assuming spherical wave propagation through a horizontal atmospheric path, the log-amplitude variance is calculated by [6]

$$\sigma_\chi^2(d) = \min\left\{0.124k^{7/6} C_n^2 d^{11/6}, 0.5\right\} \tag{11.3}$$

where k is the wave number and C_n^2 is the refractive index structure constant and the minimum is taken to consider the case of saturated scintillation. The validity of log-normal model under weak turbulence conditions has been confirmed by numerous experiments [6].

In this chapter, we consider an aggregated channel model where both path loss and turbulence-induced fading are considered [1]. Let A denote the channel coefficient of an optical link given by

$$A = |\alpha|^2 l(d) \tag{11.4}$$

where $l(d) = L(d)/L(d_{S,D})$ is the normalized path loss with respect to the distance of the direct link between the source and the destination, i.e., $d_{S,D}$.

11.3 Outage performance

Let us define $C(\cdot)$ the instantaneous capacity function which is monotonically increasing with respect to instantaneous electrical signal-to-noise ratio (SNR) γ. Since $C(\cdot)$ is monotonically increasing with respect to γ, the outage probability at a targeted transmission rate of R_t is given by [8]

$$\begin{aligned} P_{\text{out}}(R_t) &= \Pr(C(\gamma) < R_t) \\ &= \Pr(\gamma < \gamma_{\text{th}}) \end{aligned} \tag{11.5}$$

where $\gamma_{\text{th}} = C^{-1}(R_t)$ is the minimum acceptable SNR above which the quality of service, i.e. the desired transmission rate, is satisfied.

In the following sections, we provide the outage probability expressions for the aforementioned relaying techniques.

11.3.1 Serial DF relaying

In serial DF relaying, an outage occurs when any of the intermediate single-input single-output (SISO) links fails. Therefore, the corresponding outage probability of the end-to-end link can be calculated as

$$P_{\text{out}} = \Pr\left(\bigcup_{i=1}^{N+1} \{\gamma_i < \gamma_{\text{th}}\}\right) = 1 - \prod_{i=1}^{N+1} (1 - \Pr(\gamma_i < \gamma_{\text{th}})) \tag{11.6}$$

where γ_i $(1 \leq i \leq N+1)$ denotes the SNR of the ith intermediate SISO link. Assuming a signal-independent additive white Gaussian channel with zero mean and variance of $\sigma_n^2 = N_0/2$, γ_i can be obtained by [1]

$$\gamma_i = R^2 T_s^2 P^2 A_i^2 / N_0. \tag{11.7}$$

In (11.7), R is the responsitivity of the photodetector, T_s is the duration of the signal/non-signal slots, P is the average transmitted optical power per transmit aperture, and A_i is the channel gain of the ith intermediate SISO link with length d_i.

In serial DF relaying, P is related to the total transmit power by $P = P_t/(N+1)$. The end-to-end outage probability of the serial DF relaying scheme can be derived from (11.7) as [1]

$$P_{\text{out}} = 1 - \prod_{i=1}^{N+1} \left(1 - Q\left(\frac{\ln(l(d_i)P_M/(N+1)) + 2\mu_\chi(d_i)}{2\sigma_\chi(d_i)}\right)\right) \tag{11.8}$$

where P_M denotes power margin and is defined as $P_M = \sqrt{P_t^2 R^2 T_s^2 / N_0 \gamma_{th}}$.

11.3.2 Parallel DF relaying

To calculate the outage probability of parallel DF relaying, let us first define the decoded set as the set of relays having successfully decoded the signal consisting of 2^N possibilities. In parallel relaying, an outage occurs if either the decoded set is empty or multiple-input single-output (MISO) link between the relays in the decoded set and the destination fails. Assuming $S(i)$ denotes the ith possible set, and $\bar{d}_{S(i)}$ denotes the set of all distances between the destination and the relays in the decoded set, i.e., $d_{j,D} \in \bar{d}_{S(i)}, \forall j \in S(i)$, the corresponding outage probability can be expressed as [1]

$$P_{\text{out}} = \sum_{i=1}^{2^N} P_{\text{out-p}}(S(i)) = \sum_{i=1}^{2^N} \left[\prod_{j \in S(i)} \left(1 - Q\left(\frac{\ln\left(\frac{l(d_{S,j})P_M}{2N}\right) + 2\mu_\chi(d_{S,j})}{2\sigma_\chi(d_{S,j})} \right) \right) \prod_{j \notin S(i)} Q\left(\frac{\ln\left(\frac{l(d_{S,j})P_M}{2N}\right) + 2\mu_\chi(d_{S,j})}{2\sigma_\chi(d_{S,j})} \right) \right] Q\left(\frac{\ln\left(\frac{P_M e^{\mu_\xi}}{2N}\right)}{\sigma_\xi(\bar{d}_{S(i)})} \right) \tag{11.9}$$

where $\mu_\xi(\bar{d}_{S(i)})$ and $\sigma_\xi^2(\bar{d}_{S(i)})$ are given by

$$\mu_\xi(\bar{d}_{S(i)}) = \ln \sum_{i \in S(i)} l(d_{i,D}) - \sigma_\xi^2(\bar{d}_{S(i)})/2 \tag{11.10}$$

$$\sigma_\xi^2(\bar{d}_{S(i)}) = \ln\left(1 + \sum_{i \in S(i)} l^2(d_{i,D})\left(e^{4\sigma_\chi^2} - 1\right) \Big/ \left(\sum_{i \in S(i)} l(d_{i,D}) \right)^2 \right) \tag{11.11}$$

11.3.3 Optimization of relay location

In this section, we formulate optimization problems based on the end-to-end outage probability obtained in (11.8) and (11.9) to determine the optimal relay locations for serial and parallel DF relaying schemes. We then quantify performance improvements obtained through optimal relay placement.

In order to find the optimal relay locations for serial relaying, (11.8) needs to be minimized with respect to the length of the intermediate SISO links. Let $d_1, d_2, \ldots$ and d_{N+1} denotes these lengths to be optimized and define the following functions

$$h(d_1, d_2, \ldots, d_{N+1}) = \prod_{i=1}^{N+1} \Phi(f(d_i)), \tag{11.12}$$

$$g(d_1, d_2, \ldots, d_{N+1}) = \sum_{i=1}^{N+1} d_i, \tag{11.13}$$

where $\Phi(x) = 1 - Q(x)$ is the cumulative distribution function of the normal Gaussian distribution [9], and $f(d_i)$ is given by

$$f(d_i) = \frac{\ln(l(d_i)P_M/(N+1)) + 2\mu_\chi(d_i)}{2\sigma_\chi(d_i)} \tag{11.14}$$

It can be easily verified that the optimum place for each relay is on the direct path from the source to the destination [2]. Thus, the optimization problem can be stated as

$$\begin{aligned} &\max_{d_1,d_2,\ldots,d_{N+1}} \quad h(d_1, d_2, \ldots, d_{N+1}) \\ &\text{s.t. } g(d_1, d_2, \ldots, d_{N+1}) = d_{S,D} \end{aligned} \tag{11.15}$$

$h(\cdot)$ is a concave function with respect to the optimization parameters since $f(d_i)$ is a monotonically decreasing function. By applying Lagrange multiplier method, the solution of (11.15) can be found as [2]

$$d_i = d_k. \tag{11.16}$$

Therefore, the outage probability is minimized when the consecutive nodes are placed equidistant along the path from the source to the destination.

For parallel DF relaying given by (11.9) is minimized with respect to the distance of source-to-relay and relay-to-destination links. Similar to the serial relaying, it can be shown that for achieving the optimum performance, all the relays should be located along the direct path from the source to the destination. The optimization problem can be stated as

$$\begin{aligned} &\min_{d_{S,j},d_{j,D},j=1,2,..,N} \quad z\left(d_{S,1}, d_{S,2}, \ldots, d_{S,N}, d_{1,D}, d_{2,D}, \ldots, d_{N,D}\right) \\ &\text{s.t. } d_{S,j} + d_{j,D} = d_{S,D} \qquad j = 1, 2, \ldots, N \end{aligned} \tag{11.17}$$

where $z(\cdot)$ is defined as

$$\begin{aligned} &z\left(d_{S,1}, d_{S,2}, \ldots, d_{S,N}, d_{1,D}, d_{2,D}, \ldots, d_{N,D}\right) \\ &= \sum_{i=1}^{2^N}\left[\prod_{j\in S(i)}\left(1 - Q\left(u\left(d_{S,j}\right)\right)\right)\prod_{j\notin S(i)} Q\left(u\left(d_{S,j}\right)\right)\right] Q\left(v\left(\bar{d}_{S(i)}\right)\right) \end{aligned} \tag{11.18}$$

In (11.18), $u(\cdot)$ and $v(\cdot)$ are given by

$$u\left(d_{S,j}\right) = \left(\ln\left(\frac{l(d_{S,j})P_M}{2N}\right) + 2\mu_\chi\left(d_{S,j}\right)\right)/2\sigma_\chi\left(d_{S,j}\right) \tag{11.19}$$

$$v\left(\bar{d}_{S(i)}\right) = \ln\left(\frac{P_M e^{\mu_\xi}}{2N}\right)/\sigma_\xi\left(\bar{d}_{S(i)}\right) \tag{11.20}$$

This non-convex optimization problem can only be solved by numerical methods such as genetic algorithm [10] since a closed-form analytical solution is not readily available.

The numerical optimization through genetic algorithm shows that the optimum solution of (11.17) occurs if all the relays (that should be located along the direct path from the source to the destination) are placed at exactly the same point. It should be noted that, in practice, due to the size of FSO transceivers, relay terminals cannot be obviously placed at the same physical location. However, as demonstrated later in Section 11.4, placement of relay terminals (separated from each other by a distance more than the spatial coherence length) on the perpendicular line crossing the optimum location results in near-optimum performance.

Table 11.1 provides the optimized distances from the source for relay terminals for some given values of power margin (P_M), link range ($d_{S,D}$), number of relays (N), wavelengths (λ), refractive index structure constant $\left(C_n^2\right)$ and atmospheric attenuation (σ). It is observed from Table 11.1 that the optimum position slightly decreases by increasing the wavelength while it slightly increases by increasing the refractive index structure constant. It is also observed that the optimum position is somewhere close to the source, and becomes closer as the number of relays increases or for the low values of power margin.

Based on the numerical optimization results, a heuristic expression is further developed through an interpolation on logarithmic functions in [2]. This expression for optimum relay location is given by

$$d_{\text{opt}}\left(P_M, d_{S,D}, N\right) = 0.5d_{S,D} + \frac{\eta}{P_M}\ \ln\left(\varpi \mathrm{P}_M^{\kappa} + \Delta\right), N \geq 2 \tag{11.21}$$

where ϖ, κ, Δ and η are defined, respectively, as, $\varpi = -2.7 \times 10^{-5}d_{S,D} + 0.1\ \ln(N) + 0.11$, $\kappa = -6.5 \times 10^{-6}d_{S,D} + 1.19$, $\Delta = 4.5 \times 10^{-5}d_{S,D} + 0.024N + 0.9$ and $\eta = d_{S,D}(7 \times 10^{-6}d_{S,D} - 0.14\ \ln(N) - 0.5)$.

11.3.4 Multi-hop parallel DF relaying

So far, we considered serial and parallel relaying. In this subsection, we consider a mesh configuration building on a combination of serial and parallel relaying.

Table 11.1 Normalized optimal relay locations for parallel relaying $\lambda = 1550$ nm, $\sigma \approx 0.1$ and $C_n^2 = 10^{-14}m^{2/3}$. The unit of $d_{S,D}$ is kilometers. N denotes the number of relays. Since optimization has yielded the same location for all relays, only one numerical value is provided regardless of the relay number

	$N = 2$		$N = 3$		$N = 4$		$N = 5$	
P_M [dB]	$d_{S,D} = 3$	$d_{S,D} = 5$	$d_{S,D} = 3$	$d_{S,D} = 5$	$d_{S,D} = 3$	$d_{S,D} = 5$	$d_{S,D} = 3$	$d_{S,D} = 5$
	Optimal relay locations							
0	0.4037	0.3997	0.3575	0.3389	0.3191	0.3001	0.2812	0.2731
5	0.4334	0.4472	0.4012	0.4160	0.3778	0.3906	0.3582	0.3698
10	0.4466	0.4617	0.4239	0.4432	0.4104	0.4260	0.3945	0.4121
15	0.4660	0.4705	0.4489	0.4588	0.4393	0.4488	0.4360	0.4443
20	0.4781	0.4871	0.4705	0.4791	0.4686	0.4710	0.4635	0.4680

Let N_i, $i = 1, 2, \ldots K$, denote the number of relay nodes in the ith group. In multi-hop parallel relaying, transmission takes place in $K+1$ phases. In the first phase, the source transmits the same signal to N_1 relay nodes (i.e., the first group of relays). In the next phase, each relay, after direct detection, modulates it and retransmits the signal to the next group. This continues until the source's data arrives at the destination node. Let us refer the relay nodes which are engaged in forwarding as active relays. The set of active relays at the ith group is named as active relay set and denoted by W_i. Also, let D_i denotes the successful relay set, i.e., the set of active relays in W_i having successfully decoded the signal. Therefore, there exists $2^{|W_i|}$[1] possible forms for the successful set at the ith phase, $i = 1, 2, \ldots K$. Let $U_i(j)$ denotes the jth possible set for D_i and $\bar{d}_{U_i(j),n}$ denotes the set of all distances between the nth active node in the $(i+1)th$ group and the successful relays in the ith active relay set pointing out in the direction of the corresponding relay node, i.e., $d_{m,n} \in \bar{d}_{U_i(j),n}$, $m \in U_i(j)$, $1 \leq n \leq |W_{i+1}|$.

In multi-hop parallel relaying, an outage occurs if either none of the active relay nodes decodes the signal successfully or MISO link in the last phase fails. The corresponding outage probability can be then given by

$$\begin{aligned} P_{\text{out}} = &\sum_{i_1=1}^{2^{N_i}} \Pr(U_1(i_1)) \sum_{i_2=1}^{2^{|W_2|}} \Pr(U_2(i_2)) \cdots \\ &\times \sum_{i_K=1}^{2^{|W_K|}} \Pr(U_K(i_K)) P_{\text{out,MISO}}\left(\bar{d}_{U_K(i_K),D}\right) \end{aligned} \tag{11.22}$$

After some mathematical manipulations, the outage probability for multi-hop parallel scheme is obtained as [4]

$$\begin{aligned} P_{\text{out}} = &\sum_{i_1=1}^{2^{N_1}} \prod_{m_1 \in U_1(i_1)} \left(1 - F\left(d_{S,m_1}\right)\right) \prod_{m_1 \notin U_1(i_1)} F\left(d_{S,m_1}\right) \\ &\times \sum_{i_2=1}^{2^{|W_2|}} \prod_{m_2 \in U_2(i_2)} \left(1 - G\left(\bar{d}_{U_1(i_1),m_2}\right)\right) \prod_{m_2 \notin U_2(i_2)} G\left(\bar{d}_{U_1(i_1),m_2}\right) \cdots \\ &\times \sum_{i_j=1}^{2^{|W_j|}} \prod_{m_j \in U_j(i_j)} \left(1 - G\left(\bar{d}_{U_{j-1}(i_{j-1}),m_j}\right)\right) \prod_{m_j \notin U_j(i_j)} G\left(\bar{d}_{U_{j-1}(i_{j-1}),m_j}\right) \cdots \\ &\times \sum_{i_K=1}^{2^{|W_K|}} \prod_{m_K \in U_K(i_K)} \left(1 - G\left(\bar{d}_{U_{K-1}(i_{K-1}),m_K}\right)\right) \prod_{m_K \notin U_K(i_K)} G\left(\bar{d}_{U_{K-1}(i_{K-1}),m_K}\right) G\left(\bar{d}_{U_K(i_K),D}\right) \end{aligned} \tag{11.23}$$

[1] $|W_i|$ denotes the cardinality of W_i and therefore represents the number of the active relays in the ith group.

where $F(d_{i,j})$ and $G(\bar{d})$ are defined as

$$F(d_{i,j}) \underline{\underline{def}} Q\left(\frac{\ln(l(d_{i,j})P_M/(N+N_1)) + 2\mu_\chi(d_{i,j})}{2\sigma_\chi(d_{i,j})}\right), \tag{11.24}$$

$$G(\bar{d}) \underline{\underline{def}} Q\left(\frac{\ln(P_M/(N+N_1)) + \mu_\xi(\bar{d})}{\sigma_\xi(\bar{d})}\right). \tag{11.25}$$

The derived expression for the multi-hop parallel relaying includes the outage probability of serial and parallel relaying as special cases. First, assume $N = K$ which corresponds to the case of serial (multi-hop) transmission. Under this assumption, (11.23) reduces to

$$\begin{aligned} P_{\text{out-serial}} &= F(d_{S,m_1}) + (1 - F(d_{S,m_1}))F(d_{m_1,m_2}) + \ldots \\ &\quad + (1 - F(d_{S,m_1}))(1 - F(d_{m_1,m_2})) \times \ldots \\ &\quad \times (1 - F(d_{m_{K-2},m_{K-1}}))F(d_{m_{K-1},m_K}) \\ &= 1 - (1 - F(d_{S,m_1}))F(d_{m_K,D})\prod_{i=1}^{K-1}(1 - F(d_{m_i,m_{i+1}})) \end{aligned} \tag{11.26}$$

which coincides with (11.8).

Now, assume $K = 1$ corresponding to the parallel relaying scheme. Under this assumption, (11.23) reduces to

$$P_{\text{out - parallel}} = \sum_{i_1=1}^{2^N}\left[\prod_{m_1 \in U_1(i_1)} (1 - H(d_{S,m_1})) \prod_{m_1 \notin U_1(i_1)} H(d_{S,m_1})\right] G(\bar{d}_{U_1(i_1),D}) \tag{11.27}$$

which coincides with (11.9) derived in Section 11.3.2.

11.3.5 Serial AF relaying

In AF relaying, the relay nodes forward the signal to the next node without any decoding. Thus, in order to calculate the outage probability, the total received SNR at the destination should be obtained.

Assuming BPPM, the received SNR at the destination node is given by [1]

$$\gamma = \frac{\prod_{i=0}^{N} a_i^2 A_i^2}{N_0\left(\sum_{j=1}^{N}\prod_{i=j}^{N} a_i^2 A_i^2 + 1\right)} \tag{11.28}$$

where a_i is the amplification factor at ith node. Replacing (11.28) in (11.5), the end-to-end outage probability of the serial AF relaying scheme can be estimated as [1]

$$P_{\text{out}} \approx Q\left(\frac{\ln\left(P_M^2/(N+1)^2\right) - \mu_\varepsilon}{\sigma_\varepsilon}\right) \tag{11.29}$$

where μ_ε and σ_ε are defined, respectively, in (11.40) and (11.41) of [1].

11.3.6 Parallel AF relaying

For parallel relaying, the received SNR at the destination node is given by [1]

$$\gamma = \frac{\left(RT_s P \sum_{i=1}^{N} a_i^2 A_{i,D} A_{S,i}\right)^2}{N_0\left(\sum_{i=1}^{N} a_i^2 A_{i,D}^2 + 1\right)} \tag{11.30}$$

Replacing (11.30) in (11.5), the end-to-end outage probability of AF parallel relaying can be calculated using pdf of the bivariate normal distribution as

$$P_{\text{out}} = \int_{-\infty}^{\infty}\int_{-\infty}^{\omega_0} \frac{1}{2\pi\sqrt{|\Sigma|}} \times\exp\left(-\frac{\left(\sigma_2^2(\omega_1-\mu_1)^2 + \sigma_1^2(\omega_2-\mu_2)^2 - 2\sigma_{12}(\omega_1-\mu_1)(\omega_2-\mu_2)\right)}{2|\Sigma|}\right) d\omega_2 d\omega_1 \tag{11.31}$$

where the upper limit of the inner integration in (11.31) is expressed as

$$\omega_0 = 0.5\ln(2N(\exp(\omega_2)+1)/P_M). \tag{11.32}$$

In (11.31), the log-amplitude pair (ω_1, ω_2) follows a correlated bivariate normal distribution [11]. Their mean and covariance matrix are defined, respectively, as

$$\mu_1 = \ln\sum_{i=1}^{N} a_i l(d_{S,i}) l(d_{i,D}) - \sigma_1^2/2 \tag{11.33}$$

$$\mu_2 = \ln\sum_{i=1}^{N} a_i^2 l^2(d_{i,D})\exp(4\sigma_\chi^2(d_{i,D})) - \sigma_2^2/2 \tag{11.34}$$

$$\Sigma = \begin{bmatrix} \sigma_1^2 & \sigma_{12} \\ \sigma_{12} & \sigma_2^2 \end{bmatrix} \tag{11.35}$$

where σ_1^2, σ_2^2, and σ_{12} are given by

$$\sigma_1^2 = \ln\left(1 + \frac{\sum_{i=1}^{N} a_i^2 l^2(d_{S,i}) l^2(d_{i,D})(\exp(4(\sigma_\chi^2(d_{S,i}) + \sigma_\chi^2(d_{S,D}))) - 1)}{\left(\sum_{i=1}^{N} a_i l(d_{S,i}) l(d_{i,D})\right)^2}\right) \tag{11.36}$$

$$\sigma_2^2 = \ln\left(1 + \frac{\sum_{i=1}^{N} a_i^4 l^4(d_{i,D})(\exp(24\sigma_\chi^2(d_{S,D})) - \exp(8\sigma_\chi^2(d_{S,D})))}{\left(\sum_{i=1}^{N} a_i^2 l^2(d_{i,D}) \exp(4\sigma_\chi^2(d_{S,D}))\right)^2}\right) \tag{11.37}$$

$$\sigma_{12} = \ln\left(1 + \frac{\sum_{i=1}^{N} a_i^3 l(d_{S,i}) l^3(d_{i,D})(\exp(12\sigma_\chi^2(d_{S,D})) - \exp(4\sigma_\chi^2(d_{S,D})))}{\left(\sum_{i=1}^{N} a_i^2 l^2(d_{i,D}) \exp(4\sigma_\chi^2(d_{S,D}))\right)\left(\sum_{i=1}^{N} a_i l(d_{S,i}) l(d_{i,D})\right)}\right) \tag{11.38}$$

Unfortunately, a closed-form expression of (11.31) is not available although it can be numerically calculated through multidimensional integration routines such as Gauss–Hermite quadrature formula.

11.4 Performance results of AF and DF relaying

In this section, we present the outage performance for serial, parallel and multi-hop relaying schemes for an FSO system with $\lambda = 1,550$ nm operating in clear weather conditions with a visibility of 10 km. We assume the end-to-end link range is $d_{S,D} = 5$ km, the atmospheric attenuation is 0.43 dB/km (i.e., $\sigma \approx 0.1$), and refractive index structure constant is $C_n^2 = 10^{-14}\text{m}^{2/3}$. The log-amplitude variance is therefore calculated as $\sigma_\chi^2 = 0.38$ below the saturation regime.

In Figure 11.4, we present the outage performance for serial DF relaying assuming $N = 3$ relays. Following configurations for relay locations are considered.

- Scenario 0 (Optimized): $d_1 = d_2 = d_3 = d_4 = d_{S,D}/4$
- Scenario 1: $d_1 = d_4 = d_{S,D}/8, d_2 = d_3 = 3d_{S,D}/8$
- Scenario 2: $d_1 = d_2 = d_3 = d_{S,D}/5, d_4 = 2d_{S,D}/5$
- Scenario 3: $d_1 = d_2 = d_3 = d_{S,D}/6, d_4 = d_{S,D}/2$

It is observed from Figure 11.4 that relay location optimization substantially improves the performance. At an outage probability of 10^{-6}, performance improvements of 7.2, 8.2, and 13.4 dB are achieved, respectively, with respect to the non-optimized scenarios 1, 2, and 3.

In Figure 11.5, we present the optimized outage performance for parallel DF relaying assuming $N = 2, 3, 4, 5$ relays along with the equi-distant relay cases. The

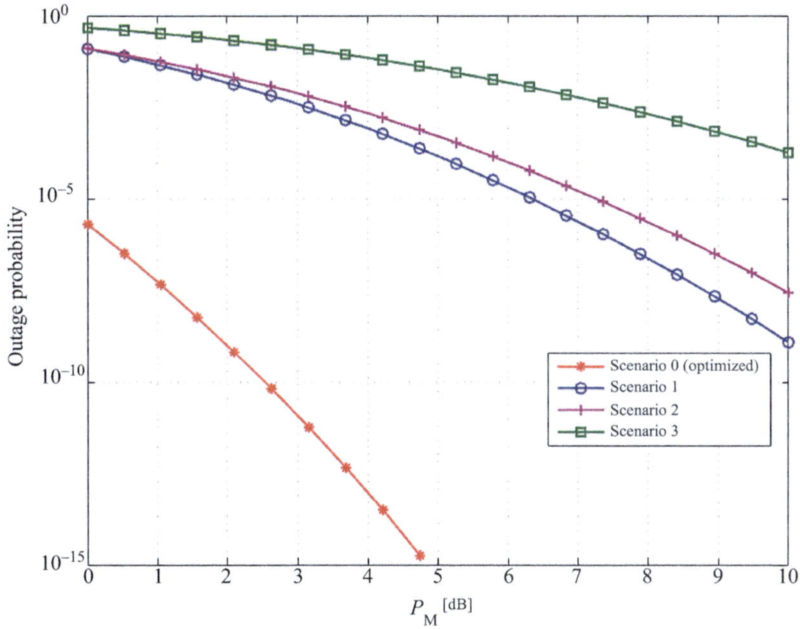

Figure 11.4 Outage probability of the serial FSO relaying scheme for different scenarios

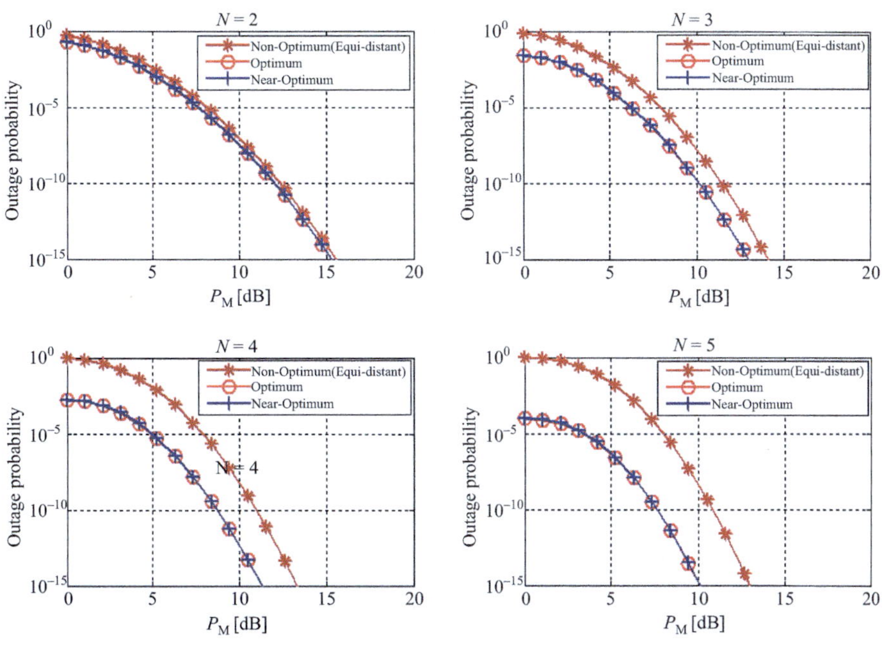

Figure 11.5 Outage probability of the parallel FSO relaying scheme for different number of relays and scenarios

optimized relay locations are already given in Table 11.1. To reflect the performance in practical settings, we have also included the performance when the relay terminals are placed on the perpendicular line crossing the optimum location and separated from each other by a distance of 50 m. These two cases are labeled as "optimum and "near-optimum." At a target outage probability of 10^{-6}, performance improvements of 1.5, 1.7, 2.8, and 4 dB, respectively, for $N=2$, 3, 4, and 5 with respect to equidistant located relays are observed.

Furthermore, it is observed that separation of the relay terminals results in a negligible performance degradation, i.e., within the line of thickness of outage probability plots. It is also demonstrated in Figure 11.5 that the performance improvements of parallel relaying are less than those observed in serial relaying. This is as a result of the nature of parallel relaying in which there are only two hops that can be optimized regardless of relay numbers.

For multi-hop parallel relaying, we assume a total of $N = 6$ relays and present the outage performance for a number of different configurations:

- Configuration 1 (C1): In this configuration, we have three groups of relays (i.e., $K=3$) each of which consists two relays (i.e., $N_1 = N_2 = 2$)
- Configuration 2 (C2): In this configuration, we have two groups of relays (i.e., $K=2$) each of which consists three relays (i.e., $N_1 = N_2 = 3$)
- Configuration 3 (C3): In this configuration, we have a single group of six relays located at the vertical line of the midway point between the source and the destination.

These configurations based on the mix use of parallel and multi-hop relaying are illustrated in Figure 11.6. As benchmarks, we also consider some configurations in which only the use of either parallel or serial relaying is allowed for the same number of relays. Specifically, Figure 11.7(a). and (b), respectively, illustrate the serial and parallel relaying schemes for $K=3$ and $N_1=N_2=2$. These will be used as benchmarks for C1. Similarly, Figure 11.8(a). and (b), respectively, illustrate the serial and parallel relaying schemes which will be used as benchmarks for C2. Figure 11.9 illustrates the benchmarking serial relaying scheme for C3.

Figures 11.10 and 11.11 present the outage performance of C1 and C2 with the aforementioned benchmarks. As clearly seen from these figures, the combined use of parallel and serial relaying substantially improves the performance with respect to stand-alone uses of serial and parallel relaying.

Particularly, for an outage probability of 10^{-6}, we observe performance improvements of 6.8 dB and 9.8 dB for $K = 2$ and 3 with respect to the stand-alone parallel relaying. On the other hand, with respect to stand-alone serial relaying, performance gains of 3.8 dB and 9.2 dB are, respectively, observed for $K = 2$ and 3. It should be noted that these performance gains are a result of the nature of multi-hop parallel relaying which smartly exploits the distance-dependency of the log-amplitude variance more than the stand-alone versions.

In Figure 11.12, the outage performance of C3 along with the benchmark is illustrated. We note that for $K=1$ the performance of serial and parallel relaying coincide as expected. For a target outage probability of 10^{-6}, this configuration

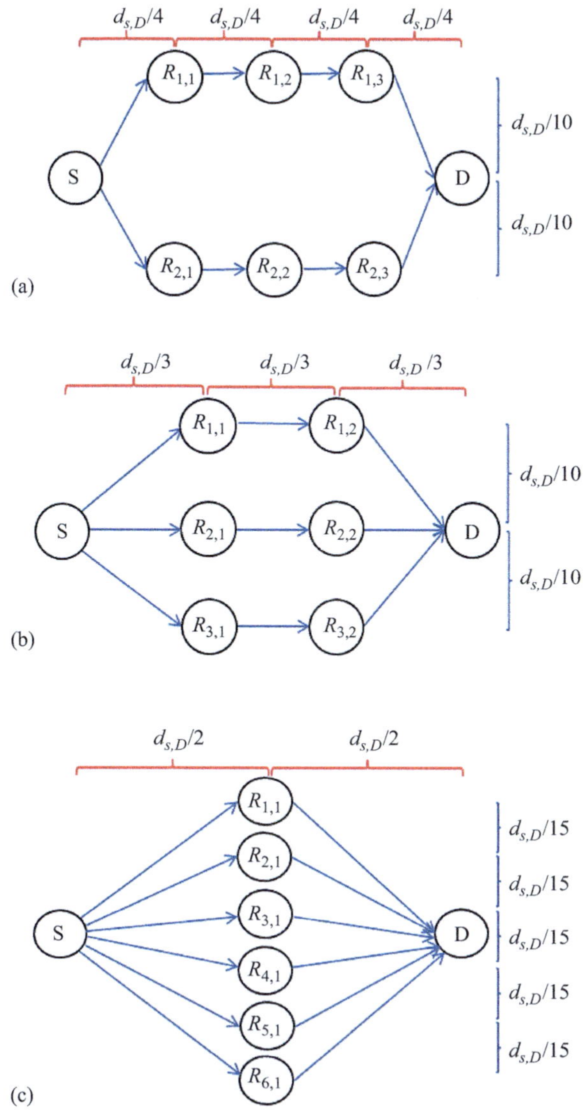

Figure 11.6 Scenarios under consideration: (a) Scenario 1: one group of relays, (b) Scenario 2: two groups of relays, and (c) Scenario 3: three groups of relays

achieves a performance improvement of 8.3 dB with respect to the stand-alone version.

In Figure 11.13, we compare the performance of C1, C2 and C3 and investigate the effect of the number of groups for a fixed number of relays. We observe from Figure 11.13 that increasing the number of groups results in performance

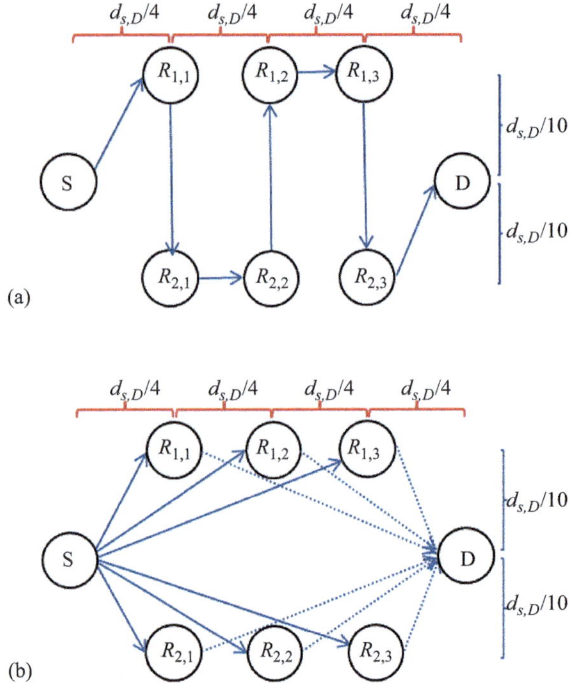

Figure 11.7 (a) Multi-hop and (b) parallel relaying benchmarking schemes for C1

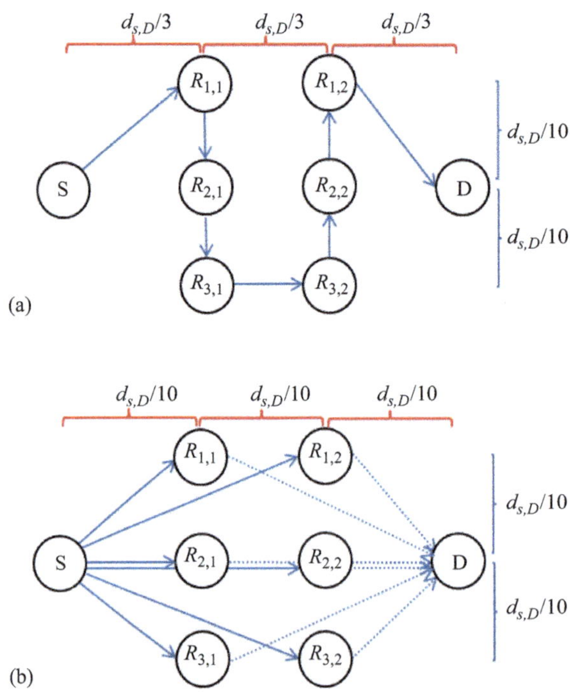

Figure 11.8 (a) Multi-hop and (b) parallel relaying benchmarking schemes for C2

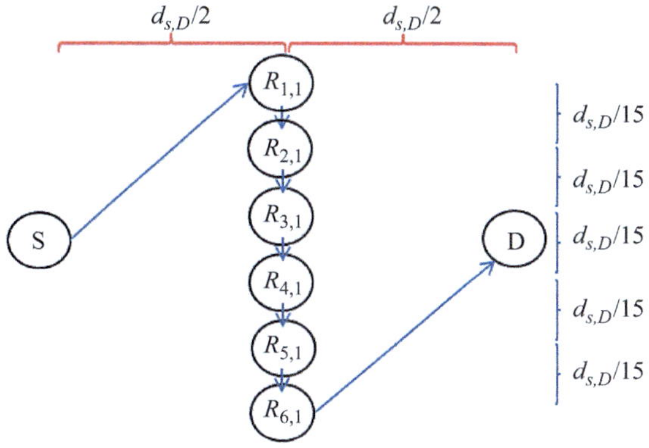

Figure 11.9 Multi-hop relaying benchmarking scheme for C3

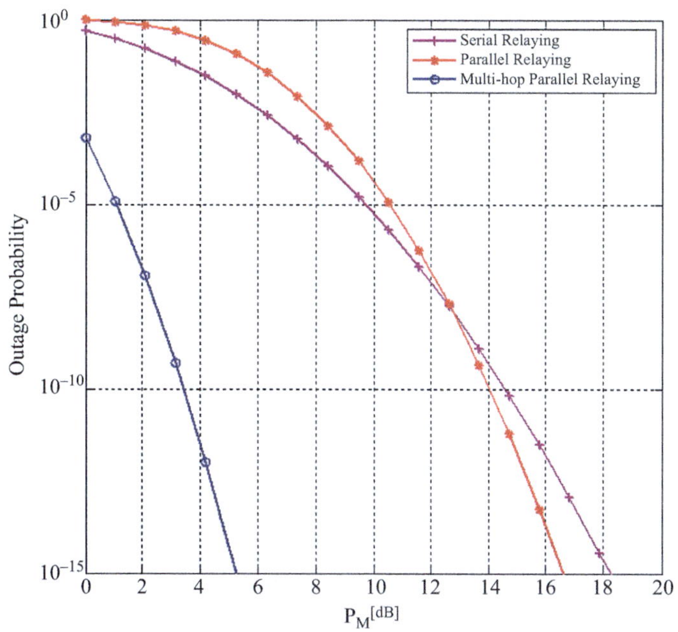

Figure 11.10 Performance comparison of multi-hop parallel, serial, and parallel relaying for C1

improvement. For a target outage probability of 10^{-6}, the performance gains for $K = 3$ are 7.5 dB and 2.9 dB with respect to $K = 1$ and 2. This is as a result of increasing hops exploiting the distance-dependency of the log-amplitude variance to a greater extent.

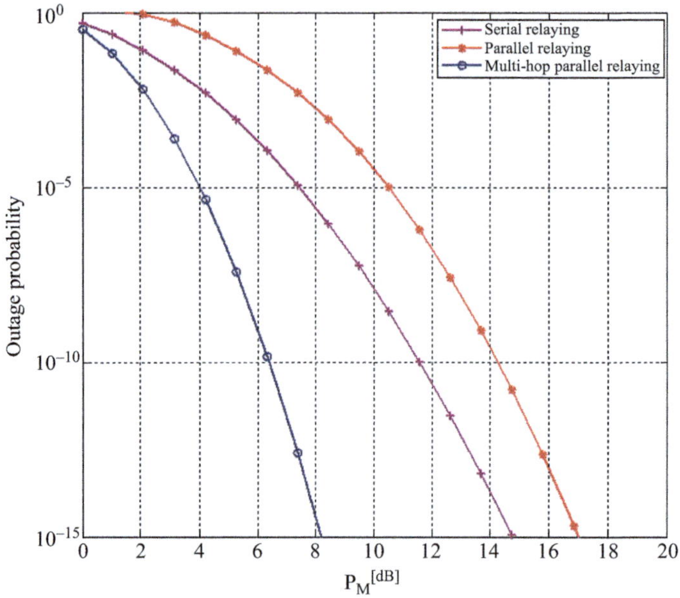

Figure 11.11 Performance comparison of multi-hop parallel, serial, and parallel relaying for C2

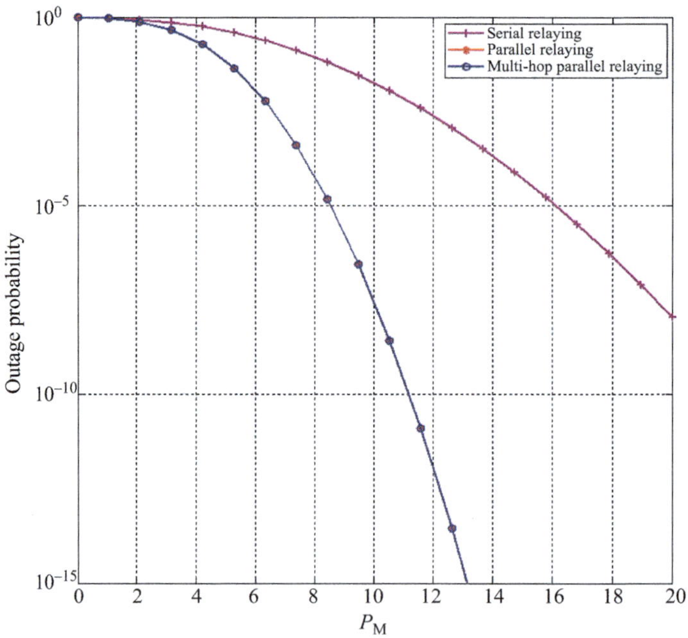

Figure 11.12 Performance comparison of multi-hop parallel, serial, and parallel relaying for C3

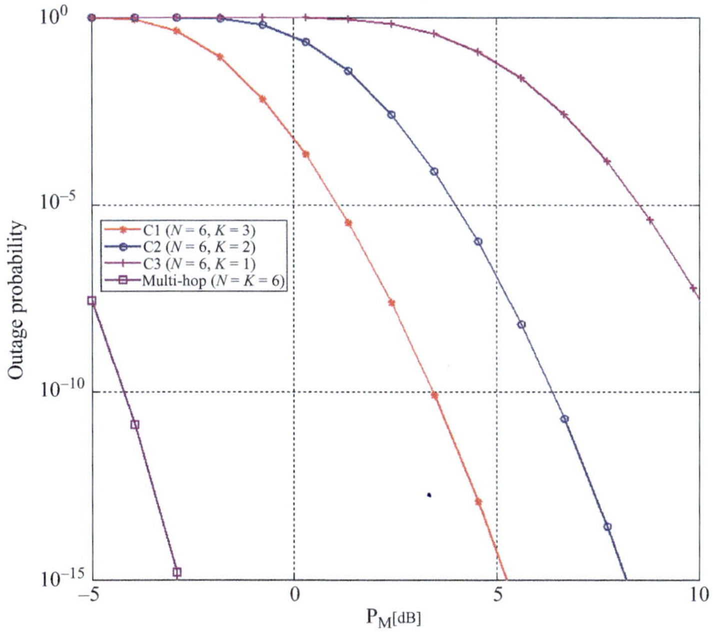

Figure 11.13 Effect of different number of groups on the outage performance

In Figures 11.14 and 11.15, we compare the performance of AF and DF relaying assuming $N = 1$, 2, and 3. For serial relaying, we assume the consecutive nodes are equidistant along the path from the source to the destination. In parallel relaying, the relays are located on the halfway point. It is observed that DF relaying outperforms its AF counterpart. However, AF relays enjoy a lower complexity in comparison with DF counterparts since it does not require any decoding process. In particular, for an outage probability of 10^{-6} and serial relaying, DF relaying obtains performance improvements of 6.3 dB, 7.7 dB and 8.2 dB for $N = 1$, 2, and 3 with respect to the AF relaying. For parallel relaying, the performance improvements are 1.4 dB and 3.2 dB for $N = 1$, 2, and 3.

11.5 All-optical AF relaying system

The actual advantage of AF relaying over the DF counterpart emerges if its implementation avoids the requirement for high-speed (in the order of GHz) electronics and electro-optics. This becomes possible with all-optical AF relaying where the signals are processed in optical domain and the relay requires only low-speed electronic circuits to control and adjust the gain of amplifiers [12]. Therefore, EO/OE domain conversions are eliminated, allowing efficient implementation.

In this section, as an alternative implementation, we consider an FSO system with all-optical AF relaying and investigate its outage performance considering the

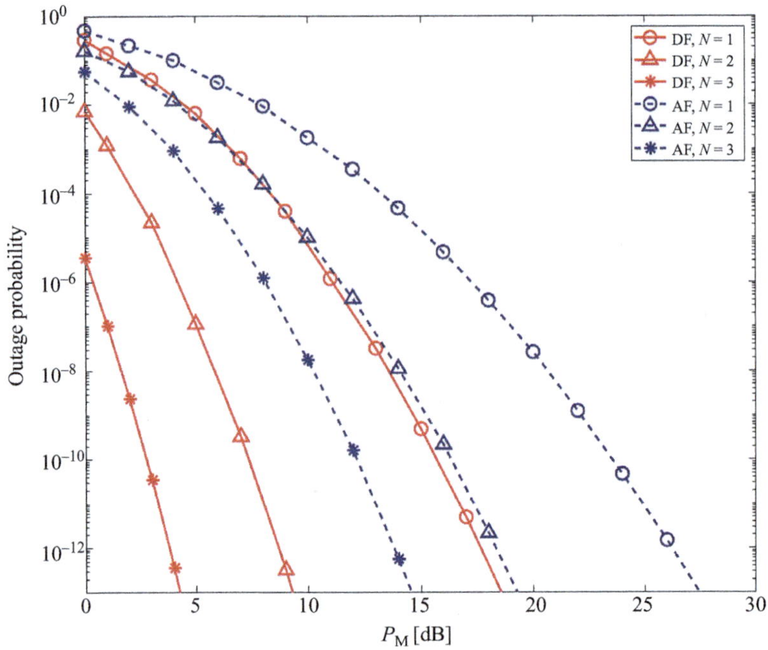

Figure 11.14 Performance comparison of serial AF and DF relaying

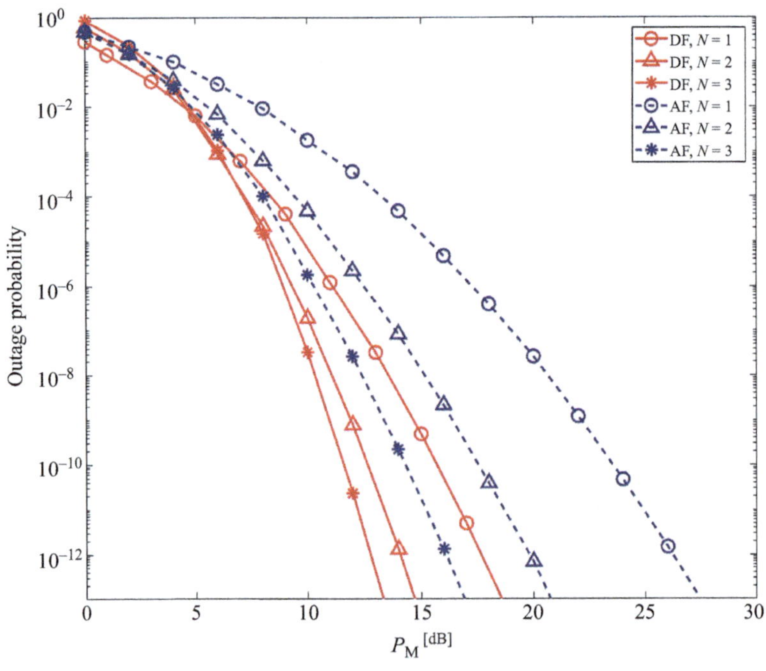

Figure 11.15 Performance comparison of parallel AF and DF relaying

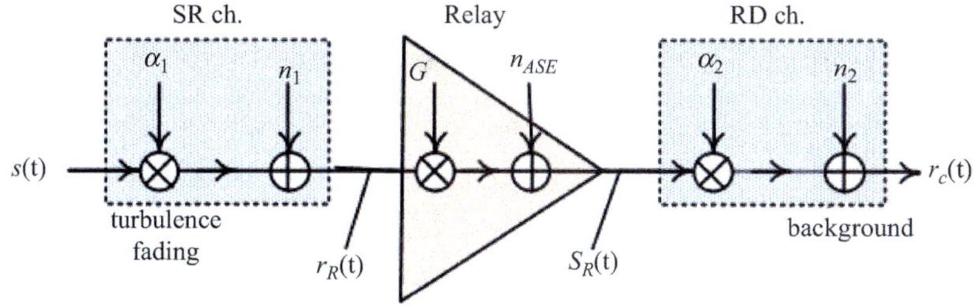

Figure 11.16 Block diagram of an all-optical relay terminal

effects of amplified spontaneous emission (ASE) noise and optical degree-of-freedom (DoF). ASE is the main source of noise in doped fiber amplifiers (DFAs) and will inevitably affect the performance of a relay. DoF quantifies the ratio of optical filter bandwidth to the electrical bandwidth and can be on the order of 1,000 unless narrowband optical filtering is employed [13].

For the sake of presentation, we consider a dual-hop intensity-modulation direct-detection FSO transmission system with BPPM. The source terminal modulates and transmits the intensity-modulated signal. At the relay terminal (see Figure 11.16), the received signals are optically filtered and amplified. In practice, EDFAs are typically employed for signal amplification.

Under the assumption of a high gain EDFA (~30 dB) at the destination, a shot-noise limited system is considered in which the effect of thermal noise can be neglected and only shot noise caused by background radiation is dominant. Therefore, the noise at the destination is modeled by additive noise. Photon counting methodology is employed to derive closed form expressions for the end-to-end SNR and the outage probability.

Let us assume that source terminal transmits photons with rate $s(t)$ (with an average of m_1) which obeys a Poisson distribution. The photon rate at the relay can be expressed as [14]

$$r_R(t) = \alpha_1 s(t) + n_1(t) \tag{11.39}$$

where α_1 and n_1 denote the atmospheric fading and background noise, respectively, for the source-to-relay (SR) channel. The relay transmits the photons to the destination with an average of m_2. The photon count due to background radiation is modeled by a negative binomial distribution (also known as Bose–Einstein) with an average of m_R [15]. The averages m_1, m_2, and m_R are related to the physical characteristics of the transmitter and the channel such as wavelength, transmit power, modulation rate, and the atmospheric conditions. Their exact calculations can be found in [13]. The total photon rate at the destination is given by

$$r_D(t) = G\alpha_2\alpha_1 s(t) + G\alpha_2 n_1(t) + \alpha_2 n_{\mathrm{ASE}}(t) + n_2(t) \tag{11.40}$$

where α_2 and n_2 denote the atmospheric fading and the background noise of the relay-to-destination (RD) channel and G is the gain of the all-optical AF relay.

Recall that a log-normal distribution for the statistics of the atmospheric turbulence is assumed, therefore α_i follows the log-normal distribution.

The ASE is modeled by an additive noise (similar to the background noise) with a count of $m_{ASE} = n_{sp}(G-1)$ and M DOF [13]. The value of M depends on both spatial and temporal filtering at the relay terminal. Assuming only one spatial mode, we have $M = 2B_o/B_e$ where B_o and B_e denote the equivalent optical and electrical bandwidths of optical filters, respectively. The factor n_{sp} denotes the spontaneous emission factor of the amplifier and can be approximately taken equal to unity for most practical purposes.

The photon count statistics (denoted by k) at the destination obeys Laguerre distribution [16]. Therefore, the probability of k counts is given by

$$\begin{aligned} p_{\text{count}}(k) &= Lag(k, s, n, M) \\ &= \frac{n^k}{(1+n)^{k+M}} \exp\left(-\frac{s}{1+n}\right) L_k^{M-1}\left(-\frac{s}{n(1+n)}\right) \end{aligned} \tag{11.41}$$

where s and n denote the average photon counts, respectively, for the desired signal and the total background noise. In (11.41), $L_k^M(x)$ is the generalized Laguerre polynomial of degree k defined as [17]

$$L_k^M(x) = \sum_{i=0}^{k} \frac{(-x)^i (k+M)!}{i!(k-i)!(M+i)!} \tag{11.42}$$

In our case, based on (11.40), we have $a = G\alpha_1\alpha_2 m_1$ and $b = m_R + G\alpha_2 m_R + \alpha_2 m_{ASE}$. In practical situations, the average photon counts due to the background noise m_R is much smaller than the average photons received at the relay (or destination), i.e., $m_R << \overline{\alpha_i} m_i, m_{ASE}, i = 1, 2$ where $\overline{\alpha_i} = E(\alpha_i)$ denotes the statistical average of the atmospheric fading. Hence, we have $b \cong \alpha_2 m_{ASE}$.

The end-to-end SNR of the system under consideration is given by $\gamma_T = a^2/\sigma_n^2$ where σ_n^2 is the variance of the photon counts at the receiver and can be expressed as $\sigma_n^2 = s + Mn(1+n) + 2sn$ for Laguerre distribution. In full-CSI relaying, the relay has the full knowledge of the instantaneous SR channel condition and sets the amplifier gain to $G = m_2/(\alpha_1 m_1 + m_R) \cong m_2/\alpha_1 m_1$. In semi-blind relaying, the gain of the amplifier is fixed to $G = m_2/E(\alpha_1 m_1 + m_R) \cong m_2/m_1 E(\alpha_1)$. These choices of the amplifier gain aim to invert the fading effect of the source-to-relay channel while limiting the output power of the relay if the fading amplitude of the first hop, α_1, is low [18]. Replacing G in γ_T, we have [3]

$$\gamma_T^{-1} = \begin{cases} 2\gamma_1^{-1} + M\gamma_1^{-2} + \gamma_2^{-1}, & \text{CSI-assisted} \\ M\gamma_1^{-2} + \overline{\gamma_1}\gamma_1^{-1}\gamma_2^{-1} + 2\gamma_1^{-1}, & \text{Semi-blind} \end{cases} \tag{11.43}$$

where $\widehat{\gamma}_i = m_i\alpha_i, i = 1, 2$ denotes the conditional SNR of link i.

The instantaneous SNR for the ith link is given by $\widehat{\gamma}_i = \gamma_{avg,i}\exp(2\chi_i)$ where $\gamma_{avg,i} = m_i\beta_i$. It obeys a log-normal distribution with pdf which is denoted by $p(\widehat{\gamma}_i)$.

The outage probability can be then expressed as

$$P_{\text{out}} = \Pr(\gamma_T \le \gamma_{\text{th}}) = \int_0^{\infty} \Pr(\gamma_T \le \gamma_{\text{th}} | \gamma_1) f(\gamma_1) d\gamma_1. \tag{11.44}$$

For full-CSI relaying, this yields [3]

$$\begin{aligned} P_{\text{out}} &= \int_0^{\gamma^*} \Pr\left(\gamma_2 \ge \left(\gamma_{th}^{-1} - 2\gamma_1^{-1} - M\gamma_1^{-2}\right)^{-1} | \gamma_1\right) f(\gamma_1) d\gamma_1 \\ &\quad + \int_{\gamma^*}^{\infty} \Pr\left(\gamma_2 \le \left(\gamma_{th}^{-1} - 2\gamma_1^{-1} - M\gamma_1^{-2}\right)^{-1} | \gamma_1\right) f(\gamma_1) d\gamma_1 \\ &= 1 - Q\left(\frac{0.5 \ln\left(\gamma^* / \gamma_{avg,1}\right) - \mu_1}{\sigma_1}\right) \\ &\quad + \int_{\gamma^*}^{\infty} \left(1 - Q\left(\frac{0.5 \ln\left(\gamma_{TH,CSI} / \gamma_{avg,2}\right) - \mu_2}{\sigma_2}\right)\right) f(\gamma_1) d\gamma_1 \end{aligned} \tag{11.45}$$

where $\gamma_{TH-CSI}^{-1} = \gamma_{th}^{-1} - 2\widehat{\gamma}_1^{-1} - M\widehat{\gamma}_1^{-2}$. In (11.45), the lower limit on the integral γ^* is the first positive (nonzero) solution of $\gamma_{th}^{-1} - 2\widehat{\gamma}_1^{-1} - M\widehat{\gamma}_1^{-2} = 0$ for γ_1 and given by $\gamma^* = M / \left(\sqrt{1 + M\gamma_{th}^{-1}} - 1\right)$.

For semi-blind relaying, we have [3]

$$\begin{aligned} P_{\text{out}} &= \int_0^{\gamma^*} \Pr\left(\gamma_2 \ge \left(\frac{\bar{\gamma}_1 \gamma_1^{-1}}{\gamma_{\text{th}}^{-1} - 2\gamma_1^{-1} - M\gamma_1^{-2}}\right) | \gamma_1\right) f(\gamma_1) d\gamma_1 \\ &\quad + \int_{\gamma^*}^{\infty} \Pr\left(\gamma_2 \le \left(\frac{\bar{\gamma}_1 \gamma_1^{-1}}{\gamma_{\text{th}}^{-1} - 2\gamma_1^{-1} - M\gamma_1^{-2}}\right) | \gamma_1\right) f(\gamma_1) d\gamma_1 \\ &= 1 - Q\left(\frac{0.5 \ln\left(\gamma^* / \gamma_{\text{avg},1}\right) - \mu_1}{\sigma_1}\right) \\ &\quad + \int_{\gamma^*}^{\infty} \left(1 - Q\left(\frac{0.5 \ln\left(\gamma_{\text{TH,SBR}} / \gamma_{\text{avg},2}\right) - \mu_2}{\sigma_2}\right)\right) f(\gamma_1) d\gamma_1 \end{aligned} \tag{11.46}$$

where $\gamma_{\text{TH-SBR}}^{-1}$ is given by

$$\gamma_{\text{TH-SBR}}^{-1} = \frac{\gamma_{\text{th}}^{-1} - 2\widehat{\gamma}_1^{-1} - M\widehat{\gamma}_1^{-2}}{\overline{\widehat{\gamma}}_1 \widehat{\gamma}_1^{-1}}. \tag{11.47}$$

In the following, we present the outage performance results for all-optical AF relaying system under consideration. We assume that the relay is placed at equal

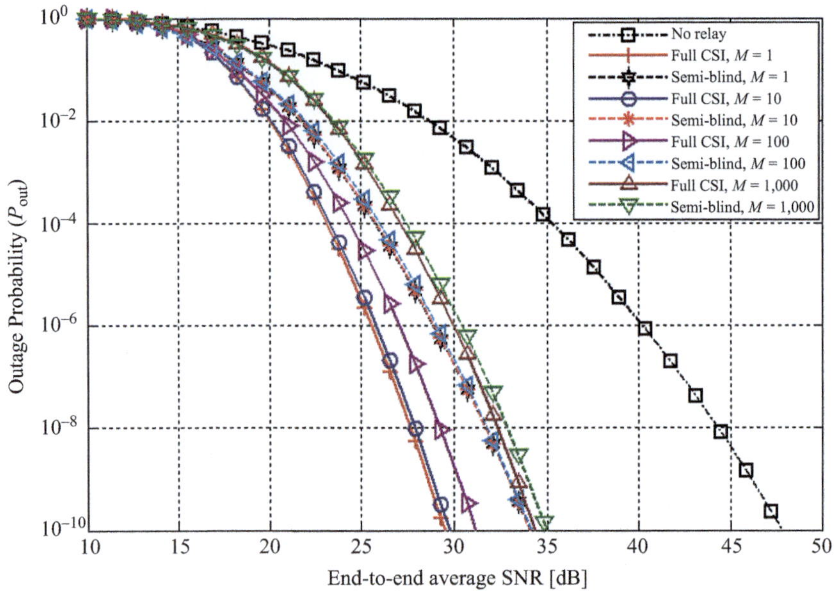

Figure 11.17 Outage probability of all-optical AF relaying

distance from both source and destination, and the links are balanced (i.e., $\overline{\gamma_1} = \overline{\gamma_2}$). A visibility of 10 km and an attenuation factor of 0.1 per km are assumed. The turbulence fading variance of the overall channel, i.e., direct link between the source and destination, is given by $\sigma_\chi^2 = 0.3$ and SNR threshold is set as $\gamma_{th} = 15$ dB.

Figure 11.17 illustrates the outage probability for full-CSI and semi-blind all-optical AF relaying based on the derived expressions in (11.45) and (11.46). As a benchmark, the performance of direct transmission is provided. We also illustrate the performance for different values of DOF, i.e., $M = 1$, 10, 100, and 1,000. Note that $M = 1$ is the ideal case where the effect of the DOF is not taken into account, while practical values of M range from 100 to 1,000 [13]. For the ideal case (i.e., $M = 1$), at a target outage probability of 10^{-6}, relay-assisted transmission with full-CSI relay results in a performance gain of 14.7 dB in comparison to the direct transmission. This impressive improvement comes from the fact that fading variance is distance-dependent in FSO channel and relay-assisted transmission takes advantage of the resulting shorter hops. The improvement reduces to 14.4 dB, 13.2dB and 10.3 dB for $M = 10$, 100, 1,000, respectively. Noting that practical values of M are in the range of 100 to 1,000, a significant performance gain over direct transmission is still maintained. We further observe that for small values of M, systems with full-CSI relay outperform those with semi-blind relay. Specifically, for $M = 1$ (i.e., ideal case), the performance gap at $P_{out} = 10^{-6}$ is 3.4 dB. For more practical values, this gap decreases and semi-blind relaying yields nearly the

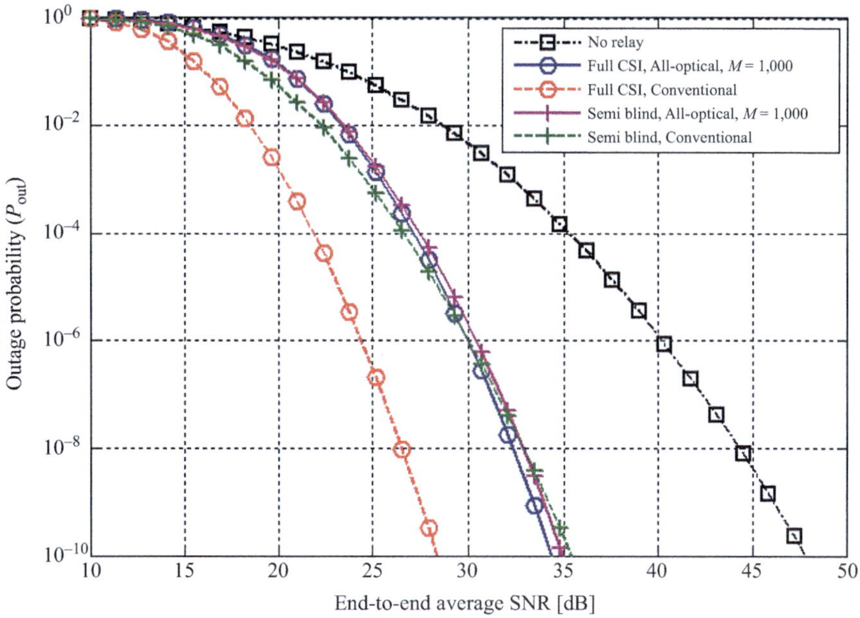

Figure 11.18 Outage probability for CSI-assisted and semi-blind relaying with optical and electrical amplification

same performance as its full-CSI counterpart. For example, for $M = 1{,}000$, the performance difference between semi-blind and full-CSI is merely 0.4 dB.

In Figure 11.18, we present a comparison between all-optical AF relaying with $M = 1{,}000$ and conventional AF relaying with OE/EO conversions. The outage probability of conventional AF relaying system can be obtained by applying the main results of [19–21] to log-normal atmospheric channels. Under the assumption of CSI-assisted relay, conventional relaying outperforms the all-optical counterpart. For semi-blind relay deployment, the performance of all-optical relaying is slightly worse than the conventional counterpart for low to medium SNR region and becomes nearly identical for high SNRs. It should be emphasized here that these results have been obtained for a relatively large value of DOF (i.e., $M = 1{,}000$). The performance of all-optical relaying can be further improved by decreasing the value of M which is possible, in practice, by a better filter design. Our comparison demonstrates that semi-blind all-optical relaying presents a favorable trade-off between complexity and performance, and can be used as a low-complexity solution.

11.6 Summary

Despite the major advantages of FSO technology and variety of its application areas, its widespread use has been hampered by its rather disappointing link reliability particularly in long ranges due to atmospheric turbulence-induced fading. Relay-assisted systems have been introduced as an effective method to extend coverage and mitigate the effects of fading in FSO links.

In this chapter, we have analyzed and investigated the outage performance of relay-assisted FSO links with AF and DF relays. For serial relaying, it has been demonstrated that the outage probability is minimized when the consecutive nodes are placed equidistant along the path from the source to the destination. For parallel relaying, it has been shown that all of the relays should be located at the same place (along the direct link between the source and the destination) closer to the source and the exact location of this place depends on the system and channel parameters.

Multi-hop parallel relaying which is the combined use of serial (multi-hop) and parallel relaying for FSO mesh networks has been also studied. Our analysis yields that multi-hop parallel relaying smartly exploits the distance dependency of the fading variance in FSO systems and bring substantial improvements with respect to standalone uses of multi-hop and parallel relaying.

As an alternative way of implementation, all-optical relaying has been also considered. Unlike the earlier relaying schemes, the signals are processed in optical domain and therefore the requirement of OE and EO conversions is avoided. Comparisons between conventional and all-optical relaying demonstrate that the latter presents a favorable trade-off between complexity and performance and can be used as a low-complexity solution.

References

[1] Safari M, Uysal M. Relay-assisted free-space optical communication. IEEE Transactions on Wireless Communications. 2008;7(12):5441–5449.

[2] Kashani MA, Safari M, Uysal M. Optimal relay placement and diversity analysis of relay-assisted free-space optical communication systems. IEEE/OSA Journal of Optical Communications and Networking. 2013;5(1):37–47.

[3] Kashani MA, Rad MM, Safari M, Uysal M. All-optical amplify-and-forward relaying system for atmospheric channels. IEEE Communications Letters. 2012;16(10):1684–1687.

[4] Kashani MA, Uysal M. Outage performance and diversity gain analysis of free-space optical multi-hop parallel relaying. Journal of Optical Communications and Networking. 2013;5(8):901–909.

[5] Gagliardi R, Karp S. Optical Communications, Wiley-Interscience, New York, NY; 1995.

[6] Osche GR. Optical detection theory for laser applications. Optical Detection Theory for Laser Applications, by Gregory R Osche, pp 424 ISBN 0-471-22411-1 Wiley-VCH, July 2002. 2002; p. 424.

[7] Haas SM, Shapiro JH. Capacity of wireless optical communications. IEEE Journal on Selected Areas in Communications. 2003;21(8):1346–1357.

[8] Acampora AS, Krishnamurthy SV. A broadband wireless access network based on mesh-connected free-space optical links. IEEE Personal Communications. 1999;6(5):62–65.

[9] Proakis JG. Digital communications fourth edition, 2001. McGraw-Hill Companies, Inc., New York, NY; 1998.

[10] Goldberg DE, Holland JH. Genetic algorithms and machine learning. Machine Learning. 1988;3(2):95–99.

[11] Forbes C, Evans M, Hastings N, Peacock B. Statistical distributions. John Wiley & Sons, Hoboken, NJ; 2011.

[12] Olsson NA. Lightwave systems with optical amplifiers. Journal of Lightwave Technology. 1989;7(7):1071–1082.

[13] Razavi M, Shapiro JH. Wireless optical communications via diversity reception and optical preamplification. IEEE Transactions on Wireless Communications. 2005;4(3):975–983.

[14] Rosenberg S, Teich M. Photocounting array receivers for optical communication through the lognormal atmospheric channel. 2: Optimum and suboptimum receiver performance for binary signaling. Applied Optics. 1973;12(11):2625–2635.

[15] Desurvire E. Erbium-doped fiber amplifiers: Principles and applications. Wiley-Interscience; 2002.

[16] Perina J. Superposition of coherent and incoherent fields. Physics Letters A. 1967;24(6):333–334.

[17] Gradshteyn IS, Ryzhik IM. Table of integrals, series, and products. Academic Press, San Diego, CA; 2014.

[18] Laneman JN, Wornell GW. Energy-efficient antenna sharing and relaying for wireless networks. In: Wireless Communications and Networking Conference, 2000. WCNC. 2000 IEEE. vol. 1. IEEE; 2000. pp. 7–12.

[19] Da Costa DB, Assa S. End-to-end performance of dual-hop semi-blind relaying systems with partial relay selection. IEEE Transactions on Wireless Communications. 2009;8(8): 4306–4315.

[20] Hasna MO, Alouini MS. A performance study of dual-hop transmissions with fixed gain relays. In: Acoustics, Speech, and Signal Processing, 2003. Proceedings (ICASSP'03). 2003 IEEE International Conference on. vol. 4. IEEE; 2003. p. IV–189.

[21] Hasna MO, Alouini MS. End-to-end performance of transmission systems with relays over Rayleigh-fading channels. IEEE Transactions on Wireless Communications. 2003;2(6):1126–1131.

Chapter 12

Experimental test of maximum likelihood thresholds based on Kalman filter estimates in on–off keyed laser communications in atmospheric turbulence

William Brown[1], Bruce Wallin[1], Daniel Lesniewski[1], David Gooding[1], James Martin[1] and Arun K. Majumdar[1]

12.1 Introduction

This chapter presents a new Kalman filter approach for tracking the turbulence probability moments for updating optimum thresholds in optical communications. The improvement in the performance of on–off keyed (OOK) data transmissions in an atmospheric laser link is shown by using recursive Kalman estimators to provide timely updates to the optimum threshold determinations. Experimental bit error rates (BERs) for a number of different approaches are compared. The central experimental result of this chapter is a 41% improvement in the BER when Kalman tracking is used. The value of the laser link study in this chapter is in the new techniques suggested for improving other low and high data rate optical links, at all strengths of atmospheric turbulence.

Laser communications offers many advantages over microwave communications and has been studied since the invention of the laser. Optical links have a potential data-carrying capacity that is extremely high (multi-gigabits/s or higher). Laser communications also is not subject to spectral or wavelength regulation or restrictions. Laser systems have the additional advantages of lower cost and quick implementation.

Laser communication links operating in the atmosphere face several significant disadvantages. First, a clear line of sight between the transmitter and the receiver is required, to prevent the laser signal from being scattered or absorbed. Second, the turbulence of the atmosphere produces beam-steering and scintillation effects that inhibit the steady, reliable transfer of laser photons from the transmitter to the receiver.

[1]Department of Physics, Colorado State University-Pueblo, USA

As streams of digital data are sent through the atmosphere, scintillation causes serious, random dropouts in the data stream. Analytical studies of this phenomena have been done over many years [1–10], and experimental data have also been collected [11–15]. Some of the mitigating approaches consist of designing optical receivers to process the incoming data in ways to approach statistical optimum methodologies. Others have suggested multiple receivers to capture multiple streams of data signals that are statistically uncorrelated.

For binary data transmissions and an OOK modulation format of the laser signal, the core of the problem of maintaining acceptable performance of the laser link is to find a threshold with which to compare incoming pulsed signals. In an OOK signal format, the laser is turned "ON" for a brief period when a "one" is sent, and the laser is turned "OFF" as much as possible for a similar brief period when a "zero" is sent. If the statistical parameters of the signals resulting from the random influences of the atmosphere were constant in time, the optimum threshold also remains always a constant in time. A single threshold to produce the minimum number of errors in the data transmission and reception would be easily calculated and would never require modification.

Unfortunately, the statistical parameters characterizing atmospheric turbulence effects are time varying, and mitigating techniques must be taken to compensate for the degrading actions of the atmosphere. Some of the possible techniques are the following. First, the laser transmitter power can be increased to compensate for the signal losses, but this is not always practical or cost-effective. Second, a forward error correction coding can be utilized with its attendant redundancy and a lower effective data rate. Third, multiple receivers can be employed as mentioned above. This may not necessarily be a practical solution when the additional cost and complexity of the additional components and receiver processing are considered. Fourth, an optimum threshold may be calculated on the basis of prior data and recalculated as frequently as possible. Recalculating as often as possible to keep up with the time-varying changes in the atmosphere must be weighed against the necessary complexity and speed of the required receiver processor. Fifth, the slowly time-varying statistical parameters may be estimated at some future time, in anticipation of forthcoming data [16]. Sixth, a small number of a priori values may be used to constantly update the time-varying optimum threshold just prior to comparing the current sample with the threshold. Seventh, a running or recurring estimate of the current state of the statistical parameters with which to determine the current optimum threshold can be updated on a pulse-to-pulse basis without using all of the prior data history. This approach is one that applies running Kalman filters to estimate the current normalized irradiance mean value and standard deviation, and then estimates the optimum threshold periodically. This approach represents an efficient compromise between having timely updates as the signal parameters change in time and a limited amount of computation required in the receiver.

It is this last approach that has been applied in this chapter, and the concept has been verified by recent experimental results using a nearly horizontal laser communication channel. A comparison of the effectiveness of several different possible approaches is shown in terms of the BERs.

12.2 Principle of the method of maximum likelihood thresholds based on Kalman filter estimates

12.2.1 Probabilistic nature of the propagating signals through atmospheric turbulence

Probability models facilitate the processing of laser communications data degraded by the turbulent atmosphere. They are often assumed to be Gaussian, as in Van Trees [17] because of their ease of use. Often the Central Limit Theorem is invoked to justify such a choice. But in atmospheric laser links, Gaussian models are not always the best fit to the experimental data. Models of the conditional probability densities used in this study are characterized by their mean values and their variances. The conditional probability density function (pdf) for the signal when the laser is "ON" is given by p (x/H1), where x is the measured voltage value given that the laser was "ON" or that hypothesis one is true. The conditional pdf for the signal when the laser is "OFF" is given by p (x/H0), where x is the measured voltage value given that the laser was "OFF" or that hypothesis zero is true

12.2.2 Maximum likelihood thresholds

Using the above pdf models, the object of this study is to find an easily computable threshold based on maximum likelihood to minimize the probability of error in decoding each bit of information transmitted on a timely basis. The BER is given as the following for analog signals.

$$BER = \int_{\eta}^{\infty} p(x/H_0)dx + \int_{-\infty}^{\eta} p(x/H_1)dx \tag{12.1}$$

For the discrete case, the form of (12.1) is the following

$$BER = \sum_{i=1}^{N} x_{0i>\eta} + \sum_{i=1}^{M} x_{1i<\eta} \tag{12.2}$$

Two sets of samples $\{x_i\}_0$, where $i = 1, \ldots, N$ and $\{x_i\}_1$, where $i = 1, \ldots, N$ are collected over a period of time given that a zero is sent and given that a one is sent, respectively. In the zero set of size N, some members will have values that exceed η. In the one set of size M, some members will fall short of η, which results in the other kind of error, called a miss. The total number of detection errors will be the sum of the numbers of elements in the "zero" set producing a false alarm plus the number of elements in the "one" set that result in a miss. The BER then will be this total number of detection errors divided by N. In this equation, n represents the ordinality or the number of elements of the sub sets of the "zero" and the "one" sets in the parentheses.

The threshold is η, and the object is to find one based on maximum likelihood that will minimize the BER in (12.2). For maximum likelihood, the two conditional pdf models p (x/H1) and p (x/H0) are combined to form a likelihood ratio function as seen in (12.3).

$$\Lambda(x) \equiv \frac{p(x/H_1)}{p(x/H_0)} \tag{12.3}$$

Wherever a sample point x has a corresponding value of the likelihood function $\Lambda(x)$ exceeding 1, then it is decoded as a "1." If a sample point x has a corresponding value of $\Lambda(x)$ less than 1, then it is decoded as a "0." If the likelihood ratio function is exactly equal to 1 for some sample value, then the sample can be decoded as either a "1" or a "0". The intersection point of p (x/H1) and p(x/H0) where the two conditional pdfs are equal will occur at some sample value $x = \eta$, which is the threshold dividing the decision space and represents the threshold to minimize the probability of error. If a sample point falls below η, then the sample point is decoded as a "0." If a sample point falls above η, then the sample point is decoded as a "1." The regions of the two pdfs that give rise to errors are the area where the two pdfs overlap.

For a perfectly stationary process, the minimum probability of error threshold η will not vary in time, as the turbulence moments are constant. If the moments are changing in time, the optimum, minimum probability of error η will also have to change if it is to keep up with the turbulence, and this will require its frequent recalculation. The frequency of recalculation will depend on the correlation times of the variations of the atmospheric turbulence. If η is fixed as a constant while the statistical moments of the atmospheric turbulence are varying in time, the BER measured will be far from optimum. If η is frequently recalculated, the BER can be expected to more closely approximate the optimum as conditions change.

12.2.3 Turbulence-tracking Kalman filter

Practically, there are a number of possible ways to combat the fact that the atmospheric turbulence parameters are time varying. First, in an optical communications link, it is possible to process and to retrieve the transmitted data post facto or in an a posteriori way. The data sample points for the entire segment of the communication transmission are collected with which the moments for the conditional densities are determined, the appropriate pdf models are applied, the optimum threshold is calculated and then the entire data record is decoded bit by bit. The advantage of this approach is that the turbulence moments can be calculated once for the entire data record. But this approach will only be valid if the turbulence moments have not changed over the entire time the data record was obtained. The disadvantages of this approach are that the decoded information in the data transmission will not be obtained in real-time and a large quantity of data will have to be processed in a single step. Second, η can be re-calculated frequently in time periods approximating the correlation time of the atmospheric turbulence. Third, the optimum threshold can be calculated in terms of the mean and the variance that are expected in the relatively slowly changing atmospheric turbulence at some time in the future. These are predictive thresholds that will function successfully for large numbers of bits in a high-data rate communications system. This experimental study has taken variations of all of these approaches and compared the results with a new technique that constantly updates the expected values of the moments of p(x/H1) on a block-by-block basis. The new technique employs Kalman filters to track the time-varying moments in a way that is not computationally intense and can also be easily done in near real-time.

The literature is filled with many detailed discussions of the Kalman filter. See References [18] and [19], for example. The Kalman filter is implemented in three steps. There are the initial estimates, the time updates and then the measurement updates.

12.2.3.1 Initial estimates

The Kalman equations reduce to very simple forms, as a result of estimating the simple scalar state of the m_L and $\sigma_{L.}$ The conditional mean and the conditional standard deviation of p(*x*/H1) are m_L and σ_L. The Kalman process in these two cases of estimation are given by

$$\hat{y}_k = \hat{y}_{k-1} + w_k \tag{12.4}$$

where *y* is either the mean or the variance of the laser signal. The *w* variable represents the process noise in Kalman filter theory, which is assumed to be Gaussian with a zero mean and a variance of *Q*. The measurement of the laser signal mean and standard deviation follow

$$z_k = y_k + v_k \tag{12.5}$$

where *v* is assumed to be the measurement noise in Kalman filter theory, also assumed to be Gaussian with a zero mean and a variance of R. Q and R are adjusted to "tune" for best results or best convergence.

12.2.3.2 Time update equations

The time update equations used are the following:

$$\hat{y}_k^- = \hat{y}_{k-1} \tag{12.6}$$

$$P_k^- = P_{k-1} + Q \tag{12.7}$$

In these equations, the negative superscript implies that the variable is an a priori value. P represents the estimate error variance. The value of *Q* used in the Kalman filter is 1×10^{-3} and $R = 0.1$

12.2.3.3 Measurement update equations

The measurement update equations used are the following:

$$K_k = P_k^- / \left(P_k^- + R\right) \tag{12.8}$$

$$\hat{y}_k = \hat{y}_k^- + K_k\left(z_k - \hat{y}_k^-\right) \tag{12.9}$$

$$P_k = (1 - K_k)P_k^- \tag{12.10}$$

where *K* is the gain factor, and (12.8) is the actual estimate in terms of the a priori values.

12.3 Experimental procedure and results

The set of experimental runs used for this study were performed in Pueblo, Colorado, in July 2004. A laser link was set up between the campus of Colorado State University-Pueblo and its observatory located at distance of 9.3 km. The laser transmitter consisted of an 8-mW laser diode coupled to a beam expander that can be zoomed from a nearly collimated beam to a strongly diverging beam, facilitating beam acquisition and alignment at the receiver end of the link at the observatory. For the most collimated beam, the measured horizontal divergence is 4.6×10^{-4} Radians, and the measured vertical divergence is 6.9×10^{-4} Radians. A programmable PIC 16F84 microcontroller modulated the Coherent 635 nm elliptical laser diode. The processor was programmed to produce square-wave and repeating random bit patterns from the laser. The square wave frequency produced was 3,145 Hz.

The receiver consists of a Celestron C-14 telescope with a 0.36 m aperture, with a Schmidt-Cassegrain central obscuration of 8.9 cm diameter and which is equipped with a Hamamatsu R1463 photo multiplier detector operated in the non-photon counting mode. Its output is channeled to a PCI-DAS6035 Measurement Computing analog to a digital converter card in a computer to collect data. The sampling rate was varied from 200 Ksamples per second to 10 Ksamples per second, but all data used in this chapter were sampled at 20 Ksamples per second.

The path of the laser beam is generally over urban and undeveloped areas. The altitude of the transmitter is 1,510 m and the observatory altitude is 1,472 m. The height of the beam over the length of the propagation path varies significantly but has an average height of 73.4 m. The path is generally over an area starting from the top of a two-story building and ending in a telescope mounted on a tripod 9.3 km distant. The two-story building is approximately 6 m high and the telescope tripod is approximately 1.5 m high and was situated on the top of a localized mesa where the university's astronomical observatory is located.

Generally, the link was set up and aligned at dusk and operated after dark. Because of the distance, the low power of the laser and the poor beam quality of the laser diode, the BER performance was generally low. Experimental data are shown in Figures 12.1 and 12.2.

Figure 12.2 shows the data samples and also the numerical derivatives of the signal used for clock synchronization by locating the rising edges of the received pulses.

The data were collected and processed and analyzed in MATLAB®. The data acquisition and conditioning steps taken are shown in Table 12.1.

The experimental distribution of p (x/H0) is shown in a histogram Figure 12.3. The normalized and shifted mean value m_0 is 0.157, and the normalized and shifted standard deviation σ_0 is 0.0269. The distribution looks much like a Gaussian distribution except for the skewness to the right.

Although not a perfect fit, this conditional density is approximated by the following Gaussian probability density having the same mean and the same

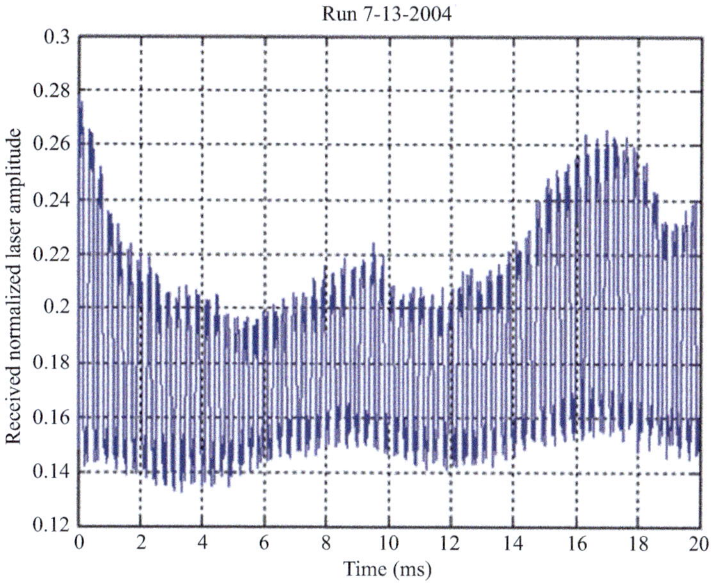

Figure 12.1 Typical laser receiver data record showing the slow time-variation of the high-photon count signal envelope

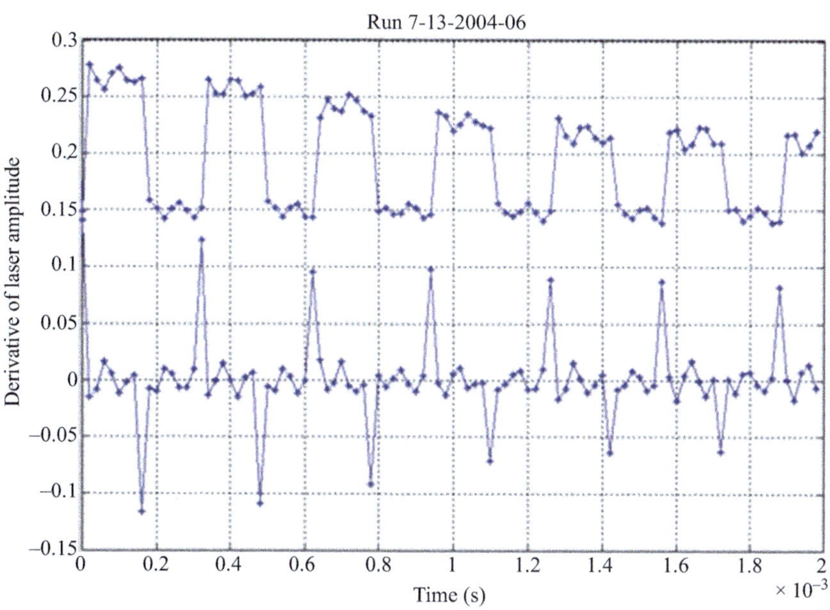

Figure 12.2 Data samples and numerical derivative

Table 12.1 The data acquisition and conditioning of the laser data

	Data acquisition procedure		
	Input	**Process**	**Output**
1	Received square-wave photomultiplier output	Analog-to-digital conversion at 50 Ksamples/s	Raw time sequence sample set
2	Raw time sequence sample set	Shift to positive values by subtracting the minimum value in set	Shifted raw sample set
3	Shifted raw sample set	Normalize by dividing each sample by the maximum value in set	Normalized, shifted sample set
4	Normalized, shifted sample set	Numerical differentiation to find rising edge of pulses	Differentiated sample set
5	Differentiated sample set and normalized, shifted sample set	Sub-sample normalized, shifted sample set 5 points beyond the leading edge for a sample given that a "1" was sent and to sub-sample 12 points beyond the rising edge for a sample given that a "0" was sent if the value of the derivative is greater than a threshold of 0.065	Preliminary sample sets of zero and one samples
6	Preliminary sample sets of zero and one samples	Eliminate outliers	Zero sample set $\{x_i\}_0$ One sample set $\{x_i\}_1$

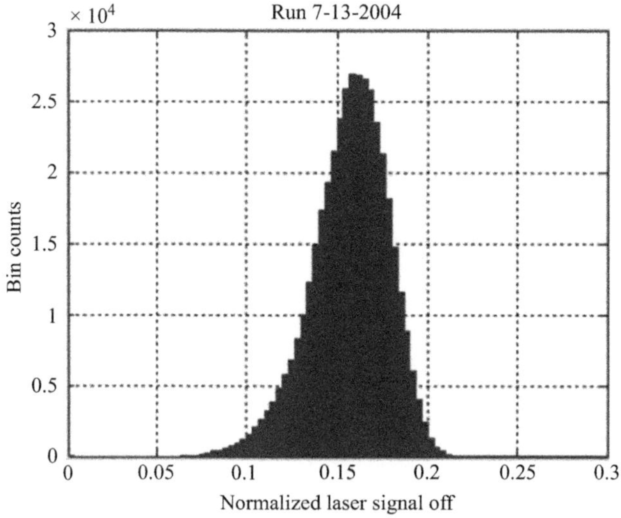

Figure 12.3 Laser signal histogram with laser off and 408,122 samples collected over 131 s at a sample rate of 50 Ksamples/s

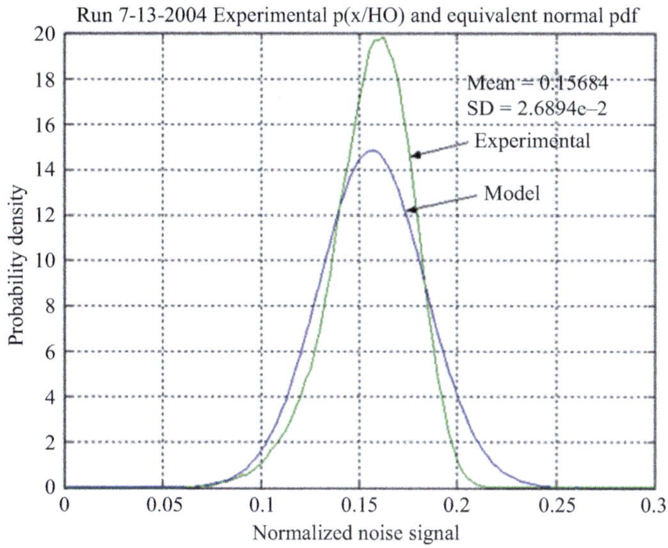

Figure 12.4 Comparison of the experimental and model p (x/H0)

variance. When the laser is off as best a finite extinction ratio will allow, the signal in the PMT will be shot noise limited because of the very high electronic gain. Shot noise is modeled as a Poisson process, but if the mean value is large, a Gaussian approximation is reasonable. A comparison of the empirical pdf and the approximating Gaussian pdf are shown in Figure 12.4.

$$p(x/H_0) = \frac{1}{\sqrt{2\pi}\sigma_0} e^{-\frac{(x-m_0)^2}{2\sigma_0^2}} \tag{12.11}$$

The condition probability density when the laser is "ON" is seen in the form of a histogram in Figure 12.5. It also is skewed, but to the left in this case. The mean value m_L in this case is 0.341 and the σ_L is 0.0907. The value of the strength of turbulence, $C_n{}^2$, corresponding to the measured normalized variance of irradiance fluctuations, $\sigma_I{}^2 = \frac{<I^2>-<I>^2}{<I>^2}$ for the propagation length of 9.5 km and aperture diameter of 0.36 m was estimated to be: $C_n{}^2 = 1.02 \times 10^{-16}$ $m^{-2/3}$. Aperture averaging was taken into account in estimating the strength of the turbulence parameter. These experiments were conducted at dusk and at night leading to this reasonable value. The assumption that the experiment was carried out in weak to moderate turbulence levels is validated. The equivalent Gaussian pdf (with the same mean and variance) and the experimental distribution are compared in Figure 12.6.

For analysis purposes, the conditional density is approximated by the following lognormal probability density, which is appropriate when the signal received consists of many photons and the strength of the turbulence is weak to moderate.

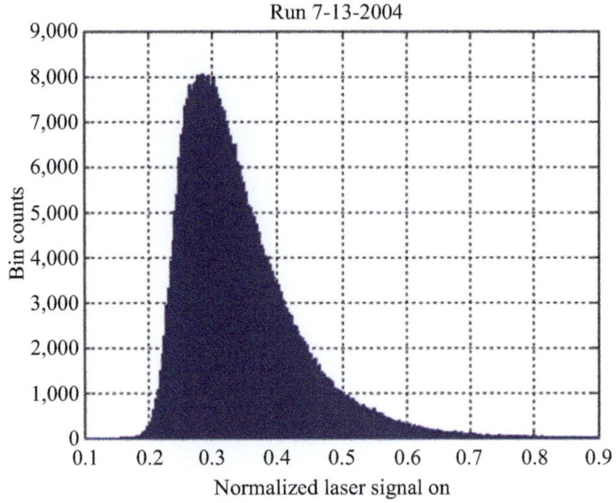

Figure 12.5 Laser signal histogram with laser ON with 408,122 samples collected over 131 s at a sample rate of 50 Ksamples/s

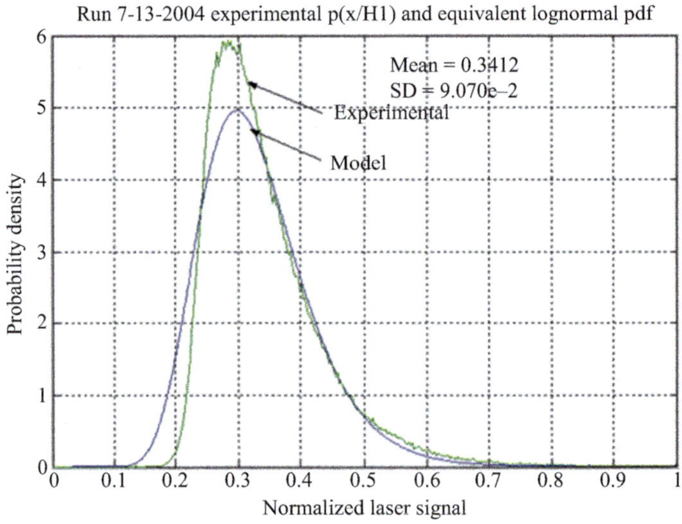

Figure 12.6 Comparison of the experimental and model p (x/H1)

The strength of the received signal in this case will be randomized by the atmospheric turbulence itself.

$$p(x/H_1) = \frac{1}{x\sigma_{1g}\sqrt{2\pi}} e^{-\frac{\left[\ln\left(x/m_{1g}\right)\right]^2}{2\sigma_{1g}^2}} \tag{12.12}$$

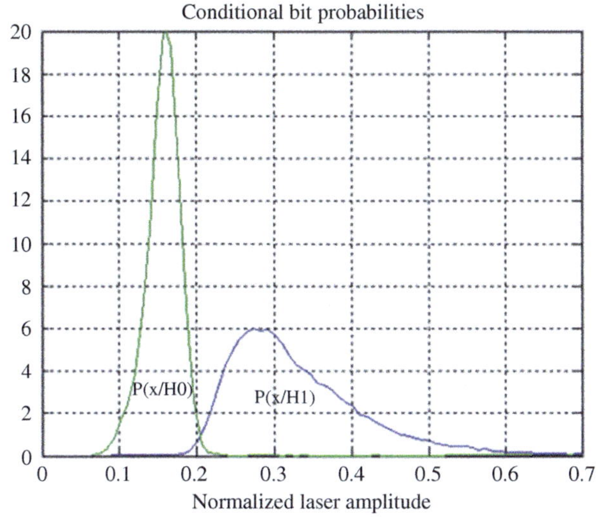

Figure 12.7 The two experimental probability densities and their overlap

where m_{1g} and σ_{1g} are the log irradiance mean and the log irradiance standard deviation. The log irradiance is assumed to be Gaussian. The relations between the lognormal and the log irradiance moments are the following. "L" subscripts refer to lognormal, and "g" refers to the Gaussian log irradiance.

$$\sigma_{1g}^2 = \ln\left(1 + \frac{\sigma_L^2}{m_L^2}\right) \tag{12.13}$$

$$m_{1g} = \frac{m_L^2}{\sqrt{m_L^2 + \sigma_L^2}} \tag{12.14}$$

The lognormal moments that are measured in the experimental distributions are σ_L and m_L. To use the lognormal pdf above, the lognormal moments are calculated in (12.13) and (12.14) from the measured experimental moments. The equivalent conditional lognormal probability density, with the equivalent m_{1g} and σ_{1g}, and the experimental distribution are compared in Figure 12.6.

Figure 12.7 shows the two experimental conditional pdfs and how they overlap. It is this overlapping area that gives rise to detection errors. The smaller the overlap area can be made, the lower the BER will be.

The Kalman filter that tracks m_L is seen is Figure 12.8.

The filter is seen generally to follow the signal but not to respond to fluctuations immediately. The advantage of a Kalman filter is that it does slowly change its estimated state. The filter estimate will maintain a best estimate based mostly on the past history of the turbulence process and to a much lesser degree on the latest changes. The Kalman estimate displays a desirable stability in the presence of atmospheric fluctuations. The correlation time of the signal fluctuations of much of

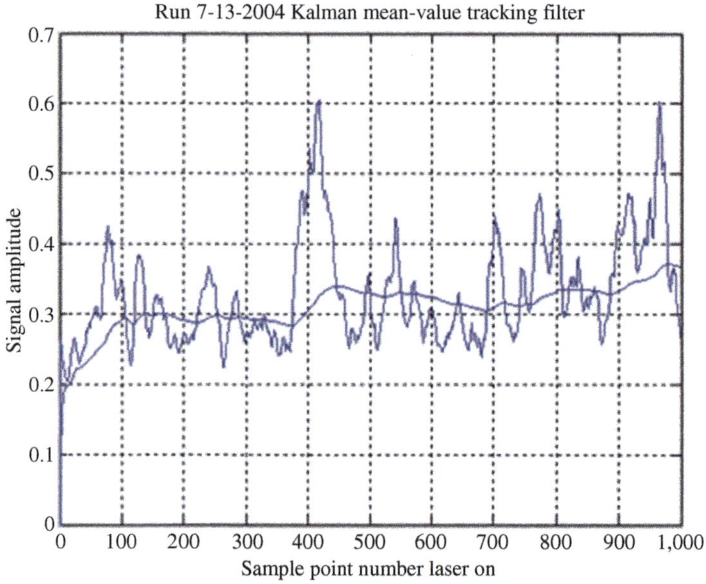

Figure 12.8 Laser signal mean Kalman tracking filter example

the experimental data is on the order of 31 ms, which corresponds to a frequency of 32.3 Hz. As a result, the latest Kalman estimate will generally provide a basis for calculating an optimum threshold that will suffice for many data symbols being transmitted at much higher rates than 32.3 Hz.

The Kalman filter that tracks σ_L is shown in Figure 12.9.

The standard deviation is calculated for each 50-sample set of points or for every 1 ms and then estimated from period to period using the Kalman filter defined above. This Kalman filter also displays the same desirable stability as shown in Figure 12.8.

The method by which the moments of p(x/H0) where calculated is shown in Figure 12.10. The entire data run was used in this estimation and is equivalent to characterizing the detector and receiver without any incoming laser signal before the actual testing would be done.

The Kalman estimation of the moments for this study is shown in Figure 12.11.

12.4 Comparison of threshold approaches

In order to compare the various threshold techniques, the total number of errors in each data run must be counted. Since this counting was not done in an electronic processor as the data arrived at the laser receiver, it was done after all of the data were collected. This post-experimental counting was done entirely in MATLAB as if it been had been done by a laser receiver processor as the data arrived. The post-arrival processing was done in accordance with the steps shown in Figure 12.12.

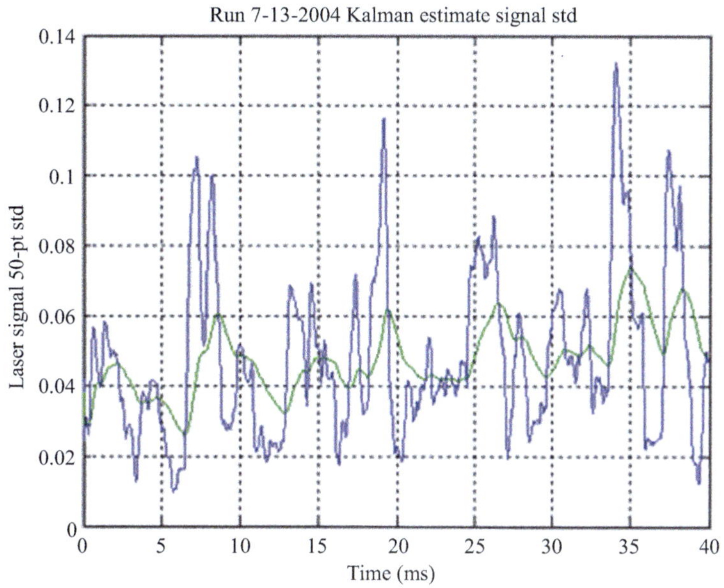

Figure 12.9 Laser signal σ_L tracking filter

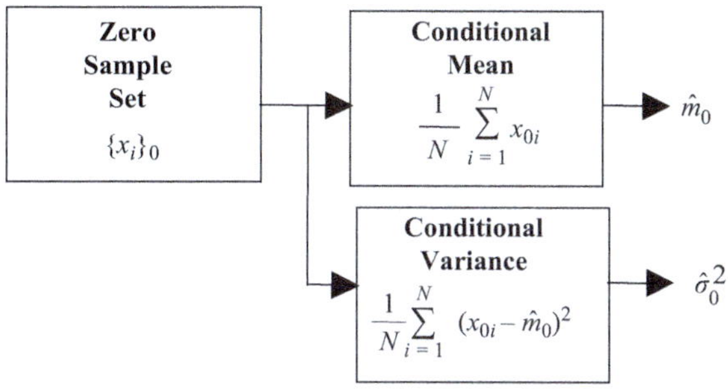

Figure 12.10 Calculating the noise moments

Processing routines were developed to estimate the BER for five different threshold approaches and the results are given in Table 12.2.

In approach 1, the entire data set of 408,122 sample points are used to calculate the means and variances of p(x/H0) and p(x/H1). This is the a posteriori approach that decodes the data at a later time than it was received. All of the data samples are taken and then the threshold decoding is done. The results of the approach show that having complete statistical knowledge over the entire 131-second data record yields a BER of 8.19×10^{-3}.

In approach 2, the updates of η are done in non-real time at each threshold decision based upon the future fifty data points to occur. For each data point compared with η, the processor has the benefit of anticipating the atmospheric

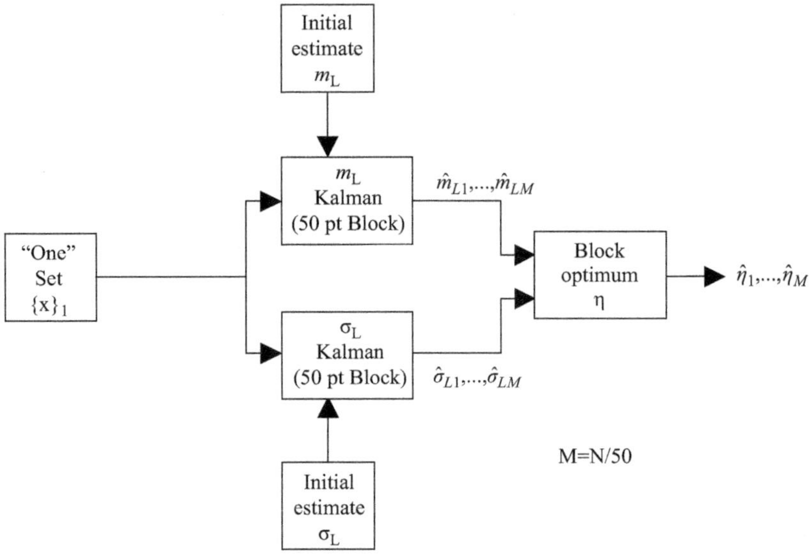

Figure 12.11 The recursive Kalman estimation of the moments of p(x/H1) and the thresholds

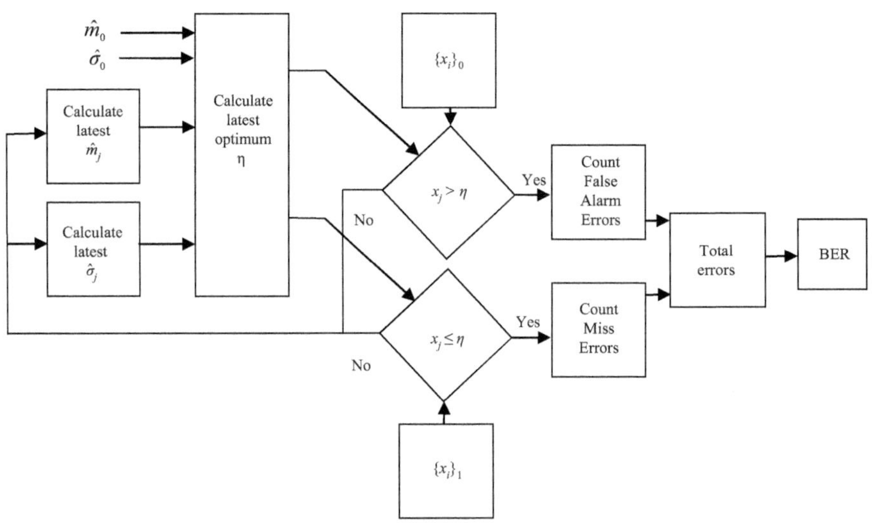

Figure 12.12 MATLAB post-processing and error counting of the Kalman estimation approach to compare with others

moments to come in the future. It is a method that is computationally intense since it requires processing fifty future data points in the record at each threshold test. This approach will also result in the unavoidable delay of fifty sampling periods. This approach provides the very best BER of 1.85×10^{-3}.

Table 12.2 Experimental results and comparison of approaches

	Processing	Source of data m_L	Source of data signal σ_L	Number of decoding errors	BER	Improvement over baseline %
1	Post	Past 408,122 samples	Past 408,122 samples	3,344	8.19×10^{-3}	16.9
2	Post	Future 50 samples	Future 50 samples	755	1.85×10^{-3}	81.2
3	Real time	Past 50 samples	Past 50 samples	4,026	9.86×10^{-3}	–
4	Real time	Recursive Kalman filter 50-pts prior to 50 threshold tests	Past 50 samples	2,392	5.86×10^{-3}	40.6
5	Real time	Recursive Kalman filter 50-pts prior to 50 threshold tests	Recursive Kalman filter 50-pts prior to 50 threshold tests	1,683	4.12×10^{-3}	58.2

Approach 3 is the baseline methodology to which all other approaches are compared. It is the baseline approach because it is a reasonable compromise when considering the amount of data to be processed in the updates of η and that processing can be done in near real-time. In this method, 50 samples occurring prior to the threshold decision of each data points are used to calculate the optimum threshold for the most recent atmospheric conditions. It is a method that is computationally intense since it requires processing fifty prior data points before the threshold decision is made for each and every data point. This approach provides the worst BER of 9.86×10^{-3}.

Approach 4 is the first instance of using a Kalman filter. In this case, m_L for p (x/H1) is recursively estimated and then used to update η for the next block of 50 sample points. As a result, the processing required is on the order of one fiftieth of the approaches recalculating η at each and every sample point test using either the past or future 50 sample points. The BER of 5.86×10^{-3} shows a significant improvement over the baseline approach.

Approach 5 uses two dual Kalman filters to recursively estimate both m_L and σ_L in 50-point blocks. An even larger improvement over the baseline is seen with a BER of 4.12×10^{-3}.

12.5 Conclusions

A new approach for mitigating atmospheric turbulence effects on free-space laser communication performance is presented. The method is based on evaluating Maximum Likelihood Thresholds using Kalman filter estimates in on–off keyed laser communications in atmospheric turbulence. Experimental results presented in this chapter clearly demonstrate the usefulness of the method. Although the results are shown for low data rate of 6.3 Kbps, the concept is valid for much higher data rate taking advantage of the higher bandwidth capability of laser communications.

When looking at the table of the comparison of various threshold approaches, it is shown clearly that the anticipatory a posteriori processing of 50 data points provides a very good bit error performance, if the extra computational burden of processing 50 prior data points for every likelihood ratio test at every sample point is acceptable. The next best approach is the use of the dual Kalman filters which yields a BER that is within a factor of two of a posteriori approach, but without having to reprocess 50 prior data points before each threshold decision. Clearly, this is an acceptable alternative to the a posteriori approach with its more extensive processing and its unavoidable delays. Because the required update rates of η is determined by the slowly varying atmospheric turbulence, high data rate links using the dual Kalman approach will allow many more times the number of data samples and bits to be tested against and validated by the most up-to-date η. The dual Kalman approach is a new approach that is expected to give better BER performance in low and high data rate atmospheric optical links and to simplify the necessary processing in the receiver. The central issue addressed in this chapter is whether Kalman Filtering is helpful in dealing with atmospheric turbulence, and this chapter shows that it is within the bounds of statistical approximations.

References

[1] P. Titterton and J. Speck, "Probability of Bit Error for an Optical Binary Communication Link in the Presence of Atmospheric Scintillation: Poisson Case," Applied Optics, 12(2), 425–426(1973).

[2] W.E.Webb and J.T. Marino, "Threshold Detection in On-Off Binary Communications Channel with Atmospheric Scintillation," Applied Optics, 14(6), 1414–1417 (1975).

[3] S.S. Gasparyan and R.A. Kazaryan, "Discrete Reception of Optical Signals in the Atmosphere," Soviet Journal of Quantum Electronics, 7(11), 1373–1375(1977).

[4] J.H. Churnside and Charles M. Mc Intyre, "Averaged Threshold Receiver for Direct Detection of Optical Communications through the Lognormal Atmospheric channel," Applied Optics, 16(2), 2669–2676(1977).

[5] R.L. Phillips and L.C. Andrews, "Effects of Atmospheric Turbulence on an Optical Communication System using a Receiver with memory," Applied Optics, 22(23), 3833–3836(1983).

[6] Y.S. Yur and Y. Weissman, "Turbulence Effects on the performance of a High order PPM Optical communication Channel Using an APD-Bases receiver," SPIE Vol. 756, Optical Technologies for Space Communications Systems, 756, 62–69(1987) in "Optical technologies for space communication systems," Proceedings of the Meeting, Los Angeles, CA, Jan. 15, 16, 1987. Bellingham, WA, Society of Photo Optical Instrumentation Engineers. 62–69(1987).

[7] D.A. Fares, "How Does Atmospheric Turbulence Affect Threshold Detection in Optical channels?" Microwave and Optical Technology Letters. 5(3), 138–141(1992).
[8] D.A. Fares, "Threshold Detection on Nonorthogonal Signaling in Turbulent Optical channels," Microwave and Optical Technology Letters. 6(2), 138–3141(1993).
[9] D.A. Fares, "How does Threshold Detection Limit Atmospheric in Number State Optical Channels?" Microwave and Optical Technology Letters. 7(8), 375–378(1994).
[10] S. Arnon and N.S. Kopeika, "Free-Space Optical communication: Detector Array Aperture for Optical Communication through thin Clouds," Optical Engineering. 34(2), 518–522(1995).
[11] G.R. Ochs, R.R. Bergman and J.R. Snyder, "Laser Beam Scintillation over Horizontal Paths from 5.5 to 145 Kilometers," Journal of the Optical Society of America. 59(2), 231–234(1969).
[12] B. Daino, M. Galeotti and D. Sette, "Error Probability of Binary Optical Communications in Turbulent Atmosphere Experimental Results," Alta Frequenza. XLII(2),80,E-83(1973).
[13] D.R. Wisely, M.J. McCullagh, P.L. Eardley, P.P. Smyth, D. Luthra, R. Couto De Miranda and R. Cole, "4 km Terrestrial line-of-Sight Optical free-Space Link Operating at 144 Mbit/s," SPIE Vol. 2123 Free-Space Laser communication Technologies VI 108–118(1994).
[14] K.E. Wilson and A. Biswas, "Effect of Aperture Averaging on a 570 Mbps 42 km Horizontal Optical Link," SPIE Vol. 2471 Free-Space Laser Communication Technologies, 62–69(1995).
[15] P.Stroud, "Statistics of Intermediate Duration Averages of Atmospheric Scintillation," Optical Engineering. 35(2), 543–548(1996).
[16] W.C. Brown, "Optimum Thresholds for Optical On-Off keying Receivers Operating in the turbulent Atmosphere," SPIE Vol. 2990 Free-Space Laser Communication Technologies, 254–261(1997).
[17] H.L.Van Trees, Detection, Estimation and Modulation Theory, Part I, Wiley, New York (1968).
[18] R.G. Brown and P.Y.C. Hwang, Intro to Random Signals and Applied Kalman Filtering, Wiley, New York (1992).
[19] M.S. Grewal and A.P. Andrews, Kalman Filtering: Theory and Practice Using Matlab, Wiley, New York (2001).

Chapter 13

Signal encryption strategies based on acousto-optic chaos and mitigation of phase turbulence using encrypted chaos propagation

Monish R. Chatterjee[1], Fares S. Almehmadi[1], Fathi H. Mohamed[1], and Ali A. Mohamed[1]

13.1 A-O Bragg diffraction of profiled optical beams

The phenomenon of acousto-optic (A-O) diffraction, first studied extensively in the late 1920s and 1930s [1,2], is used in many areas of signal processing, although this behavior is complex, and despite extensive generalized analyses and applications, comprehension of the phenomenon in its entirety is still incomplete [3]. A-O diffraction refers to the interaction of light and sound waves, and it is used to controllably diffract light beams. The behavior of an A-O cell depends on several system parameters, and, in particular, the thickness of the crystal L and the wave numbers of both sound (K) and light (k). These quantities are summarized as a figure-of-merit by the Klein–Cook parameter (Q) which is used to characterize the regimes of A-O operation [4]. For strict Bragg operation, which finds the most applications for these devices in practice, Chatterjee and Chen showed that Q should be larger than 8π [5]. In this regime, under perfect Bragg-matching, there is only one *diffracted* order. If Q is much smaller than one, the mode of operation is called the Raman–Nath regime, which is characterized by multiple diffracted orders with intensities given by Bessel functions [6,7].

Weak interaction theory is used in the analyses of AO diffraction, and this theory rests upon the assumption of uniform plane waves of sound and light. These assumptions, though not physically realistic, allow for tractable analyses and lead to observable results. In 1979, a plane wave theory of A-O interaction was developed by Korpel and Poon, in which the light and sound waves are represented by plane wave decomposition [7]. This theory (for uniform light and profiled sound) leads to well-known expressions for Bragg and Raman–Nath diffraction. However, there is much that is unknown about A-O diffraction with profiled beams, which are physically more realistic. Extensive research has been conducted to analyze special

[1]Department of Electrical and Computer Engineering, University of Dayton, USA

cases such as 2D and 3D sound profiles, and higher-order or strong interactions [8,9]. Nevertheless, examining A-O scattering under arbitrary beam profiles remains a complex problem. Beginning with the multiple plane wave theory due to Korpel and Poon [7], Chatterjee *et al.* obtained a transfer function formalism for evaluating scattered output profiles for arbitrary input profiles [10].

The transfer function approach utilizes a plane wave angular spectrum of the field distribution (valid for small deviations from the exact Bragg angle), which allows the scattered fields to be represented by Fourier integrals in the angular domain. This makes it possible to apply the FFT algorithm to numerically generate the scattered fields of arbitrary inputs. Transfer function expressions for both Bragg orders are developed and may be readily applied in the Fourier transform domain. These expressions are convenient for modeling the effects of various parameters (such as phase shift and Q), as well as arbitrary input profiles.

13.2 Transfer function formalism (TFF) for arbitrary optical profiles

Figure 13.1 illustrates the standard geometry of an A-O Bragg cell, showing two scattered orders created by an arbitrary input profile, assuming upshifted (i.e., a first-order beam with a frequency higher than that of the incident beam by the amount of the acoustic frequency in the cell) operation at the (exact) Bragg incidence. We assume that the profiled beam is nominally incident at the Bragg angle (i.e., the "ray" (or wave vector) corresponding to the center of the profiled beam is incident at the Bragg angle). In this figure, r and r' represent the transverse radial coordinates with respect to the direction of the incident field and the diffracted field, respectively. $E_0(r)$ and $E_1(r')$ represent the zeroth- and first-order scattered outputs, δ is the angular deviation from the nominal Bragg angle $\phi_B(\approx K/2k)$, and $\overline{K}$ is the acoustic wave vector [10].

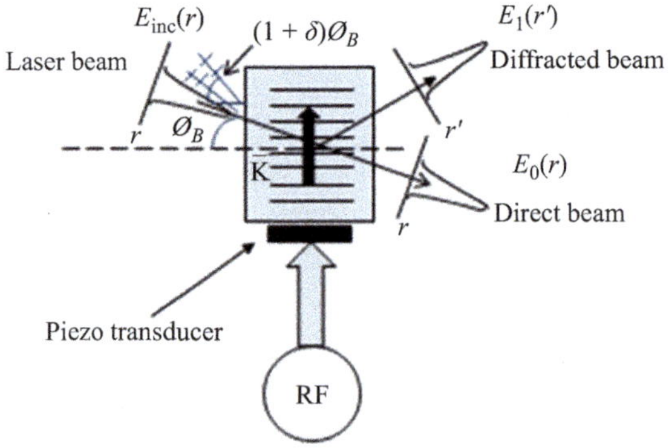

Figure 13.1 Bragg diffraction with an arbitrary incident beam profile

Both diffracted orders can be described by a set of coupled differential equations given by [7]:

$$\frac{d\tilde{E}_n}{d\xi} = -j\left(\frac{\widehat{\alpha}}{2}\right)\left[e^{\left\{-j\left(\frac{1}{2}\right)\left[\frac{\phi_{inc}}{\phi_B}+(2n-1)\right]Q\xi\right\}}\tilde{E}_{n-1} + e^{\left\{j\left(\frac{1}{2}\right)\left[\frac{\phi_{inc}}{\phi_B}+(2n+1)\right]Q\xi\right\}}\tilde{E}_{n+1}\right]. \tag{13.1}$$

In this equation, $\widehat{\alpha}(=kC|A|L/2)$ is the peak phase delay, $\xi(=z/L)$ is the normalized propagation distance in the sound cell, ϕ_{inc} is the incident angle corresponding to a uniform plane wave input, ϕ_B is the Bragg angle of the sound cell, and Q is the Klein–Cook parameter [10]. This general equation leads to the two scattered orders $\tilde{E}_0$ and $\tilde{E}_1$ for near-Bragg diffraction by setting $\phi_{\text{inc}}(=-(1+\delta)\phi_B)$. With this substitution for near-Bragg incidence with an angular deviation $\delta\phi_B$, the coupled equations reduce to the following:

$$\frac{d\tilde{E}_0}{d\xi} = -j\left(\frac{\widehat{\alpha}}{2}\right)e^{-jQ\xi\delta/2}\,\tilde{E}_1 \tag{13.2}$$

$$\frac{d\tilde{E}_1}{d\xi} = -j\left(\frac{\widehat{\alpha}}{2}\right)e^{jQ\xi\delta/2}\,\tilde{E}_0 \tag{13.3}$$

The TFF due to Chatterjee *et al.* [10] is a direct consequence of the solutions of the above coupled equations. As may be shown, the solutions for the zeroth- and first-orders under arbitrary angular deviations from the Bragg angle, when normalized relative to the incident beam angular spectrum, yield the following two transfer functions for the fields in the δ domain. Thus, we have [10]:

$$\tilde{H}_0(\delta) = \frac{\tilde{E}_0(\zeta)_{\xi=1}}{\tilde{E}_{inc}} = -\frac{e^{-j\frac{\delta Q}{4}}}{\sqrt{\left(\frac{\delta Q}{4}\right)^2+\left(\frac{\hat{\alpha}}{2}\right)^2}}\left(\sqrt{\left(\frac{\delta Q}{4}\right)^2+\left(\frac{\hat{\alpha}}{2}\right)^2}\right.$$

$$\left.\times\cos\left(\sqrt{\left(\frac{\delta Q}{4}\right)^2+\left(\frac{\hat{\alpha}}{2}\right)^2}\right)+j\frac{\delta Q}{4}\sin\left(\sqrt{\left(\frac{\delta Q}{4}\right)^2+\left(\frac{\hat{\alpha}}{2}\right)^2}\right)\right) \tag{13.4}$$

$$\tilde{H}_1(\delta) = \frac{\tilde{E}_1(\zeta)_{\xi=1}}{\tilde{E}_{inc}} = -j\left(\frac{\widehat{\alpha}}{2}\right)\frac{e^{j\frac{\delta Q}{4}}}{\sqrt{\left(\frac{\delta Q}{4}\right)^2+\left(\frac{\widehat{\alpha}}{2}\right)^2}}\left(\sin\left(\sqrt{\left(\frac{\delta Q}{4}\right)^2+\left(\frac{\widehat{\alpha}}{2}\right)^2}\right)\right). \tag{13.5}$$

Both scattered beams may now be found by applying an inverse Fourier transform to the product of the incident spectrum and the corresponding transfer function, $\tilde{E}_{\text{inc}}(\delta)H(\delta)$ [10]. This process is shown in the following equation, where

$E_{out}(r)$ is either first- or zeroth-order output, depending on the transfer function used, and $\tilde{E}_{inc}(\delta)$ is the angular spectrum of the incident profiled beam:

$$E_{out}(r) = \int_{-\infty}^{\infty} \tilde{E}_{inc}(\delta)\, H(\delta)\, e^{-j\frac{2\pi}{\lambda}\delta\phi_B r}\left(\frac{\phi_B}{\lambda}\right) d\delta. \qquad (13.6)$$

In the brief discussion of the results reported, the angular spectrum of the incident light is inserted into this equation, and the output fields are computed by numerically solving the same. These fields are functions of the peak phase delay $\widehat{\alpha}_0$ and the Klein–Cook parameter Q.

13.3 Examination of the nonlinear dynamics under profiled beam propagation

Using uniform input beams, researchers have studied the hybrid closed-loop system created by detecting the first-order scattered output with a photodetector and feeding the resulting signal into the (external) bias input of the RF generator. The resulting nonlinear dynamics exhibit mono-, bi-, multistable and also chaotic behavior [11–13]. These properties have been exploited for a variety of signal processing applications including signal encryption and decryption [14]. The understanding of the nonlinear dynamics thus far has been limited by the assumption of uniform input beams. Practical optical beams are more likely to be nonuniform in profile, and it has been shown that significant, unexpected deviations in the first-order output occur for open loop A-O systems [15]. Since chaos is extremely sensitive to amplitudes, it becomes necessary to examine the consequences of specific profiled light beams upon the feedback system.

If the first-order diffracted light is picked up by a photodetector, and its output is amplified and fed back into the acoustic driver, then a familiar closed-loop system is created as shown in Figure 13.2. The resulting nonlinear system displays complex bistability behavior which was first observed in 1978 [13]. Under the assumption of a uniform plane wave input, the output from the photodetector follows the nonlinear equation:

$$I_1(t) = I_{inc}\sin^2\left\{\frac{1}{2}\left[\widehat{\alpha}_0(t) + \tilde{\beta}\, I_1(t - TD)\right]\right\}. \qquad (13.7)$$

In this equation, I_{inc} is the amplitude of the incident light, $\widehat{\alpha}_0$ is the peak phase delay, $\tilde{\beta}$ is the feedback gain, and TD is the feedback time delay. Depending on the values of these parameters, this equation leads to *monostable*, *bistable,* and *multistable* behavior, as well as chaos. Simulations of such behavior are shown in Figures 13.3 and 13.4, and consist of plots of intensity versus the optical phase shift at a fixed value of the feedback gain. Figure 13.3 illustrates bistability and the beginning of chaos with a feedback gain of 2.42, which is just above the nominal threshold value between bistability and chaos under the assumption of a uniform

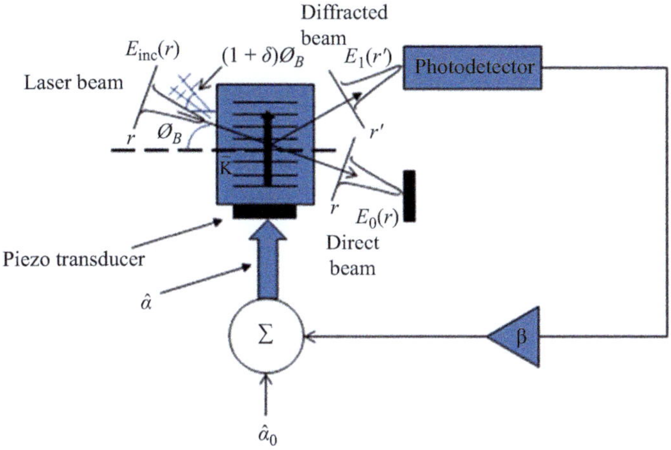

Figure 13.2 A-O closed-loop hybrid system with an arbitrary incident beam profile

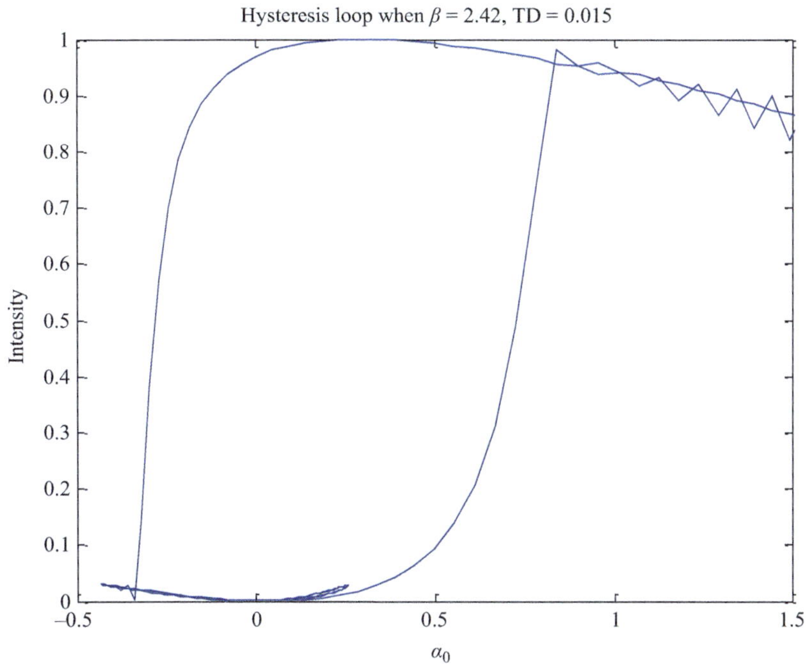

Figure 13.3 Transition from bi- to multi-stable oscillations for uniform plane wave input

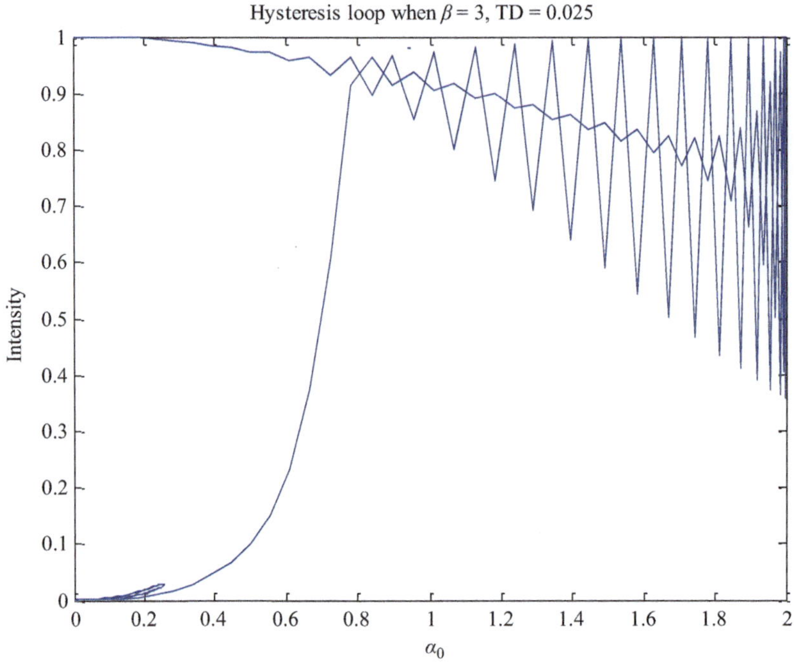

Figure 13.4 Chaotic oscillations for uniform plane wave input when $\tilde{\beta}$ increased to 3

plane wave input. Figure 13.4 illustrates stronger chaotic behavior of the closed loop system for feedback gain $\tilde{\beta} = 3$.

Most of the literature on nonlinear assumes a uniform plane wave input in order to make the analysis tractable. Since practical optical beams are more likely to be nonuniform in profile, and nonlinear dynamics are extremely sensitive to amplitudes, it becomes necessary to examine the consequences of specific profiled light beams upon the feedback system under examination. Therefore, the transfer function technique described earlier is used to model the scattered fields created with a Gaussian input profile [10]. Differences in the scattered fields, as previously discussed, are observed between the Gaussian assumption and uniform assumption. This causes the above equation for photodetector current (derived from the first-order optical intensity) to become invalid for Gaussian input profiles. Figure 13.5 displays the results for the Gaussian case, showing the output intensity profile along the peak phase delay $\widehat{\alpha}_0$ for three different values of Q.

As is clearly shown in Figure 13.5, unexpected deviations from the standard theory occur at higher Q values. The curve for $Q = 20$ appears consistent with uniform plane wave inputs, although the impact is significant in the closed-loop system (primarily due to the nonuniform output amplitudes) as is discussed in the results. As seen, the $\widehat{\alpha}_0$ intensity profile begins to deviate from the standard $\sin^2$

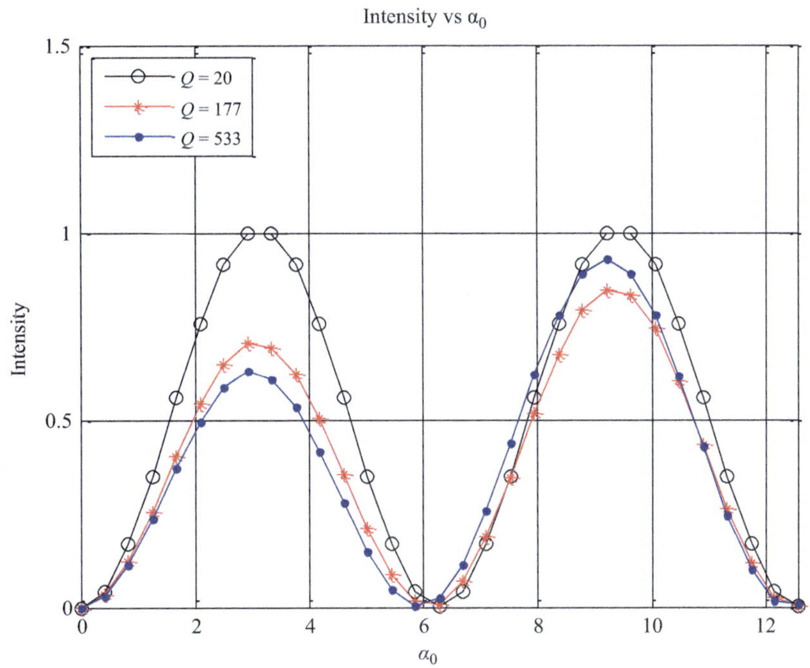

Figure 13.5 First-order intensity Bragg diffraction versus the optical phase shift for $Q = 20$, 177, and 533

behavior as Q increases; thus, the curve for $Q = 533$ is clearly no longer similar to uniform plane wave input results. The effect of nonuniform inputs on nonlinear dynamics has not been previously studied.

As one would expect, changes in the scattered beam caused by a profiled input create corresponding changes in the nonlinear dynamics for the close-loop system. The threshold value of the feedback gain $\tilde{\beta}$, at which the system become monostable, bistable, and chaotic, changes for profiled inputs. There are also changes in the Lyapunov exponents and bifurcation maps which will have implications for encryption and decryption of RF signals applied via the bias loop. These differences are described and shown in the results. To generate these characteristics, the simulated first-order diffracted beam collected by the photodetector (which creates a current $I_{\mathrm{ph}}(t)$) is applied to the feedback system as a set of numerical data derived via the open-loop analysis instead of the standard analytic expression known for the uniform profile case. This photon current is amplified and time-delayed, and then fed back into the acoustic driver, as is done in the standard case. The following equation represents the photodetector current, where the function f represents the observed output for a Gaussian input profile (which is not the same as the sine-profile that occurs under the standard analysis), where the argument of f represents one-half of the equivalent sound pressure applied to the Bragg cell through the feedback. There is no closed form expression for f in this case, since it is related to Q and effective beam width in a complicated way, as was numerically examined for the open-loop. Thus,

$$I_{\text{ph}}(t) = \left| f\left(\frac{1}{2}\left[\widehat{\alpha}_0(t) + \tilde{\beta}\left(I_{\text{ph}}(t - TD)\right)\right]\right)\right|^2. \tag{13.8}$$

One of the main effects of a nonuniform input profile is that the threshold value of $\tilde{\beta}$ for transition between bistability and chaos becomes a function of the Klein–Cook parameter Q. As Q increases, the threshold value of $\tilde{\beta}$ also increases. This is demonstrated by Figure 13.6 that shows plots of the intensity versus the peak phase shift (immediately prior to reaching chaos) for two values of Q, where the value of $\tilde{\beta}$ in each case is chosen to be just below the threshold of transition between bistability and chaos. In Figure 13.6, this subthreshold value of $\tilde{\beta}$ is 1.28 corresponding to $Q = 20$. At this low value of Q, even though the scattered output appears to be $\sin^2$, this value of $\tilde{\beta}$ for the hybrid cell is significantly different from that for the uniform case. This is most likely due to the deviations in the scattered output profile, as was demonstrated in Figure 13.5.

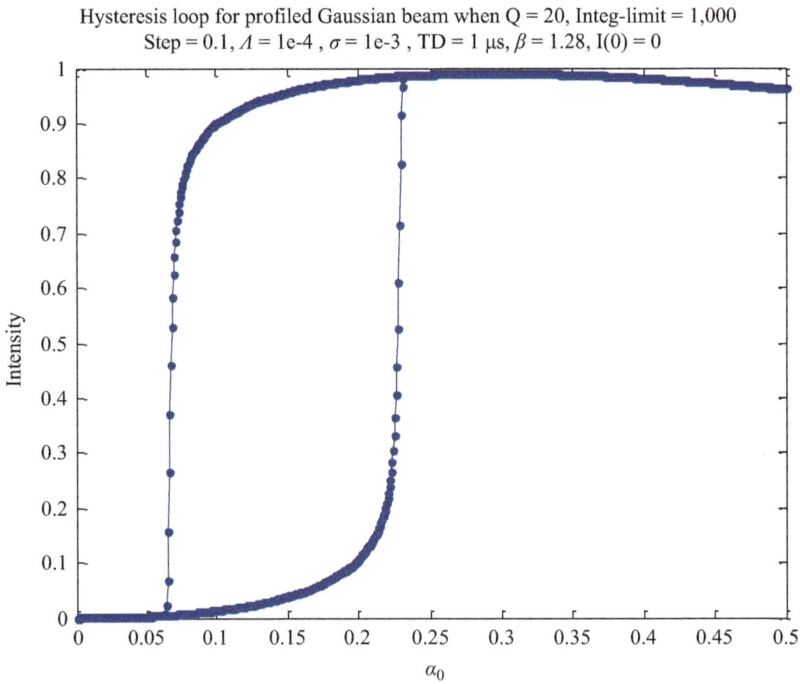

Figure 13.6 Hysteresis loop with an arbitrary Gaussian incident beam profile for $Q = 20$

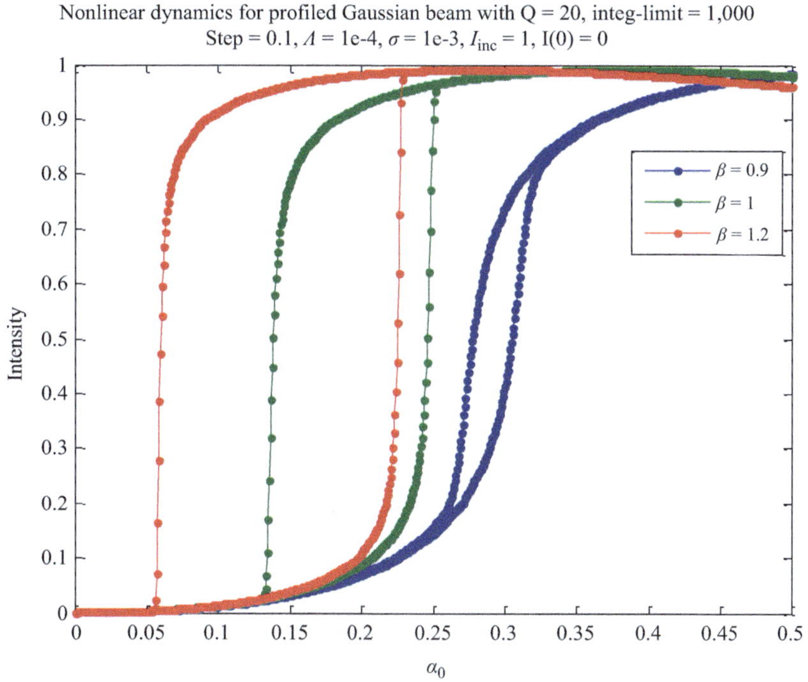

Figure 13.7 Nonlinear dynamics with three different values of the effective feedback gain for $Q = 20$

Figure 13.7 illustrates the intensity versus the peak phase shift with three different values of $\tilde{\beta}$ {0.9, 1, 1.2}, all below the threshold value of 1.28. These demonstrate the feedback gain tuning sensitivity of this hybrid closed-loop system.

As $\tilde{\beta}$ is increased above the threshold, chaotic behavior occurs as shown in Figures 13.8 and 13.9. Figure 13.8 uses a $\tilde{\beta}$ value of 1.6 where the transition to chaos is evident. When $\tilde{\beta}$ is increased to 2.2 as in Figure 13.9, chaos is strongly present. The chaotic thresholds, which are of special interest for encryption applications, are significantly lower due to the nonuniform profiled beam. Additionally, as can be observed in Figures 13.8 and 13.9, the chaotic bands appear to migrate along the peak phase delay into the originally bistable/hysteretic regions. For encryption applications, it is necessary to predict where the chaotic bands occur; this is explored in the next subsection.

13.4 Examination of dynamical behavior based on both Lyapunov exponent and bifurcation maps

The theory of Lyapunov exponents (LEs), developed by Ghosh and Verma to predict the nonlinear dynamics of the A-O feedback system [16], is briefly outlined here, beginning with the discretization of the photodetector current. Ghosh and

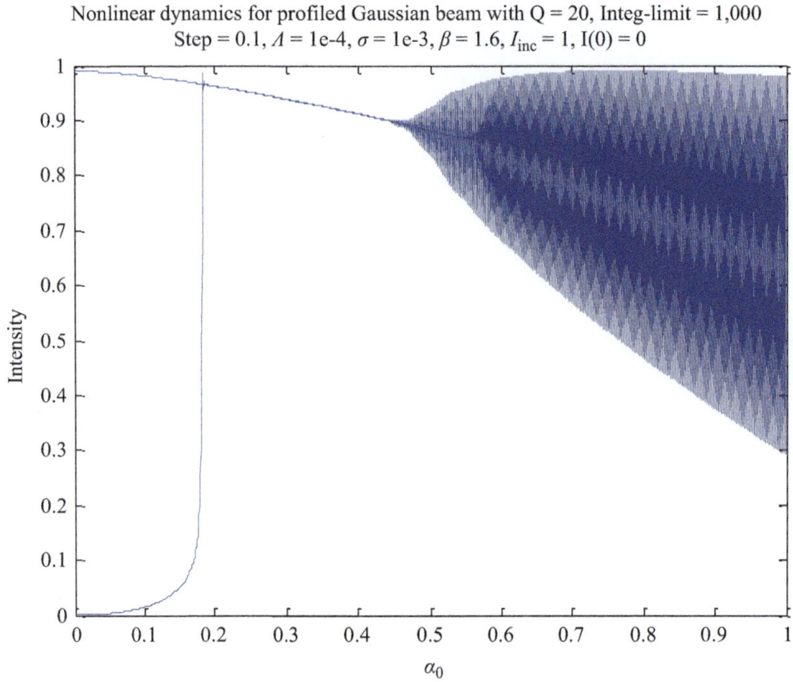

Figure 13.8 Nonlinear dynamics versus optical phase shift with a profiled Gaussian incident beam when $\tilde{\beta}$ increased to 1.6 for $Q = 20$

Verma showed that the incremental change in the feedback intensity after n iterations can be represented by an exponential function [16]:

$$\frac{\Delta I_1}{\varepsilon} \rightarrow e^{n\lambda} \tag{13.9}$$

where ε represents a small random perturbation to the initial condition and λ is the LE. This equation is true in the limit as $n \rightarrow \infty$ and $\varepsilon \rightarrow 0$. This leads to the fact that if $\lambda > 0$, the iterations diverge leading to chaotic behavior and if $\lambda < 0$, chaos does not occur. Ghosh and Verma also derived a necessary but not sufficient condition for chaos, given by $|\tilde{\beta} I_{\text{inc}}| > 1$. The behavior of the LE as a function of the feedback system parameters has been explored numerically with the assumption of uniform input beams [14,16]. In the current work, we explore the effect of non-uniform Gaussian input beams on the LE and chaos, a necessary and novel development because realistic input beams are not truly uniform.

In order to use an A-O feedback system for encryption and decryption of data, it is necessary to characterize the chaotic behavior of such systems using the Lyapunov exponent or bifurcation maps. Both approaches depend on the nonlinear system parameters, including bias, gain, input intensity and $I_{\text{ph}}(0)$. Chaos can also be explored using plots of the photodetector output versus gain or peak phase delay

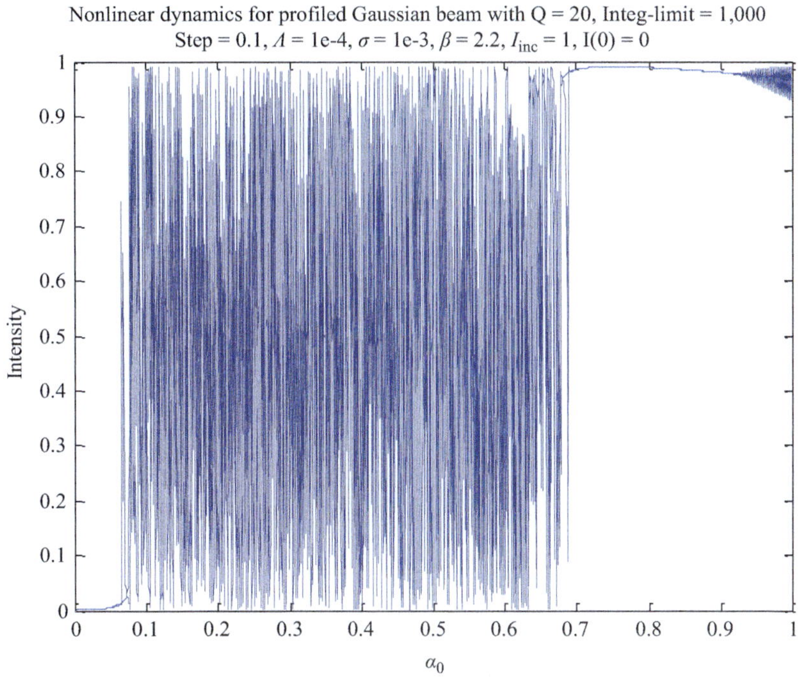

Figure 13.9 Nonlinear dynamics versus optical phase shift with a profiled Gaussian incident beam when $\tilde{\beta}$ increased to 2.2 for $Q = 20$

while holding other parameters constant. These plots, known as bifurcation maps, are generated from the simulation and can be compared to the behavior predicted by the LE theory. Bifurcation maps illustrate sudden changes in the dynamic behavior as a system parameter crosses a threshold.

Figures 13.10 and 13.11 consist of bifurcation maps, which show photodetector intensity versus bias $\widehat{\alpha}_0$, alongside the Lyapunov exponent characteristics. These two types of results independently describe the state of the system as a function of system parameters, and they are seen to be in close agreement.

In Figure 13.10, for $\tilde{\beta} = 1$, the LE is negative for all $\widehat{\alpha}_0$ except at 1.7, where the LE is zero. The corresponding bifurcation map shows a period-doubling at this peak phase shift value. In Figure 13.11, where $\tilde{\beta}$ is 1.2, there are multiple locations where the LE is zero and the corresponding bifurcation map shows consistent period-doubling at each location. As $\tilde{\beta}$ is increased above the threshold for chaos, the LE becomes positive for some bands of peak phase delay. This is clearly shown in Figure 13.12, with $\tilde{\beta} = 2$, where the bifurcation map shows bands of chaos whenever the LE is positive. The locations and strengths of these bands change with $\tilde{\beta}$, as evident from Figure 13.12.

Figure 13.13 uses LE and bifurcation maps to explore the dynamic behavior as a function of the feedback gain for fixed peak phase delays. It uses $\widehat{\alpha}_0 = 1$ and illustrates bands of chaos along the gain dimension.

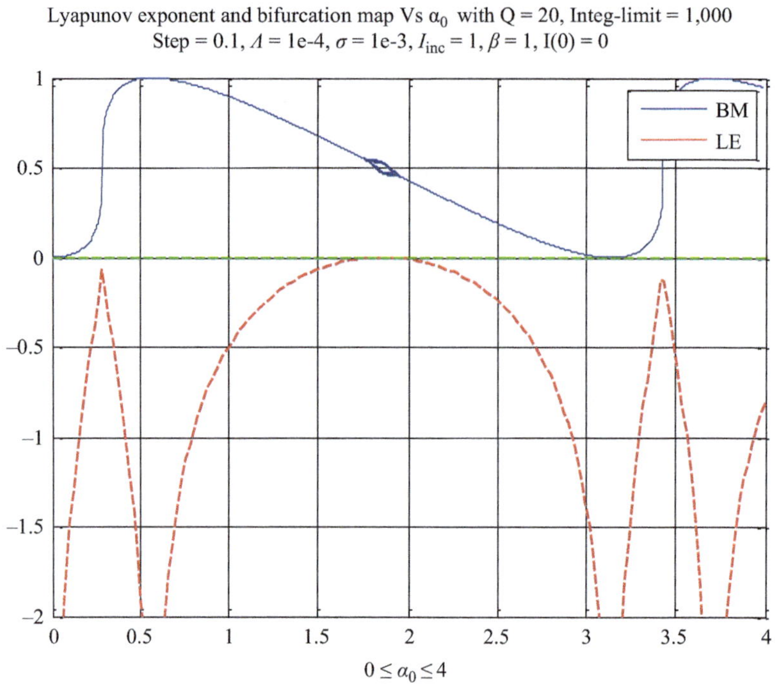

Figure 13.10 Lyapunov exponent and bifurcation maps versus the optical phase shift when $\tilde{\beta} = 1$

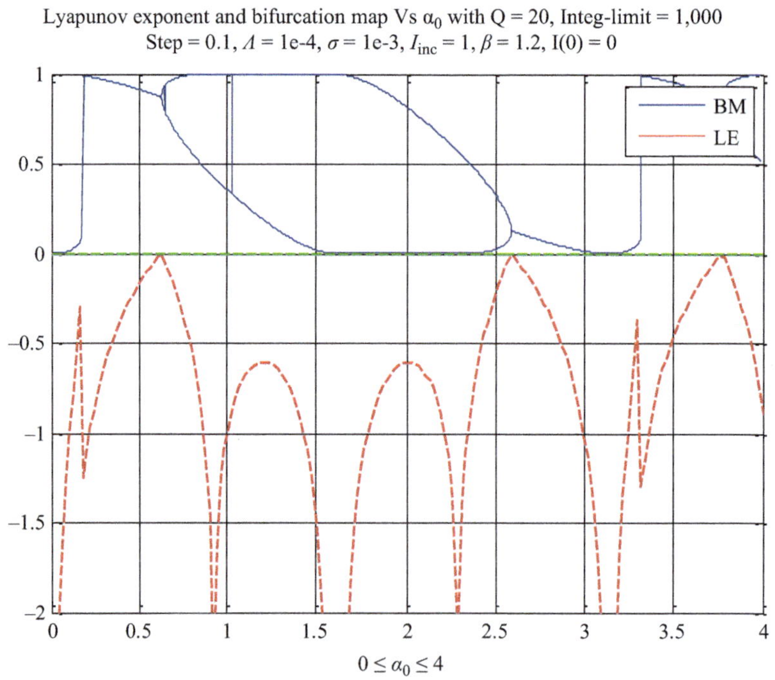

Figure 13.11 Lyapunov exponent and bifurcation maps versus the optical phase shift when $\tilde{\beta} = 1.2$

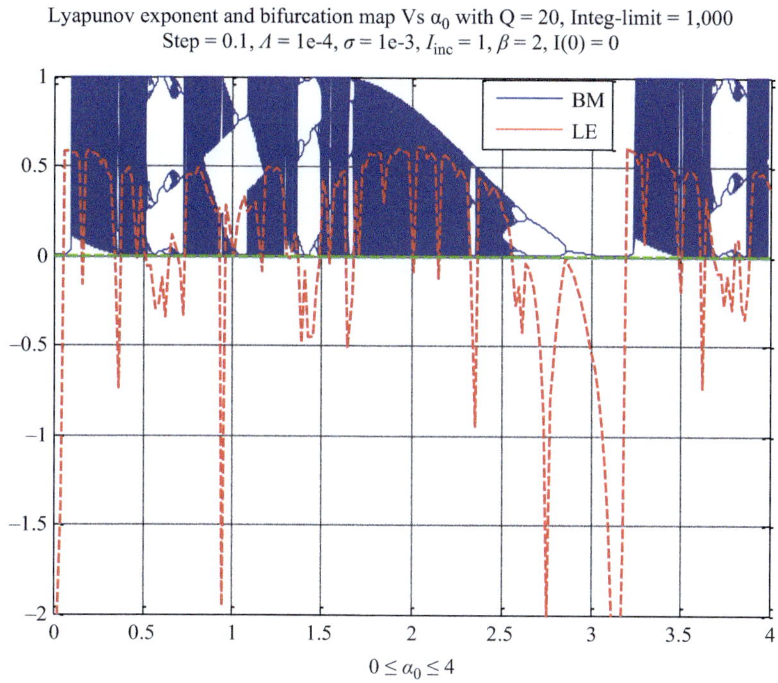

Figure 13.12 Lyapunov exponent and bifurcation maps versus the optical phase shift when $\tilde{\beta} = 2$

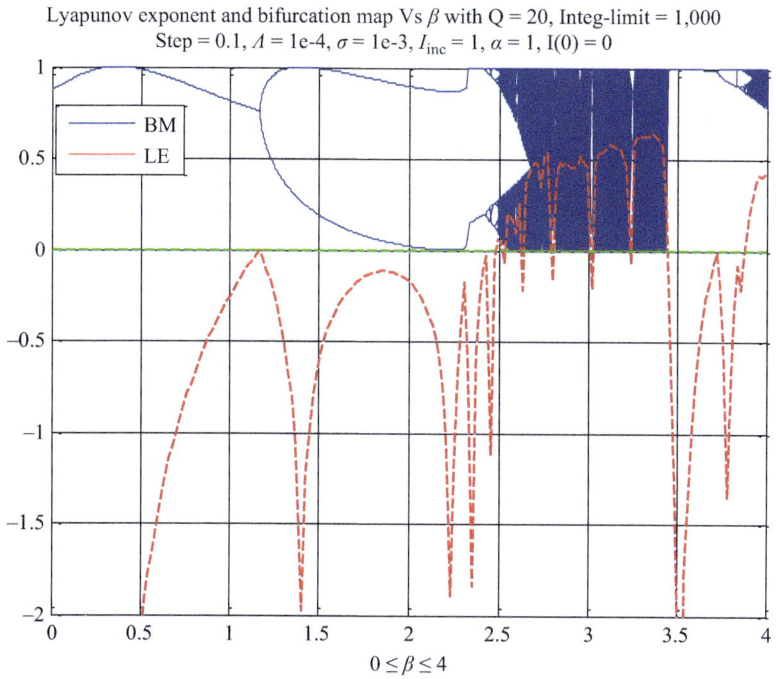

Figure 13.13 Lyapunov exponent and bifurcation maps versus the effective feedback gain when $\widehat{\alpha} = 1$

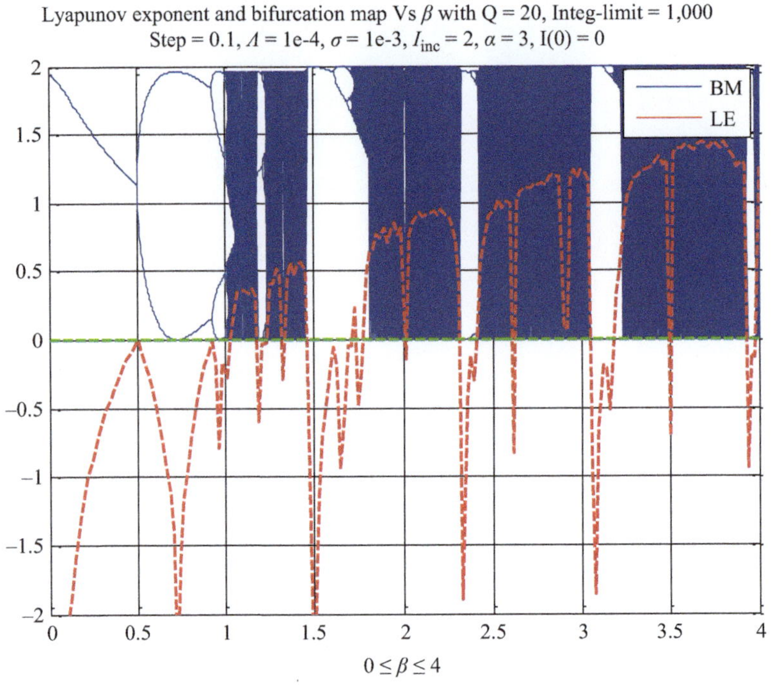

Figure 13.14 Lyapunov exponent and bifurcation maps versus the effective feedback gain when $\widehat{\alpha} = 3$ and $I_{inc} = 2$

For $\widehat{\alpha}_0 = 3$, and for a doubled input intensity, the location of the chaotic bands are as shown in Figure 13.14, indicating a sensitivity to input intensity as well as feedback gain, peak phase delay and initial condition.

13.5 Chaotic encryption and decryption in hybrid acousto-optic feedback (HAOF) devices

In order to utilize chaos as a means of encrypting and securely transporting a signal waveform, we need to apply suitable AC signal to the bias driver. A complete block diagram of the transmitter-heterodyne receiver scheme is shown in Figure 13.15. From our simulation, we may interpret the chaos waveform as an approximately sinusoidal carrier, albeit chaotic, and therefore somewhat random in nature. From A-O, we may readily show that the chaos amplitude is related to the bias voltage amplitude ($\widehat{\alpha}_{tot}$) via the RF oscillator. Therefore, by incorporating a time varying signal with the bias, we expect the chaos waveform to be effectively amplitude modulated, and because of its chaotic nature, the signal will now be encrypted in the chaos.

At the receiver end, the signal is recovered (in the manner of standard heterodyne detection) as follows. First, a local chaos wave is generated using a second Bragg cell with matched parameters $\widehat{\alpha}$, $\tilde{\beta}$, and T_d. This is shown in Figure 13.15.

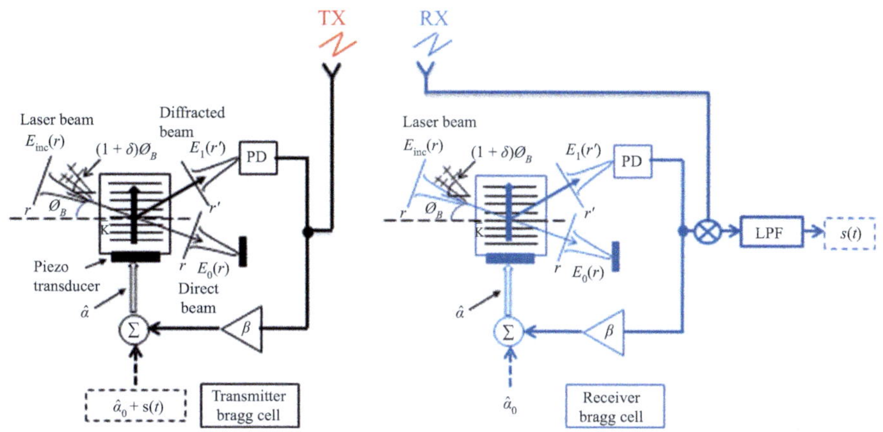

Figure 13.15 Heterodyne scheme for encrypting and decrypting using A-O chaos

The local chaos (which is a photo-detected RF current corresponding to the first-order light) is then multiplied with the incoming photo-detected modulated chaotic signal. The product waveform is then passed through a low-pass filter (LPF) with cutoff frequency adjusted to accommodate the signal bandwidth.

13.6 Preliminary results for chaotic encryption and decryption

To apply chaos as a means of encrypting a signal waveform $s(t)$, the signal is applied to the bias driver such that, as shown in the transmitter of Figure 13.15, the peak phase delay has the form of $\widehat{\alpha} = \widehat{\alpha}_0 + s(t)$. The constant offset $\widehat{\alpha}_0$ is chosen to be at the center of a chaotic passband, and the range of $s(t)$ does not exceed the width of the passband. The resulting chaotic photodetector current is then viewed as a modulated version of the input signal. After transmission through a channel, $s(t)$ is recovered in the manner of standard heterodyne detection, as shown in the receiver of Figure 13.15, where a local chaos wave is generated using a second Bragg cell parameters matched to the transmitter cell. The local chaos is multiplied with the modulated signal, and the product is low-pass filtered to recover $s(t)$.

If the HAOF parameters are not chosen correctly, the signal $s(t)$ will modulate the chaos as standard AM, and its shape appears in the envelope [14]. With the proper choice of parameters, $s(t)$ is completely encrypted, and a simulation of this case is shown in Figure 13.16 using a digital signal. The chaos contains no apparent hint of the original signal, but by using matched parameters in the receiver, it is recovered with no bit errors [17].

Bit errors are examined in previous works as a function of channel noise and parameter mismatch between the transmitter and receiver HAOFs [17]. For the channel noise case, marginally higher error rates are measured relative to systems that provide no encryption, showing a low energy cost of encryption [18].

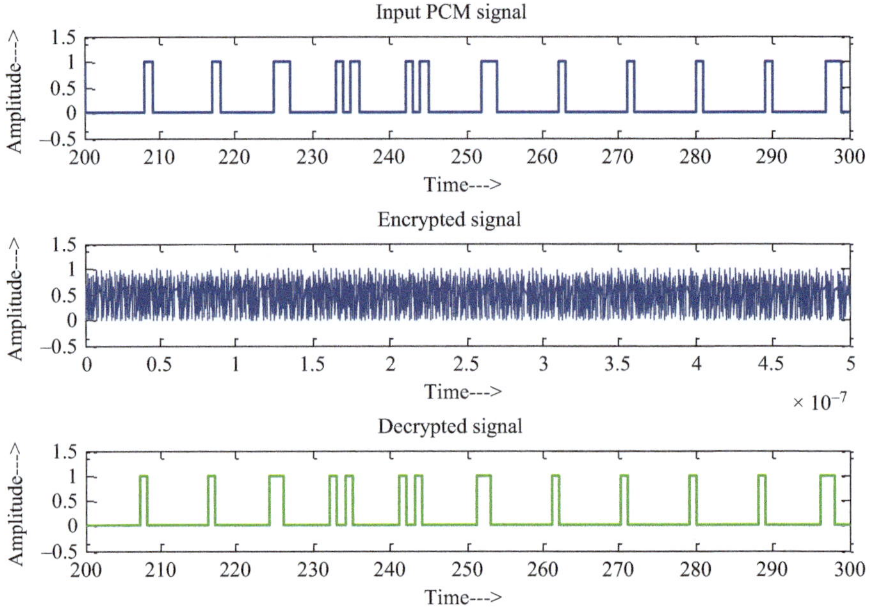

Figure 13.16 Encryption and recovery with matched keys; $\tilde{\beta} = 3$, $TD = 0.05\ \mu s$, $\widehat{\alpha}_0 = 2$

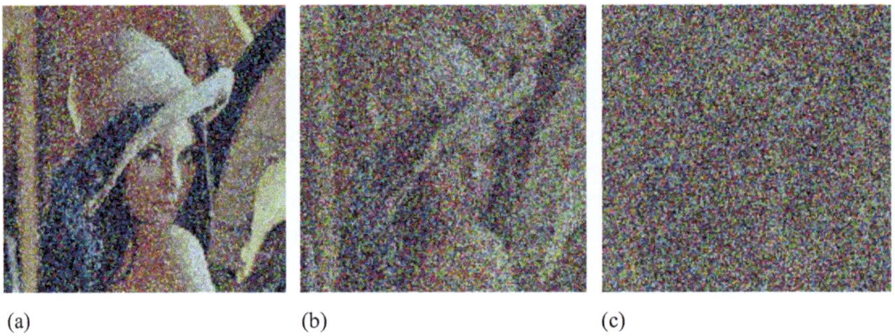

Figure 13.17 Recovered lens images with three levels of mismatch in the $\tilde{\beta}$ *parameter: (a) 0.1% mismatch; (b) 0.2% mismatch; and (c) 0.4% mismatch*

Additionally, it was found that the error rates are highly sensitive for parameter mismatch, indicating that the parameters form a robust encryption key. This encryption key strength is due to the fact that profiled beams are used in the simulation. Encryption results with uniform beam simulations indicated significantly weaker encryption [17]. To illustrate the sensitivity to mismatch achieved using nonuniform input beams, Figure 13.17 shows digital images recovered with different levels of mismatch in a single parameter. The original image is a color version of the standard Lena test image, with 512×512 pixels and 24 bits per pixel, and mismatches of 0.1%, 0.2%, and 0.4% are used in the gain parameter. The 0.1% mismatch causes a significant number of bit errors, but enough pixels remain in the recovered image that the original can visually be recognized. For 0.2% mismatch, the original image is barely recognizable, and fine details are lost. For 0.4%, the image appears as noise, and there is no hint of the original.

13.7 Propagation of a profiled beam through MVKS type phase turbulence

13.7.1 An overview

Atmospheric turbulence has a significant impact on the quality of free-space EM propagation over long distances. The Earth's atmosphere can be described as a locally homogeneous medium in which its properties vary with respect to temperature, pressure, wind velocities, humidity and other factors. Inhomogeneities in the temperature and pressure of the atmosphere lead to variations of the refractive index along the transmission path. These variations of the refractive index can cause fluctuations in both amplitude and phase of the received signal (in some instances, the random index fluctuations may also cause the phenomenon of scintillation), which increase the bit errors in a digital communication link; additionally, the above lead to distortion and other degradations of the transmitted signal or message. In order to quantify the performance limitations, a better understanding of the effect of turbulence-induced intensity fluctuations on the received signal under different turbulence categories is needed. Three major parameters characterizing atmospheric turbulence are: the refractive index structure parameter C_n^2, and the turbulence inner (ℓ_0) and outer (L_0) scales. The refractive index structure parameter was first introduced in turbulence theory by Kolmogorov and Obukhov [19–21]. This theory is commonly referred to as the Kolmogorov turbulence theory. The refractive index structure parameter describes the strength of the (spatial) refractive index fluctuations and marks the first major principle on which the development of the classical Kolmogorov atmospheric turbulence theory depends. When a laser beam propagates through the atmosphere, the randomly varying spatial distribution of refractive index that it encounters causes a number of effects.

Propagation through the turbulent atmosphere often consists of a laser beam propagating through the otherwise clear atmosphere with very small changes in the refractive index present. These small changes in refractive index, which are typically on the order of $\frac{\Delta n}{n} \sim 10^{-6}$ are related primarily to the small variations in temperature (on the order of 0.1–1°C), which are produced by the turbulent motion of the atmosphere, and also typical thermal gradients between the earth's surface and the atmosphere [22]. Since the turbulence is irregular and random, typically any information transmitted will suffer amplitude and phase distortions upon propagation through the turbulence. There may be multiple means of trying to mitigate the effects of turbulence in the wave communication. One such, which is the potential use of an encrypted chaotic carrier wave which carries the information and then transmitted through the turbulence. Since chaos itself has a certain degree of randomness in its characteristics, the latter problem therefore becomes more complex, involving two different random effects, of which the turbulence possess one or more spatial models; however, the chaos is not equally amenable to mathematical modeling, and in the work presented here, occurs only in time and not in space.

13.7.2 The von Karman spectrum

The power spectrum density of the von Karman spectrum (also called the MVKS in the form shown) is given by [23]:

$$\Phi_n(k) = 0.033C_n^2 \frac{\exp\left(-\frac{k^2}{k_m^2}\right)}{\left(k^2 + k_o^2\right)^{\frac{11}{6}}}, 0 \leq k \ll \infty \tag{13.10}$$

where C_n^2 is the medium structure parameter, $k_m = 5.92/\ell o$ is an equivalent wavenumber related to the inner scale, $k_0 = 2\pi/L_o$ is a wavenumber related to the outer scale, and k is the unbounded nonturbulent wavenumber in the medium. In the above equation, $\Phi_n(k)$ represents the socalled power spectral density (PSD) of the *refractive index* of the medium. Additional models such as the Modified Atmospheric Spectrum also exist in the literature [23]. The Modified Atmospheric model has not been considered since it is more complicated than the MVKS.

13.7.3 Thin-phase screen generation

In this subsection, we discuss the generation of a phase screen to mimic the statistical behavior of the phase fluctuations due to a turbulent atmosphere using a discrete grid and generating the phase screen from the given spectrum based on fast Fourier transform (FFT) techniques. The purpose of a phase screen is to simulate the random phase perturbations resulting from random index fluctuations in narrow and extended atmospheric turbulence [24]. The generated random phase screen (either planar or extended) in this research is characterized by several different parameters: C_n^2 (or Fried parameter r_0), inner and outer scales, ℓo and L_0, respectively, and the incremental spatial frequencies $\Delta k_x, \Delta k_y$.

The procedure for phase screen generation is as follows: beginning with the MVKS model with given parameters as mentioned, and by using a standard scheme based on Fourier transform generation, a set of random complex numbers (following a Gaussian distribution) is generated on the chosen grid. Following (13.10), the random numbers are multiplied by the square root of the phase power spectrum (PPS) wherefrom an inverse Fourier transform produces the phase screen. The real part of the result is taken to be the random phase function $\varphi(x, y)$ due to atmospheric fluctuations based on the MVKS model. The discrete phase distribution in 2D is given as:

$$\varphi_{ij} = Re\left\{IFFT\left((a + jb)\sqrt{\Phi_p \Delta k_x \Delta k_y}\right)\right\} \tag{13.11}$$

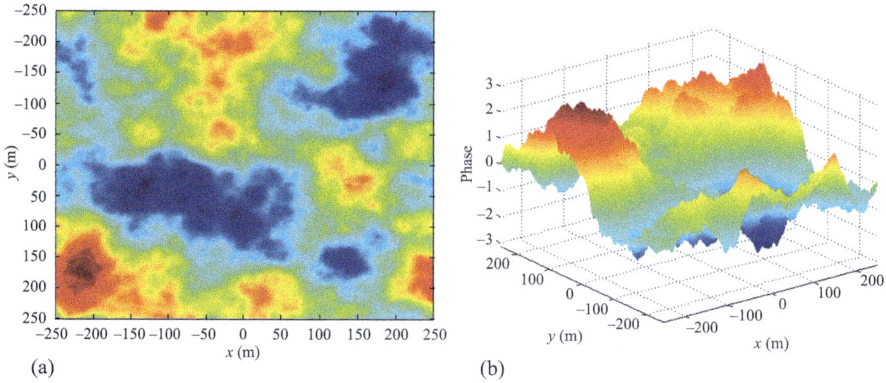

Figure 13.18 Random phase screen distribution profile: (a) 2D and (b) 3D

where *IFFT* represents the inverse *fast Fourier transform* operation, $\Delta k_x, \Delta k_y$ is the incremental spatial frequencies, Φ_p is the power spectral density (given below) evaluated in the transverse plane, and (a and b) are random numbers generated in order to appropriately mimic the random noise-like characteristics of the von Karman phase. Correspondingly, the MVKS model in the spatial domain is expressed as:

$$\Phi_p(k) = 0.23 r_0{}^{-\frac{5}{3}} \frac{\exp\left(-\frac{k^2}{k_m^2}\right)}{\left(k^2 + k_o^2\right)^{\frac{11}{6}}}. \tag{13.12}$$

Propagation through thin phase screens specified by (13.10) is numerically explored using the split-step algorithm, whereby the incident EM wave is alternately transmitted across the screen and thereafter diffracted over incremental distances before encountering a possible subsequent sequence of phase screens (representing an extended phase turbulence), or being propagated directly to a receiving or image plane using standard Fresnel–Kirchhoff diffraction. The 2D and 3D random phase screen distribution profiles are shown in Figure 13.18.

13.8 Spectral approach to the propagation of a (*non-chaotic*) EM wave through turbulence using SVEA and Fourier transforms

In this section, we examine the propagation of (non-chaotic) EM waves through atmospheric turbulence and their characteristics in the image plane. Starting with a profiled beam represented by a time-harmonic carrier:

$$E(x, y, z, t) = E_o(x, y, z) cos\,(\omega_o t - kz) \tag{13.13}$$

with $\bar{k} = k_z \hat{a}_z$, and

$$E_o(x, y, z) = A \frac{w_o}{w(z)} e^{-r^2/w^2(z)} \tag{13.14}$$

where $w^2(z)(= w_o^2(1 + \frac{z^2}{z_o^2}))$ defines the beam spot size; $r^2 = x^2 + y^2$; ω_o is the carrier frequency; w_0 is the beam waist, and z_o is the depth of focus of the (Gaussian) beam [25,26].

By substituting (13.14) in (13.13), we get

$$E(x, y, z, t) = A \frac{w_o}{w(z)} e^{-r^2/w^2(z)} cos\,(\omega_o t - kz). \tag{13.15}$$

The (AM) modulated EM wave may be expressed as:

$$E_{AM}(x, y, z, t) = E_o(x, y, z)[1 + ms(t)] cos\,(\omega_o t - kz) \tag{13.16}$$

where m is the modulation index and $s(t)$ is the modulating signal (message). The above then represents a directly modulated optical carrier.

By substituting (13.16) in (13.15), we get

$$E_{AM}(x, y, z, t) = A \frac{w_o}{w(z)} e^{-r^2/w^2(z)} [1 + ms(t)] cos\,(\omega_o t - kz). \tag{13.17}$$

Note that the term $A \frac{w_o}{w(z)} e^{-r^2/w^2(z)} [1 + ms(t)]$ is a time-varying spatial envelope defined by the signal waveform.

As mentioned above, in our approach we use the SVEA approximation where the envelope varies slowly compared with the rate of change of the optical carrier (i.e., envelope BW B $\ll \omega_o$). This method involves expressing the optical carrier via phasor representation wherein the envelope (containing the signal waveform) is slowly varying in time. The analysis then proceeds with the application of the diffraction integral acting upon the above complex spatial phasor, and the resulting propagation is tracked numerically to the receiver or "image" plane.

Accordingly, we next propagate the modulated EM wave ($E_{AM}(x, y, z, t)$) converted to the phasor domain (while retaining the SVEA envelope) by applying the Fresnel–Kirchhoff diffraction integral along the propagation path. An inverse phasor transform restores the optical carrier. A photodetector recovers the modulated envelope by retaining the intensity profile (which removes the carrier). For a directly modulated optical carrier, the signal is recovered by using appropriate filtering and scaling. On the other hand, if the optical carrier has an envelope containing an RF modulated carrier (such as a modulated chaos wave), recovery would require additional use of heterodyne AM demodulation followed by filtering. At the receiver, the scalar output is proportional to the PD current at the point location (say at 0, 0, z_i for on-axis detection) where the PD intercepts the input field:

$$I_{PD}(t) \propto |E_{out}(0, 0, z_i)|^2 [1 + ms(t)]^2, \tag{13.18}$$

where $E_{out}(0, 0, z_i)$ is equivalent to the spatial Fourier transform of $E_o(x_o, y_o)e^{-jk_z z}g(x_o, y_o)$ in the far-field on-axis with $E_o(x_o, y_o)$ being the profile of the input wave and $g(x_o, y_o)$ being the aperture function through which the wave is transmitted at the "object" plane.

The message signal may be retrieved from the photocurrent via appropriate electronics. This establishes the fact that under homogeneous propagation, a modulated carrier may be demodulated directly using time-domain strategies without having to consider spatial diffraction and other effects. On the other hand, when the intervening medium is turbulent (in phase or otherwise), there is a likelihood that the random spatial effects would affect the signal riding on the carrier wave; hence, one of the goals of this study is to examine the effects of such atmospheric turbulence on the transmitted signal.

13.9 A uniform (nonturbulent) propagation prototype

To test the behavior under turbulence, we first present below results from a simple homogeneous, diffraction-limited propagation of a modulated (optical) EM wave through the use of the Kirchhoff–Fresnel diffraction integral. Accordingly, a Gaussian profile beam (Figure 13.19(a)) is modulated by a *sinc*-type pulse (Figure 13.19(b)), and is received as a diffracted Gaussian (Figure 13.19(c)) with a circular symmetry (Figure 13.19(d)) and a broadened profile (Figure 13.19(e)). Figure 13.19(f) shows the final recovered signal after carrying out the necessary demodulation operation using the photodetector output.

The above figures demonstrate typical spatial transmission and recovery of an EM wave through a homogeneous medium whereby the profiled EM wave (here Gaussian) undergoes diffractive spread; however, the time signal embedded within the optical carrier is simply recovered using standard detection electronics (see Figure 13.19(b) and (f)). We include this relatively trivial result here simply to enable further understanding as to the consequences on the homogeneous prototype when medium turbulence is incorporated into the problem.

13.9.1 Propagation through weak *turbulence*

For weak turbulence, we choose a Fried parameter $r_0 = 10$ mm with structure parameter $C_n^2 = 1.067 \times 10^{-18}\ m^{-2/3}$. Note that turbulence is typically modeled as a spatial effect; however, in reality it also is subject to random time variations. In our analysis, we incorporate the temporal variation of turbulence by repetitive insertion of the thin phase screen over pre-selected time intervals which thereby defines an equivalent mean (temporal) turbulence frequency.

13.9.1.1 Propagation through *weak* turbulence with mean frequency $f_T = 20$ Hz

In this section, we repeat the procedure described in Section 13.9 for a narrow, weakly turbulent medium. To represent the time fluctuations of the medium phase, the narrow phase screen is inserted into the propagation algorithm (at the half-way

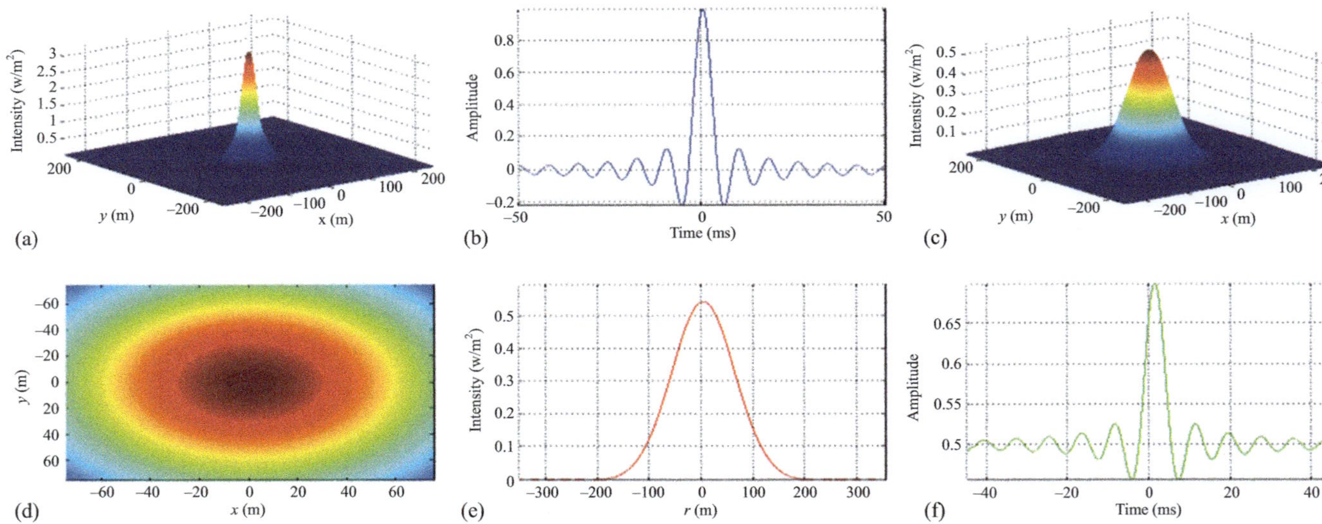

Figure 13.19 *Propagation of EM waves through uniform medium: (a) 3D Gaussian beam input ($w_0 = 30$ mm); (b) modulating signal; (c) 3D Gaussian beam output ($w_0 \approx 80$ mm); (d) cross-section of the output Gaussian intensity; (e) 2D output Gaussian profile; and (f) recovered signal*

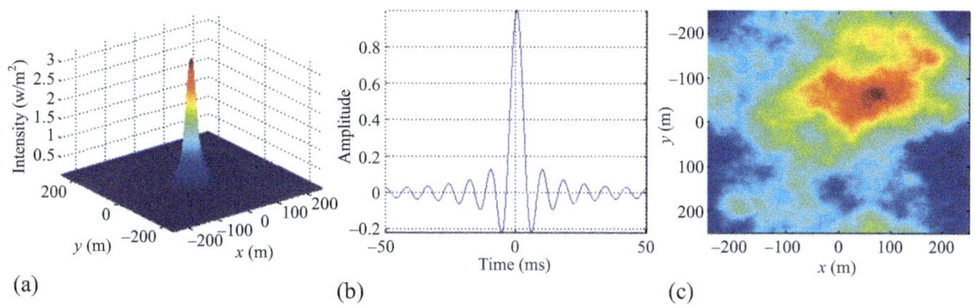

Figure 13.20 Propagation of EM waves through weak turbulence ($f_T = 20$ Hz). (a) 3D Gaussian beam input ($w_0 = 30$ mm); (b) modulating signal; and (c) random phase screen distribution profile

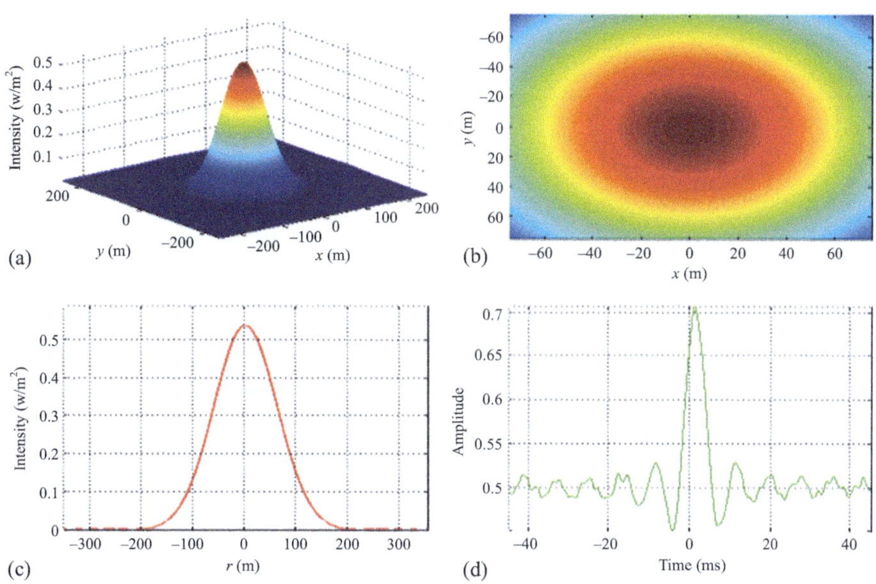

Figure 13.21 Propagation of EM waves through weak turbulence ($f_T = 20$ Hz). (a) 3D Gaussian beam output ($w_0 = 80$ mm); (b) cross-section of the output Gaussian intensity; (c) 2D output Gaussian profile; and (d) recovered signal

distance) repeatedly every (1/20) second, thereby generating random results at a mean turbulence frequency of 20 Hz.

The results (see Figures 13.20 and 13.21) show that under weak phase turbulence, the recovered signal upon averaging the repeated time profiles for each pass over the 20 Hz turbulence is beginning to exhibit some distortion (Figure 13.21(d)).

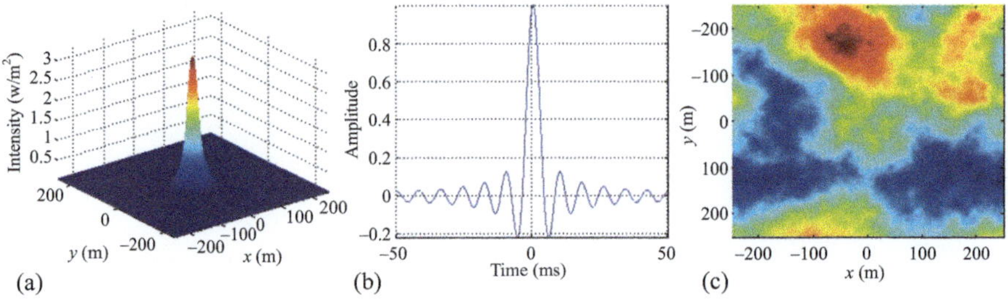

Figure 13.22 Propagation of EM waves through weak turbulence ($f_T = 50$ Hz): (a) 3D Gaussian beam input ($w_0 = 30$ mm); (b) modulating signal; and (c) random phase screen distribution profile

13.9.1.2 Propagation through *weak* turbulence with mean frequency $f_T = 50$ Hz

In this section, we repeat the same procedure for 50 Hz turbulence and the simulation results are shown in the following two figures.

We note from Figures 13.22 and 13.23 that when the turbulence frequency increases, the received Gaussian profile begins to undergo some peak amplitude splitting (Figure 13.23(c)); additionally, the recovered superposed signal waveform shows even greater distortion (Figure 13.23(d)).

13.9.1.3 Propagation through *weak* turbulence with mean frequency $f_T = 100$ Hz

Here, we increase the turbulence frequency to 100 Hz. The numerical results are shown in Figures 13.24 and 13.25.

Again, from Figures 13.24 and 13.25, we observe that as the turbulence frequency increases, the Gaussian profile nominally broadens according to the Kirchhoff–Fresnel diffraction model, but begins to exhibit a small splitting in the transverse direction (Figure 13.25(c)). Moreover, the recovered signal likewise begins to get more distorted (Figure 13.25(d)). This directly demonstrates the damaging effect of turbulence (with higher mean frequency) on a modulated EM wave even when the turbulence is relatively weak.

13.9.2 Propagation through strong *turbulence*

In this section, we examine the EM wave propagation for the case where the turbulence is made several orders of magnitude stronger. Accordingly, we choose $r_0 = 0.01$ mm with $C_n^2 = 1.067 \times 10^{-13}\ m^{-2/3}$.

13.9.2.1 Propagation through *strong* turbulence with mean frequency $f_T = 20$ Hz

In this case, we propagate the modulated EM waves along the propagation path with (mean) turbulence frequency 20 Hz and the simulation results are shown in Figures 13.26 and 13.27.

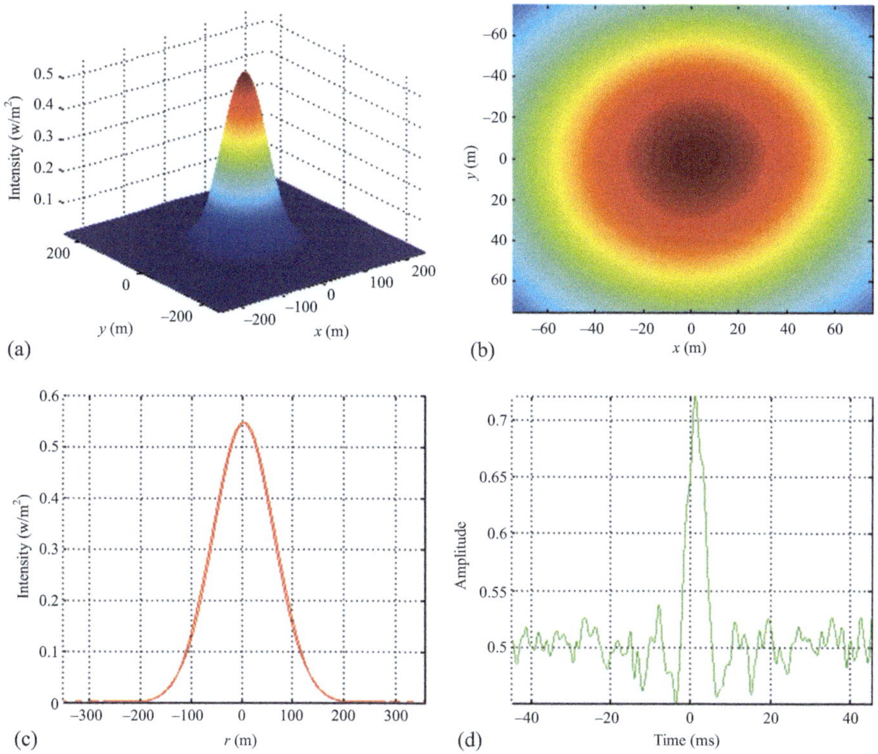

Figure 13.23 *Propagation of EM waves through weak turbulence ($f_T = 50$ Hz). (a) 3D Gaussian beam output ($w_0 \approx 80$ mm); (b) cross-section of the output Gaussian intensity; (c) 2D output Gaussian profile; and (d) recovered signal*

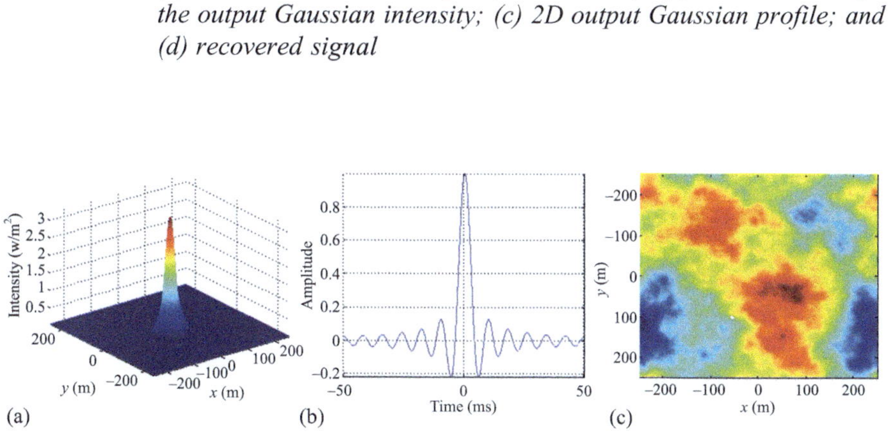

Figure 13.24 *Propagation of EM waves through weak turbulence ($f_T = 100$ Hz). (a) 3D Gaussian beam input ($w_0 = 30$ mm); (b) modulating signal; and (c) random phase screen distribution profile*

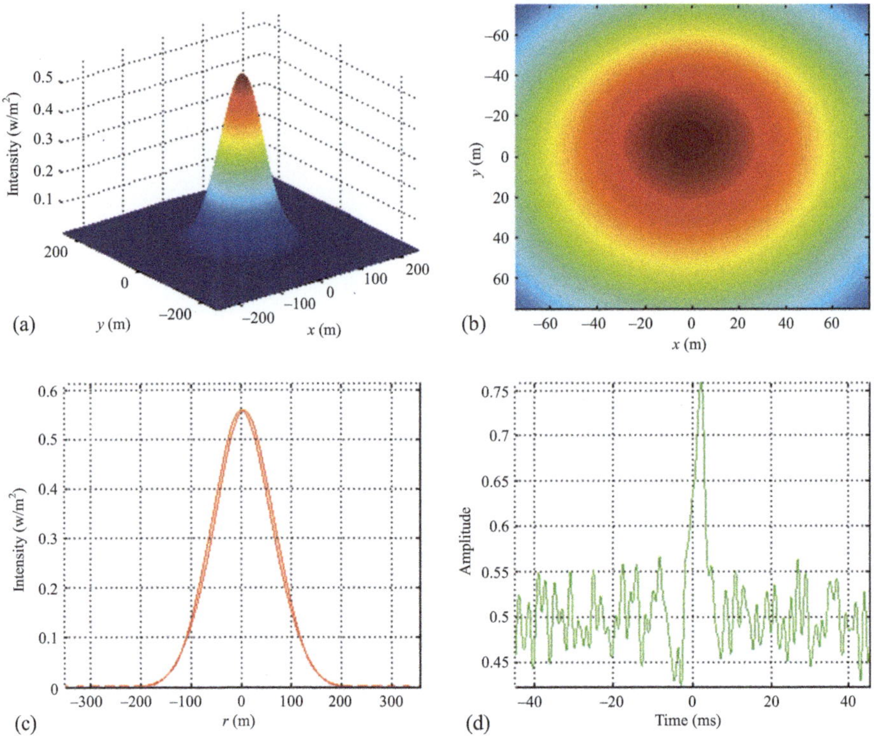

Figure 13.25 Propagation of EM waves through weak turbulence ($f_T = 100$ Hz): (a) 3D Gaussian beam output ($w_0 \approx 80$ mm); (b) cross-section of the output Gaussian intensity; (c) 2D output Gaussian profile; and (d) recovered signal

From Figure 13.26, we now notice the impact of the random phase screen (representing strong turbulence) on the field profile *before* and *after* traversal across the screen. As seen in Figure 13.26(d), the profile has undergone some amplitude distortion around the peak. In Figure 13.27, the EM profile undergoes even more radial splitting (Figure 13.27(d)). Another interesting feature to be noted is that following traversal through turbulence (whereby a peak distortion had occurred), the subsequent diffraction broadening has likely "smoothened" out the peak distortion features immediately following the turbulence region (see Figure 13.27(a)). Also, the recovered signal suffers even greater amplitude distortion compared with the weak turbulence under the same turbulence frequency, as illustrated in Figure 13.27(e).

13.9.2.2 Propagation through *strong* turbulence with mean frequency $f_T = 50$ Hz

In the next series of simulation results, the mean turbulence frequency is increased to 50 Hz. The simulation results are shown in Figures 13.28 and 13.29.

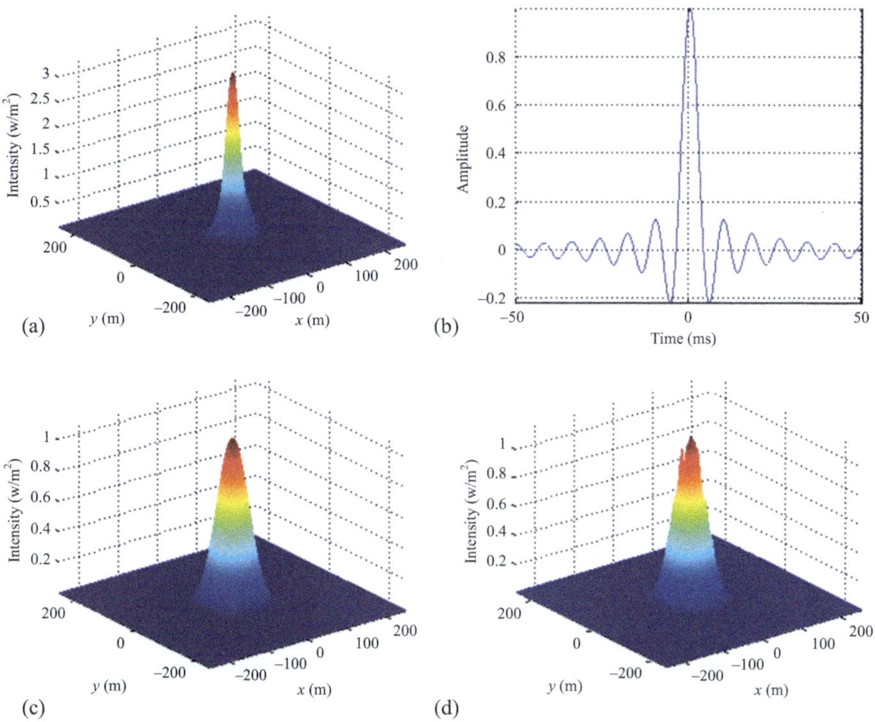

Figure 13.26 Propagation of EM waves through strong turbulence ($f_T = 20$ Hz): (a) 3D Gaussian beam input ($w_0 = 30$ mm); (b) modulating signal; (c) 3D field before the phase screen; and (d) 3D field after the phase screen

The profile of the Gaussian beam gets more distorted now following the phase screen compared with the profile before the screen in a manner similar to the 20 Hz case (Figure 13.28(c) and (d)). In addition, we observe that the received Gaussian once again exhibits smoothening near the peak area; however, there occurs some additional distortion in the radially outward portion of the waveform (Figure 13.29(a)). Also, the recovered signal distortion and Gaussian profile splitting scenarios indicate progressively greater magnitude when the mean turbulence frequency increases, as shown in Figure 13.29(d) and (e). Further details are available in [27].

13.9.2.3 Propagation through *strong* turbulence with mean frequency $f_T = 100$ Hz

Here we examine the effect of the propagation of modulated EM wave through strong turbulence with mean frequency 100 Hz. The significance of the damaging effects of strong phase turbulence on the propagating Gaussian beam is self-evident from Figures 13.30 and 13.31. Thus, there is even greater peak distortion following

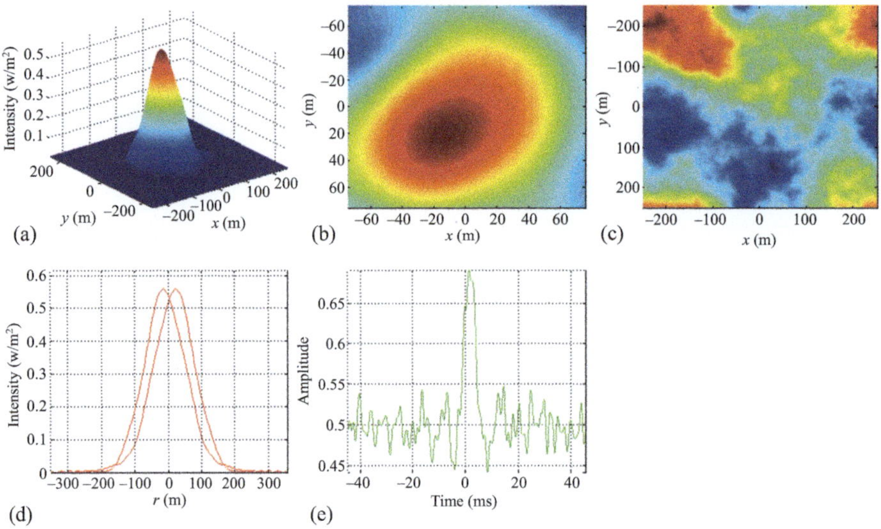

Figure 13.27 Propagation of EM waves through strong turbulence ($f_T = 20$ Hz): (a) 3D Gaussian beam output ($w_0 \approx 80$ mm); (b) cross-section of the output Gaussian intensity; (c) random phase screen distribution profile; (d) 2D output Gaussian profile; and (e) recovered signal

passage across the phase screen (Figure 13.30(d)); similarly the received Gaussian beam now is split into spatially separated humps with peak smoothening and radial distortion (Figure 13.31(a) and (d)); finally, the *sinc*-signal waveform is distorted almost beyond recognition and is essentially pure amplitude noise (Figure 13.31(e)).

13.10 Spectral approach to encrypted *chaotic* wave propagation through turbulence using SVEA and Fourier transforms

In this section, we propagate a modulated *chaotic* wave through atmospheric turbulence. The AM modulation using chaos frequency may be expressed as [28,29]:

$$S_{\mathrm{ch}}(t) = A_c[1 + ms(t)]cos\,\omega_{\mathrm{ch}}t \tag{13.19}$$

where m is the modulation index, and ω_{ch} is an equivalent chaos frequency. It has been verified that in an A-O Bragg cell with first-order feedback, the time-dependent chaos wave generated is approximately amplitude modulated when the signal wave is applied via the RF (sound) input [28].

We note that the chaotic waveform in (13.19) is essentially RF in nature, and manifests itself in the feedback loop as a current in the PD [28]. However, the chaos wave in turn rides on the optical carrier at the output of the Bragg cell (here assumed to be in the first order), and is therefore manifested as *envelope modulation* of the optical carrier.

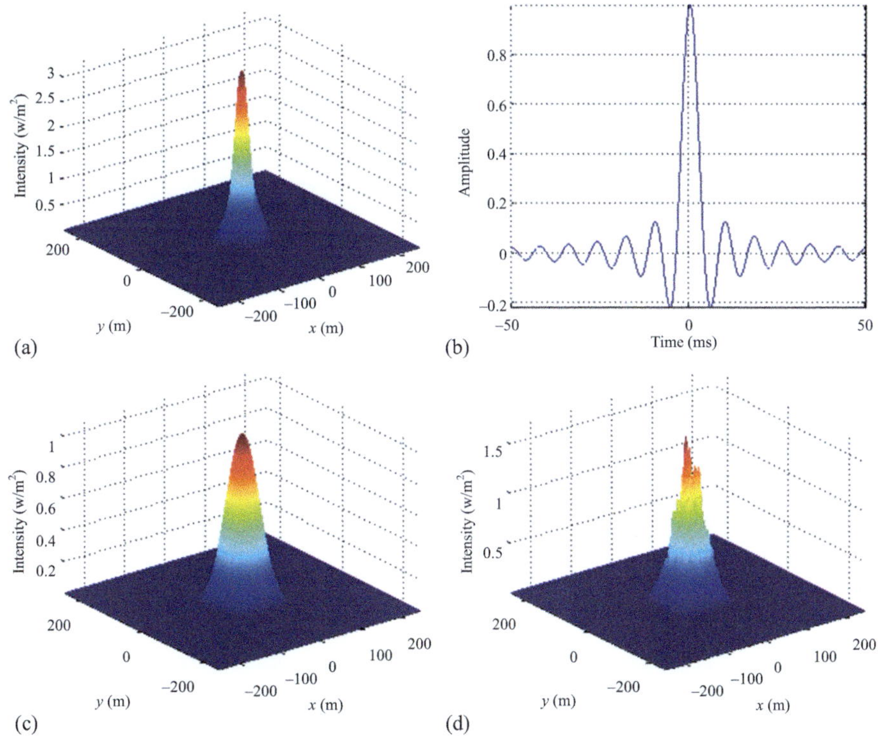

Figure 13.28 Propagation of EM waves through strong turbulence ($f_T = 50$ Hz): (a) 3D Gaussian beam input ($w_0 = 30$ mm); (b) modulating signal; (c) 3D field before the phase screen; and (d) 3D field after the phase screen

The modulated optical wave at the output of the Bragg cell may therefore be approximately written as (assuming AM/envelope modulation):

$$E_{\text{AM}}(x,y,z,t) = E_o(x,y,z)[1 + \tilde{m}S_{\text{ch}}(t)]\cos(\omega_o t - kz) \tag{13.20}$$

where $\tilde{m}$ is the optical modulation index, ω_o is the optical frequency, $E_o(x,y,z)$ is the spatial profile of the optical wave as discussed earlier, and k in the unbounded wavenumber in the medium of propagation.

By substituting (13.14) and (13.19) in (13.20), we get (assuming Gaussian optical profile):

$$E_{\text{AM}}(x,y,z,t) = A\frac{w_o}{w(z)}e^{\frac{-(x^2+y^2)}{w^2(z)}}\{1 + A_c\tilde{m}[1 + ms(t)]\cos\omega_{\text{ch}}t\}\cos(\omega_o t - kz). \tag{13.21}$$

As mentioned earlier, we have here two levels of AM modulation. We will follow the same process by propagating the chaos-modulated optical carrier along

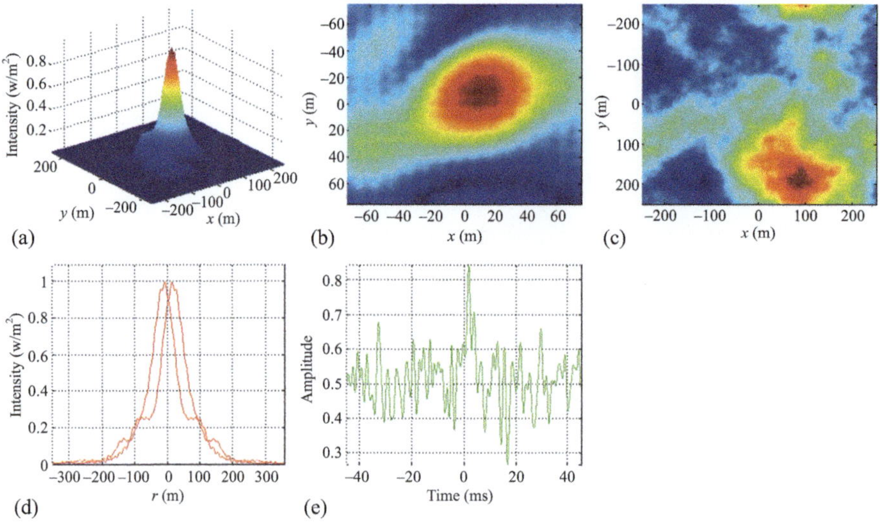

Figure 13.29 Propagation of EM waves through strong turbulence ($f_T = 50$ Hz): (a) 3D Gaussian beam output ($w_0 \approx 80$ mm); (b) cross-section of the output Gaussian intensity; (c) random phase screen distribution profile; (d) 2D output Gaussian profile; and (e) recovered signal

the propagation distance from the transmitter to the receiver, and include the random phase screen located mid-way when incorporating turbulence. Also, we will use SVEA approximation because the optical carrier is orders of magnitude faster than either the chaotic or the signal waveforms. As described, a *sinc*-type message waveform is therefore embedded onto the chaotic carrier at the output of a Bragg cell. The resulting numerical data for the encrypted chaos wave is used to amplitude-modulate the optical carrier. It must be noted that for the case where information is used to directly modulate the optical carrier (as described in Section 13.11), photo detection automatically transfers the intensity corresponding to the envelope, and the optical carrier is eliminated. Thereafter, signal recovery involves simple electronic processing. For the case of a *chaotic* envelope riding the optical carrier, however, photo detection will at first result in a scaled version of the modulated chaos to be detected via the photocurrent. Eventually, the heterodyne receiver strategy with appropriate cut-off frequency may be used to retrieve the baseband (message) signal.

13.10.1 Numerical simulations, results, and interpretations

13.10.1.1 A uniform (nonturbulent) propagation prototype

In this case, we examine the propagation of chaotic EM waves through a uniform medium. Normally, transmission and recovery of time-modulated EM wave through uniform or homogeneous media is fairly standard and well-established, and does not require any examination as such. However, the purpose of including this propagation

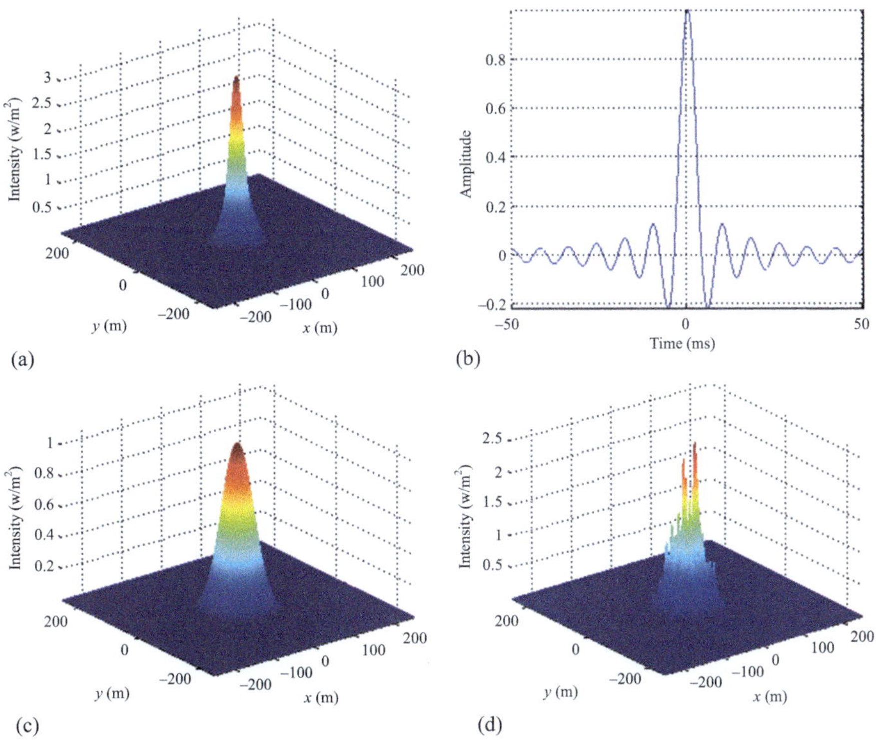

(a) (b) (c) (d)

Figure 13.30 Propagation of EM waves through strong turbulence ($f_T = 100$ Hz): (a) 3D Gaussian beam input ($w_0 = 30$ mm); (b) modulating signal; (c) 3D field before the phase screen; and (d) 3D field after the phase screen

problem prior to introducing turbulence is threefold. First, the time-modulated signal is being propagated over a homogenous region wherein we incorporate the diffractive effects along the path. Since diffraction is essentially a spatial phenomenon, our interest is to ascertain the effect of spatial diffraction on the time signal carried within the modulated chaotic carrier. Second, we wish to determine whether standard demodulation strategies will enable extraction of the message from the photo-detector response at the receiver. Third, the recovery in this case relates to a message carried within a chaotic wave; retrieval of the message therefore needs special consideration vis-à-vis the properties of the chaos wave as opposed to a standard sinusoidal carrier. The simulation results are shown in the following figures.

From Figure 13.32, we observe that the profile of the EM beam is broadened due to diffraction during passage through the (homogenous) medium (Figure 13.32 (a) and (d)). Also, Figure 13.32(f) shows that the received signal is recovered without any distortion using a simple heterodyne receiver used to filter out the chaos frequency and its harmonics in the photocurrent, thereby enabling recovery of the message signal [28,29].

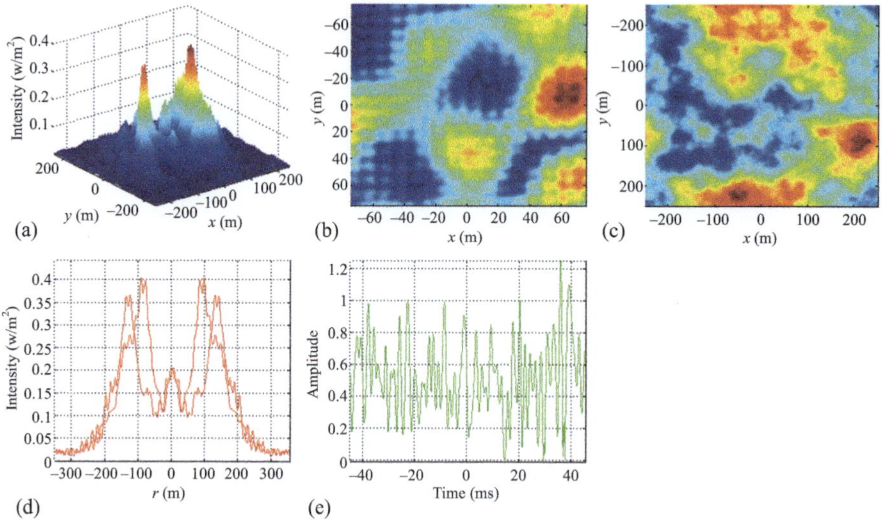

Figure 13.31 *Propagation of EM waves through strong turbulence (f_T = 100 Hz): (a) 3D Gaussian beam output; (b) cross-section of the output Gaussian intensity; (c) random phase screen distribution profile; (d) 2D output Gaussian profile; and (e) recovered signal*

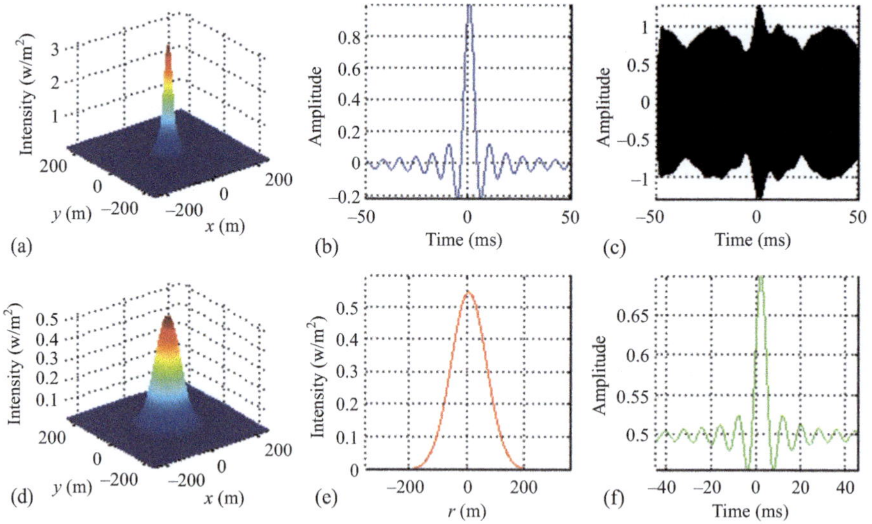

Figure 13.32 *Propagation of modulated chaotic waves through a uniform medium: (a) 3D Gaussian beam input (w_0= 30 mm); (b) modulating signal; (c) modulated signal; (d) 3D Gaussian beam output; (e) 2D output Gaussian profile; and (f) recovered signal*

13.10.1.2 Chaotic propagation through *weak* turbulence with mean frequency $f_T = 50$ Hz

Here, we propagate the chaotic EM wave through narrow and weak turbulence with mean frequency of 50 Hz. For weak turbulence, we choose $r_0 = 10$ mm with $C_n^2 = 1.067 \times 10^{-18}\ m^{-2/3}$; the corresponding simulation results are illustrated in Figure 13.33.

In the presence of turbulence (here weak), we find that the diffracted beam profile (Figure 13.33(d) and (e)) begins to exhibit radial splitting (rather small for the weak case; see Figure 13.33(e)). As may be recalled, even under weak turbulence, transmission of a directly modulated EM wave results in significant distortion of the recovered signal (see Figure 13.33(d)). When the message is embedded in a chaotic carrier, it is seen that even after passage though relatively weak chaos, the message is recovered with noticeably higher integrity (see Figure 13.33(f)). Although this finding is relatively limited in scope, it nevertheless holds out the promise that a message wave may be secured and protected from a turbulent environment by packaging it inside a chaos wave via the RF encryption methodology which has been developed by our group. As is shown later, when the turbulence becomes stronger, the recovery of the message from the chaos wave is subject to some degree of distortion, and is not entirely distortion-free. However,

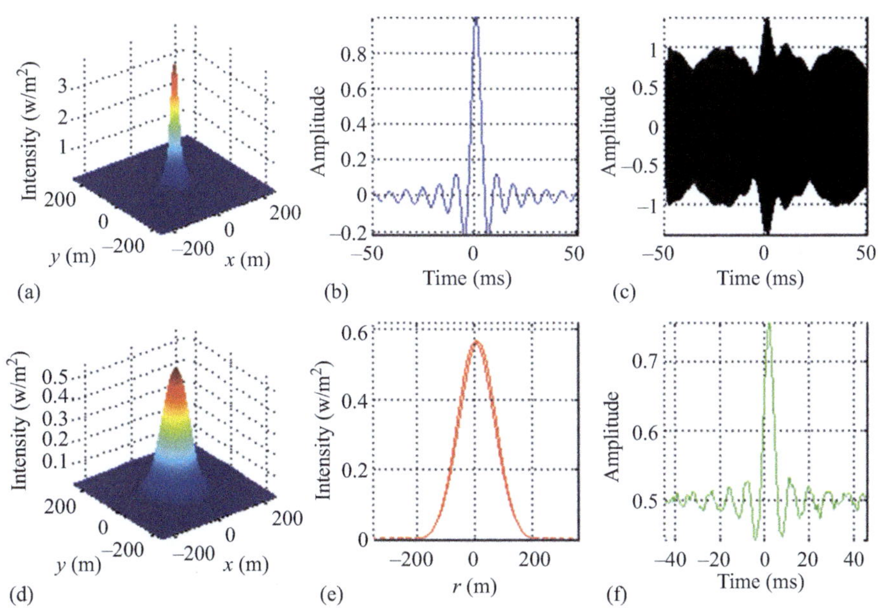

Figure 13.33 Propagation of modulated chaotic waves through weak turbulence ($f_T = 50$ Hz): (a) 3D Gaussian beam input ($w_0 = 30$ mm); (b) modulating signal; (c) modulated signal; (d) 3D Gaussian beam output; (e) 2D output Gaussian profile; and (f) recovered signal

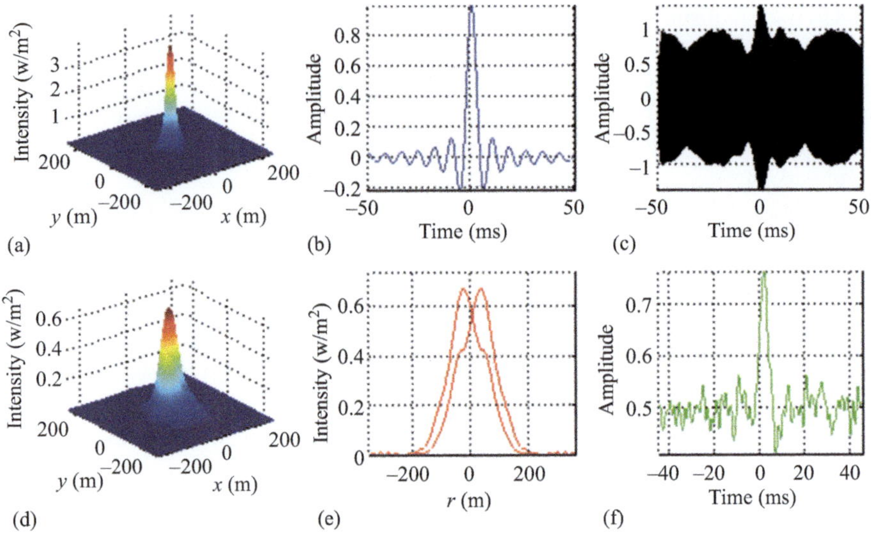

Figure 13.34 Propagation of modulated chaotic waves through strong turbulence ($f_T = 50$ Hz): (a) 3D Gaussian beam input ($w_0 = 30$ mm); (b) modulating signal; (c) modulated signal; (d) 3D Gaussian beam output; (e) 2D output Gaussian profile; and (f) recovered signal

the results indicate that packaging a message in a chaos wave does still offer some immunity from even stronger turbulence compared with non-chaotic transmission.

13.10.1.3 Chaotic propagation through *strong* turbulence with mean frequency $f_T = 50$ Hz

Under strong turbulence, we choose $r_0 = 0.01$ mm with $C_n^2 = 1.067 \times 10^{-13}\ m^{-2/3}$.

Figure 13.34 shows the simulation results for propagation of the modulated chaotic wave over a medium under strong turbulence. It is evident that under strong turbulence, the diffracted profiled is even further split in the radial direction (see Figure 13.34(e)). For this case, the signal recovered using the heterodyne strategy is noticeably distorted (Figure 13.34(f)); however, the degree of distortion is considerably lower than that obtained for non-chaotic propagation (see Figure 13.34(e) for comparison). The above further reinforces the observation that chaotic encryption likely immunizes a message waveform from the damaging effects of turbulence.

13.11 Propagation through phase turbulence using altitude-dependent structure parameter without and with A-O chaos

13.11.1 Hufnagel-Valley (HV) model

For relatively low-altitude propagation, the strength of turbulence along the propagation path is described by the structure parameter $C_n^2(h)$, where h is the physical altitude at the point of measurement relative to the ground plane. There are several profiles that have been established to estimate $C_n^2(h)$. The most commonly used profile is the HV model [30,31]:

$$C_n^2(h) = Ae^{\left(-\frac{h}{100}\right)} + 5.94 \times 10^{-53} \times \left(\frac{v}{27}\right)^2 h^{10} e^{\left(-\frac{h}{1,000}\right)} + 2.7 \times 10^{-16} e^{\left(-\frac{h}{1,500}\right)}. \tag{13.22}$$

The parameter A represents the turbulence strength at ground level while v is the high altitude wind speed. The most commonly used standard is the HV5/7 which sets $A = 1.7 \times 10^{-14}\ m^{-2/3}$ and $v = 21$ m/s. We note that for $h = 0$, $C_n^2(0)$ becomes $A = 2.7 \times 10^{-16}$, which is approximately equal to the ground level turbulence strength. The C_n^2 profile may be broken into a finite number of layers (i) with individual $C_{n_i}^2$. The seeing cell size is commonly called the Fried parameter; the Fried parameter for the ith layer is given by [30,31]:

$$r_{o_i} = 0.185\left[\frac{4\pi^2}{k^2 C_{n_i}^2 \Delta z_i}\right]^{3/5}, \text{ and} \tag{13.23}$$

$$r_0 = r_{0\text{eff}} = 0.185\left[\frac{4\pi^2}{k^2 C_n^2 L}\right]^{3/5}. \text{ Also,} \tag{13.24}$$

$$r_0 = \left[\sum_{i=1}^{n} r_{o_i}^{-5/3}\right]^{-3/5} \tag{13.25}$$

where k is the wave number; z_i and Δz_i are the locations of the phase screen and the thickness of the ith layer, respectively. L is the length of the propagation path through turbulence. Another important parameter in defining atmospheric turbulence in an optical system is log amplitude variance, which is defined as a common measure of scintillation [32]:

$$\sigma^2 = 1.33k^{-5/6}L^{5/6}\sum_{i=1}^{n} r_{o_i}^{-\frac{5}{3}}\left(1 - \frac{z_i}{L}\right)^{5/6}. \tag{13.26}$$

The parameter σ^2 captures the variance in the amplitude of the wavefront due to the atmosphere. It is used as a reliable measure of the strength of scintillation in optics. It turns out that in standard applications, the variance typically takes the range of values as: < 0.3 for weak turbulence; $0.3 < \sigma^2 < 4$ for moderate turbulence, and $\sigma^2 > 4$ for strong turbulence [31]. The final turbulence parameter to be addressed is the *isoplanatic* angle θ_0 defined as the largest field angle over which turbulence can be accurately estimated.

$$\theta_0 = \left[6.8794L^{5/3}\sum_{i=1}^{n} r_{o_i}^{-\frac{5}{3}}\left(1 - \frac{z_i}{L}\right)^{5/3}\right]^{-3/5}. \tag{13.27}$$

It turns out that the angle θ_0 is typically quite small (around the microradians range); hence, most slanted propagation tends to require the anisoplanatic model for $C_n^2(h)$.

We report here on numerical simulations of imaging using a plane EM wave via two approaches: (1) propagating the wave through a transparency (taken as a chessboard pattern) over a sufficiently long path (including a region of turbulence), and imaging the 2-D pattern through a thin positive lens in its back focal plane; and (2) digitizing the transparency function using pixels encoded into PCM data, modulating the optical carrier with this data, propagating the modulated (wave over the same region (with turbulence), and finally recovering and reconstructing the image using photodetection and D-A conversion with thresholding.

13.11.2 Plane EM wave propagation through a transparency-thin lens combination with turbulence

We present a model for simulating the effects of turbulence using a technique similar to Roggemann's model [31]. The technique implements the commonly used split-step propagation method with four phase screens optimally placed along the optical path. In this part, we present the case for an unmodulated Gaussian-profiled EM wave propagated over a thin positive lens with a 2-D image-pattern transparency placed at a distance sufficiently greater than *2f* (where *f* is the focal length) in front of the lens. The effect of imaging and other related received transverse field patterns under propagation through a turbulent layer along a slanted path is examined. We present a numerical model (refer to Figure 13.35(a)) by setting the value of L_T (equal to L_D) at 3.5 km. Note that L_T is the horizontal distance corresponding to the turbulence layer and L_D is a region of propagation strictly under diffraction. Also, the angle of elevation has been chosen as 8°. This gives a total optical slant range of $L = 7,068.8$m with a receiver height (max. altitude $h_{\max}$) above the ground of 983.78 m. Figure 13.35(b) shows a schematic of the numerical model where L_1 *is* the propagation distance with turbulence and L_2 is the propagation distance without turbulence along the slant path.

Let the nonturbulent distance $L_2 = z_1$ where $z_1 \gg f$ the distance in front of a positive lens. Then by standard geometrical optics, the image occurs at the image focal plane of the lens. Hence, the image plane is at $z_2 \approx f$ behind the lens, where $\frac{1}{f} = \frac{1}{z_1} + \frac{1}{z_2}$. We represent the complex field immediately behind the last phase screen by $U_4(\xi, \eta)$ and its image $U_i(u, v)$ via the integral [33,34]:

$$U_i(u,v) = \iint_{-\infty}^{\infty} h(u,v;\xi,\eta) U_4(\xi,\eta) d\xi d\eta \qquad (13.28)$$

The finite extent of the lens can be accounted for by associating with the lens a pupil function $p(x, y)$ defined by:

$$p(x,y) = \mathrm{circ}\left(\frac{\sqrt{x^2+y^2}}{w}\right), w = d/2 \text{ and } d \text{ is the } \textit{lens} \text{ diameter} \qquad (13.29)$$

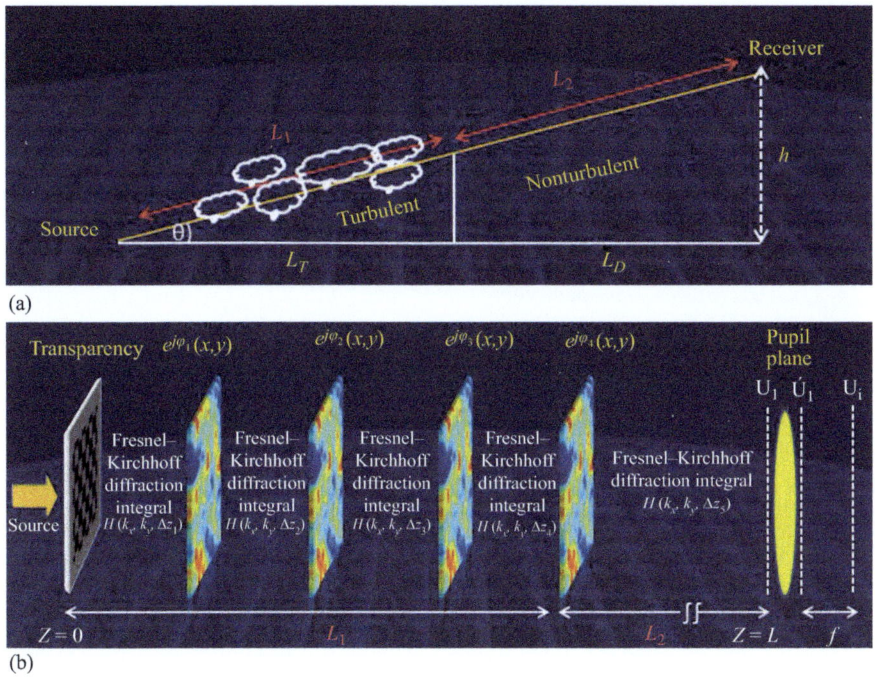

Figure 13.35 (a) Numerical model for slant path propagation and (b) Schematic illustration and physical interpretation of propagation through turbulence

The coherent transfer function is:

$$H(f_u,f_v) = \text{circ}\left(\frac{\sqrt{(\lambda z_2 f_u)^2 + (\lambda z_2 f_v)^2}}{w}\right) = \text{circ}\left(\frac{\sqrt{f_u^2 + f_v^2}}{f_0}\right) \quad (13.30)$$

where the coherent cutoff frequency $f_0 = \frac{w}{\lambda z_2}$; λ is the wavelength; f_u *and* f_v are the spatial frequencies. To find the spatial impulse response h, let the input be a spatial δ function; then the point-spread function introduced by diffraction is defined as [30,33]:

$$\tilde{h}(u,v) = \iint_{-\infty}^{\infty} p(\lambda z_2 \tilde{x}, \lambda z_2 \tilde{y}) e^{\{-j2\pi[u\tilde{x}+v\tilde{y}]\}} d\tilde{x} d\tilde{y} \quad (13.31)$$

where $\tilde{h}, \tilde{x}$ *and* $\tilde{y}$ the final set of coordinate normalizations such that: $\tilde{h} = \frac{1}{|M|} h, \tilde{x} = \frac{x}{\lambda z_2}, \tilde{y} = \frac{y}{\lambda z_2}$, and M is the magnification of the system equal to $-\frac{z_2}{z_1}$. For simulating coherent imaging on the computer is based on (13.7) can be implemented as [33,34]:

$$U_i(u,v) = \mathfrak{I}^{-1}\{H(f_u,f_v)\mathfrak{I}\{U_g(u,v)\}\} \quad (13.32)$$

where $H(f_u,f_v) = J\{\tilde{h}(u,v)\}$ and $U_g(u,v) = \frac{1}{|M|} U_4\left(\frac{u}{M}, \frac{v}{M}\right)$.

The diffracted EM wave through turbulence finally traverses a distance f past the lens with this transmission function incorporated. In this manner, the image of the transparency is reconstructed in the focal plane.

13.11.3 Fixed L_T and L_D distances for different turbulence strengths

We generate weak, moderate, and strong turbulence for fixed L_{T} and L_{D} (= 3.5 km) by setting the effective $C_n^2 = 1 \times 10^{-17} m^{-2/3}$, $C_n^2 = 3.5021 \times 10^{-15} m^{-2/3}$ and $C_n^2 = 1 \times 10^{-13} m^{-2/3}$, respectively. We use (13.23) to find r_0, giving $r_0 = 59.83$ cm, 1.78 cm, and $r_0 = 0.23$ cm, respectively. For the case of weak, we need to find minimization for each r_{0_i} that gives $C_n^2 = 1 \times 10^{-17}$ or ($r_0 = 0.5983$ m) and $\sigma^2 << 0.3$ (applying (13.24) and (13.25)) while for case of strong $\sigma^2 = 28$. We use MATLAB code fmincon. We note that when the HV model and Roggeman algorithm (as outlined earlier) are applied to the system shown in Figure 13.35, the computed equivalent C_n^2 happens to fall in the moderate turbulence range (i.e., about $3.5021 \times 10^{-15} m^{-2/3}$). Therefore, in order to numerically simulate the weak and strong cases, a separate set of calculations is needed. This entails the following strategy. First, an equivalent C_n^2 (say $1 \times 10^{-17} m^{-2/3}$ for the weak case and $1 \times 10^{-13} m^{-2/3}$ for the strong) is assumed and plugged into (13.25) whence an equivalent Fried parameter r_0 is computed. Thereafter, in conjunction with (13.24) and (13.25), the four segmented r_{0_i} values are calculated, from which the corresponding phase screens along the propagation path may be constructed. The resulting values of r_0 and each r_{0_i}, log-amplitude variance, and the isoplanatic angles are listed in Table 13.1. The results show strong turbulence leads to the highest distortion in the recovered image. Weak turbulence, on the other hand, primarily displays diffraction limited behavior, while moderate turbulence displays limited image distortion. Cross correlations (CCs) are computed for the three cases. To do this, the image is transmitted through the medium assuming no turbulence (diffraction limit only) and thereafter transmitted under the three turbulence conditions. For each case, image CCs are computed relative to the nonturbulent case. For weak turbulence, the CC value is 0.9555 and decreases for moderate to 0.9513. For strong turbulence, the value decreases to 0.8719. All of these computations are at fixed L_1 and L_2 of 3.5 km. Note that the CC declines noticeably for strong turbulence. The results for the three cases are shown in Figure 13.36(d)–(f), with corresponding phase screens in Figure 13.36(a)–(c).

Table 13.1 Turbulence parameters $L_{\mathrm{T}} = L_{\mathrm{D}}$ (= 3.5 km)

	$r_{0_1}(m)$	$r_{0_2}(m)$	$r_{0_3}(m)$	$r_{0_4}(m)$	$r_{0,\mathrm{eff}}(cm)$	σ^2	$\theta_0(\mu\mathrm{rad})$
Weak	1.2	1.4	1.45	1.5	59.83	0.0022	82.37
Moderate	0.027	0.032	0.069	0.12	1.78	0.9632	2.01
Strong	0.0034	0.0048	0.0085	0.0110	0.23	28	0.262

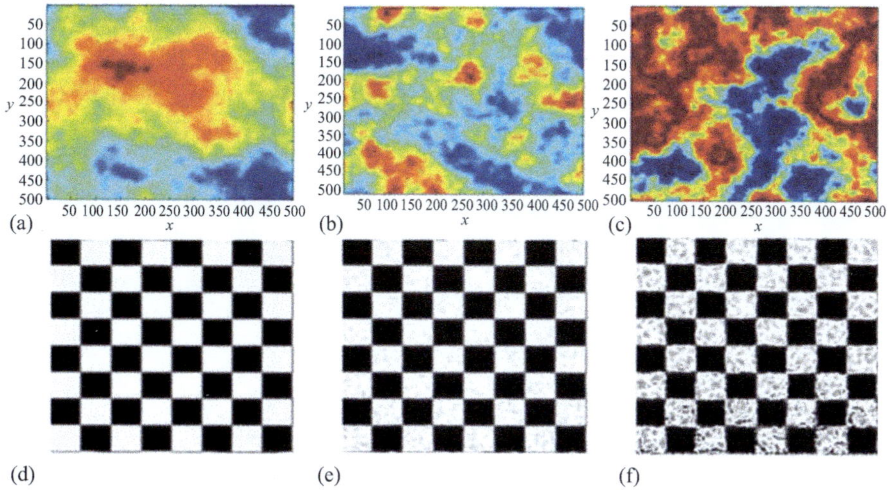

Figure 13.36 *Thin phase screen for (a) weak, (b) moderate, (c) strong turbulence. Simulator image output under turbulence conditions (d) weak, (e) moderate, (f) strong. Only the first of four phase screens is displayed in each case*

13.11.4 Fixed C_n^2 and L_T for three different (nonturbulent) distances L_D

In this part, we increase the L_D from 3.5 km to 10 km and 14 km, and test the effect of increasing L_D under fixed C_n^2 for weak, moderate, and strong turbulence. For all cases, the value of L_T is fixed at 3.5 km. Figure 13.37(a)–(c) shows that the image distortion increases as the (nonturbulent) propagation distance increases for all three cases. Thus, the image distortion is seen to be highest in Figure 13.37(c). The CC values are computed for each L_D at different turbulence strengths. The results are shown in Figure 13.38(a). In the figure, L_D is increased from 3.5 to 14 km for weak, moderate, and strong turbulences, and the CC products are plotted with different legends. We find from the plots that the CCs consistently decrease for all cases as L_D increases; additionally, the products are consistently lower as the turbulence strength increases. Thus, the cross-correlation decreases from 0.9555 to 0.9317 for the weak case; from 0.9513 to 0.9126 for the moderate case, and from 0.8719 to 0.6448 for the strong case. In summary, then, image distortion under turbulence is controlled by two effects, viz., (1) the (nonturbulent) propagation distance beyond the region of turbulence; and (2) the turbulence strength.

13.11.5 Fixed C_n^2 and L_D, for three different turbulence distances L_T

Since there is a relation between $r_{o,\text{eff}}$ and L_T (propagation distance along the path with turbulence), an increase in L_T requires a corresponding re-calculation of $r_{o,\text{eff}}$ in

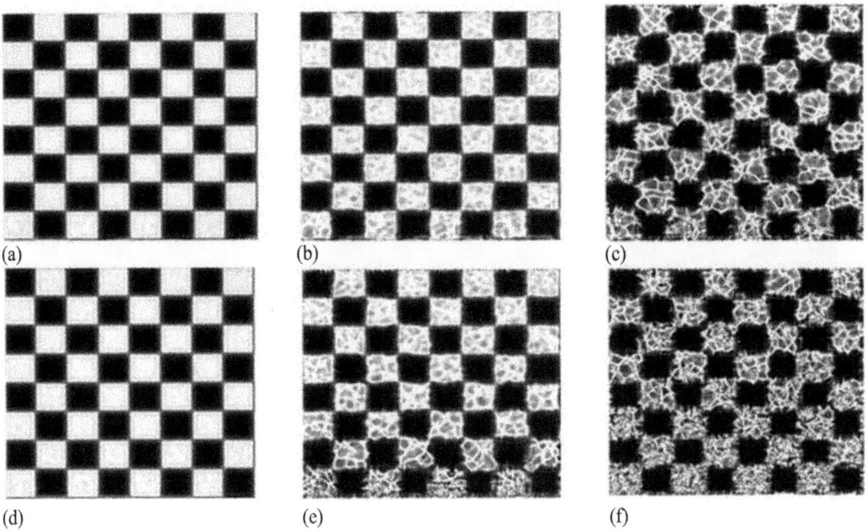

Figure 13.37 *Simulator output with increasing (a)–(c)* $L_D = 14$ km *weak, moderate, and strong turbulence; (d)–(f)* $L_T = 14$ km *weak, moderate, and strong turbulence*

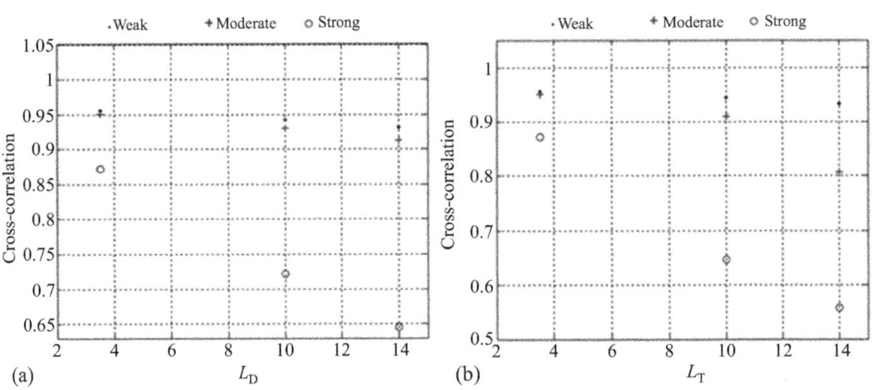

Figure 13.38 *Cross-correlation for weak, moderate, and strong turbulence, (a)* L_D *3.5, 10 and 14 km,* $\frac{\Delta CC}{\Delta L_D}\Big|_{strong} = -0.022$*;*
(b) L_T *3.5, 10 and 14 km,* $\frac{\Delta CC}{\Delta L_T}\Big|_{strong} = -0.03$

order to keep C_n^2 fixed while satisfying (13.23). We note in this part L_D is fixed at 3.5 km. The effect of increasing L_T for different C_n^2 strengths is examined Table 13.2.

Figure 13.37(d)–(f) shows recovered images for weak, moderate, and strong turbulence, respectively, at $L_T = 14$km. We note L_D is fixed at 3.5 km for all cases. The cross-correlation products corresponding to the above cases are shown in Figure 13.38(b). For weak turbulence, the CC value decreases from 0.9555 to

Table 13.2 Re-computed Fried parameters for $L_T = 10$ and 14 km corresponding to weak, moderate, and strong turbulence

	Weak	Moderate	Strong
$L_T = 10\,\text{km}$	$r_0 = 0.3187\,\text{m}$	$r_0 = 0.0095\,\text{m}$	$r_0 = 0.0013\,\text{m}$
$L_T = 14\,\text{km}$	$r_0 = 0.2604\,\text{m}$	$r_0 = 0.0077\,\text{m}$	$r_0 = 0.0010\,\text{m}$

0.905; for moderate it decreases from 0.9513 to 0.8076, and for strong from 0.8719 to 0.5571, respectively. Overall, it is clear that for similar increases in the distances L_D and L_T, the rate of decrease in the CC values dramatically increases for higher L_T distances than L_D distances for all turbulence cases. This implies that image or signal distortion increases significantly when the effective longitudinal range of the turbulence is increased.

13.11.6 Modulated EM wave (non-chaotic and chaotic) with a digitized image pattern

In this section, we examine the propagation of (modulated) EM waves through atmospheric turbulence and their characteristics upon recovery in the image plane. The AM-modulated EM wave in the presence of (in this case, digitized) information signal may be expresses as:

$$E_{AM}(x,y,z,t) = A\frac{w_0}{w(z)}e^{-r^2/w^2(z)}[1 + ms(t)]\cos(\omega_0 t - kz). \tag{13.33}$$

For the case of a (modulated) chaotic wave, the resulting waveform may be approximately expressed as:

$$s_{ch}(t) = A_C[1 + ms(t)]cos(\omega_{ch}t). \tag{13.34}$$

The modulated optical wave at the output of the Bragg cell may therefore be approximately written as (assuming AM/envelope modulation):

$$E_{AM}(x,y,z,t) = A\frac{w_0}{w(z)}e^{-r^2/w^2(z)}[1 + \tilde{m}s_{ch}(t)]\cos(\omega_0 t - kz) \tag{13.35}$$

In the above process, there are therefore two levels of AM modulation. We will follow the same process by propagating the chaos-modulated optical carrier along the propagation distance, and include the four random phase screens when incorporating turbulence. Also, we will use SVEA approximation because the optical carrier is orders of magnitude faster than either the chaotic or the signal waveforms. Photo detection automatically transfers the intensity corresponding to the envelope, and the optical carrier is eliminated. Thereafter, signal recovery involves simple electronic processing. For the case of a chaotic envelope riding the optical carrier, however, photo detection will at first result in a scaled version of the modulated chaos to be detected via the photocurrent. Eventually, a *heterodyne* receiver strategy with appropriate filter cut-off frequency may be used to retrieve the baseband (message) signal following additional filtering as needed.

13.11.7 Fixed L_T and L_D distances for different turbulence strengths under a modulated EM wave propagation

In this section, we will examine the propagation of a modulated EM wave (non-chaotic and chaotic) through a turbulent medium (weak, moderate, and strong) with effective temporal frequency (related to the mean time fluctuations within the turbulence) f_T (assumed 50 Hz here; typically, this number varies from about 50–200 Hz), where L_T and L_D are fixed at 3.5 km. The digitized image corresponding to the spatial chessboard pattern converted to pixelated grey scale values (On–off NRZ is used for encoding through the MATLAB routine) is shown in Figure 13.39(a). Also, the encrypted chaotic image signal is shown in Figure 13.39(b). We note that the encrypted chaos is propagated through the 4-segment turbulent layer as described before, and eventually the transmitted image is reconstructed using photodetection, heterodyning and two levels of filtering. The final filtered images are not shown here. The CC values (computed using pixel-by-pixel data from the two images for the nonturbulent and turbulent cases) for both non-chaotic and chaotic propagation are computed for the three cases and the results plotted in Figure 13.39(c). As expected, weak turbulence has the highest cross-correlation which decreased monotonically from moderate to strong turbulence. When the image is embedded in a chaotic carrier, it is seen that the CC is greater than for the non-chaotic case for each of the different turbulence strengths. Quantitatively, use of a chaotic carrier leads to a cross-correlation or image improvement of about 0.3 dB under strong turbulence and 0.4 dB under weak turbulence.

13.11.8 Fixed C_n^2 and L_T for three different destination distances L_D

In this case, we examine the propagation of chaotic and non-chaotic EM waves with increasing the nonturbulent distance L_D. We increase L_D from 3.5 km to 10 km

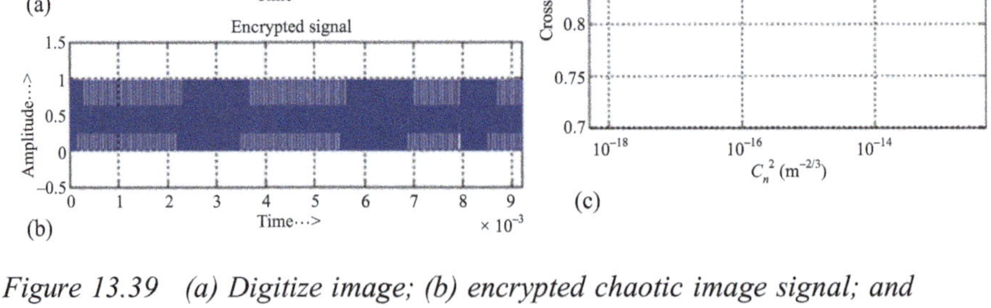

Figure 13.39 (a) Digitize image; (b) encrypted chaotic image signal; and (c) cross-correlation for non-chaotic vs. chaotic under weak, moderate, and strong turbulence and $L_T = L_D$ at 3.5 km

and 14 km, and examine the effect of increasing L_D under fixed C_n^2 for strong turbulence. The recovered digital image signals under chaotic and nonchaotic conditions for $L_D = 14$ km are shown in Figure 13.40(a) and (c), respectively. In Figure 13.40(b) and (d), thresholding the recovered signal over half of the overall amplitude helps compensate for the ringing noise derived from turbulence and undesirable signal components which emerge at the output upon filtering. The resulting reconstructed/decoded images for both cases are shown in Figure 13.41. We find that for modulated non-chaotic propagation under strong turbulence, the CC decreases from 0.8567 at 3.5 km to 0.7398 at 14 km, while in the modulated chaotic case, the CC decreases from 0.9205 to 0.8704 as shown in Figure 13.42(a). In Figure 13.42(b), we plot the computed mean square error (MSE) between the original and recovered images in order to measure image distortion due to different nonturbulent distances L_D under non-chaotic and chaotic transmission. It is obvious from Figure 13.42(a) that chaotic transmission maintains greater image integrity against turbulence for various L_D distances. Similar observations may be made for propagation of non-chaotic and chaotic signals through strong turbulence over different propagation distances as shown in Figure 13.42(b).

13.11.9 Fixed C_n^2 and L_D for three different destination distances L_T

In this section, we intend to propagate a modulated chaotic and non-chaotic waves through atmospheric turbulence for different turbulence distances L_T. We increase L_T from 3.5 km to 10 km and 14 km, and examine the effect of increasing L_T under fixed C_n^2 for strong turbulence. Figure 13.43 shows the simulation results for the received image signal under chaotic and non-chaotic modulation over a medium with strong turbulence. The recovered digital image signals under chaotic and non-chaotic for $L_T = 14$ km and strong turbulence are shown in Figure 13.43(a) and (c),

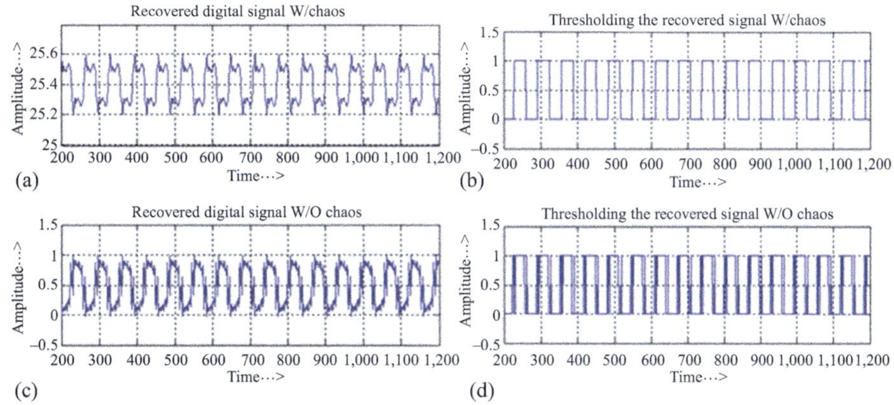

Figure 13.40 The received image signal: (a) under chaotic transmission; (b) thresholding the recovered signal in (a). (c) Under non-chaotic propagation; and (d) thresholding the recovered signal in (c). $L_D = 14$ km and strong turbulence

Figure 13.41 The received image under (a) non-chaotic and (b) chaotic propagation

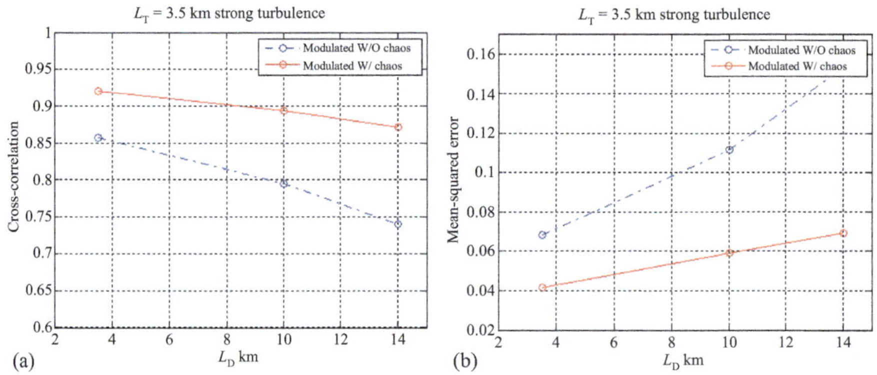

Figure 13.42 (a) Cross-correlation and (b) mean square error for non-chaotic vs. chaotic with ($L_D = 3.5$, 10 and 14 km) strong turbulence

respectively. In Figure 13.43(b) and (d), recovered signals via thresholding applied to Figure 13.43(a) and (c) are shown. It is evident that under strong turbulence, the degree of distortion under chaotic propagation (Figure 13.43(b)) is considerably lower than that obtained for non-chaotic propagation (Figure 13.43(d)). Figure 13.44 shows the received images for non-chaotic vs. chaotic transmissions, clearly indicating a substantial improvement in recovery. We plot also the computed CC and MSE values between the original and recovered images in order to measure the image distortion due to different L_T distances under non-chaotic and chaotic transmission. We observe that when L_T increases to 14 km, the CC decreases from 0.8567 to 0.70 for non-chaotic and from 0.9205 to 0.8026 for chaotic (Figure 13.45(a)). The above further reinforces the observation that chaotic

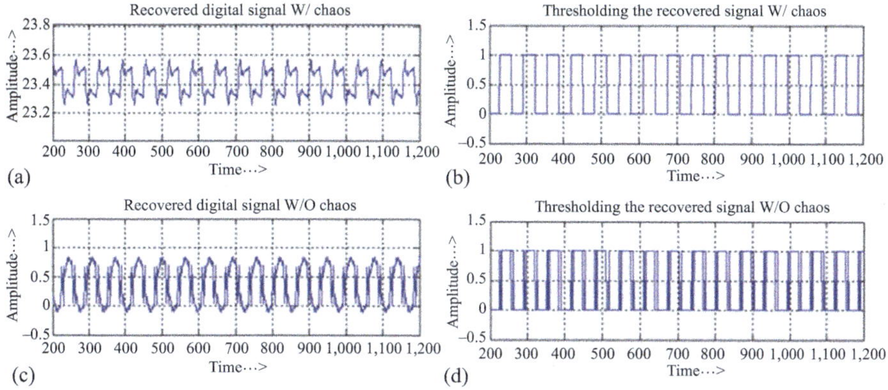

Figure 13.43 The received image signal: (a) with chaotic transmission; (b) thresholding the recovered signal in (a). (c) with non-chaotic transmission; (d) thresholding the recovered signal in (c). L_T = 14 km and strong turbulence

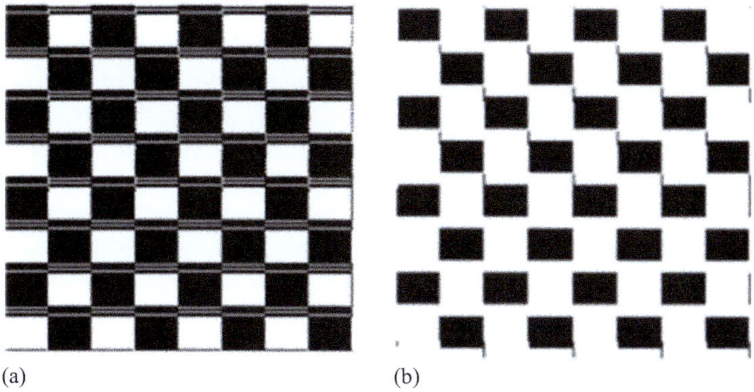

Figure 13.44 The received image: (a) with non-chaotic transmission; (b) with chaotic transmission

encryption reduces turbulence-induced message waveform distortion compared with non-chaotic transmission.

Propagation through MVKS-type phase turbulence has been examined under a variety of conditions. The problem has been set up as an electromagnetic wave propagation along a slanted path (at a slant angle sufficiently higher than the isoplanatic angle) made up of a layer of turbulence (L_T) and one of non-turbulence (L_D). The goal was to conduct a numerical simulation of the diffraction plus turbulence problem for a modulated EM wave (consisting of a digitized 2D image) and examine the recovered signal at the destination in comparison with non-turbulent (purely diffraction limited) propagation. Cross-correlation (CC) products and MSEs are computed as performance measures relative to nonturbulent recovery. It is observed

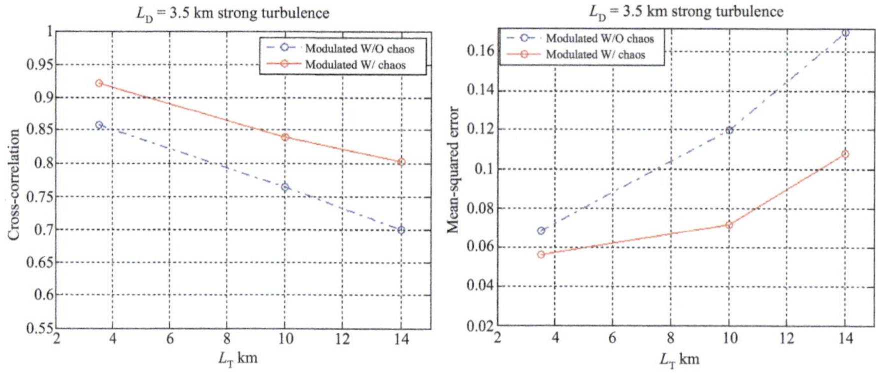

Figure 13.45 (a) Cross-correlation and (b) mean square error for non-chaotic vs. chaotic with (L_T = 3.5, 10 and 14 km) with strong turbulence

that the CC consistently decreases for higher turbulence strengths, and also for increasing L_D and L_T distances (with a higher loss of CC for L_T increases at each level of turbulence strength). A second series of simulations consisted of "packaging" the information signal (digitized 2D image) inside an acousto-optic chaos wave which in turn modulates the optical carrier (the first-order laser beam). This twice-modulated laser beam is then propagated across the slanted path under conditions identical to those for the non-chaotic cases. A case-by-case examination reveals that packaging a signal inside chaos consistently improves the performance of the system (lower signal distortion or higher CC product) in each case. A third series consists of propagating the EM carrier through a spatial 2D image (consisting of a 2D transparency) placed at an input aperture, then propagating the resulting wave through a similar L_T plus L_D combination as was done for the modulated wave. The "image wave" is then passed through a positive thin lens placed at a distance sufficiently greater than $2f$ (where f is the focal length) and finally recovered in the back focal plane of the lens. This recovered image is once again quantitatively examined relative to the non-turbulently propagated and recovered images (primarily via the CC products) for different turbulence strengths, and L_D, L_T distances. A special note regarding the use of strong turbulence in the MVKS-based simulations reported in this chapter. It turns out that generally the use of a specific turbulence model requires compatibility or admissibility under one or more cumulative figures of merit. One such is the so-called Rytov variance which takes into account the different turbulence parameters in the simulation to determine if the assumptions in a model will apply. For the strong turbulence regimes (as presented here), sometimes the MVKS model does not satisfy the Rytov variance condition. The strong turbulence results presented here are likely on the margins of acceptable Rytov variances [35]. It is found overall that while many of the imaging characteristics are similar between the transparency-lens vs. the modulated wave problems, the latter approach generally appears to provide better performance than the purely spatial transmission and recovery using lens-based optics. Some related results appear in Ref. [36].

References

[1] L. Brillouin, "Diffusion of light and X-rays by a transparent homogeneous body," Annales de Physique 17, 88–122 (1922).
[2] C.V. Raman and N.S. Nagendra Nath, "The diffraction of light by high frequency sound waves: Part I," Proceedings of the Indian Academy of Sciences 2, 406–412 (1935).
[3] A. Korpel, "Acousto-Optics," 2nd ed., Marcel Dekker, New York (1997).
[4] W.R. Klein and B.D. Cook, "Unified approach to ultrasonic light diffraction," IEEE Trans. Sonics Ultrason. SU-14, 123–133 (1967).
[5] S.-T. Chen and M.R. Chatterjee "A numerical analysis and expository interpretation of the diffraction of light by ultrasonic waves in the Bragg and Raman–Nath regimes multiple scattering theory," IEEE Trans. on Education 39, 56–68 (1996).
[6] C. Webb and J. Jones, "Handbook of Laser Technology and Applications – Vol. II," 1st ed., IOP, Cornwall, UK (2004).
[7] A. Korpel and T.-C. Poon, "Explicit formalism for acousto-optic multiple plane-wave scattering," J. Opt. Soc. Am. 70, 817–820 (1980).
[8] R. Pieper and A. Korpel, "Eikonal theory of strong acousto-optic interaction with curved wavefronts of sound," J. Opt. Soc. Am. A 2, 1435 (1985).
[9] A. Korpel, C. Venzke, and D. Mehrl, "Novel algorithm for strong acousto-optic interaction: application to a phase profiled sound column," Proc. Ultrason. Int. 91, Le Touquet, France, July 1–4 (1991).
[10] M.R. Chatterjee, T.-C. Poon, and D.N. Sitter, Jr., "Transfer function formalism for strong acousto-optic Bragg diffraction of light beams with arbitrary profiles," Acustica 71, 81–91 (1991).
[11] J. Chrostowski and C. Delisle, "Bistable piezoelectric Fabry–Perot interferometer," Can. J. Phys. 57, 1376–1379 (1979).
[12] J. Chrostowski and C. Delisle, "Bistable optical switching based on Bragg diffraction," Opt. Commun. 41, 71–74 (1982).
[13] P.P. Banerjee, U. Banerjee, and H. Kaplan, "Response of an acousto-optic device with feedback to time-varying inputs," Appl. Opt. 31, 1842–1852 (1992).
[14] M.A. Al-Saedi and M.R. Chatterjee, "Examination of the nonlinear dynamics of a chaotic acousto-optic Bragg modulator with feedback under signal encryption and decryption," Opt. Eng. 51, 018003 (2012).
[15] M.R. Chatterjee and F. Almehmadi, "Numerical analysis of first-order acousto-optic Bragg diffraction of profiled optical beams using open-loop transfer functions," Opt. Eng. 53(3), 036108 (2014).
[16] A.K. Ghosh and P. Verma, "Lyapunov exponent of chaos generated by acousto-optic modulators with feedback," Opt. Eng. 50, 017005 (2011).
[17] F.S. Almehmadi and M.R. Chatterjee, "Improved performance of analog and digital acousto-optic modulation with feedback under profiled beam propagation for secure communication using chaos," Opt. Eng. 53, 126102 (2014).
[18] J. Durnin, J. J. Miceli, and J. Eberly, "Diffraction-free beams," Phys. Rev. Lett. 58, 1449–1501 (1987).

[19] A. N. Kolmogorov, "The local structure of turbulence in incompressible viscous fluid for very large Reynolds numbers," Proceedings: Mathematical and Physical Sciences 434, 9–13 (1991).
[20] A. N. Kolmogorov, "A refinement of previous hypotheses concerning the local structure of turbulence in a viscous incompressible fluid at high Reynolds number," J. Fluid Mech. 13, 82 (1962).
[21] A. M. Obukhov, "On the distribution of energy in the spectrum of turbulent flow," Dokl. Akad. Nauk SSSR 32, 22–24 (1941).
[22] J.W. Strobehn (ed.), "Topics in Applied Physics," Vol. 25, Laser Beam Propagation in the Atmosphere. Springer-Verlag, New York (1978).
[23] L.C. Andrews and R.L. Phillips, "Laser Beam Propagation through Random Media," 2nd ed., Bellingham: SPIE Press (2005).
[24] M.C. Roggemann and B.M Welsh, "Imaging Through Turbulence," CRC Press, Boca Raton, FL (1996).
[25] P.P. Banerjee and T.-C. Poon, "Principles of Applied Optics," Aksen Associates, Inc., USA (1991).
[26] J.M. Jarem and P.P. Banerjee, "Computational Methods for Electromagnetic and Optical Systems," 2nd ed., CRC Press, NY (2011).
[27] M.R. Chatterjee and F.H.A. Mohamed, "Split-step approach to electromagnetic propagation through atmospheric turbulence using the modified von Karman spectrum and planar apertures," Opt. Eng. 53(12), 126107 (December 2014).
[28] M.R. Chatterjee and M. Alsaedi, "Examination of chaotic signal encryption and recovery for Secure communication using hybrid acousto-optic feedback," Opt. Eng. 50, 055002-1-14, (2011).
[29] M.R. Chatterjee and F.S. Almehmadi, "Numerical examination of acousto-optic Bragg interactions for profiled light waves using a transfer function formalism," SPIE Photonics West, San Francisco, CA, Oct. 2013.
[30] J. Power, "Modeling anisoplanatic effects from atmospheric turbulence across slanted optical paths in imagery," Master's Thesis, University of Dayton, 2016.
[31] M.C. Roggemann and B.M. Welsh, "Imaging through Turbulence," 1st ed., CRC Press, New York, NY (1996).
[32] J.D. Schmidt, "Numerical Simulation of Optical Wave Propagation with Examples in Matlab," SPIE Press, Bellingham, Washington (2010).
[33] J.W. Goodman, "Introduction to Fourier Optics," 2nd ed., McGraw-Hill, NY (1996).
[34] D. Voelz, "Computational Fourier Optics (A MATLAB Tutorial)," SPIE Press, Bellingham, WA (2011).
[35] A. Mohamed and M.R. Chatterjee, "Non-chaotic and chaotic propagation of stationary and dynamic images through MVKS turbulence," revised and submitted for publication in J. Mod. Opt. (2019).
[36] M.R. Chatterjee and A. Mohamed, "Anisoplanatic electromagnetic image propagation through narrow or extended phase turbulence using altitude-dependent structure parameter," Frontiers in Optics, OSA Technical Digest, paper # JW4A.90, Rochester, NY (October 2016).

Chapter 14

Distributed sensing with free space optics

Timothy J. Brothers[1] *and Arun K. Majumdar*[2]

14.1 Introduction

A distributed sensing system utilizes multiple geographically separated sensors to observe the world. The sensor systems can process data and then transmit it, receive data and then process it, or some combination of the two. For this chapter, we will examine a distributed radio frequency (RF) sensor system set on mobile platforms. The main challenges of any distributed system are localization, synchronization, and processing capabilities. This paper presents a novel solution that utilized a free space optical (FSO) link for communication between nodes. The FSO link will aid in localization and synchronization while also providing a high-speed communications path between sensor units to allow maximum flexibility in processing optimization. An example of FSO communication link between unmanned air vehicles (UAVs) is given to show the viability of establishing multi-Gbit/s optical communication links in presence of atmospheric turbulence. Optical communication offers the advantages of sensor information exchanges at high data rates as well as secure communications needed for a number of tactical applications.

Before we discuss about the distributed system, a brief introduction into signals, systems, and signal processing is provided.

14.2 Signals

Signals are defined mathematically as a function that maps a domain to a range [1]. Continuous time, discrete time, and frequency are all examples of domains, where a range could be voltage, current, temperature, etc. For the basis of this work we will focus on sinusoidal signals that map time and frequency to voltage. The sinusoid will be defined by frequency, wavelength, and velocity. Frequency is the rate of oscillation of the wave defined in cycles per second or Hertz (Hz). Wavelength is the measured length in meters of a single cycle. Velocity is the rate of propagation of the signal in units of meters per second. Equation (14.1) is the relation between

[1]Georgia Institute of Technology, Georgia Tech Research Institute, USA
[2]Department of Physics, Colorado State University, USA

the frequency (f), wavelength (λ), and velocity (v). For RF signals, the velocity is the speed of light (c) shown in (14.2).

$$f = v/\lambda \tag{14.1}$$

$$f = c/\lambda \tag{14.2}$$

Equation (14.2) represents a single RF sinusoid. More complex signals can be represented as a sum of sinusoids. Specifically, the Taylor Series Expansion states that any periodic signal can be represented by a sum of sinusoids (14.3).

$$f(x) = \sum_{n=-\infty}^{\infty} A_n \cos(2\pi f_n t + \theta_n) \tag{14.3}$$

where each sinusoid in this summation contains a unique frequency. The representation of each frequency component contains an amplitude (A_n) and a phase offset (θ_n). It is normal for many of the values of A_n to be zero resulting in a finite sum of terms that would fully describe the signal. If a plot is created where the finite sum of frequencies are represented on the independent axis and the amplitude is presented on the dependent axis the plot of the spectrum of a signal can be obtain, see Figure 14.1.

The bandwidth of a signal is the RF region where the signal exists. Signals with wider bandwidth scan contain more complex signals. For a communication signal the increased bandwidth can provide increased communication capacity. In practice the bandwidth is usually limited by the hardware or legal restriction.

Different frequency regions are beneficial for specific applications. For instance, a radar system at lower frequencies (HF, VHF, UHF, and L-band) is better

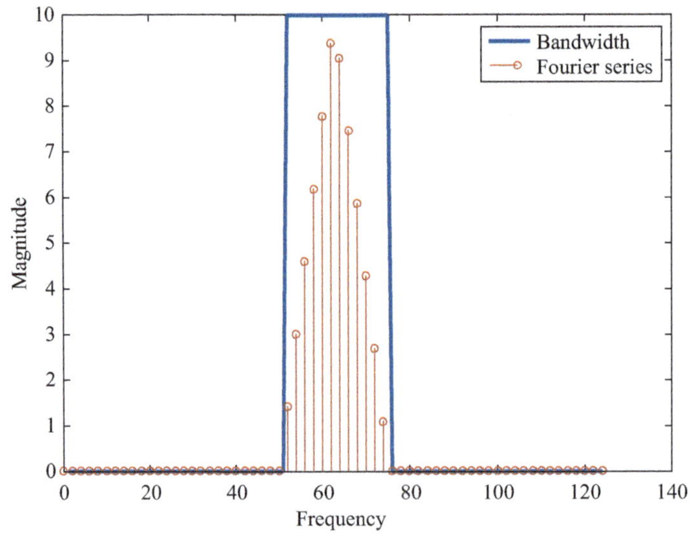

Figure 14.1 Example bandwidth with Fourier series

Table 14.1 Radar frequency bands [2]

Frequency	Frequency range	Example application
High frequency (HF)	3–30 MHz	Ground-penetrating radar
Very high frequency (VHF)	30–300 MHz	Foliage-penetrating radar
Ultrahigh frequency (UHF)	0.3–1 GHz	Airborne surveillance radar
L-Band	1–2 GHz	Air traffic control radar
S-Band	2–4 GHz	Naval surface radar
C-Band	4–8 GHz	Weather radar
X-Band	8–12 GHz	Air interceptor radar
Ku-Band	12–18 GHz	Surface-moving target indication radar
K-Band	18–27 GHz	None
Ka-Band	27–40 GHz	Missile seekers radar
Millimeter wave (mmw)	40–300 GHz	Automotive radar

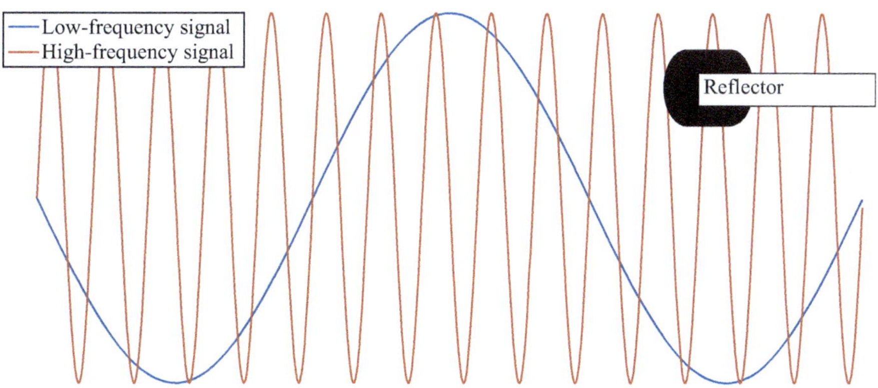

Figure 14.2 High- and low-frequency signals

at foliage penetration and long-range sensing, see Table 14.1. This is due to the low atmospheric loss and the long wavelengths. The size of object that a radar can detect is inversely proportional to the wavelength due to the scattering reflection off the object. RF signals are scattered off sharp edges or flat surfaces where the incident angle is equal to the reflected angle. The higher the frequency the greater the probability the incident angle will reflect off a small object, see Figure 14.2. This is why low frequency radars are used for foliage penetration and Missile Seeker Radars utilize high frequency signals.

Signal energy propagation, multi-path, and Doppler are three of the major channel considerations when trying to receive an RF signal. Energy propagation is the natural spreading of an RF signal over distance. Imagine a stone that is dropped into a lake. The ripples caused by the stone propagate in a radial fashion from the origin of the stone drop. For an RF signal the signals will propagate in a spherical manner. With a reduction in signal amplitude equivalent to the area of the wavefront.

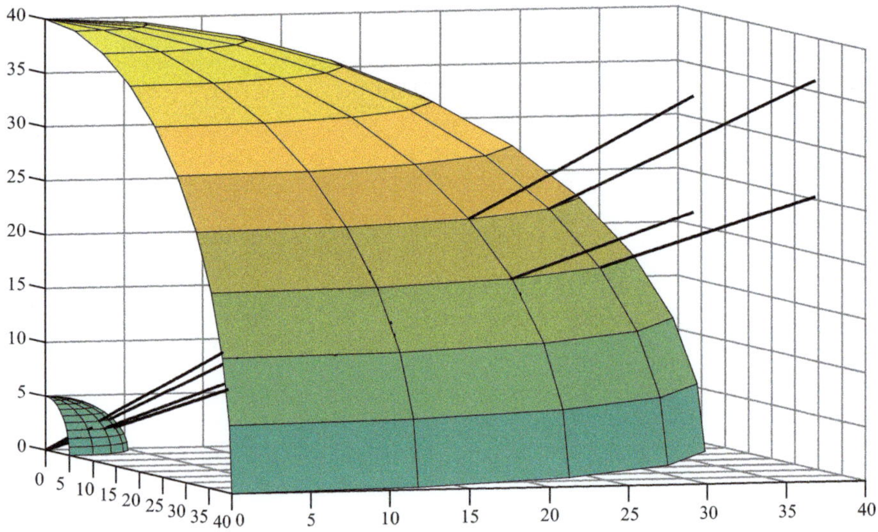

Figure 14.3 Propagation of RF energy

Specifically, the totally energy is conserved and the amplitude decreased appropriately (if atmospheric attenuation is neglected). This can be seen in Figure 14.3. Equation (14.4) represents the power density (S_u in watts per meter squared) given the radius (R in meters) from the transmitter and the power of the transmitter (P_s in Watts). The significant term in this equation is the squared distance.

$$S_u = \frac{P_s}{4\pi R^2} \tag{14.4}$$

Multi-path is related to propagation in that it is the result of the wavefront reflection off a surface (such as the ground) and creating a secondary path to between the transmitter and the receiver. This will cause the multiple time delayed copies of the signal to be received, see Figure 14.4.

The third major consideration for the signal channel is Doppler. Doppler occurs when the transmitter or receiver is moving. The motion causes the signal to be compressed spatially. This results in higher frequency on the receive side if the transmitter and receiver are moving toward one another and a lower frequency if they are moving away, see Figure 14.5.

For a traditional system where an array of antenna is located at a single location these effects can be considered a constant over all of the receiving elements. However, for a distributed system these effects can be very different from sensor to sensor. Let us consider a group of four sensors that are collectively receiving a single signal from a moving emitter as shown in Figure 14.6.

For this example, assume an ideal omnidirectional antenna pattern. Each sensor will receive a signal with slightly different frequencies due to the Doppler shift

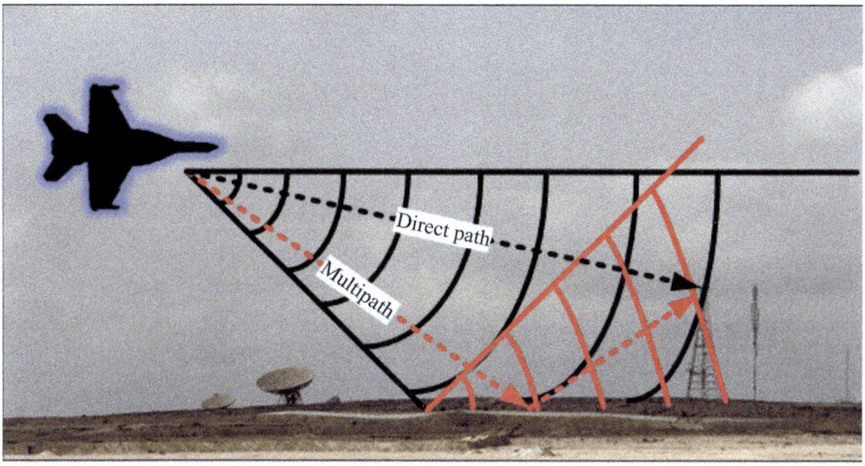

Figure 14.4 Multipath illustration

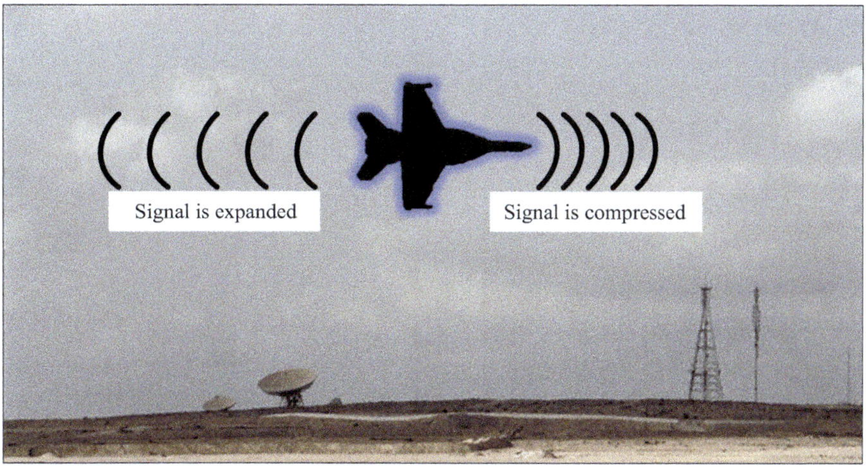

Figure 14.5 Doppler

as seen by each receiver. In a traditional antenna array all of the receiver elements are closely located so it can be assumed that all of the signals have the same Doppler. In addition to the frequency shift there is also a difference in the receiver time due to the signal propagation. Again, traditional systems are collocated so it simplifies this problem by assuming the signal source is in the far field. Which means that the source is of sufficient distance from the receiver that the signal is no longer represented by a spherical propagation model, but instead the radius of the sphere is so large that the portion of the signal observed can be assumed to be linear. For this distributed system the signal is not only in the near field it could be

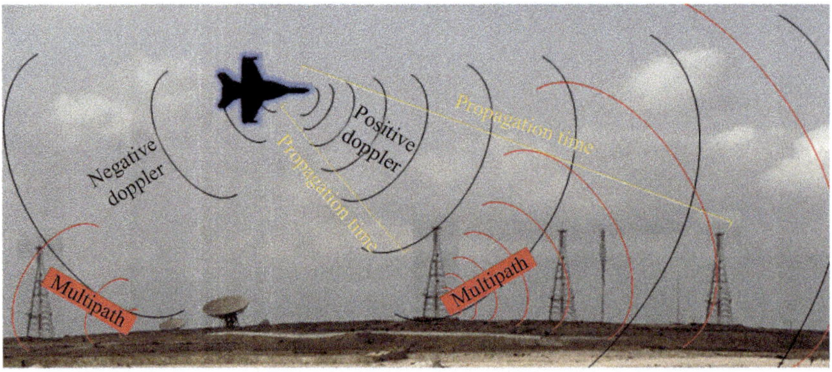

Figure 14.6 Distributed sensors and channel considerations

encircling the transmitter. One advantage of the distributed nature of the sensors in this example is that the received signals are all statistically independent due to the large geographical separations. This will allow for each received signal to be treated as a totally independent equation when trying to process the data for additional information.

For distributed sensing system the signal characteristics must be considered prior to doing collaborative signal processing. The next section will discuss how a distributed system can actually take advantage of the statistical independence of the Doppler and energy propagation in a collaborative manner.

14.3 Distributed sensing systems

A pure distributed system, as defined by most academics, would not contain a central processing node. A pure distributed system operates on a set of rules and the behavior is not directly controlled resulting in what is called emergent behavior. For most practical system it is very beneficial if not essential to have a central processing node that will also direct the members of the distributed system. Figure 14.7 shows all of the components of a distributed sensing system.

The sensors are the physical interface to the surrounding world. The sensors could be any type of sensor or a combination of multiple sensors such as visible imagery, infrared (IR) imagery, audio, RF, etc. Localization is also required at each sensor. One of the most common and inexpensive localization systems is the global positioning system (GPS). The localization system is one of the largest contributors of error in a distributed system. Another component of the sensing node is the distributed processing system. This layer of processing takes the raw sensor data and reduces it to a level that can be transmitted over the communication link to other sensor nodes or to the central processing node. The central processing node preforms the final processing on the data streams from the sensor nodes and presents information to the user. Each of these systems is interrelated and the design of the distributed sensor system is a trade space between the capabilities and

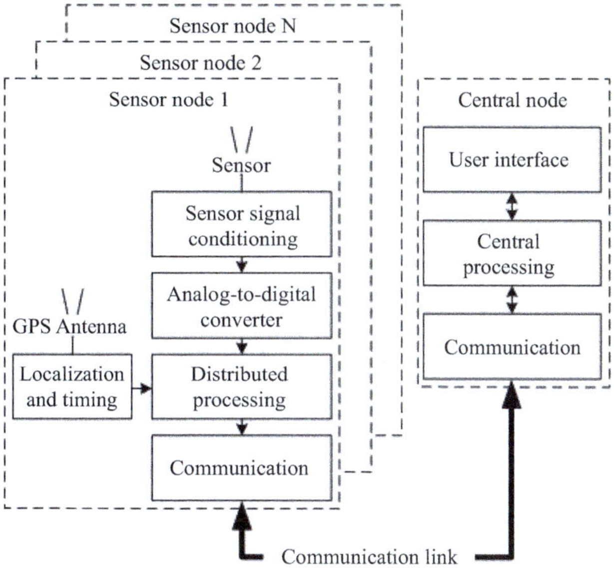

Figure 14.7 Distributed sensor components

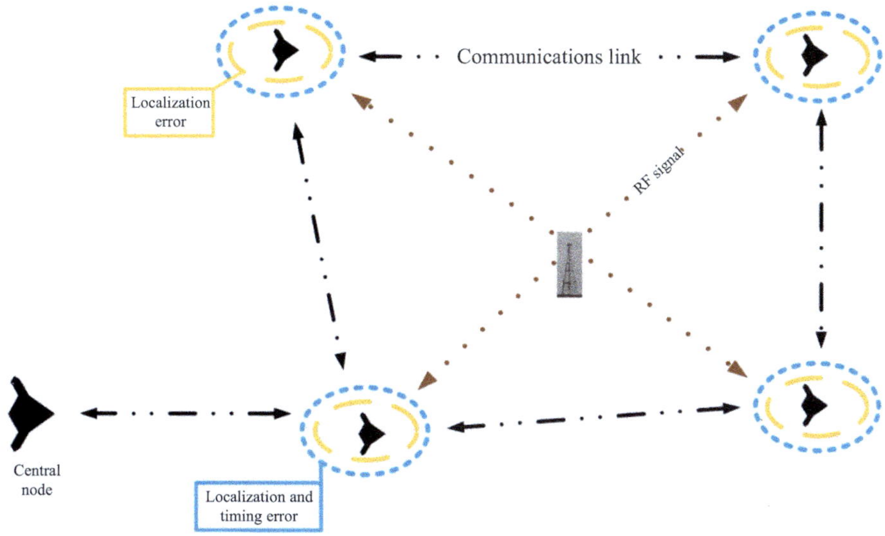

Figure 14.8 Distributed sensor error

limitation of the communication, processing, and localization. Take, for example, a distributed communication system shown in Figure 14.8 where four UAVs compose a distributed sensor system that is receiving from a transmitter and sending data to a central node.

The error shown in Figure 14.8 is represented by the circles around the sensors. Using GPS results in a large region of uncertainty when compared to the signal wavelength. There is also a smaller uncertainty due to the clock synchronization error between nodes. However, as we have already discussed there is also a degree of independence between each sensor due to the energy propagation of the signal of interest, so we already know by the physics of the environment that all of the sensors will not receive the signal simultaneously as they would in a traditional antenna array. Therefore we must do some signal processing to align the signals in time. Consider a traditional digital beamformer where the signal inputs are shifted in phase so a beam is formed. One way of implementing a traditional beamformer is to insert a very small-time delay at each antenna tap to steer the beam, see Figure 14.9(a). However, for the distributed system the signals are in the near field so instead of forming a beam the distributed system will form a point, see Figure 14.9(b). To do point forming in a distributed system all of the signals must first be time aligned to some reference. This is done in much the same way as a traditional beamformer, except the delay is much larger. Please note the simulation in Figure 14.9 assumed ideal omnidirectional antenna patterns. Given a realistic antenna pattern the multiple beams in Figure 14.9(a) will be reduced and the point former will be more irregular.

The area of the point former is related to the cumulative error of the nodes in the distributed system (see Figure 14.8) reduced by the number of sensors. The exact error reduction is based on the processing conducted. However, the error declines exponentially given the increased number of nodes as shown in Figure 14.10(a). However as the error decreased the throughput requirement on the communications link increases linearly without bound.

To obtain the error decrease as shown in Figure 14.10(a), the signal must undergo a correction. This can be done at the central processor or on the distributed processing elements. For this work it is shown on the distributed processing elements because the processing requirements are minimal as compared to the increase flow of communication required if it were done on the central node.

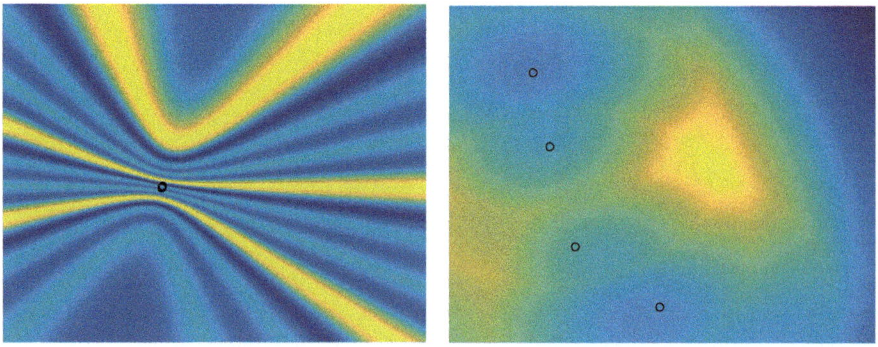

Figure 14.9 (a) Traditional beamformer pattern (left) and (b) distributed sensor point former (right)

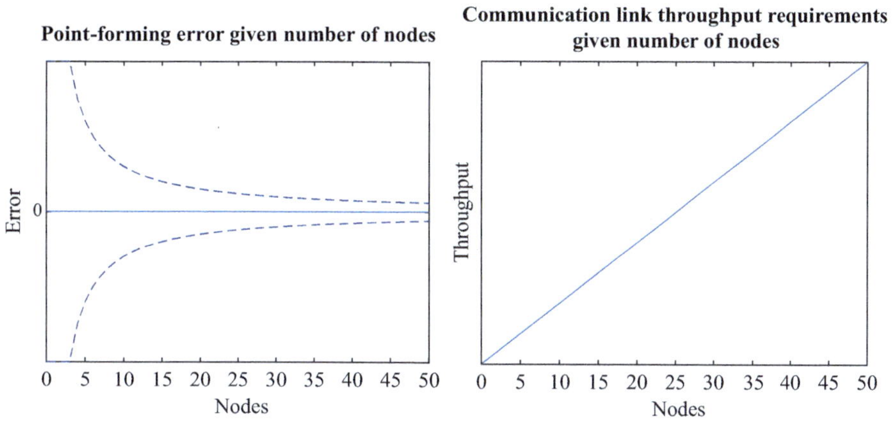

Figure 14.10 (a) Point forming error given number of distributed nodes (left) and (b) communications link throughput requirements given number of nodes(right)

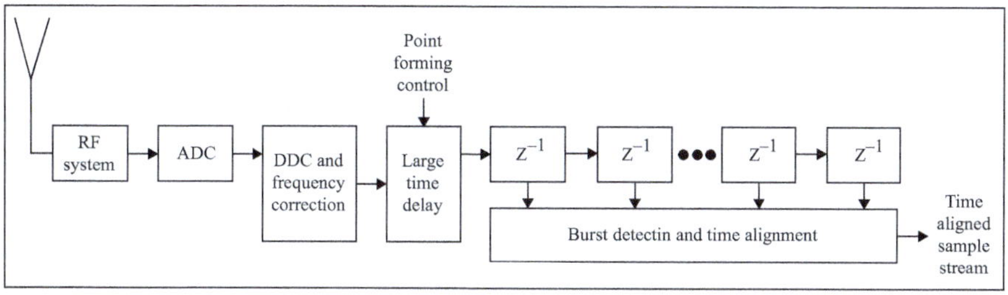

Figure 14.11 Block diagram for point forming

Figure 14.11 shows the block diagram for a point-forming processing system where ADC is the Analog-to-Digital Converter, DDC is the digital down converter, and Z^{-1} represents a single clock time delay. The size of the large time delay block will be dynamically adjusted to move the location of the Point in space. The delay chain (represented by Z^{-1} blocks) must be of sufficient length to correct for the errors of the localization system and the time system. This length can be quantified by (14.5).

$$\text{Delay chain length} = \left(\frac{\text{Localization}_{\text{Error}}}{c} + \frac{1}{\text{Time Sync}_{\text{Error}}}\right) f_s \tag{14.5}$$

where the localization error is in meters, the speed of light (c) is in meters per second, the time sync error is in Hertz, and the sampling frequency (f_s) is in samples per second. The delay chain combined with the burst detection and time alignment block can be considered the auto calibration system. Where the burst detection and time alignment block is search for a known sequence in the data and

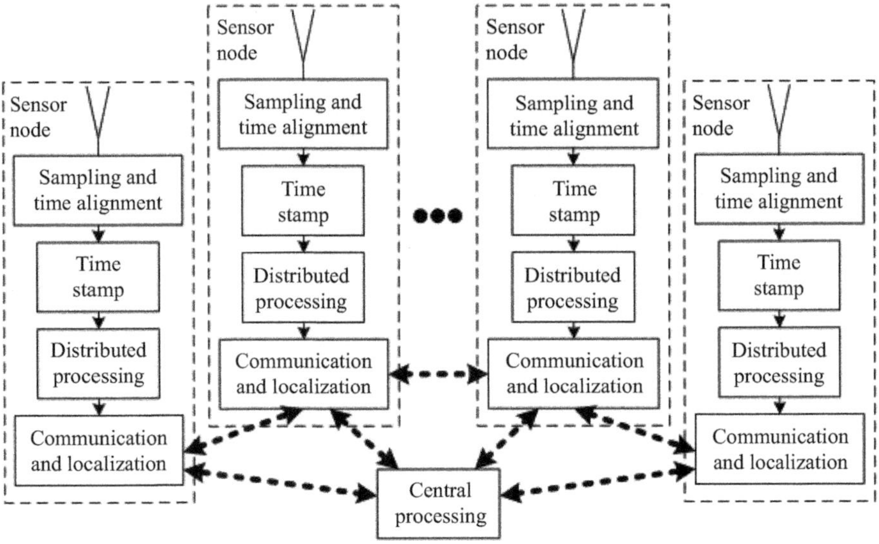

Figure 14.12 Distributed processing block diagram

when it finds this sequence it is selecting the correct time delay tap to correct for the localization and timing errors. The output of this block will be time stamped and then sent over the communication system for additional processing.

The additional processing will be determined by the application of the distributed sensing system. Until this point we have simply stated that the system will be a RF sensor. For the sake of explanation let us assume the system will be a communications receiver, but it could just as easily be applied to radar, communications transmitter, or most any other RF system. Figure 14.12 shows an example of how a distributed sensor system could be connected for the next stage of processing. In this example every sensor node is connected to its next neighbor and also to the central processing node. The nodes can be connected in many different ways depending on the requirements of the systems and the capability of the communication system. There is a design trade space between the processing and the communications systems. Consider a system that is processing 100 MHz of RF bandwidth. In order to obtain alias free digital signals the ADC must sample at 200 M samples per second. Assuming 12 bit in-phase/quadrature (IQ) would result in a total data throughput of 4.8G bits per second (bps). If there are 8 sensor nodes this would result in a total of 38.4 Gbps! As this example shows the data requirements on the communication links can quickly become unreasonable. This is why the distributed processing the essential.

Consider the same example system, with the additional information that the signal of interest is a frequency hopping quadrature phase shift keying (QPSK) signal that has a chip rate of 10 MHz. If the demodulation and decoding is done on each node the data throughput transmitted to the central node would be reduced to

20 Mbps per node. The total throughput to the central node would be 160 Mbps. If the distributed nodes are only doing the frequency hop recovery and still sending IQ data to the central node the throughput would be 480 Mbps per node resulting in a total of 3.84 Gbps cumulative data to the central node. All of these approaches have the advantage of statistically independent receive signals from the geographical diversity. This will be advantageous in minimization of non-Gaussian noise that would be compounded in a traditional system.

14.4 Summary of a distributed system

A distributed sensor network utilizes multiple geographically separated sensors connected by a high-speed data link. One advantage of the distributed sensor network is the ability to operate in the near field to produce point forming solutions. The second major advantage of a distributed network is the ability of the system to be resistant to non-Gaussian noise.

The major disadvantage of the system is the required correction of the errors associated with the localization and signal propagation error. The correction of these errors required a higher degree of digital signal processing capability. Along with the errors associated with the localization and propagation the system will also have an increased data throughput between the processing nodes.

To minimize the localization error and increase the throughput the distrusted processing system will utilize a FSO link. The FSO link will be utilized for angle measurements and distance measurements between nodes. The FSO link will also be able to support high data throughput. The combination of FSO and advanced digital signal processing will allow *the next generation of sensing systems*. The following section will detail the FSO communication link.

14.5 Free space optical communication between two UAVs: BER and adaptive beam divergence analysis

There is an increasing interest in UAVs for critical civil and military applications especially in the area of surveillance of defined area with a variety of sensors. Using multiple UAVs in cooperative swarm mode allow to survey a larger area, collect a huge amount of data in short duration, and increase the robustness by sharing information between UAVs and sending information back to the command center on the ground. Thus, UAVs require high data rate connectivity. Data rates in the range of 100 Mbit/s to larger than 1 Gbit/s will be required to handle multiple sensor information in real time and in parallel. UAVs equipped with FSO can have the technological capabilities to meet this high data rate requirement [3,4]. With proper designs, the data rates can even go 2.5 Gbit/s or much higher using recent technological devices and optical networks. Advanced imaging sensors capable of handling high bandwidth and large amount of data fast and in real-time need communication links to transmit more data between UAVs. Free-space optical communications (FSOC) is the only technology available which can meet this demand for transferring data at required data rate. FSOC between UAVs is also secure because of transferring a narrow optical beam along the line-of-sight (LOS) between the UAVs which additionally also provide other advantages such as immunity to detection, interference, interception or jamming which are very important in critical missions. Compared to RF system, FSO use extremely compact size, weight, and power (SwP), fast and are perfect to be mounted on UAVs. This chapter discusses the Free-space laser communication link analysis between two UAVs and communication performance analysis in terms of achievable BER in the presence of atmospheric turbulence in between the UAVs. Finally, the effects of varying beam divergence to keep the connectivity between the UQAV platforms are also described to suggest for developing adaptive beam divergence for maintain the communication links.

14.6 Technical issues for mobile UAV FSO communication

Some of the main challenges involve in establishing and maintaining the LOS between two UAV terminals are the effects of atmosphere and the effective divergent loss consisting of geometrical and pointing error losses for inter-UAV communication links. The communication channel for UAV FSO communication is the atmosphere where the beam attenuation due to the atmosphere on one side and the alignment of the FSO units on the other side. Since in order to achieve LOS communication link, the alignment becomes a technical challenge in tracking and acquisition in presence of the atmosphere. There are three basic FSO communications link scenarios with UAV platform. These are (i) ground-to-UAV, (ii) UAV-to-ground, and (ii) UAV-to-UAV or between UAVs (UAV Swarm). This paper will address and analyze the communication system performance for a link between two UAVs separated by a fixed distance. The key point here is to develop the UAVs capabilities to handle multiple sensor information in real-time and in parallel so that a large amount of data can be transferred in real-time with a goal to achieve a rate of 1 Gbps or higher. An optical communication terminal receives a command from ground via RF links which also receives continuous updates of the GPS information collected by the UAV GPS receiver. Simultaneously the UAV provides the optical communication terminal with its GPS information. The optical communication terminal uses updated information from both the UAV GPS receiver and the ground location to blind point the gimbal by sending a beacon signal to initiate the communication signal. Data communication starts when the signal terminal tracks on the beacon signal.

14.6.1 Atmospheric and turbulence effects

For different UAV scenarios, the atmosphere plays different roles. Optical signals propagating in the earth's atmosphere experience degradation due to absorption, scattering and turbulence. Absorption and scattering are caused by the interactions of optical waves with atmospheric gases and particulates such as aerosols and fog and result mainly in the attenuation of the signal. Turbulence is caused by the random variations of the refractive index at optical wavelength, and when an optical beam propagates through a turbulent medium atmospheric turbulence causes irradiance fluctuations, beam wandering, and loss of spatial coherence of the optical wave. The horizontal link described in this paper is defined when both of the communication terminals communicate via horizontal path in space.

14.6.2 Atmospheric models related to UAV FSO communication links

Based upon the measurements, empirical and parametric models of turbulence strength parameter, C_n^2 have been derived and several different models are commonly used to represent the effects of atmospheric turbulence, and are described below; h is altitude (meter) [3]:

14.6.2.1 Hufnagel-Valley (HV) model

$$C_n^2(h) = 0.00594\left(\frac{v}{27}\right)^2\left(10^{-5}\,h\right)^{10}\exp\left(-\frac{h}{1000}\right) + 2.7\times10^{-16}\exp\left(-\frac{h}{1500}\right) + A\exp\left(-\frac{h}{100}\right) \tag{14.6}$$

where v is the RMS wind speed. Typical value of the parameter, $A = 1.7\times10^{-14}\ \mathrm{m}^{-2/3}$.

14.6.2.2 Modified Hufnagel-Valley (MHV) model

$$C_n^2(h) = 8.16\times10^{-54}\,h^{10}\exp\left(-\frac{h}{1000}\right) + 3.02\times10^{-17}\exp\left(-\frac{h}{1500}\right) + 1.90\times10^{-15}\exp\left(-\frac{h}{100}\right) \tag{14.7}$$

14.6.2.3 SLC-Day model

$$\begin{aligned} C_n^2 &= 0 && 0\,\mathrm{m} < h < 19\,\mathrm{m}\\ &= 4.008\times10^{-13}\,h^{-1.054} && 19\,\mathrm{m} < h < 230\,\mathrm{m}\\ &= 1.300\times10^{-15} && 230\,\mathrm{m} < h < 850\,\mathrm{m}\\ &= 6.352\times10^{-7}\,h^{-2.966} && 850\,\mathrm{m} < h < 7{,}000\,\mathrm{m}\\ &= 6.209\times10^{-16}\,h^{-0.6229} && 7{,}000\,\mathrm{m} < h < 20{,}000\,\mathrm{m} \end{aligned} \tag{14.8}$$

14.6.2.4 CLEAR1 model

Note: Here, h is the altitude in kilometers above the mean sea level (MSL).

$$\begin{aligned}
&1.23 < h \leq 2.13\\
&\log_{10}(C_n^2) = A + Bh + Ch^2\\
&\text{where } A = -10.7025,\ B = -4.3507,\ C = +0.8141\\
\\
&2.13 < h \leq 10.34\\
&\log_{10}(C_n^2) = A + Bh + Ch^2\\
&\text{where } A = -16.2897,\ B = +0.0335,\ C = -0.0134\\
\\
&10.34 < h \leq 30\\
&\log_{10}(C_n^2) = A + Bh + Ch^2 + D\exp\left\{-0.5[(h - E)/F]^2\right\}\\
&\text{where } A = -17.0577, B = -0.0449, C = -0.0005\\
&\qquad D = 0.6181, E = 15.5617, F = 3.4666
\end{aligned} \tag{14.9}$$

HV model turbulence profile is generally used. For a specific modulation scheme, such as OOK modulation, and knowing the parameters such as wavelength, UAV height, transmitter divergence angle, and a data rate, the UAV communication performance BER can be computed [3]. For the horizontal case the value of C_n^2 remains constant along the propagation path. The value of C_n^2 at the altitude of the UAVs needs to be accounted for when estimating intensity fluctuations or beam wander.

When the UAVs have to operate under atmospheric scattering conditions, scattering of optical wave by aerosol particulates and fog can be important. The proposed model for short distance link is the Kruse model [5] which relates the attenuation to visibility V (in km) for a given wavelength (in nm). In visible and near IR wavelength up to about 2.5 μm, the attenuation is given by [3]:

$$\gamma(\lambda) \approx \beta_a(\lambda) = \frac{\ln(\tau_{\mathrm{TH}})}{V}\left(\frac{\lambda}{550\ \mathrm{nm}}\right)^{-q} = \frac{3.912}{V}\left(\frac{\lambda}{550\ \mathrm{nm}}\right)^{-q} \tag{14.10}$$

where τ_{TH} is the transmission.

The attenuation a_{dB} over the link path distance d_{link} can be calculated from the measured transmission τ or the extinction coefficient $\gamma(\lambda)$ (in km^{-1}) using the following relation:

$$a_{\mathrm{dB}} = 10\log\left(\frac{1}{\tau}\right) = \frac{10}{\ln(10)}\gamma(\lambda)d_{\mathrm{link}} \tag{14.11}$$

14.6.3 Alignment and tracking of a FSO communications link to a UAV

For establishing a successful FSO communication link between two UAVs, the most important criteria to start with is to make sure that the mechanical gimbal can accurately track the moving UAV in presence of atmospheric turbulence and absorption. From UAV side, it is important that the minimum transmitting beam divergence is such that the probability of fading of the signal reaching the receiver due to beam wandering caused by atmospheric turbulence is below a required threshold. The repeatability and the accuracy of the gimbal to align and track a ground-to-UAV FSO link needs to be verified. Divergence of the transmitting beam is one technique to help with alignment and tracking of FSO link which will be discussed in this paper. For short-range inter-UAV FSO optical communication link, generally a fixed single beam divergence is employed in the link which may limit the transmission distance between UAVs. Furthermore, in order to establish and maintain the LOS between two UAVs, the optical system requires pointing ability. Adjusting the transmitting beam divergence by adaptive means may help in solving this problem and will be described later.

14.7 FSO optical communication system performance in turbulence: BER and SNR calculation

In this chapter, we are using weak turbulence approximations for signal fading caused by atmospheric turbulence to calculate the BER for communication link between two horizontally separated UAVs. The FSO link consists of transmitter, channel and receiver. For simplicity, an ON–OFF modulation scheme (OOK) is considered. The transmitter's optical signal from one UAV is transmitted over FSO channel which is assumed to have weak turbulence at the altitude of the UAV under consideration. The optical signal is then received by the receiver located on the other UAV. The receiver has a PIN (or APD) detector and the detected signal which contain the information provided by the sensor is stored and/or retransmitted to another UAV if needed. This section includes an analysis of the effect of atmospheric turbulence on FSO performance by referring to the two criteria, namely, BER and signal-to-noise ratio (SNR) between two UAVs.

We have assumed a collimated optical beam transmitted from one UAV and for this FSO system (plane wave) the weak and symmetrical atmospheric turbulence the so-called log-irradiance variance is given by [6]

$$< \chi^2 >= 0.31\, C_n^2\, k^{7/6}\, L^{11/6} \tag{14.12}$$

Assuming the only dominant noise source is the atmospheric turbulence between the UAVs, χ represents the fluctuations of the log of the amplitude of the field, $A_n(r)$ is the amplitude of turbulence-induced noise, and $A_0(r)$ is the amplitude of the laser beam without turbulence.

$$\chi = \ln\left[\frac{A_n(r)}{A_0(r)}\right] = \ln(1+\varepsilon) \tag{14.13}$$

where $\varepsilon = \frac{A_n(r)}{A_0(r)}$

The performance of the communication system between UAVs is evaluated generally in terms of the two parameters, SNR and BER. The object is to increase the average received power and to reduce the effects of atmospheric turbulence by applying a number of various techniques for the optimum design. The SNR in presence of turbulence is given by

$$SNR = \frac{< A_0(r)^2 >}{< A_n(r)^2 >} = \left[< \varepsilon^2 >\right]^{-1} \tag{14.14}$$

At the UAV altitude, we assume a weak turbulence model, ε is therefore very small, and

$$\ln(1+\varepsilon) \approx \varepsilon$$

The SNR can be therefore be written as for weak turbulence,

$$SNR = < \chi^2 >^{-1} = \left(0.31\, C_n^2\, k^{\frac{7}{6}}\, L^{\frac{11}{6}}\right)^{-1} \tag{14.15}$$

For the FSO link between UAVs with OOK modulation scheme, the BER can be written as [7]

$$BER = \frac{exp^{\left(-\frac{SNR}{2}\right)}}{(2\pi SNR)^{0.5}} \tag{14.16}$$

From (14.15) and (14.16), the UAV FSO communication systems can be evaluated for a typical turbulence strength, C_n^2, optical wavelength, λ (and thus wave number $k = 2\pi/\lambda$) and path length, L.

If the beam spreading effect is included in the SNR calculation, the effective SNR can be defined as [6,8,9]

$$SNR_{\text{eff}} = \frac{SNR_0}{\left[1 + 1.33\sigma_{lnIR}^2\left(2L/kw_L{}^2\right)^{5/6}\right] + F\cdot\sigma_{lnIR}^2\cdot SNR_0} \tag{14.17}$$

where σ_{lnIR}^2 is the log amplitude variance, $\chi = 0.5\,\sigma_{lnIR}$, F is the aperture averaging factor which is the ratio of normalized intensity variance of the signal at a receiver with diameter D to that of a point receiver $D = 0$ [6]:

$$F = \frac{\sigma_{lnIR}^2(D)}{\sigma_{lnIR}^2(D=0)} \tag{14.18}$$

The value of the aperture averaging factor depends on the wavelength, turbulence strength, path length, and the wave number defined earlier.

14.8 Data rate

It was mentioned earlier that for many missions UAVs data rates in the range of 100 Mbit/s to larger than 1 Gbit/s will be required to handle multiple sensor information in real time and in parallel. The data rates can even go 2.5 Gbit/s or much higher using recent technological devices and optical networks. Advanced imaging sensors capable of handling high bandwidth and large amount of data fast and in real-time need communication links to transmit more data between UAVs.

Given a laser transmitter power P_t, with transmitter divergence of θ_t, receiver telescope area A, transmit and receive optical efficiency τ_{opt}, the achievable data rate R can be obtained from the following equation [4]:

$$R = \frac{P_t \cdot \tau_{opt} \, \tau_{ATM} \cdot A}{\pi(\theta_t/2)^2 \cdot L^2 \cdot E_p \cdot N_b} \tag{14.19}$$

where $E_p = hc/\lambda$ is the photon energy and N_b is the receiver sensitivity in # photons/bit. τ_{ATM} is the value of the atmospheric transmission at the laser transmitter wavelength, τ_{ATM} may be written in terms of the atmospheric attenuation factor α, given by $-10\,\log(\tau_{ATM})/L$.

If we write τ_{ATM} in terms of atmospheric attenuation factor, α(dB/km) at the wavelength, λ, the τ_{ATM} is then given by:

$$\tau_{ATM} = 10^{(-\alpha \cdot L/10)}$$

We have estimated the achievable data rates using the following scenario: $L = 175$ m, $D = 10$ cm, $P_t = 50$ mW, $\theta_t = 10$ mrad. The receiver sensitivity for this example was taken to be $N_b = 1{,}568$ photons/bit [4] at the wavelength $\lambda = 1{,}550$ nm. The data rates were calculated for two atmospheric conditions: clear atmosphere $\alpha = 0.2$ dB/km, and haze condition $\alpha = 4$ dB/km. When we included the atmospheric turbulence conditions, the τ_{ATM} was multiplied by the turbulence-induced.

Factor τ_{TURB}, where $\tau_{TURB} = 10^{(-X/10)}$. The values of X were taken as turbulence-induced loss and were taken in this example as $X = 6$ dB (weak turbulence), and $X = 10$ dB (moderately strong turbulence). The data rates were calculated for these different atmospheric conditions and are shown in Table 14.2.

Note that from Table 14.2, for clear atmosphere data, rates between two UAVs can be achieved to be 453.3 Gbit/s which is reduced to 113.86 Gbit/s and 45.33 Gbit/s including some atmospheric turbulence-induced losses. For haze conditions, the achievable data rate is 389 Gbit/s, and if atmospheric turbulence-induced losses are taken into account, the data rates are 97.6 Gbit/sec and 38.9 Gbit/s, respectively, for weak and moderate turbulence conditions. The corresponding data rates for 4 UAV separations are also shown in the same table. In conclusion, because of the short distance between two UAVs, very high data rates required for distributed sensing. Even if we use a much lower power transmitter down to even 10 mW, the data rates in the order of a few Gbit/s can be expected!

14.9 Beam divergence effects for inter-UAV FSO communication

For establishing inter-UAV FSO communications, in general the beam divergence present in the FSO link should be sufficient to offset any error introduced into the alignment and tracking algorithm by the gimbal with a very low probability of signal fade for the UAV optical link.

Table 14.2 Data rates for different atmospheric conditions between UAVs

Data rate, R			
Communication range = 175 m (between two UAVs)			
Clear atmosphere: $\alpha = 0.2$ dB/km		Haze: $\alpha = 4$ dB/km	
With no turbulence-induced loss	With turbulence-induced loss	With no turbulence-induced loss	With turbulence-induced loss
	(a) 6 dB loss (b) 10 dB loss		(a) 6 dB loss (b) 10 dB loss
453.3 Gbit/s	(a) 113.86 Gbit/s (b) 45.33 Gbit/s	389 Gbit/s	(a) 97.6 Gbit/s (b) 38.9 Gbit/s
Communication range = 700 m (4 UAV separation)			
Clear atmosphere: $\alpha = 0.2$ dB/km		Haze: $\alpha = 4$ dB/km	
With no turbulence-induced loss	With turbulence-induced loss	With no turbulence-induced loss	With turbulence-induced loss
	(a) 6 dB loss (b) 10 dB loss		(a) 6 dB loss (b) 10 dB loss
27.6 Gbit/s	(a) 6.93 Gbit/s (b) 2.76 Gbit/s	14.99 Gbit/s	(a) 3.76 Gbit/s (b) 1.499 Gbit/s

14.9.1 Adaptive beam divergence technique

New technique based on adaptive beam divergence is presented [10] for inter-UAV FSO under varying distance conditions. A single beam divergence employed in the link of optical communications may limit the transmission distance between UAVs. An adaptive beam divergence can improve the free space communication link performance and provides advantages over a fixed beam divergence in the inter-UAV FSO. The general link equation can be written as:

$$P_{rx} = P_{tx} \cdot L_{rx} \cdot G_{tx} \cdot L_p \cdot L_R \cdot L_{\mathrm{atm}} \cdot G_{rx} \cdot L_{rx} \tag{14.20}$$

where P_{rx} is the received power, P_{tx} = transmit power, G_{tx} = transmit optics efficiency, L_p = transmit gain, L_R = pointing loss, L_{atm} = atmospheric loss, G_{rx} = receiver gain, and L_{rx} = receiver optics loss. If we combine the transmit gain, range loss, and receiver gain into a single term, geometrical loss, L_{geo}, then (14.19) can rewritten as:

$$P_{rx} = P_{tx} \cdot L_{\mathrm{geo}} \cdot L_p \cdot L_R \cdot L_{\mathrm{atm}} \cdot L_{rx} \tag{14.21}$$

$$L_{\mathrm{geo}} = [a_{rx}/(\theta_{\mathrm{div}} \cdot R)]^2 \tag{14.22}$$

$$L_p = \exp\left[-8 \cdot (\theta_{\mathrm{err}}/\theta_{\mathrm{div}})^2\right] \tag{14.23}$$

where a_{rx} = receiver aperture diameter, θ_{div} = beam divergence, R = communication range, and θ_{err} = pointing error. When two UAVs are trying to establish a communication link, knowledge of each other's location is required provided either by a ground control relay or through the swarm control channel. An exact instantaneous position for each UAV is difficult because of its inaccuracy in its on-board positioning system and causes a spherical error region whose radius depends on the confidence value of error probability [10]. Viewed from another UAV, this region becomes a 2D circular region (*uncertainty area*) around the measured UAV position coordinates and actual UAV position can be anywhere within the uncertainty area, from the center to the edge of the area [10,11]. The size of the uncertainty area is described by its diameter d_{uca}. Platform jitter, θ_{jitter} can be neglected if it is very small compared to the uncertainty area angular size, i.e.,

$$\theta_{\text{div}} \geq \theta_{\text{uca}} + \theta_{\text{jitter}} \tag{14.24}$$

An optimum beam divergence is to be determined now in order to deliver more beam power to the edge of the uncertainty area so that UAV at that location will receive sufficient beam power to continue communications. From (14.21) and (14.22), a larger beam divergence will increase the geometrical loss, the pointing loss will be reduced for a fixed pointing error. From (14.21) and (14.22) we can find the optimum beam divergence to send the most beam power to the edge of the uncertainty area while satisfying the condition, (14.23). The distance between the two communicating UAVs is continuously changing which affects the maximum angular pointing error, which is the maximum mispointing of beam when the UAV is the edge of the uncertainty area defined by [10]:

$$\theta_{\text{max-err}} = 0.5 \cdot \left(d_{\text{uca}}/R + \theta_{\text{jitter}}\right) \tag{14.25}$$

Instead of increasing the receiver aperture size or transmit power, a more efficient method is to adopt an adaptive beam divergence to mitigate the loss due to the distance under the constraint of (14.23). Thus, the loss due to the geometrical and pointing loss can be reduced by constantly changing the beam divergence according to the distance. The amount of loss at the edge of the uncertainty area is given by [10]:

$$EdgeLoss = 0.36 \cdot \left\{a_{rx}/\left[1.4 \cdot \left(d_{\text{uca}} + \theta_{\text{jitter}} \cdot R\right)\right]\right\}^2 \tag{14.26}$$

For a Gaussian beam, collimated beam diameter is given by [11]:

$$\theta_{\text{div}} = (4 \cdot \lambda)/(\pi \cdot d_{\text{out}}) \tag{14.27}$$

where λ is the transmitter wavelength, and d_{out} is the collimated output beam diameter. The collimated beam output can therefore be altered to provide adaptive beam divergence adjustment in order to improve the FSO communication performance under varying distance conditions. If high amplitude jitter is present, the adaptive beam divergence method can be used to suppress the jitter.

14.10 Results and discussions

This section shows the results from the above discussions for calculating SNR and BER as well as analyzing the effects of divergent angles on the inter-UAV communication system performance. The UAV scenario is the following:

There is a number of distributed sensors located in an UAV array of size N, UAV spacing is 175 m from each other in the array. The UAVs are located at an altitude of 5 km from the ground. The FSO system will be used for inter-UAV communication and to transmit and send the collected data. As mentioned earlier, the atmospheric turbulence strength, C_n^2 varies with altitude either estimated from the accepted atmospheric models or from measurements. At 5 km altitude, the turbulence strength is weak. In the present calculation of BER and SNR, C_n^2 is assumed to be in the range of 10^{-17} to 10^{-13} $\mathrm{m}^{-2/3}$ [12,13]. The following inputs were assumed for the calculation of BER and SNR:

Communication range, $L = 175$ m (separation between two UAVs), wavelength, $\lambda = 1{,}550$ nm, receiver diameter, $D = 10$ cm, laser transmitting power $= 50$ mW, the aperture average factor at the wavelength, receiver diameter and range was taken to be $F = 0.02$ (estimated from [14]), transmitting beam size, $w_0 = 2$ cm.

Figure 14.13 shows the BER performance calculated for the inter-UAV optical communication in presence of atmospheric turbulence at the height of the UAV (5 km). The BER is calculated using the effective SNR_{eff}. From the calculations, we have also found that for BER, there is marginal improvement for SNR compared with SNR_{eff}. In other words, there is no appreciable effect of beam spreading at this short range. The strength of turbulence, C_n^2 was assumed from the range of 2.10^{-17}

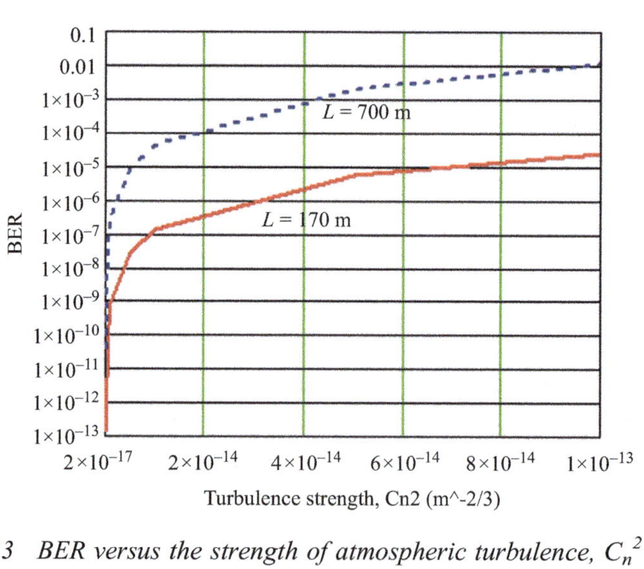

Figure 14.13 BER versus the strength of atmospheric turbulence, C_n^2 at the height of the UAV for two UAV separations of 175 and 700 m (4 UAV separations)

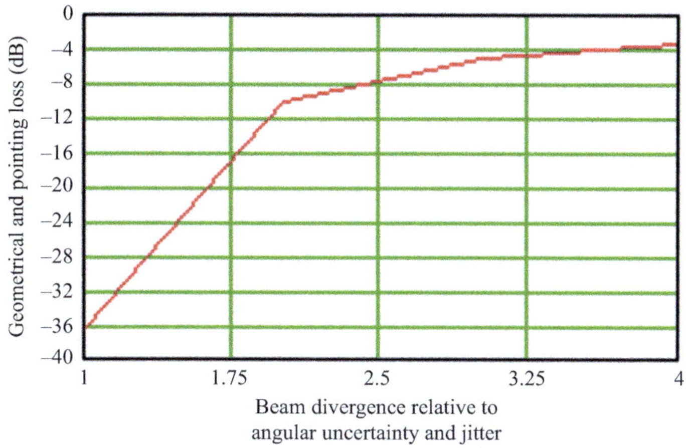

Figure 14.14 Combined geometrical and pointing loss at the edge of the uncertainty area due to different beam divergence relative to the angular size of uncertainty area and jitter. The ratio of angular divergence and angular uncertainty, i.e., $x = (\theta_{div}/\theta_{uca})$ is plotted along the x-axis, and combined geometric and pointing loss is plotted along the y-axis

to 10^{-13} $m^{-2/3}$, which is still in the weak turbulence regime so that (14.12) and (14.14) could be applied under the approximation described in the text to calculate SNR and then BER. The results of BER are shown for the two communication ranges of 175 m and 700 m. It is interesting to observe from the figure that for such a short range of 175 m between two UAVs, the expected BER is very small from 10^{-13} to 3×10^{-5} within the C_n^2 range. This means that there will be enough received signal to continue communications with an acceptable communication. For a longer communication range of 700 m, the BER will increase little from 10^{-12} for very weak turbulence to about 10^{-2} for somewhat moderate turbulence strength. The typical BER of 10^{-6} can be easily accomplished for weak turbulence strength of about 5×10^{-17}.

From Figure 14.14, for a value of $x = 1$ (relative beam divergence of 1), the combined loss is about -36 dB and increasing x to 2.5 or 3.5, the losses can be less about -8 and -4 dB, respectively. This loss can be compensated either by adjusting the beam divergence adaptically or sometimes increasing the laser input power.

14.11 Conclusions and future research

This paper presents a novel solution that utilized a FSO communications link combined with advanced digital signal processing between nodes to establish a distributed sensing system. The advanced digital signal processing presented in this paper enables the correction of errors associated with localization and timing.

Combining the digital signal processing with inter-UAV FSO communications has shown that with weak atmospheric turbulence the throughput of the communications link is more than adequate to sustain the signal processing. The simulation results show that an acceptable BER performance can be achieved with a reasonable transmitting laser power in one UAV. The results are also presented for four times UAV separation and show the achievable BER. Other results show the combined geometric and pointing loss versus beam divergence relative to angular uncertainty area and jitter. The acceptable communication capability can be improved if automatic adaptive beam divergence is implemented for inter-UAV free space optical communications.

Future research can address automatic alignment between UAVs by retro-reflectors installed in the two UAVs which will require no mechanical tracking and pointing system. An electronically steerable flash lidar (ESFL) [15] can be very useful for auto-tracking and exact locating moving vehicles such as UAVs for maintaining successful communications between UAVs. The auto-tracking and exact localization will also reduce errors in the signal processing resulting in the enabling of higher frequency signal transmission/reception on the distributed sensing system.

References

[1] C. Chen, *Signals and Systems*, Oxford University Press, New York, NY, 2004.

[2] W. Melivin and J. Scheer, *Principles of Modern Radar,Vol III: Radar Applications*, SciTech Publishing, Edison, NJ, 2014.

[3] A. K. Majumdar, *Advanced Free Space Optics (FSO): A systems Approach*, Springer, New York, 2015.

[4] A. K. Majumdar and J. C. Ricklin, *Free-Space Laser Communications: Principles and Advances*, Springer, New York, 2008.

[5] Ch. Chlestil, E. Leitgeb, S. Sheikh Muhammad, *et al.*, "Optical Wireless on Swarm UAVs for High Bit Rate Applications," Proceedings of IEEE Conference, CSNDSP, 19th–21st July, Patras, Greece, 2006.

[6] L. C. Andrews, R. L. Phillips, and C. Y. Hopen, *Laser Beam Scintillation with Applications*. Philadelphia, PA: SPIE Press, 2001.

[7] X. Guoliang, Z. Xuping, W. Junwei, and F. Xiaoyong, Influence of atmospheric turbulence on FSO link performance, Proc. SPIE, Optical Transmission, Switching, and Subsystems, Vol. 5281, pp. 816–823, 2004.

[8] L. C. Andrews and R. L. Phillips, *Laser Beam Propagation through Random Media*, SPIE Optical Engineering Press: Bellingham, Washington, 2005.

[9] I. Toseli, L. C. Andrews, R. L. Phillips, and V. Ferrero, Free space optical system performance for a Gaussian beam propagating through non-Kolmogorov weak turbulence, *IEEE Trans. Antennas Propag*, vol. 57, pp. 1783–1788, 2009.

[10] K. H. Heng, N. Liu, Y. He, W. D. Zhong, and T. H. Cheng, "Adaptive Beam Divergence for Inter-UAV Free Space Optical Communications," IEEE Conference IPGC, 2008.

[11] S. G. Lambert and W. L. Casey, *Laser Communications in Space*, Artech House, Boston, MA, 1995.

[12] J. C. Ricklin, S. M. Hammel, F. D. Eaton, and S. L. Lachinova, Atmospheric channel effects on free-space laser communication, in the book by A. K. Majumdar, J. C. Ricklin, *Free-Space Laser Communications: Principles and Advances*, Springer, New York, 2008.

[13] S. Hagelin, E. Masciadri, and F. Lascaux, Optical Turbulence: The influence of the atmosphere on ground-based astronomy, from a publication in a study founded by a Marie Curie Excellence Grant (MEXT-CT-2005-023878).

[14] J. C. Ricklin, Free-Space Laser Communication using a Partial Coherent Source Beam, Ph.D. dissertation, The Johns Hopkins University, March 2002.

[15] C. Weimer, T. Ramond, I. Burke, Y. Hu, and M. Lefsky, An Electronically Steerable Flash Lidar (EFSL): https://www.researchgate.net/publication/2667220787

Chapter 15

Quantum-based satellite free space optical communication and microwave photonics

Arockia Bazil Raj[1], *Vishal Sharma*[2], *and Subhashish Banerjee*[2]

15.1 Introduction to spread spectrum techniques

Spread-spectrum techniques are widely used in radio communication and telecommunication. Any signal like acoustic, electrical, and electromagnetic signal produced with a specific bandwidth is spread in frequency hence results in a wider bandwidth. Spread spectrum techniques are deployed in telecommunication because of many significant advantages, e.g., to achieve secure communications, to detect the eavesdropping, to resist natural interference, to bound power flux density for satellite down links and resistance to noise and jamming. In spread spectrum technique, frequency hopping (FH) is used as a basic modulation technique by which any telecommunication signal can be transmitted on a wider bandwidth (radio bandwidth) as compared to frequency value of the original signal. Spread spectrum techniques deploy FH, direct sequence (DS), or mix of both methods so that it can be used for multiple access and reduces the interference to other receivers to get the overall privacy. At the receiver side, the received signals are correlated to extract the original information being sent. The two main motivation behind spread spectrum are: to create anti-jamming for unauthenticated person and to provide low probability of interception.

Spread spectrum technique includes chirp spread spectrum (CSS), direct-sequence spread spectrum (DSSS), frequency-hopping spread spectrum (FHSS), and time-hopping spread spectrum (THSS). In all these methods pseudorandom number sequences are used throughout the bandwidth to confirm and to adjust the spreading pattern. IEEE 802.11 wireless standard uses DSSS or FHSS in radio communication. Spread spectrum principle makes use of noise like carrier waves, bandwidth is wider than used in simple point to point communication at the same data rate. DS is capable to resist continuous-time narrowband interference (jamming). FH is good at resisting the pulse interference. In DS spread spectrum, the

[1]Department of Electronics Engineering, Defence Institute of Advanced Technology, India
[2]Department of Physics, Indian Institute of Technology Jodhpur, India

narrow band jamming degrades the signal quality because jamming power concentrates on the whole signal bandwidth [1]. Signal power is spread over a large bandwidth hence it is lower than noise PSD (power spectral density). Hence it is difficult for eavesdroppers to find out meaningful information. Spread spectrum systems need same amount of energy per bit before spreading as narrow band systems.

The classical data transmission can also be achieved by using quantum communication protocols like dense coding [2,3], quantum key distribution (QKD) [4], and quantum teleportation [2,5]. Most of these protocols use quantum states of light to transmit the information through optical fibers [6]. The problem of channel access become apparent when number of users increases. In classical communication networks, when many users want to use the same channel at the same time, many multiple access methods are used. On the other side, quantum communication networks also deploy frequency and wavelength division multiple access techniques, hence each user make use of same channel at different frequencies and different time instance. In addition to this, many multiple access techniques employ orbital angular momentum of single photons and coherent states. Quantum spread spectrum multiple access scheme described here for optical fibers could be used for free-space data transmission. In this scheme, multiple users send their photons via same optical fiber, sharing their path, time window, and frequency band. A number of techniques have been proposed previously for data transmission for quantum optical communication which provides heavy losses while combining and separating the transmitted data from different users. The described scheme is developed to follow classical spread spectrum methods, and also follows add-drop architecture with simple extraction and combination points.

15.1.1 Spread spectrum scheme

Spread spectrum technique is used to share channel in cellular communication networks, in which a modulated signal $d(t)$ of bandwidth W is transformed to make it a spread signal $s(t)$ of higher bandwidth SW. Here, S is known as spreading factor. We describe DSSS and its use in code division multiple access (CDMA). In CDMA, out of N users, every user U_i is assigned a code C_i. The codes C_i is considered as vectors 1 and -1. The codes are selected based on the orthogonal condition $c_i \cdot c_j^T = \delta_{ij}$, to be orthogonal, with $c_i \cdot c_j^T \leq m$ for $i \neq j$, for smallest integer value m.

Figure 15.1 represents despreading and spreading methods. Spreading can be estimated as a second modulation in which the code C_i is multiplied with the signal being transmitted. On the other side, despreading is achieved at the receiver end by multiplying the data signal with the code C_i. Selection of proper codes being used for multiplication help in signal separation from each user even if they share the same bandwidth at a time. Spread spectrum technique enhances the number of users with minimum noise probability. It also depends on the number of orthogonal codes of certain length. DSSS technique is helpful in extending the photon's wavefunction and can be recovered at the receiver end. The despreading approach

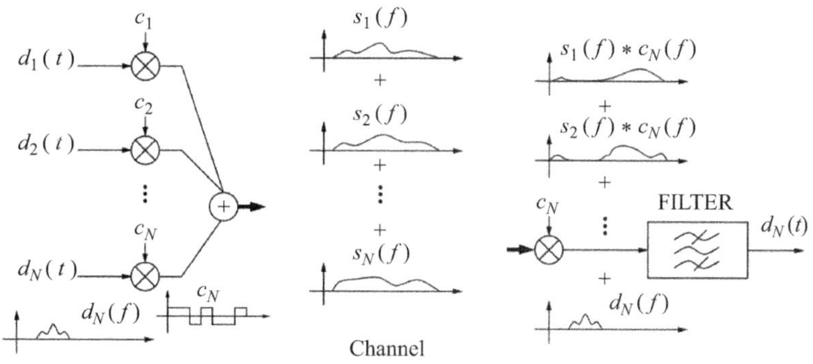

Figure 15.1 DSSS: Each c_i code is multiplied with the corresponding data signal $d_i(t)$ and sent via the same channel. At the receiver side, we can obtain any data signal sent from the transmitter side. To recover $d_N(t)$, the total signal is multiplied by c_N. All other signals will spread except the original data signal $d_N(t)$. In case of orthogonal signals, a BPF (band-pass filter), cancel out the interference of other signals. In case of non-orthogonal signals, energy of other signals will spread throughout the SW bandwidth, but a fraction $\frac{1}{S}$ will affect the data signal being recovered

recovers the single photon's wavefunction to its original bandwidth and noise is filtered out at the receiver end.

15.1.2 Basic building block for quantum spread spectrum

In this particular technique of spread spectrum-based multiple access, we need three elements: optical modulators, circulators and fiber Bragg gratings (FBG). The signal waveform is altered and controlled by a control signal with the use of an electro-optic modulator [7]. How to perform the same operation in quantum domain is described in [8]. The function of the optical modulator is to change the photon's phase at different time instances. In addition to this, the phase shift of amount $\pm\frac{\pi}{2}$ can be inserted to different time instances of the photon's wavefunction [9]. The time length T signal can be divided into S segments. Further, the code c_i corresponding to each element ensures the phase shift $\frac{\pi}{2}$ or $-\frac{\pi}{2}$ according to the presence of element in time segment 1 or -1, respectively. As an alternate, we can also perform this operation by using an optical modulator which provides phase 0 and π for an element 1 and -1, respectively.

Further, we need to combine the spread photons of all the users in the same optical fiber cable. This can be achieved by using circulators and FBGs, as shown in Figures 15.2 and 15.3, respectively.

15.1.3 Incoming data signals

In classical spread spectrum technology, frequency modulation is preferred for encoding. Here we follow the same model for our incoming data signals (qubit)

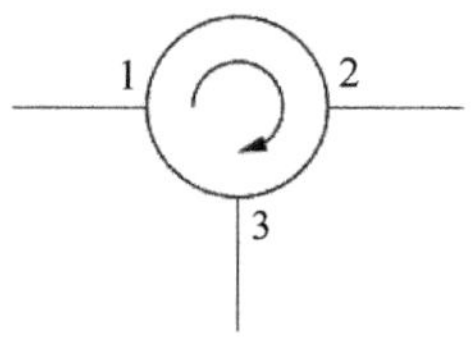

INPUT PORT	OUTPUT PORT
1	2
2	3
3	1

Figure 15.2 Working principle of circulator and its input–output action at different ports [10]

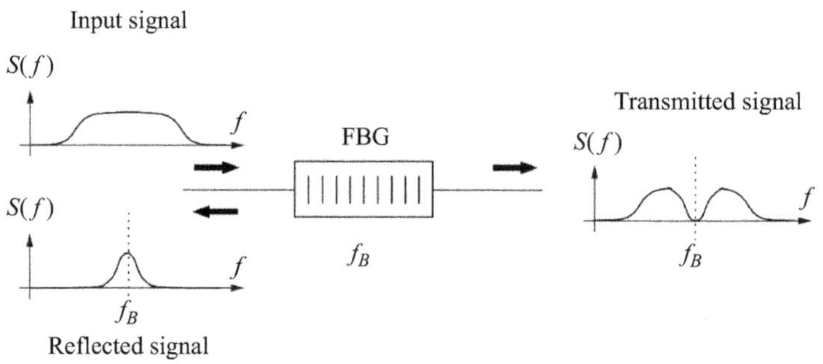

Figure 15.3 FBG works as a band-stop filter. It splits the power spectral density of the incoming signal into two parts. One part of the incoming signal is reflected around a frequency f_B and the remaining input signal is transmitted [10]

encoding, in which frequency qubits, $|0\rangle$ and $|1\rangle$ denote different wavefunctions at different frequencies. Also, phase modulation can be used on the sidebands of a strong carrier. Time bin encoding can also be used to encode qubit information. For this, encoding of $|0\rangle$ is achieved by sending a photon with a wavefunction bounded in a time window $(0, T_0)$. On the other side, $|1\rangle$ encoding is performed by introducing a delay in time window $(T_0, 2T_0)$ of the considered wavefunction. Among encoding schemes, time-bin qubit scheme is the preferred one for applications like optical fibers transmission and experimental QKD. Here, codes are constructed for T_0 time period and used for two times. It is done to avoid interference otherwise, two different users will give same spread signals for the two orthogonal codes with the same first or second half of time length.

15.2 Laser satellite communication

Lasers have the capability to transmit information effectively in space and improves satellite communication. The satellites as communication relays collect the surrounding images using laser and transmit them to ground station in form of a concentrated beam. The laser-based satellite system accomplishes this task in a few minutes. On the other hand, communication system based on radio waves takes hours to transmit images. The laser system enhances the transmission rate by enabling the spacecraft circling the ground station and its surroundings to transmit vertically to a newly launched node. This becomes much easier to approach because it is placed at a higher altitude than other satellites. Hence, laser satellite communication, transmits strong beams which are not disruptive.

For disaster relief, laser-based communication system provides fast relief, because it quickly transmits earth images and further rescue operations can be taken effectively for those areas on earth which are severely under the grip of natural calamities. Under the effects of matter, sheer volume of space, noise or disruption, any communication signal becomes weaker [11]. Radio waves degrade in space and radio signals become weaker under these effects. In addition to this, the radio signals need more complex relay systems even if operating from ground. In case of radio waves which require much power, they get confronted with the vastness of space, hence radio waves are inadequate. In such situations, laser satellite communication becomes efficient by transmitting encoded data through a narrow beam.

Radio wavelengths are much busy for data transmission due to ever increasing data transmission demands. Hence, laser communication is a better and effective technique which achieves a higher volume of data transmission than radio. Comparatively smaller antennas are needed for laser communication due to their narrower waves. As an estimate, an average movie can be downloaded in 639 h with the help of currently available S-band communication model, but the laser technology reduces this time period to eight times. Laser satellite communication plays a vital and significant role in many areas such as in delivering videos, images and other data from satellites, aircrafts, different space stations, and unmanned aerial vehicles (UAVs), permitting fast and complete communications in the event of disaster or natural hazards. Also, the laser-based satellite communication system can be used for environment monitoring, weather forecasting, agriculture, maritime surveillance and natural disaster.

In the field of quantum-based satellite FSO communication, QKD [12–15,16] is an advanced and effective approach to perform secure global communication between a satellite and earth-based stations. Although quantum repeaters are capable of enhancing the communication distance by deploying a number of repeaters per station with many qubits, then cannot achieve true global distances. On the other side optical fibers are limited to only a few kilometers due to presence of optical and material losses.

Quantum key distribution is the first mature and secure cryptographic technology which has been experimentally tested both in optical fibers and in free space [12–14]. The optical fibers are not the good choice for establishing a global quantum network, because of their inherent losses. In addition to this, if we talk about ground to ground communication, quantum repeaters are deployed to enhance the communication distance, which cannot be achieved by simple optical fibers. Other techniques such as optical fiber-based quantum repeaters [17] and error correction-based methods are unable to establish a true global distance. A large amount of classical and quantum resources in quantum repeaters make the overall quantum communication system costly and suffers from poor performance due to losses at each successive node. Finally, it is essential to look for other methods such as quantum-based satellite communication which overcomes all the mentioned drawbacks and provides true global reach for real field applications.

There are three main techniques to establish a secure quantum-based satellite communication. The first approach deals with an entangled source which is itself mounted on the satellite and distributes photons to the two ground stations, separated hundreds of kilometers apart, at the same time. These can be a lower earth orbit (LEO) satellite, in which correlation property is examined to confirm the secure communication. The beauty of this approach arises due to a unique property of quantum mechanics known as entanglement, which is classically absent. Here, random detection of arrival photons is stored for extracting the final secure key. Hence, any autonomous satellite system with a mounted entangled source can make the source functional, which is essential for such kind of research where quantum-based satellite communication plays an important role in achieving true global distances.

Attenuated laser pulses with low optical power are the alternate source which emits single photons that reduce the occurrence of photon number splitting attack (PNS). In addition to this, decoy states must be used to avoid side channel attack due to multi photons per pulse. In the third scenario, both the transmitter and receiver are located at the ground and satellite station in space. In such a scenario, photons are allowed to propagate from Earth to space. In such an arrangement, the quantum source is located at the ground which is adaptive according to the requirement during the complete mission. This is the only unique arrangement which accomplishes both the foundational tests of quantum cryptography and quantum mechanics. Here, we simulate such kind of arrangements in our current study.

Many studies and experiments were proposed to set up a quantum-based satellite communication for establishing a global quantum network. Some of these schemes are based on FSO communications. In this context, free-space QKD suffers from atmospheric turbulences which degrade the system performance and becomes a major issue for improvement of the security, data rate and communication distance [1,11,18].

QKD implementation in free space is possible when a direct line of sight is not available. Such a scheme can be deployed for quantum cryptography applications, where security comes from Heisenberg's uncertainty principle and No-Cloning principle. Such a QKD scheme outperforms in free space communications and is free from traditional optical and material losses present in optical fiber cables. This plays a vital role in establishing a global free space optical network for sharing secret messages. For this, it is significant to study the impact of atmospheric effects on traveling photons emitted from the laser source, both in case of uplink and downlink communication. For the detailed analysis of different parameters such as bit error rate, communication distance, and selection of optimum photon number, we consider two QKD protocols, that is, the BB84 and the SARG04 protocols.

15.3 Free space quantum optical satellite link

In free space optical communication, three effects mainly contribute to the channel attenuation (represented as $\delta \in [0, 1]$), that is, receiver efficiency, atmospheric propagation, and diffraction. Hence, the total channel attenuation can be obtained using

$$\delta = \delta_{\text{diff}}\delta_{\text{atm}}\delta_{\text{rec}}. \tag{15.1}$$

The above equation for total attenuation (δ) is represented in dB (dB is calculated as $10\log_{10}(\delta) = 10\log_{10}(\delta_{\text{diff}}) + 10\log_{10}(\delta_{\text{rec}})$). In the above equation δ_{diff}, δ_{atm}, and δ_{rec} represent attenuation due to geometrical losses, atmospheric losses, and losses due to receiver inefficiency, respectively. In this study, we can represent the basic building block for a quantum-based LEO satellite FSO communication as shown in Figure 15.4.

Here, we consider Cassegrain type telescope architectures both at the transmitter and receiver ends and laser beams of Gaussian type. Attenuation is caused by diffraction and obstruction; see for a detailed description.

$$\begin{aligned}
\delta_{\text{diff}} &= \left(e^{-2\gamma_t^2\alpha_t^2} - e^{-2\alpha_t^2}\right)\left(e^{-2\gamma_r^2\alpha_r^2} - e^{-2\alpha_r^2}\right), \\
\gamma_t &= \frac{b_t}{R_t}, \gamma_r = \frac{b_r}{R_r}, \alpha_t = \frac{R_t}{\omega_t}, \alpha_r = \frac{R_r}{\omega_r}, \\
\omega_t &= R_t, \omega_r = \frac{\sqrt{2}\lambda L}{\pi R_t}.
\end{aligned} \tag{15.2}$$

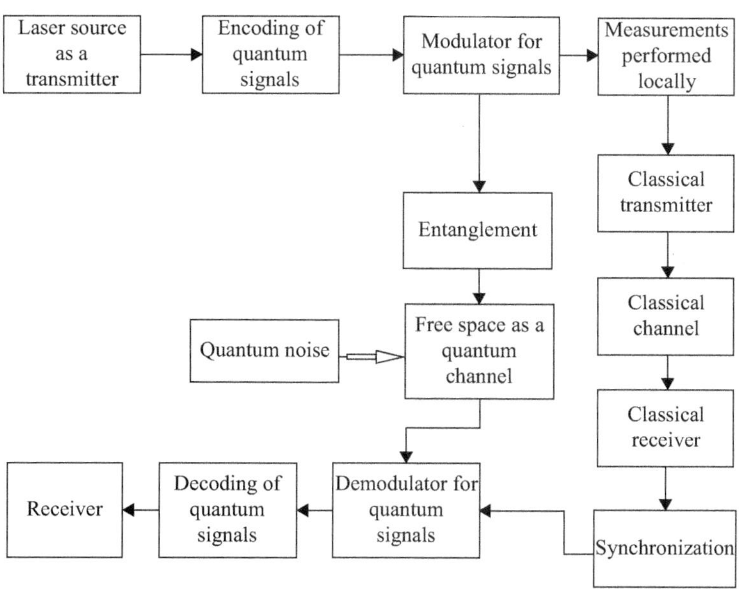

Figure 15.4 Quantum-based LEO satellite FSO communication system

Here, b_t, b_r and R_t, R_r represent radii of the secondary (b) and primary (R) mirrors at transmitter (t) and receiver (r), respectively; L is the distance between the telescopes (also known as the link distance), λ represents the wavelength, and $\omega_{t,r}$ denotes the beam radius at transceiver ends. The atmospheric attenuation δ_{atm} is caused by various phenomena such as turbulence, scattering, absorption, rain, haze, and fog. Hence, this can be represented as $\delta_{\text{atm}} = \delta_{\text{scatt}}\delta_{\text{abs}}\delta_{\text{turb}}\delta_{\text{rain}}$, where each quantity represents the corresponding attenuation for the considered phenomena.

In this context, taking into account the total attenuation mentioned above and excluding the rain effects, various experiments have been performed for QKD in FSO link and optical fibers. The received P_R and transmit P_T power at the receiver and transmitters ends, respectively, are

$$P_R = P_T O \alpha_{\text{geo}} e^{-\sigma L}. \tag{15.3}$$

Here,

$$\sigma = \left(\frac{3.912}{V}\right)\left(\frac{\lambda}{550}\right)^{-q'}, \tag{15.4}$$

$$\alpha_{\text{geo}} = \frac{d_2^2}{[d_1 + (L\theta)]^2}. \tag{15.5}$$

O stands for optical losses, d_2 is the diameter of receiver aperture in meters, d_1 is the diameter of transmitter aperture in meters, θ is the beam divergence in mrad, L is the link range in meters, λ represents the considered wavelength (nm), V is the visibility (km), and q' represents the size distribution of diffusing particles. To examine the FSO links under harsh atmospheric conditions, power equation at the receiver side is

$$P_R(\text{dBm}) = P_T(\text{dBm}) - O(\text{dB}) - \alpha_{\text{geo}}(\text{dB}) - \alpha(\text{dB}) - \alpha_{\text{haze/cloud}}(\text{dB}), \tag{15.6}$$

$$\alpha(\text{dB}) = 10\log_{10}\left(\exp^{-\sigma L}\right) + L\left(aR^b\right). \tag{15.7}$$

In the above, R denotes the rain rate and a, b are rain coefficients which are dependent on rain characteristics and the wavelength under consideration of the FSO link. $O(\text{dB})$, $P_T(\text{dBm})$, $P_R(\text{dBm})$, $\alpha_{\text{geo}}(\text{dB})$ and $\alpha(\text{dB})$ are optical losses, transmitted power, received power, geometric attenuation, and atmospheric attenuation coefficient in dB, respectively. Signal at the receiver can be recovered if the received signal power P_R is more than the minimum power level which depends on the sensitivity of the receiver.

Geometric loss is directly proportional to the link range. In a free space optic model, the geometric loss can be minimized by considering the low value of divergence angle of the laser beam. Under geometric attenuation light beam diverges from the transmitter to receiver, hence most of the light beam does not reach the receiver's telescope and signal loss occurs. It is required to increase the receiver aperture area, so that the geometric losses can be controlled by collecting more signals at the receiving telescope ends.

15.4 Analysis of secure key generation rate

In this section, we will describe BB84 and SARG04 QKD protocols, with and without decoy states. Moreover we calculate many important parameters such as quantum bit error rate (QBER), mutual information shared between legitimate users and Eve, and secure key generation rate of these quantum key distribution protocols.

15.4.1 The BB84 QKD Protocol

In the category of quantum key distribution protocols, BB84 is the basic quantum cryptography protocol which was proposed in [12,19,20]. The practical QKD schemes need attenuated laser pulses which are described by the coherent states and it is also required to avoid PNS. This has been taken into account for the analysis of key generation rate, mutual information shared between the communicating parties, later in this chapter. The output pattern of the lasers follow the Poisson distribution

$$|\alpha\rangle = e^{\frac{-|\alpha|^2}{2}} \sum_{n=0}^{\infty} \frac{\alpha^n}{\sqrt{n!}} |n\rangle. \tag{15.8}$$

Here, $|\alpha| = \sqrt{\mu}$, μ is the average photon number of a pulse. The probability for n photons in a pulse is given by

$$p_n = |\langle n|\alpha\rangle|^2 = e^{-|\alpha|^2} \frac{|\alpha|^{2n}}{n!}. \tag{15.9}$$

In quantum key distribution, the optical pulses are emitted from the laser source. These optical pulses are the information content in the form of bit streams and transmitted via a quantum channel [2,3]. These optical pulses are known as beam intensity μ (mean photon number) which ranges from 0.1 to 0.5. The value 0.1 for mean photon number indicates 1 photon every 10 pulses. The polarization encoding is used in BB84 protocol. In addition to this, polarization filters are used to change the photon polarization. The mutual information $I(A:B)$ and $I(B:E)$, shared between, Alice-Bob and Bob-Eve are calculated in bits/pulse [13,21]. Here,

$$I(A:B) = \sum_{n=0}^{\infty} (1 - (1 - \delta)^n) P_n(\mu) \approx \mu\delta, \tag{15.10}$$

$$I(B:E) = \sum_{n\geq 2}^{\infty} P_n(\mu). \tag{15.11}$$

Eve's mutual Information, I_{Eve}, is written as

$$I_{\text{Eve}} \approx \frac{I(B:E)}{I(A:B)}. \tag{15.12}$$

The lowest value for the key generation rate R (in bits/pulse) is represented in [4,21]

$$R \geq q\left(-Q_\mu f(E_\mu)H_2 E_\mu + \Omega Q_\mu\left(1 - H_2\left(\frac{E_\mu}{\Omega}\right)\right)\right), \tag{15.13}$$

where Ω represents the photons, from which Eavesdroppers cannot extract any significant information, also called as untagged photons [22]. H_2 denotes Shannon entropy function. Also q denotes the efficiency of the protocol under examination, the values of q are $\frac{1}{2}$ and $\frac{1}{4}$ for BB84 and SARG04 protocols, respectively. $f(x)$ represents the bi-directional error correction efficiency, whose value is 1.22 for the Cascade protocol [23]. The quantity Y_n represents the Yield of the n-photon pulses [22].

The expected raw key rate Q_μ and quantum bit error ratio (QBER) E_μ are

$$Q_\mu = \sum_{n=0}^{\infty} Y_n P_n(\mu) \tag{15.14}$$

and $$E_\mu = \frac{\sum_{n=0}^{\infty} Y_n P_n(\mu) e_n}{Q_\mu} = \frac{Y_0}{2Q_\mu}. \tag{15.15}$$

In the above, e_n represents bit error ratio of n-photon signals, $e_n = \frac{Y_0}{2Y_n}$, and Y_0 denotes the rate of dark counts (counts/pulse). Y_n is the yield of n-photons, $Y_n = Y_0 + \delta_n$, $\delta_n = 1 - (1-\delta)^n$, here δ_n denotes attenuation for n-photon signal.

15.4.2 The Scarani–Acin–Ribordy–Gisin 2004 (SARG04) QKD Protocol

The SARG04 protocol was proposed by Scarani *et al.* [24] and is more advanced as compared to BB84 protocol against the PNS. The quantum communication phase in SARG04 protocol is similar to that in the BB84 protocol, but the difference lies in the encryption and decryption of Shannon's classical information part [13]. In this protocol, the bases are not communicated, but Alice announces one non-orthogonal state out of the four pairs $A_{\omega,\omega'} = \{|\omega x\rangle, |\omega' z\rangle\}$, where ω, $\omega' \in \{+,-\}$ and $|\pm x\rangle = 0, |\pm z\rangle = 1$ [25]. While performing attacks on the communication channel, an Eavesdropper introduces attenuation which is expressed as

$$\delta = \frac{(1-t)P_1 + P_2(\mu) + \chi}{\mu}, \; t \in [0,1]. \tag{15.16}$$

Here, χ is written as

$$\chi = \sum_{n\geq 3}^{\infty} P_n(\mu) P_{ok}(n), \tag{15.17}$$

where P_{ok} represents the probability of acceptance. The value of P_{ok} is 0.5 for BB84 protocol.

The value of the attenuation in this case can be represented as

$$\delta = \frac{(1-s)P_2(\mu) + \chi}{\mu}, \quad s \in [0, 1]. \tag{15.18}$$

15.4.3 The decoy-states protocols

Decoy states are used to detect the presence of Eavesdroppers; these are extra states other than the information bits. In this method, the signal states are used for key generation. The decoy state method was proposed in [26–28]. The mutual information shared between the communicating parties and Eve can be written as [4]

$$I(A:B) = P_1(\mu)(1-t) + P_2(\mu)(1-s) + \sum_{n\geq 3}^{\infty} P_n(\mu)P_{ok}(n), \tag{15.19}$$

$$I(B:E) = P_2(\mu)(1-s)I_2 + \sum_{n\geq 3}^{\infty} P_n(\mu)P_{ok}(n), \tag{15.20}$$

where t denotes the fraction of the single photon pulses blocked by Eavesdropper, and s represents a fraction of the two-photon pulses. I_2 is the amount of information that Eve can obtain from a single copy of the state [24]. Next we investigate the security of the quantum cryptography protocols.

15.4.3.1 BB84 QKD protocol: vacuum + weak decoy states

The key generation rate in its lower bound form [22], which uses decoy state method and entanglement distillation can be written as

$$R_{\mathrm{BB84}} \geq q(-Q_\mu f(E_\mu)H_2(E_\mu) + Q_1(1 - H_2(e_1))). \tag{15.21}$$

Here, Q_μ denotes the gain of the signal states, E_μ represents the quantum bit error ratio, Q_1 denotes the gain of the single-photon states, and e_1 is the error rate of single-photon states.

The quantity Q_1 is

$$Q_1 = Y_1 e^{-\mu}\mu. \tag{15.22}$$

The lower bound for Q_1 and upper bound for e_1 with the vacuum and a weak decoy state (ν) is [26]

$$Y_1^L = \frac{\mu}{(\mu\nu - \nu^2)}\left(Q_\nu e^\nu - Q_\mu e^\mu\left(\frac{\nu^2}{\mu^2}\right) - \frac{(\mu^2-\nu^2)}{\mu^2}Y_0\right) \leq Y_1, \tag{15.23}$$

$$Q_1^L = \mu e^{-\mu} Y_1^L \leq Q_1, \tag{15.24}$$

$$e_1^U = \frac{e_0 Y_0}{Y_1^L} \geq e_1. \tag{15.25}$$

15.4.3.2 The SARG04 QKD protocol: vacuum + two weak decoy states

In BB84 QKD protocol only a single photon state contributes to the key generation rate. On the other hand, both single and two photon states help in key generation rate in SARG04 protocol. The gain in two photon pulses is [21]

$$Q_2 = Y_2 e^{-\mu} \frac{\mu^2}{2}. \tag{15.26}$$

Three decoy states are used in SARG04 protocol, ν_0, ν_1, and ν_2, assuming that ν_0 is the vacuum (i.e., $\nu_0 = 0$), and the two weak decoy states are ν_1 and ν_2. For these decoy states, gain and quantum bit error rate are

$$Q_{\nu_i} = \sum_{n=0}^{\infty} Y_n P_n(\nu_i), \tag{15.27}$$

$$E_{\nu_i} = \sum_{n=0}^{\infty} \frac{Y_n P_n(\nu_i) e_n}{Q_{\nu_i}}. \tag{15.28}$$

The bit error ratio of the n-photon signals, which is due to the dark counts Y_0 only, is

$$e_n = \frac{Y_0}{2Y_n}.$$

Let the legitimate users (Alice and Bob) choose ν_1 and ν_2 which follows

$$0 < \nu_1 < \nu_2,\ \nu_1 + \nu_2 < \mu. \tag{15.29}$$

Now, the secure key generation rate can be written as

$$R_{\text{SARG04}} \geq q\left(-Q_\mu f(E_\mu) H_2(E_\mu) + Q_1\left(1 - H\left(\frac{Z_1}{X_1}\right)\right) + Q_2(1 - H(Z_2))\right), \tag{15.30}$$

where X_n and Z_n denote the binary random variables. The Shannon binary entropy function, $H_2(\cdot)$, can be seen in [21].

The two photon gain in its lower limit form can be written as

$$Q_2^L = \frac{Y_2^L \mu^2 e^{-\mu}}{2} \leq Q_2. \tag{15.31}$$

The upper limit of e_2 can be calculated by considering quantum bit error rate of weak decoy states

$$E_{\nu_i} Q_{\nu_i} e^{\nu_i} = e_0 Y_0 + e_i \nu_i Y_1 + e_2 \frac{\nu_i^2}{2} Y_2 + \sum_{n=3}^{\infty} e_n Y_n \frac{\nu_i^n}{n!}. \tag{15.32}$$

15.5 Design parameters and results

The simulation results performed on three different scenarios are uplink, downlink, and intersatellite links. The quantities for quantum communication link establishment are detector efficiency (δ_{rec}), satellite telescope radius ($R_{t,r}$), satellite secondary mirror radius ($b_{t,r}$), ground telescope radius ($R_{t,r}$), ground secondary mirror radius ($b_{t,r}$), dark counts (Y_0), and communication wavelength (λ) whose values are 65%, 15 cm, 1 cm, 50 cm, 5 cm, 50×10^{-6} counts/pulse and 650 nm, respectively. The value of $\lambda = 650$ nm shows an absorption window with a commercial photon detector made of silicon avalanche photo diode with high detection efficiency.

The turbulence generated attenuation in uplink scenario is calculated under two different cases, one for $\delta_{turb} = 5$ dB and other for $\delta_{turb} = 11$ dB. In downlink communication, the turbulence effects are almost negligible. A value of $\delta_{scatt} = 1$ dB is obtained for the scattering plus absorption attenuation by the Clear Standard Atmosphere model. Figure 15.5 is adapted from [4] and depicts the comparison between I_{Eve} and communication distance in km under BB84 and SARG04 protocols. This is calculated based on the link parameters described in subsequent sections. From this figure, it is observed that Eavesdropper receives more information in BB84 protocol as compared to SARG04 protocol. Hence, it can be concluded that SARG04 protocol outperforms the BB84 protocol under such conditions. The condition $I_{Eve} = 1$ is obtained for the optimum parameters (attenuation = 13 dB, $\mu = 0.1$ for BB84 QKD protocol and attenuation = 25.6 dB, $\mu = 0.2$ for the SARG04 QKD protocol) [4,24].

Figure 15.6 shows the variation in secure key generation rate with the communication distance for the protocols under study. The conversion from the bits/pulse to bits/second is achieved for the pulses emitted from the laser source.

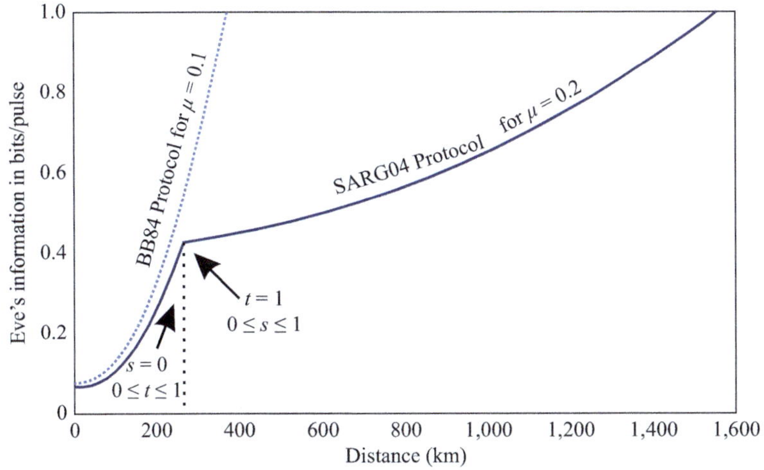

Figure 15.5 Comparison between I_{Eve} and communication distance in km for BB84 and SARG04 protocols [4]

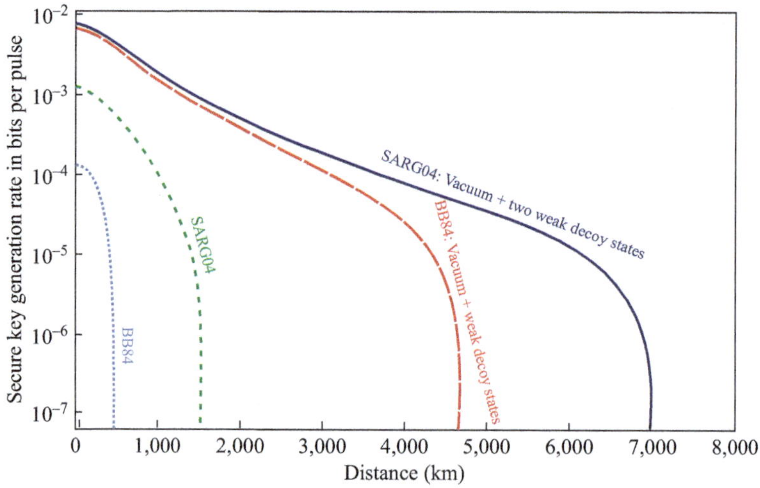

Figure 15.6 Secure key generation rate in uplink case for all the protocols [4]

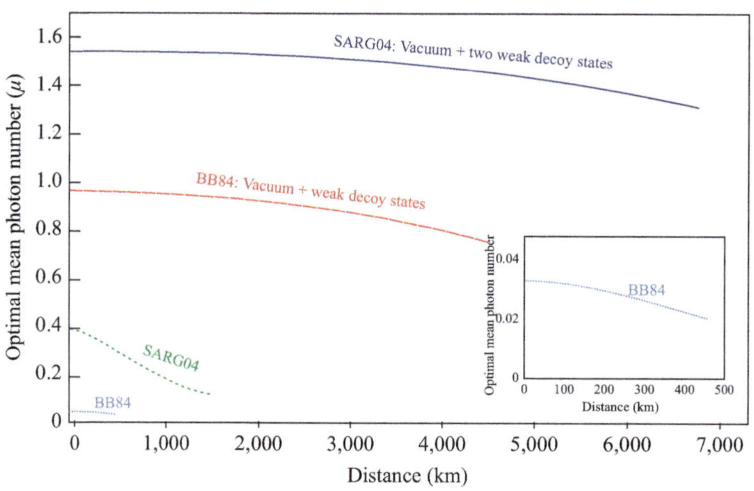

Figure 15.7 In uplink, variation in optimum mean photon number with communication distance [4]

In Figure 15.7, we have optimized $v_i^{'s}$ and μ in each protocol for both the states to receive the highest key generation rate [4].

In Figure 15.6, it is seen that the critical distance achieved for SARG04 protocol is higher than BB84 protocol, for both the cases: with and without decoy states. Also in Figure 15.5, it is shown that SARG04 is more resilient against eavesdropping than BB84 with an optimal mean photon number. The decoy state method used in BB84 protocol increases the critical distance. Decoy state method is

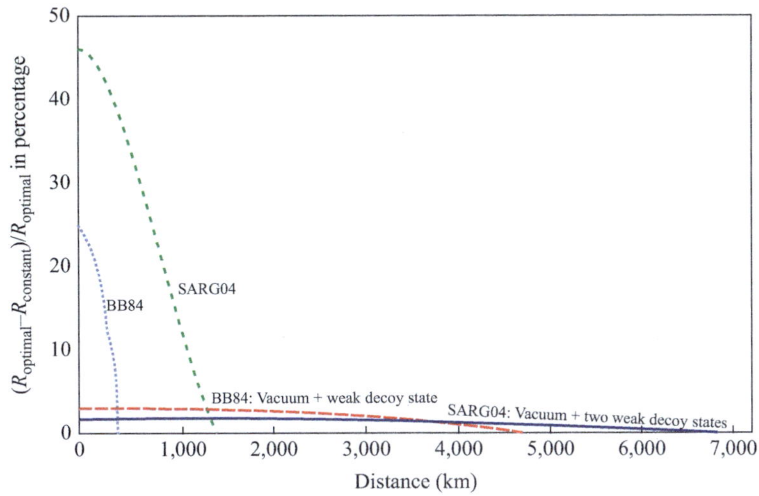

Figure 15.8 In uplink, variation in secure key generation rate (R) with communication distance at constant value of mean photon number [4]

a powerful approach that enhances both the critical distances and key generation rate for both the entangled and non-entangled-based quantum key distribution protocols.

With increasing attenuation, the number of multi photon pulses must be reduced which helps in reducing the chance of attacks performed by Eve (in such case μ must be reduced). At higher values of μ, the protocol becomes more long-lasting, as shown in Figure 15.7. With increasing mean photon number, we achieve enhanced communication distance and at the same time the considered protocols are resistant to Eve's PNS attack. Due to movement of the satellite along its orbit, its distance with the ground station varies. The value of μ has to be modified to attain the maximum secure rate, which is the challenging part of the research.

The maximum key generation rates are shown in Figure 15.8, keeping μ constant to that of optimal μ for maximum distance. It is seen that in the protocols based on decoy states the secure rate starts decreasing to a level below 3%, which means that in such conditions the variation in mean photon number with distance is not necessary. The result is opposite to that of protocols based on non-decoy states where reduction in rate occurs rapidly. This implies that the value of mean photon number should vary with distance for obtaining optimum results for secure rates. The rest of the three cases (uplink, downlink on clear weather conditions, and inter satellite links) follows the same steps.

The critical distance (in km) obtained for BB84 protocol are: 1,540 km in downlink case, 430 km in intersatellite case, 460 km in uplink case ($\delta = 5$ dB), and negligible critical distance in case of uplink with $\delta = 11$ dB. On the other side, the critical distance (in km) attained for SARG04 protocol are: 3,290 km in downlink case, 920 km in intersatellite case, 1,520 km in uplink case ($\delta = 5$ dB), and 500 km in case of uplink with $\delta = 11$ dB. Following the same method, the critical distance (in km) for BB84 protocol with vacuum state and weak decoy states received from simulations are: 9,450 km in downlink case, 2,660 km in intersatellite case, 4,650 km in uplink case ($\delta = 5$ dB), and 2,200 km in case of uplink with $\delta = 11$ dB. The critical distance (in km) attained for SARG04 protocol with vacuum state and two weak decoy state are seen to be: 14,100 km in downlink case, 3,900 km in intersatellite case, 6,980 km in uplink case ($\delta = 5$ dB), and 3,460 km in case of uplink with $\delta = 11$ dB. The maximum possible secure rate (in bits/pulse) for BB84 protocol achieved from simulations are: 1.7×10^{-2} in downlink case, 2.0×10^{-2} in intersatellite case, 1.4×10^{-4} in uplink case ($\delta = 5$ dB), and almost negligible secure rate in case of uplink with $\delta = 11$ dB. The maximum possible secure rate (in bits/pulse) for SARG04 protocol are as follows: 2.4×10^{-2} in downlink case, 2.6×10^{-2} in intersatellite case, 1.2×10^{-3} in uplink case ($\delta = 5$ dB), and 7.5×10^{-5} in case of uplink with $\delta = 11$ dB [4].

In a similar fashion, it can be shown that the maximum possible secure rate (in bits/pulse) for BB84 QKD protocol with vacuum state and weak decoy states are as follows: 4.8×10^{-2} in intersatellite case, 4.4×10^{-2} in downlink case, 5.8×10^{-3} in uplink case ($\delta = 5$ dB), and 1.4×10^{-3} in case of uplink with $\delta = 11$ dB. The maximum possible secure key generation rate (in bits/pulse) for SARG04 protocol with vacuum state and two weak decoy states are as follows: 4.6×10^{-2} in downlink case, 5.0×10^{-2} in intersatellite case, 6.5×10^{-3} in uplink case ($\delta = 5$ dB), and 1.6×10^{-3} in case of uplink with $\delta = 11$ dB. In these cases values are different but the curves follow the same steps [4].

It is clear from the figures that we achieve maximum distance in downlink communication which is due to absence of turbulence in downlink and hence no attenuation. In medium-earth-orbit (MEO) satellites, cryptography techniques can be implemented by employing SARG04 protocol with decoy states. Inter satellite links suffer from reduced telescope dimensions and hence cannot attain maximum distance. In all these operations two major obstacles are telescope dimensions and turbulence-induced attenuation which severely deteriorate the performance of the quantum-based satellite FSO communication.

Geometric attenuation is responsible for the light beam to diverge in its propagation path. To reduce these signal losses, receiver aperture area is increased to receive more light by the telescope to lessen the geometric losses. Hence SARG04 protocol with decoy states achieves highest key generation rate and maximum link range. Finally, we can claim that the optimum results are attained when we use pulses with two photons plus optimum μ.

From the above results, it can be concluded that the SARG04 protocol attains optimum result as compared to the BB84 protocol under the considered attack, for example, PNS attack. Based on these results, we can claim that two decoy states based on the SARG04 QKD protocol is the best choice for QKD-based satellite FSO communication. Here, we have optimized all the results for the optimum value of mean photon number to achieve maximum communication distance and secure key generation rate.

To achieve long distance communication, it is necessary to diminish the link losses. Actual data may be deployed to better analyze the turbulence effects and define a real propagation model that should help the receiver and transmitter design. In addition to this, new communication protocols that exploit the atmospheric turbulence as a resource can be explained. Our telescope design data could be used in future for single photons long distance free space experiments, like teleportation and QKD. This study will help to experimentally demonstrate the feasibility of Earth-space quantum-based satellite FSO links.

The uplink permits the complex quantum source to be kept on the ground while only simple receivers are in space, but suffers from high link loss due to atmospheric turbulence, necessitating the use of specific photon detectors and highly tailored photon pulses. For better performance and to increase the communication distance one could use six or more nonorthogonal states. Further, the effect of adding pointing and misalignment errors need to be taken into account for greater improvement in the overall system performance.

15.6 Introduction to microwave photonics

Photonics is the science and technology of generating, managing, and detecting photons. Photons, unlike electrons, have neither mass nor charge. Therefore, photonic systems are not affected by external electromagnetic fields and have a much greater transmission distance and signal bandwidth. The first important technical device using photons was the laser, invented in 1960. After the fiber-optic transmissions began in the 1980s, the term photonics became more common. Through the end of the twentieth century, photonics was largely focused on telecommunications. In particular, it became the basis for the development of the Internet. Currently, radio-photonics have started to replace classical telecommunications. This has led to the interdisciplinary field of radionics which is the intersection of various fields such as wave optics, microwave RF systems and optoelectronics, etc. RF photonics are used in the transmission of information using electromagnetic waves of microwave, photonic devices, and systems that can create radio-frequency waves with parameters unattainable with conventional electronics.

The cross-fertilization between photonics and microwave systems is setting a new paradigms in radio technologies, promising improved performance and new applications with strong benefits for communication systems broadly as well as for public safety specifically. In particular, the use of photonics in radar systems is leading to a new generation of multifunctional systems that can manage multiple simultaneous coherent radio signals at different frequencies; which enables multispectral imaging for advanced surveillance.

The next generation of radar systems needs to be based on software-defined radar (SDR) to adapt to variable environments with higher carrier frequencies for smaller antennas and broadened bandwidth for increased resolution. The tunability and huge bandwidth of photonics meet the urgent need of flexibility requirements for the future SDR architecture. Today's digital microwave components (direct digital synthesizer (DDS) and analog-to-digital converters (ADCs)) suffer from limited bandwidth with high noise at increasing frequencies, thus a fully digital radar systems can work up to only a few GHz and noisy analogue up/down conversions especially in higher frequencies. In contrast, photonics provide high precision and ultra-wide bandwidth, allowing both the flexible generation of extremely stable RF signals with arbitrary waveforms up to mm waves. Detection of such signals and their precise direct digitization without down conversion are also possible in photonics domain. It also guarantees the system compactness because a single shared transceiver can be used for multiband operations with the added potential of photonic integration. The increasingly demanding requirements, for remote sensing systems, are putting current electronic technologies under pressure [29–32].

Radar system applications, traditionally limited to ranging and surveillance, are becoming more multifunctional with improvement in spatial and Doppler resolution. This requires very stable RF generation and very precise signal detection and digitization. Multifunctional radar system requires reconfigurable and software-defined RF signal source capable of producing wideband waveforms with maintaining the phase stability necessary for coherent-pulse Doppler processing, target imaging and clutter rejection. Further, a radar receiver with a digital backend working in the millimeter wave range would be beneficial for system reconfigurability, reliability, and noise reduction with respect to its analog counterpart. Nevertheless, today direct generation of modulated RF signals by means of DDSs with acceptable stability is limited to a few GHz and the need of multiple upconvertors worsen the phase stability. Similarly, the precision of ADCs drop as the input bandwidth and sampling speed increase, which therefore requires downconversion stages. Consequently, the technical requirements for SDRs pose currently a serious challenge for high-speed DDS and ADC [33,34].

Over the last decade, the new field of microwave photonics is investigating how to overcome these limitations of electronic systems for futuristic radar applications. Microwave photonics aims to exploit the specific features of photonics; such as, wide bandwidth, immunity to electro-magnetic interference (EMI), low loss and low propagation distortion, low phase noise and extremely high frequency flexibility, etc., in microwave systems. These attractive and unique properties have been exploited to obtain a wide range of functionalities in microwave systems as well as in radar systems. Photonics can be used for generating RF signals in a wide range of carrier frequencies (up to the mm wave) with superior phase stability and also permits the simultaneous generation of multiple RF carriers. It enables the beam forming of wideband signals in phased-array antennas (PAAs) capable of implementing tunable filtering and beamforming operations. At the receive side, photonics-based ADC guarantees a large input bandwidth, high sampling rates with extremely low jitter, fully digital approach, independence from the RF carrier and capacity to simultaneously receive multiple signals. Extensive use of photonics in a radar system allows RF signals to be easily distributed from the transceiver to the antenna site through optical fibers that guarantees low loss, low distortion and EMI immunity [35,36]. The potential of photonics establishes a new benchmark for

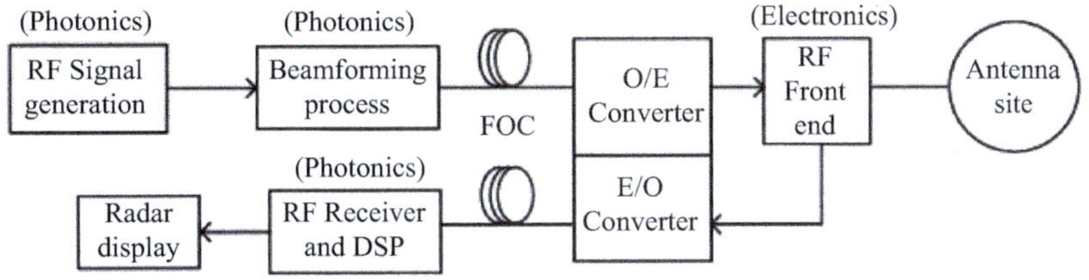

Figure 15.9 A typical method of exploiting microwave photonics in a radar system

multifunctional radar that can manage several applications; such as meteorology, environment monitoring, target detection, communications, etc., in one system. This functional integration will also enable simplified implementation of multiband radar, reduction of cost and power consumption since the same hardware is shared among different functions.

A simple illustration of photonics radar assembly is shown in Figure 15.9. In fact, while early-warning applications prefer S-band radars (strong immunity against weather clutter), target tracking is usually realized in the X-band to generate narrower beams. Moreover, different materials show different reflectivity profiles along the RF range. The flexibility of adapting the radar carrier frequency would therefore allow optimizing the performance based on the operative conditions (weather, target distance, target material, required precision, etc.). A convenient way of implementing the frequency flexibility is the realization of a multiband system and combining the information from multiple signal bands. In addition, multiband radars have a larger total bandwidth which can be used for improving the observation precision through merging the data from different bands.

15.6.1 Photonics for broadband microwave measurements

The microwave measurement refers to the acquisition of parameters of microwave signal or the properties of an object via microwave-based approaches [33,34]. Photonics provides broad bandwidth and high speed microwave measurements in the optical domain that gives better performance which may not be achievable using traditional even state-of-the-art electronics. In this section, the techniques for photonics-based broadband and high speed microwave measurements are discussed. In general, photonics-based microwave measurements can be roughly divided into two categories: (i) measurements (e.g., power, phase, phase noise, instantaneous frequency, and spectrum) and (ii) measurement of the properties of an object (e.g., position, velocity, and direction of arrival). In the above measurements, the microwave signal to be measured is first converted to the optical domain using a suitable modulator. The microwave-modulated optical signal is then sent to an optical processing module where the parameter to be measured is converted to an amplitude change. The amplitude information is finally extracted with a photodetector (PD). Few of the main radar related measurement techniques are as follows.

15.6.1.1 Microwave spectrum measurement

Photonics-based microwave spectrum measurement technique can be done by two methods. In the first method, an optical channelizer is employed to spectrally divide the microwave modulated optical signal into continuous parallel channels as shown in Figure 15.10. The spectrum of the optical microwave signal is obtained by detecting the optical signals in each channel. There, the microwave signal is first converted into an optical signal using an electro optical modulator (EOM), and then the optical microwave signal is split into various continuous parallel channels with the optical channelizer. Each channel corresponds to a particular microwave frequency. The split optical signals are then detected by an array of PDs and finally sent to an electrical processing unit to obtain the spectral information [37,38].

The key component in the system is the optical channelizer which is effectively an array of narrow band optical filters with adjacent pass bands. Although the optical channelizer can support spectrum measurement of microwave signal with a large bandwidth, its resolution is limited by the minimum achievable bandwidth of the optical filter array (typically larger than 1 GHz) which must have precise center frequencies and identical bandwidth.

In the second method, optical spatial spectral material is applied to analyze the spectra of broadband microwave signals as shown in Figure 15.11. There, the light

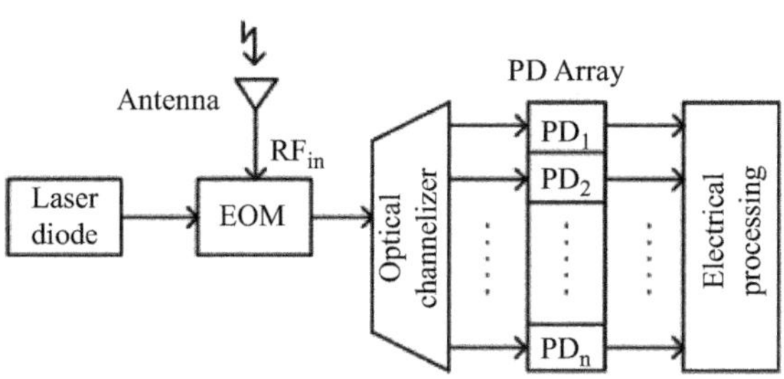

Figure 15.10 Microwave spectrum analysis system based on an optical channelizer

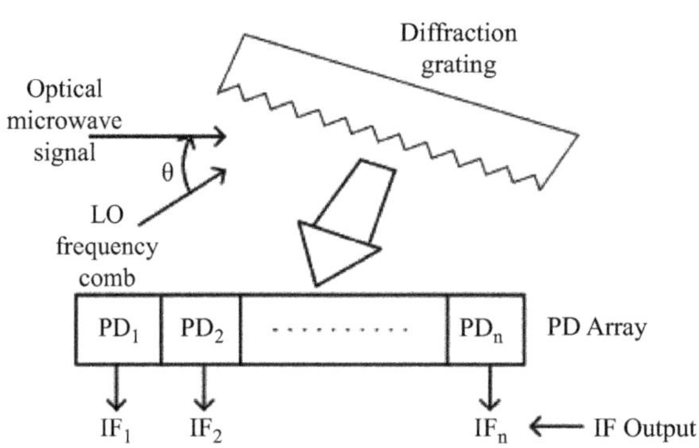

Figure 15.11 Microwave spectrum analysis using a different grating

wave modulated by the microwave signal to be measured is incident to a diffraction grating. Different frequency components are dispersed with different angles and subsequently directed onto a detector array and the position of the detector corresponds to a particular microwave frequency. An OFC with a frequency spacing equal to the channel spacing is incident to the grating at an offset angle with respect to the optical microwave signal, so every portion of the signal spectrum is translated to the same IF band and each channel can use the same post-processing electronics which greatly reduces the complexity of the processing system [37,38].

15.6.1.2 Instantaneous frequency measurement (IFM)

It is an important task particularly in radar, early warning receiver, anti-stealth defense, electronic intelligence, systems etc. Fast and accurate frequency measurement is essential for these applications. Unfortunately, conventional IFM systems are vulnerable to EMI and usually have measurement range limited to about 18 GHz due to the limited bandwidth of the electronic components. Microwave photonic techniques can be employed to extend the bandwidth of IFM to tens or even hundreds of GHz. For the IFM, as compared to the spectrum analysis, we need not to know the detailed spectrum of the microwave signal, so that the measurement can be significantly simplified while the measurement resolution and range are improved. Generally, photonics-based IFM can be implemented by monitoring microwave power, optical power or time delays since a fixed relationship between the microwave frequency and the microwave power, optical power or time-delay can be established in the optical domain.

A typical IFM scheme based on microwave power monitoring is shown in Figure 15.12. The unknown microwave signal is first converted at an EOM into an optical microwave signal consisting of two sidebands and an optical carrier, then the optical microwave signal is inserted in to an optical channel that can induce a frequency-dependent power penalty (FDPP). Beating the optical carrier with the upper sideband and the lower sideband at a PD, would generate a new microwave signal of which the power, as compared to the original power of the microwave signal, is a function of the microwave frequency [37–39]. To eliminate the effect of power fluctuation produced by the measurement system two parallel measurements using two optical channels with different frequency-dependent power penalties are usually required. The ratio of the microwave powers in the two measurements is defined as the amplitude comparison function (ACF). If the ACF has a monotonic

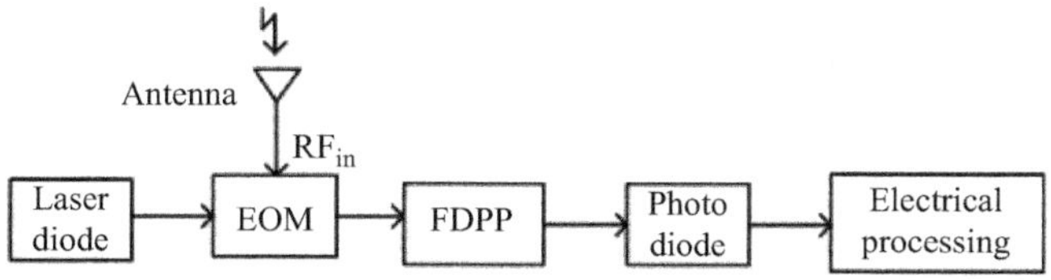

Figure 15.12 Microwave frequency measurement using FDPP

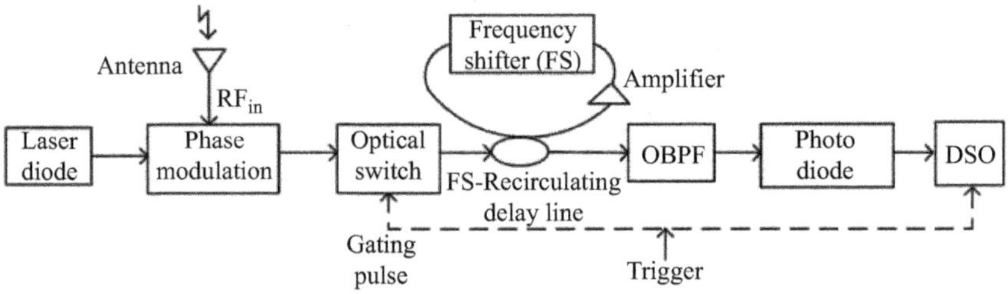

Figure 15.13 Multiple frequency measurement using FS-RDL

relationship with the frequency of the microwave signal in a certain frequency range, IFM can be realized without ambiguity.

Another method based on frequency-to-time mapping using frequency shift recirculating delay line (FS-RDL) can also be used for multiple-frequency measurement. Figure 15.13 shows the schematic diagram of a typical multiple-frequency measurement scheme based on an FS-RDL. An optical carrier from an LD is modulated by the unknown microwave signal in a phase modulator (PM), then the phase-modulated optical signal is time-gated by a high-speed on–off optical switch before being launched into an FS-RDL loop. The FS-RDL, as shown in Figure 15.13, consists of an optical frequency shifter and an optical amplifier. The output of the FS-RDL loop is connected to a narrowband optical band-pass filter (OBPF) [38,39].

The optical signal from the OBPF is detected by a PD and the obtained photo current is observed on a low-frequency oscilloscope that is synchronized with the control pulse connected to the optical switch. The principle of the frequency measurement system based on FS-RDL is illustrated in Figure 15.13. Due to the small signal modulation, we assume that the output of the optical switch contains only three optical spectral components: upper sideband, optical carrier, and lower sideband. This signal is injected into the FS-RDL loop. After each circulation, a frequency down-shift of Δf is introduced. After a certain number of circulations, the down-shifted signal would drop in the passband of the OBPF as shown in Figure 15.13. If the OBPF bandwidth is narrow enough, only one optical component is selected at a time, resulting in a square pulse being displayed on the oscilloscope. Corresponding to the relative locations in the optical spectrum, three pulses show up on the oscilloscope sequentially at the time t_{-k}, t_0 and t_k.

To avoid ambiguity, the trigger of the oscilloscope is adjusted to only display the pulse from the upper sideband. When the frequency of the beat signal falls into the PD response window, an electrical pulse can be observed on the oscilloscope. The number of circulation can be counted accurately based on the time of the pulse on the oscilloscope. Thus, the frequency of the unknown signal can be calculated if the frequency shift Δf of the FS-RDL is known. Due to the fast development of the photonic integration technique, photonic integrated chips (PICs) for IFM applications have been growing fast.

15.6.2 Photonics-based wideband RF signal generation for radar applications

In conventional systems, the RF signals are generated through multistage electronic up-conversions that exploit different electronic local oscillators (LOs). Each stage introduces non-negligible phase and amplitude noise due to the phase drifts of the LO and the use of noisy electronic mixers. Moreover, the generation of each specific RF frequency requires specific LOs, usually incoherent with one another and whose stability rapidly decreases as the RF increases. RF signal source is a key device in both the transmitter and receiver of the radar system, because, its property determines the performance of entire system. Currently, the signal generation based on pure electronic technologies is limited in the aspects of carrier frequency, signal bandwidth, and noise properties. Thus, it cannot meet the requirement of high frequency, large bandwidth, and low noise signals for the futuristic high performance radar designs. Microwave photonics, a promising technology, can generate microwave signals with the advantages of high frequency, large bandwidth, and superior noise performance. This section presents the photonics-based radar signal generation such as high-quality local oscillator signal and generation of phase-coded/linear frequency modulated (LFM) signals, etc.

As the improvement of radar systems necessitates digital approaches, photonics is becoming a solution for software defined, high frequency, and high stability RF signal generation. Several new radar concepts have been recently proposed like SDR, multi-function radar, arbitrary beamforming radar, and shared aperture radar. All these new paradigms in the radar field will require to flexibility in control for the generation of one or more RF signals and this would be conveniently managed in a digital fashion. In order to apply the digital trend to RF signals, high-speed digital electronics must be developed to surpass the need for the old analog electronics. But as the carrier frequency increases, the task becomes more and more challenging and nowadays DDS can only range up to few GHz [40,41]. Photonics is being proposed as the solution to fill the gap between the urgent need for digital solutions and the limitations of current electronics at higher frequencies. The flexibility, wide bandwidth and technological maturity of photonics in fact have led to numerous proposed architectures for the generation, manipulation, detection of RF signals and also merging the radar applications with related other fields.

Coherent radars require RF carriers with lower phase noise for improved sensitivity and higher frequency for smaller antenna. A modulation or coding of the phase of the radar pulse carrier is also necessary to implement pulse compression techniques for increased resolution without dangerous transmitted peak power. Moreover, the generation of multiple signals is sought for frequency agility and for multiband multifunctional radars. For high frequency carriers, the usual frequency multiplication processes reduce the phase stability of the original RF oscillators and frequency diversity radars are commonly realized by using two or more radar transmitters with increasing cost and power consumption [42,43]. In the last few years, photonics has been found suitable for generation of low phase-noise RF

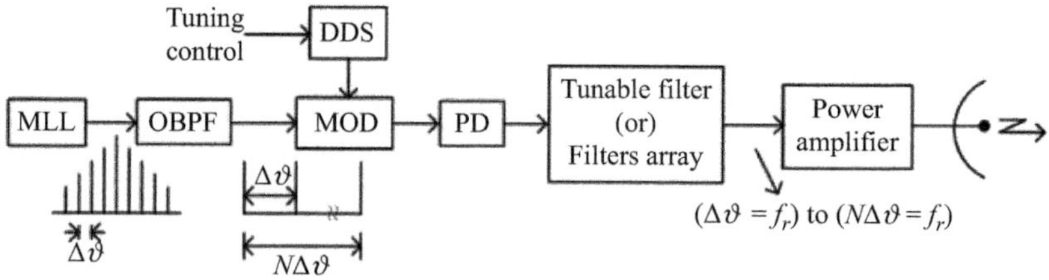

Figure 15.14 Photonic generation of RF signal by modulating a MLL at intermediate frequency

carriers. Among other techniques, the heterodyning of modes from a mode-locking laser (MLL) has proven to generate low phase noise RF carriers up to the EHF band (30–300 GHz). Wideband modulation and coding can also be applied in the photonic domain that avoids use of multiple frequency-specific electronic devices. But most of these modulation schemes exploit optical interferometric structures which are hardly suitable for coherent radars with demanding frequency agility.

A technique has been proposed for optically generating multiple phase-coded RF signals with flexible carrier frequencies and with a phase stability suitable for coherent radar systems. The proposed scheme, as sketched in Figure 15.14, modulates the spectrum of a MLL at intermediate frequency (IF), so that phase- and amplitude-modulated RF signals at any carrier frequency can be obtained by mixing a modulation sideband and a MLL mode in a PD, thus realizing a stable photonic up-conversion. Therefore, in principle a precise optical filtering device selecting only one sideband and one MLL mode would be necessary, but the filtering task in this case is easier and more efficient if realized in the RF domain. Thus, after the PD, an RF filter centered at the desired frequency allows to select the RF signal. The use of a precise DDS for generating the coding signals at variable IF allows the implementation of a SDR transmitter without losing the original phase stability of the MLL that matches the requirements of demanding surveillance systems [44].

As shown in Figure 15.14, the frequency agility can be implemented with a single MLL instead of a series of electronic oscillators. The photonic-based RF generation architecture allows to generate the desired high frequency signals either simultaneously, alternately or even to continuously change their frequency by appropriately programming the DDS and tunable RF filter. The modulation can also be changed meanwhile and thus a waveform diversity can be implemented. In this way, due to the wide optoelectronic bandwidth of available commercial photodiodes, generation of RF signals up to several hundred of GHz is easily achievable. Moreover, if one laser is modulated, the heterodyning operation transfers the modulation to the carrier frequency given by the lasers detuning. Since optical modulators are available with electrooptical bandwidth above 40 GHz, photonics allows the generation of ultra-broadband width RF signals.

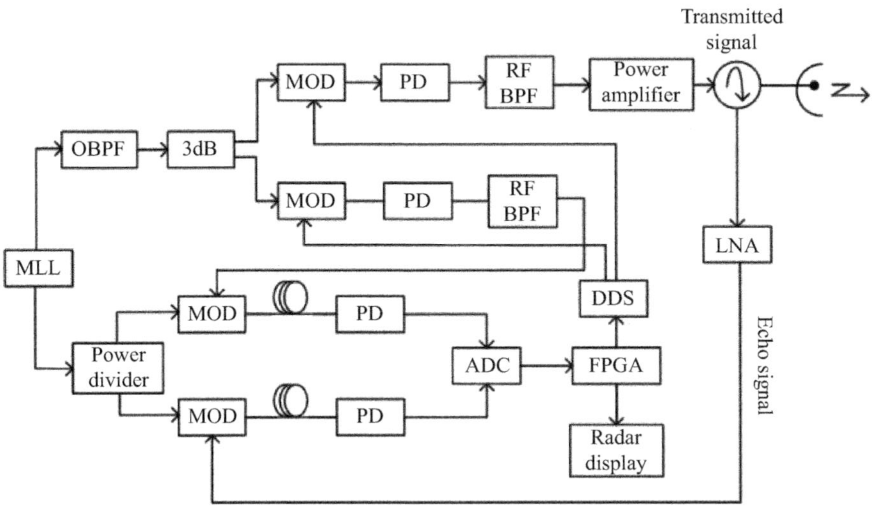

Figure 15.15 A photonics pulse radar optoelectronic assembly

15.6.3 Photonics radar system—optoelectronic assembly

The requirements of resolution, sensitivity, and flexibility of future multifunctional radar systems are pushing the development of reconfigurable and SDR transceivers that are capable of managing wideband waveforms over high-frequency carriers with the phase stability required for coherent pulse-Doppler processing. Thus, the evolution of future radars strongly depends on the progress of electronic digital components as the high speed DDSs and ADCs. Thus, bandwidth and precision of digital electronics represent the limit of current radar system performance. On the other hand, microwave photonics technologies promise to match tomorrows system radar systems requirements.

A simple architecture of the photonics radar system is shown in Figure 15.15. The photonic generation of high-quality, flexible, and high-frequency RF signals exploits the heterodyne detection of modes from an MLL. The radar waveform is generated at an IF level by a DDS and modulates the MLL signal in an electro-optical modulator. The optical signal (composed of original MLL modes and new sidebands) is then detected by a PD and a RF filter selects the desired beating frequency without resorting to electrical mixers. Hence, the carrier frequency can be arbitrarily chosen as an integer multiple of the pulses repetition rate. Thus, digitally generated and highly stable IF holds the radar waveform and guarantees a continuous tunability [45,46].

The filtered RF signal is then sent to a high-power amplifier to be transmitted from the antenna. Once the radar signal has been back-scattered from a target, it is received by the antenna, amplified, and filtered. Then, the received RF waveform is used to modulate the signal from the MLL (optical sampling), generating replicas of the detected signal spectrum as sidebands of the MLL modes. Finally, a PD detects the optical signals and performs the down-conversion of the received echo from the RF frequency back to the original IF, again without electrical mixers. An ADC eventually digitizes the down-converted signal at IF. The system shown in Figure 15.15, therefore exploits a single MLL for both the radar transmitter and receiver. Besides the radar pulses, the photonic transmitter also generates a continuous-wave reference signal which is used to implement a coherent radar deriving the targets speed.

Another scheme of a dual-band photonic radar system is shown in Figure 15.16. The transceiver is fed by a MLL which produce a pulse train whose spectrum is also shown in Figure 15.16. Two radar waveforms are generated by a single DDS at two different intermediate frequencies (IF_1 and IF_2) and up converted to their respective carrier frequencies (RF_1 e.g., S-Band and RF_2 e.g., X-band). The upconversion is obtained by modulating the MLL's light via a Mach–Zehnder modulator (MZM), thus, transferring the waveforms as lower and upper-sidebands around each mode as shown in Figure 15.16 and detecting the optical signal with a PD in which all the spectral components are heterodyned together [47,47].

Two electrical bandpass filters finally extract the replicas at the desired carrier frequencies. The same MLL is employed also in the reception of the radar echoes, which are optically under-sampled by the optical pulses. Thus, low time jitter of the laser becomes possible to precisely digitize high frequency signals using the low speed ADCs. The ADC feeds a FPGA which separates the two echoes and digitally down converts them to base-band.

To this extent, the incoming samples are multiplied with two complex numeric-controlled oscillators at two different IFs, thus, each channel is moved to base-band and the other one can be cancelled by a suitable finite impulse response (FIR) low-pass filter (LPF). In order to operate the system in a cooperative scenario, the signals from the transceiver need to be boosted before the transmission

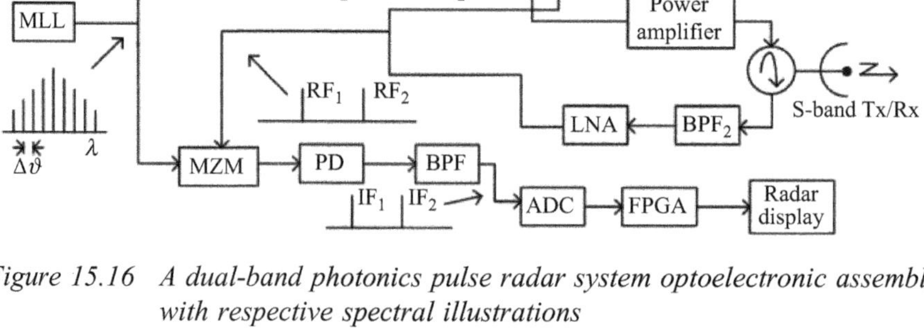

Figure 15.16 A dual-band photonics pulse radar system optoelectronic assembly with respective spectral illustrations

and the weak echoes must be amplified to drive the MZM. To this extent different-band-dedicated RF front-ends (as required) have to be employed as shown in Figure 15.16. Both the front-ends manage two parabolic antennas with required field of view and gain in both bands.

15.6.4 Broadband photonics radar system and beamforming architecture

Real-time and high-resolution target detection and imaging, such as capturing and tracking fast moving targets, are highly significant in civil and security applications which requires a radar to be operated at a high frequency and a wide bandwidth with real-time signal processing capability. This requirement creates great challenges to the state-of-the-art electronics [49–51]. On one hand, radar transmitter direct generation of Linear Frequency Modulation (LFM) signals by means of DDS is limited to a few GHz. Although this bandwidth can be expanded by multiple stages of frequency up-conversion, the signal quality would be inevitably deteriorated and eventually affects the detection performance.

On the other hand, the precision of ADC in the receiver drops rapidly as the input bandwidth and sampling rate increase, which severely restricts the resolution as well as the processing speed. As we have seen in the previous sections, microwave photonic technologies have been proposed as a promising solution for the RF signal generation, modulation, detection, processing, tracking, broadband operations, etc. Up to now, a lot of schemes for photonic generation of broadband LFM signals have been demonstrated, where a signal bandwidth over 10 GHz can be easily achieved. In this section, we propose a photonics-based real-time high-range-resolution LFMCW radar incorporating optical generation and processing of broadband signals. Proposed broadband photonics LFMCW radar experimental test-bed is shown in Figure 15.17.

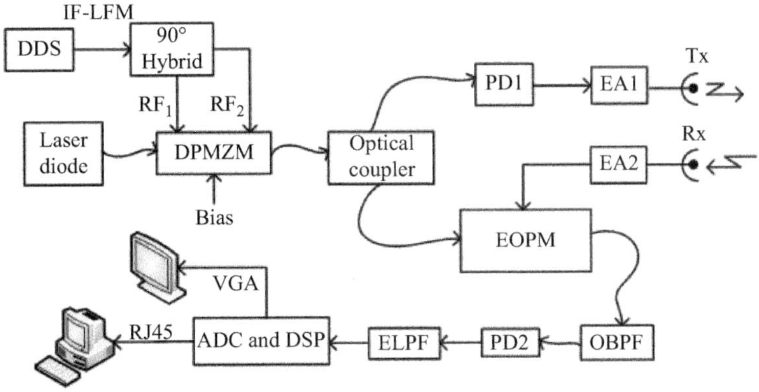

Figure 15.17 Experimental test-bed of proposed photonics linear frequency-modulated continuous wave (LFMCW) radar

A continuous-wave (CW) light from a laser diode (LD) is modulated by a Dual-Parallel Mach-Zehnder Modulator (DPMZM) which is driven by an IF-band LFM signal generated by a low-speed electrical signal generator i.e., a DDS. The instantaneous frequency of the IF-LFM signal can be expressed as $f_{IF}(t) = f_0 + kt$; where f_0 is the initial frequency and k is the chirp rate. The DPMZM consists of two MZMs embedded in each arm of the main MZM modulator. Before giving to DPMZM, the IF-LFM signal passes through an electrical 90° hybrid that produces two signals with the phase difference of 90°. These two signals are then fed into the two RF ports of the DPMZMs. By biasing the two MZMs at the maximum transmission points, a series of even-order optical sidebands are generated. If the amplitudes of the driving signals are properly controlled, only the optical carrier and $\pm 2^{nd}$ order sidebands dominate the output, since the higher sidebands have very small amplitudes. At the output of the DPMZM, only $\pm 2^{nd}$ order optical sidebands exist and thus the obtained optical signal can be expressed as

$$E_{\mathrm{DPMZM}}(t) \propto J_2(m)\cos\left[2\pi(f_c + 2f_{IF})t\right] + J_2(m)\cos\left[2\pi(f_c - 2f_{IF})t\right] \quad (15.33)$$

where f_c is the frequency of the laser diode, m is the modulation index of the two MZMs, and J_2 denotes the 2nd order Bessel function of the first kind. This modulated optical microwave signal is equally split into two branches (T_x (upper) and R_x (lower) branches) by an optical coupler. In the upper branch, the optical signal is sent to a PD1 to perform optical-to-electrical conversion [49–51]. After PD1, a frequency-quadrupled LFMCW signal is obtained which has an instantaneous frequency of $f_{\mathrm{LFMCW}}(t) = 4f_0 + 4kt$. Compared with the input IF-LFMCW signal, both the center frequency and bandwidth of the generated LFMCW signal are quadruped. Based on this principle, broadband LFMCW signals can be easily generated using low-speed electrical devices.

The generated LFMCW signal is amplified by a broadband Electrical Amplifier1 (EA1) and then transmitted to the free space through a transmit antenna for targets detection. The lower branch output of the optical coupler, is used as a reference for dechirp processing of the received radar echoes.

The echoes reflected from the targets are collected by a receive antenna, which are properly amplified by EA2. Then, the amplified signal is applied to an Electro-Optical Phase Modulator (EOPM) to modulate the reference optical signal coming from the optical coupler. Mathematically, the reference optical signal can be treated as two optical carriers at $(f_c - 2f_0 - 2kt)$ and $(f_c + 2f_0 + 2kt)$ which are phase modulated by the reflected LFMCW signal. The frequency of the first order sideband generated by phase modulating carrier at $(f_c - 2f_0 - 2kt)$ is located at $(f_c + 2f_0 + 2kt + 4k\Delta\tau)$, where $\Delta\tau$ is the time delay of the reflected LFMCW signal compared with the transmitted signal. By properly designing the parameters of the transmitted signal according to the detection range to let 4k $\Delta\tau$ be a small value, this first-order sideband is very close to the optical carrier at $(f_c + 2f_0 + 2kt)$. That means, the reflected LFM signal is dechirped to a low-frequency signal using the reference optical signal based on photonic frequency mixing. The type of implementation of photonic dechirping can directly process high-frequency and large bandwidth signals without any electrical frequency conversion. This low frequency signal can be extracted using an Optical Band-Pass Filter (OBPF). The optical signal after the filter can be written as

$$E_{\mathrm{OBPF}}t \propto J_0(m')\cos\left[2\pi(f_c + 2f_0 + 2kt)t\right]$$
$$+ J_1(m')\cos\left[2\pi(f_c + 2f_0 + 2kt + 4k\Delta\tau)t + \frac{\pi}{2}\right] \quad (15.34)$$

where m' is the phase modulation index. After the OBPF, the optical signal is sent to PD2 to perform optical-to-electrical conversion. Up to this point, photonic dechirping is implemented. The desired signal after de-chirping has a low frequency at $\Delta f = 4k\Delta\tau$, as shown in Figure 15.18, where B is the bandwidth and T is the temporal period of the LFMCW signal. It should be noted that a high frequency component at $(B - \Delta f)$ is generated at the same time, because of the temporal overlap between the received echo and the transmitted LFM signal in the next period. To remove the high frequency interference, an electrical low-pass filter (ELPF) with a proper bandwidth is applied after the PD2.

Since the frequency of the dechirped signal after the ELPF is determined by the time delay $(\Delta\tau)$ and chirp rate $(4k)$ of the echo LFMCW signal, the distance (L) of the target from the antenna pair can be calculated by

$$L = \frac{c}{2}\Delta\tau = \frac{c}{2}\frac{\Delta f}{4k} = \frac{cT}{2\beta}\Delta f \quad (15.35)$$

where c is the velocity of light in free space. By choosing a proper chirp rate of the transmitted LFM signal according to the detection range, the de-chirped signal can be controlled to have a low frequency (e.g., lower than 50 MHz) and it can be digitized by a low-speed ADC, as shown in Figure 15.17, with a high effective number of bits. Then, the digitized signal can be processed in a digital signal-processing (DSP) unit.

Suppose the de-chirped signal has M pulses in one frame and each pulse has N samples, the total samples make an $M \times N$ matrix. One-dimensional range profile is

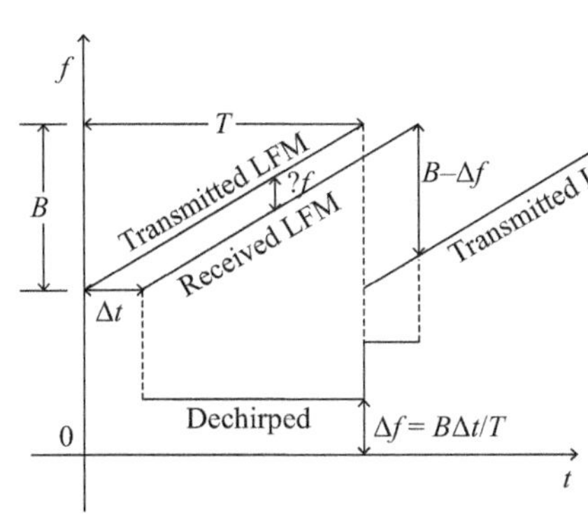

Figure 15.18 Principle illustration of LFM dechirping process

derived by performing N-point discrete Fourier transform (DFT) for all the M rows. After motion compensation, M-point inverse discrete Fourier transform (IDFT) is calculated in the cross range profile and a two-dimension image can be constructed. When performing DFT in each row, the spectral resolution is the inverse of the measurement time i.e., the minimum distinguishable spectral spacing is $\Delta f_{\min} = \frac{1}{T}$. Thus, the theoretical range resolution (L_{RES}) of the radar is

$$L_{\mathrm{RES}} = \frac{cT}{2\beta}\Delta f_{\min} = \frac{c}{2\beta} \tag{15.36}$$

and the cross-range resolution (C_{RES}) is given by

$$C_{\mathrm{RES}} = \frac{c}{2\theta f_{c_1}} \tag{15.37}$$

where f_{c_1} is the center frequency of the LFMCW signal and Î is the total viewing angle of target rotating. According to (15.36), a large signal bandwidth helps to achieve a high range resolution. The proposed radar can avoid the use of multi-stage electrical frequency conversion as well as the high-speed ADCs, thus, it enables the generation and processing of broadband radar signals. In principle, the operation bandwidth is mainly limited by the electro-optical devices, which can reach tens or even hundreds of GHz. Therefore, it is possible to achieve an ultra-high range resolution below 1 cm. The cross range resolution, determined by f_{c_1} and θ, can be chosen to be the same as or close to the range resolution. Therefore, the proposed photonics-based radar with appropriate signal processing algorithms has the potential for ultra-high resolution and real-time imaging.

Beamforming of RF signal using the suitable PAAs allow steering of the transmitted RF beam without physically moving the antenna assembly. This type of electronic beamforming is being used now a days in many applications including radars, electronic warfare, communications, etc. Common PAAs use electronic phase shifters at each antenna element (or group of antenna elements) to control the viewing angle of the array. Now a days, several adaptive array signal processing algorithms are coming up to find the optimum weight values for controlling the phase of the RF signal, at every channel, to steer the look angle of the antenna beam. The beamforming technique can also be easily implemented in photonics radar with appropriate opto-electronics devices [51–53]. The main advantages of this type of photonics-based beamforming are (i) capable of handling wide bandwidth signal, (ii) EMI insensitivity, (iii) no need of bulk RF phase-shifters, (iv) low transmission loss, (v) potential of higher beam steering resolution, etc. The optoelectronics architecture of a photonics-based beamforming is shown in Figure 15.19.

The IF/modulation electrical signal is loaded into the MZM which gets optical input from a MLL. The modulated optical microwave signal is divided by a power divider and then sends them to Tx (upper) and Rx (lower) chain, respectively. In the Tx chain, the optical coupler couples the single channel optical input-fiber optic cable (FOC) into "n" channel FOCs, that are connected to optical phase

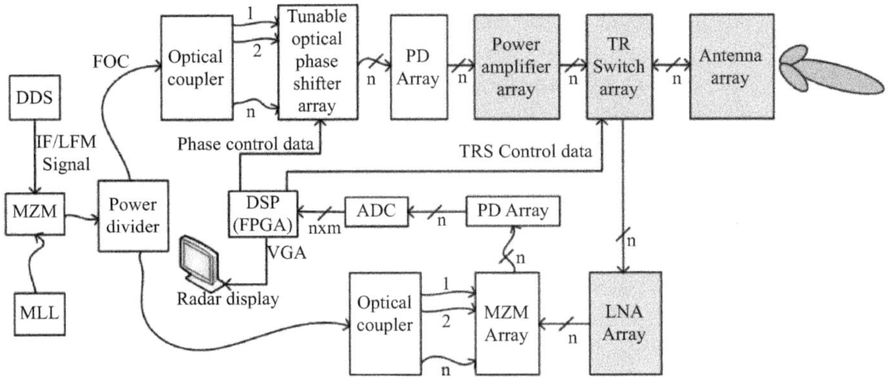

Figure 15.19 Photonics radar with beamforming optoelectronic architecture. RF stages are indicated by ash-colored boxes

shifter/controller through which the beam will be formed. The phase control, at every channel, can be done by several ways like deploying the well calculated length of fibers, optical path switching, and tunable optical phase shifters. The phase controlled-optical signals fall simultaneously on the PD array which generates the RF signal. The RF signal are amplified and applied to the antenna array via transmit–receive switches (TRS). Phased array antenna radiates the RF signal in the decided direction and then receives the echo signal. TR switches are controlled by the signal processing unit such as to route the echo signal to the LNA array. In the Rx chain, the amplified RF signal modulates the optical signal at the MZM array and the modulated optical microwave signal fall on another PD array. Since, the modulators here act as the optical mixers, their outputs are the IF signal which go into the suitable multichannel ADC. The digital data are applied, in parallel, into the FPGA where all the signal processing and computations are taking place. Finally, the FPGA displays the profile of the target(s) on the radar display. The incorporation of beamforming architecture to optically steer the beam is the near future research works.

References

[1] Vishal Sharma and Richa Sharma. Analysis of spread spectrum in matlab. *International Journal of Scientific & Engineering Research*, 5(1):1899–1902, 2014.

[2] Vishal Sharma, Chitra Shukla, Subhashish Banerjee, and Anirban Pathak. Controlled bidirectional remote state preparation in noisy environment: A generalized view. *Quantum Information Processing*, 14(9):3441–3464, 2015.

[3] Vishal Sharma, Kishore Thapliyal, Anirban Pathak, and Subhashish Banerjee. A comparative study of protocols for secure quantum communication under noisy environment: Single-qubit-based protocols versus

entangled-state-based protocols. *Quantum Information Processing*, 15(11): 4681–4710, 2016.

[4] Vishal Sharma and Subhashish Banerjee. Analysis of atmospheric effects on satellite-based quantum communication: A comparative study. arXiv preprint arXiv:1711.08281, 2017, *Quantum Information Processing*, 18(3):67, 2019.

[5] Charles H Bennett, Gilles Brassard, Claude Crépeau, Richard Jozsa, Asher Peres, and William K Wootters. Teleporting an unknown quantum state via dual classical and einstein-podolsky-rosen channels. *Physical Review Letters*, 70(13):1895, 1993.

[6] Nicolas Gisin, Grégoire Ribordy, Wolfgang Tittel, and Hugo Zbinden. Quantum cryptography. *Reviews of Modern Physics*, 74(1):145, 2002.

[7] Bahaa EA Saleh, Malvin Carl Teich, and Bahaa E Saleh. *Fundamentals of Photonics*, volume 22. Wiley, New York, 1991.

[8] José Capmany and Carlos R Fernández-Pousa. Quantum model for electro-optical amplitude modulation. *Optics Express*, 18(24):25127–25142, 2010.

[9] Chinmay Belthangady, Chih-Sung Chuu, A Yu Ite, GY Yin, JM Kahn, and SE Harris. Hiding single photons with spread spectrum technology. *Physical Review Letters*, 104(22):223601, 2010.

[10] Juan Carlos Garcia-Escartin and Pedro Chamorro-Posada. Quantum spread spectrum multiple access. *IEEE Journal of Selected Topics in Quantum Electronics*, 21(3):30–36, 2015.

[11] Vishal Sharma. Effect of noise on practical quantum communication systems. *Defence Science Journal*, 66(2):186–192, 2016.

[12] Charles H Bennett and Gilles Brassard. Quantum cryptography: Public key distribution and coin tossing. *Theoretical Computer Science*, 560:7–11, 2014.

[13] Valerio Scarani, Helle Bechmann-Pasquinucci, Nicolas J Cerf, Miloslav Dušek, Norbert Lütkenhaus, and Momtchil Peev. The security of practical quantum key distribution. *Reviews of Modern Physics*, 81(3):1301, 2009.

[14] Narayanaswami Srinatha, Srikrishna Omkar, Radhakrishna Srikanth, Subhashish Banerjee, and Anirban Pathak. The quantum cryptographic switch. *Quantum Information Processing*, 13(1):1–12, 2014.

[15] Vishal Sharma and Subhashish Banerjee Analysis of Quantum Key Distribution based Satellite Communication. arXiv:1807.07544, 2018, 9th International Conference on Computing, Communication and Networking Technologies (ICCCNT), Vol. 43488, pp. 1–5, IEEE, July 10–12, 2018, IISC, Bengaluru, India.

[16] Vishal Sharma, Shrikant Utagi, Srikanth Radhakrishna, and Subhashish Banerjee. "Decoherence can help quantum cryptographic security," arXiv preprint arXiv:1712.06519, *Quantum Information Processing*, 17(8):207, 2018.

[17] Saikat Guha, Hari Krovi, Christopher A Fuchs, *et al.* Rate-loss analysis of an efficient quantum repeater architecture. *Physical Review A*, 92(2):022357, 2015.

[18] S Omkar, R Srikanth, and Subhashish Banerjee. Dissipative and non-dissipative single-qubit channels: dynamics and geometry. *Quantum Information Processing*, 12(12):3725–3744, 2013.

[19] Charles H Bennett, François Bessette, Gilles Brassard, Louis Salvail, and John Smolin. Experimental quantum cryptography. *Journal of Cryptology*, 5 (1):3–28, 1992.

[20] Charles H Bennett and Stephen J Wiesner. Communication via one-and two-particle operators on einstein-podolsky-rosen states. *Physical Review Letters*, 69(20):2881, 1992.

[21] Thomas M Cover and Joy A Thomas. Elements of information theory, 2nd edition (Wiley series in telecommunications and signal processing). Hoboken, NJ, 2006.

[22] Hoi-Kwong Lo, Xiongfeng Ma, and Kai Chen. Decoy state quantum key distribution. *Physical Review Letters*, 94(23):230504, 2005.

[23] Gilles Brassard and Claude Crépeau. Quantum cryptography. In *Encyclopedia of Cryptography and Security*, pages 495–500. Springer, 2005.

[24] Valerio Scarani, Antonio Acin, Grégoire Ribordy, and Nicolas Gisin. Quantum cryptography protocols robust against photon number splitting attacks for weak laser pulse implementations. *Physical Review Letters*, 92(5):057901, 2004.

[25] Anthony Chefles. Unambiguous discrimination between linearly independent quantum states. *Physics Letters A*, 239(6):339–347, 1998.

[26] Xiongfeng Ma, Bing Qi, Yi Zhao, and Hoi-Kwong Lo. Practical decoy state for quantum key distribution. *Physical Review A*, 72(1):012326, 2005.

[27] Tomoyuki Horikiri and Takayoshi Kobayashi. Decoy state quantum key distribution with a photon number resolved heralded single photon source. *Physical Review A*, 73(3):032331, 2006.

[28] Xiang-Bin Wang. Beating the photon-number-splitting attack in practical quantum cryptography. *Physical Review Letters*, 94(23):230503, 2005.

[29] http://fullafterburner.weebly.com/next-gen-weapons/radio-optic-phased-array-radar-a-comprehensive-study

[30] Paolo Ghelfi, Francesco Laghezza, Filippo Scotti, Daniel Onori, and Antonella Bogoni. Photonics for radars operating on multiple coherent bands, *IEEE, Journal of Lightwave Technology*, 34(2):500–507, 2016.

[31] Paolo Ghelfi, Francesco Laghezza, Filippo Scotti, *et al.* A fully photonics-based coherent radar system. *Nature*, Nature Publishing Group, 507(7492): 341, 2014.

[32] Vishal Sharma and PC Panchariya. Experimental use of electronic nose for odour detection. *International Journal of Engineering Systems Modelling and Simulation*, 7(4):238–243, 2015.

[33] Manisha Mathur, Jaynendra Kumar Rai, and Nilakantan Sridhar. Microwave photonic network for active electronically scanned array radar. *International Journal of Microwave and Wireless Technologies*, Cambridge University Press, 9(3):543–550, 2017.

[34] Paolo Ghelfi, Francesco Laghezza, Filippo Scotti, *et al.* Photonics in radar systems: RF integration for state-of-the-art functionality. *IEEE Microwave Magazine*, 16(8):74–83, 2015.

[35] Filippo Scotti, Francesco Laghezza, Daniel Onori, and Antonella Bogoni. Field trial of a photonics-based dual-band fully coherent radar system in a maritime scenario. *IET Radar, Sonar & Navigation*, 11(3):420–425, 2016.

[36] Filippo Scotti, Francesco Laghezza, Daniel Onori, *et al.* Microwave photonics for Integrated multifrequency lidar/radar system, *Opto-Electronics and Communications Conference (OECC)*, IEEE, 1–4, 2015.

[37] Shilong Pa and Jianping Yao. Photonics-based broadband microwave measurement. *IEEE, Journal of Lightwave Technology*, 35(16):3498–3513, 2017.

[38] Xihua Zou, Bing Lu, Wei Pan, Lianshan Yan, Andreas Stöhr, and Jianping Yao. Photonics for microwave measurements. *Laser & Photonics Reviews*, Wiley Online Library, 10(5):700–700, 2016.

[39] Fangzheng Zhang and Shilong Pan. Microwave photonic signal generation for radar applications, *2016 IEEE International Workshop on Electromagnetics: Applications and Student Innovation Competition (iWEM)*, 1–3, 2016.

[40] Jayaseelan Marimuthu, Konstanty S. Bialkowski, and Amin M. Abbosh. Software-defined radar for medical imaging. *IEEE Transactions on Microwave Theory and Techniques*, 64(2):643–652, 2016.

[41] Jason D McKinney. Technology: Photonics illuminates the future of radar. *Nature*, 507(7492):310, 2014.

[42] Filippo Scotti, Antonella Bogoni, Francesco Laghezza, Daniel Onori. Tracking of a naval target with a dual-band photonic-based coherent radar system. *Radar Conference (RadarConf)*, 2016 IEEE, 1–4, 2016.

[43] Filippo Scotti, Francesco Laghezza, Giovanni Serafino, *et al.* In-field experiments of the first photonics-based software-defined coherent radar. *IEEE, Journal of Lightwave Technology*, 32(20):3365–3372, 2014.

[44] Ziyi Meng, Jianqiang Li, Chunjing Yin, *et al.* Dual-band dechirping LFMCW radar receiver with high image rejection using microwave photonic I/Q mixer. *Optics Express*, 25(18):22055–22065, 2017.

[45] Shaowen Peng, Shangyuan Li, Xiaoxiao Xue, *et al.* High-resolution W-band ISAR imaging system utilizing a logic-operation-based photonic digital-to-analog converter. *Optics Express*, 26(2):1978–1987, 2018.

[46] Filippo Scotti, Francesco Laghezza, Paolo Ghelfi, and Antonella Bogoni. Multi-band software-defined coherent radar based on a single photonic transceiver. *IEEE Transactions on Microwave Theory and Techniques*, 63 (2):546–552. 2015.

[47] Paolo Ghelfi, Francesco Laghezza, Filippo Scotti, Giovanni Serafino, Sergio Pinnai, and Antonella Bogoni. Photonic generation and independent steering of multiple RF signals for software defined radars. *Optics Express*, 21(19):22905–22910, 2013.

[48] Sasan Fathpour and Nabeel A Riza. Silicon-photonics-based wideband radar beamforming: basic design. *Optical Engineering, International Society for Optics and Photonics*, 49(1):018201, 2010.

[49] E. Ackerman, S. Wanuga, D. Kasemset, W. Minford, N. Thorsten, and J. Watson. Integrated 6-bit photonic true-time-delay unit for lightweight 3–6

GHz radar beamformer. *Microwave Symposium Digest*, 1992, IEEE MTT-S International, 681–684, 1992.

[50] Xingyuan Xu, Jiayang Wu, Thach G Nguyen, *et al.* Photonic RF and microwave reconfigurable filters and true time delays based on an integrated optical Kerr frequency comb source. arXiv preprint arXiv:1802.00057, 2018.

[51] Oded Raz, Sharon Barzilay, Ruth Rotman, and Moshe Tur. Submicrosecond scan-angle switching photonic beamformer with flat RF response in the C and X bands. *Journal of Lightwave Technology*, 26(15):2774–2781, 2008.

[52] Xingwei Ye, and Shilong Pan. Wideband Multi-Beamforming using Programmable Microwave Photonic Signal Processing Units, MTT-S Undergraduate/Pre-Graduate Scholarship Project Report, https://mtt.org/sites/default/files/project-reports/10_FR_UGSFall2015_XingWei%20Ye.pdf.

[53] Daniel Headland, Yasuaki Monnai, Derek Abbott, Christophe Fumeaux, and Withawat Withayachumnankul. Tutorial: Terahertz beamforming, from concepts to realizations. APL Photonics, AIP Publishing, 3(5):051101, 2018.

Index